Genetic Analyses of Wheat and Molecular Marker-Assisted Breeding, Volume 1

Jichun Tian · Zhiying Deng · Kunpu Zhang
Haixia Yu · Xiaoling Jiang · Chun Li

Genetic Analyses of Wheat and Molecular Marker-Assisted Breeding, Volume 1

Genetics Map and QTL Mapping

Science Press
Beijing

Springer

Jichun Tian
State Key Laboratory of Crop Biology of Agronomy
Shandong Agricultural University
Tai'an
China

Zhiying Deng
State Key Laboratory of Crop Biology of Agronomy
Shandong Agricultural University
Tai'an
China

Kunpu Zhang
Institute of Genetics and Developmental Biology
Chinese Academy of Sciences
Beijing
China

Haixia Yu
College of Life Sciences
Shandong Agricultural University
Tai'an
China

Xiaoling Jiang
Henan Institute of Science and Technology
Xinxiang
China

Chun Li
Henan Sesame Research Center
Henan Academy of Agricultural Sciences
Zhengzhou, Henan
China

ISBN 978-94-017-7388-1 ISBN 978-94-017-7390-4 (eBook)
DOI 10.1007/978-94-017-7390-4

Jointly published with Science Press, Beijing
ISBN: 978-7-03-045741-7 Science Press, Beijing

Library of Congress Control Number: 2015949319

Springer Dordrecht Heidelberg New York London

Printed on acid-free paper

Springer Science+Business Media B.V. Dordrecht is part of Springer Science+Business Media (www.springer.com)

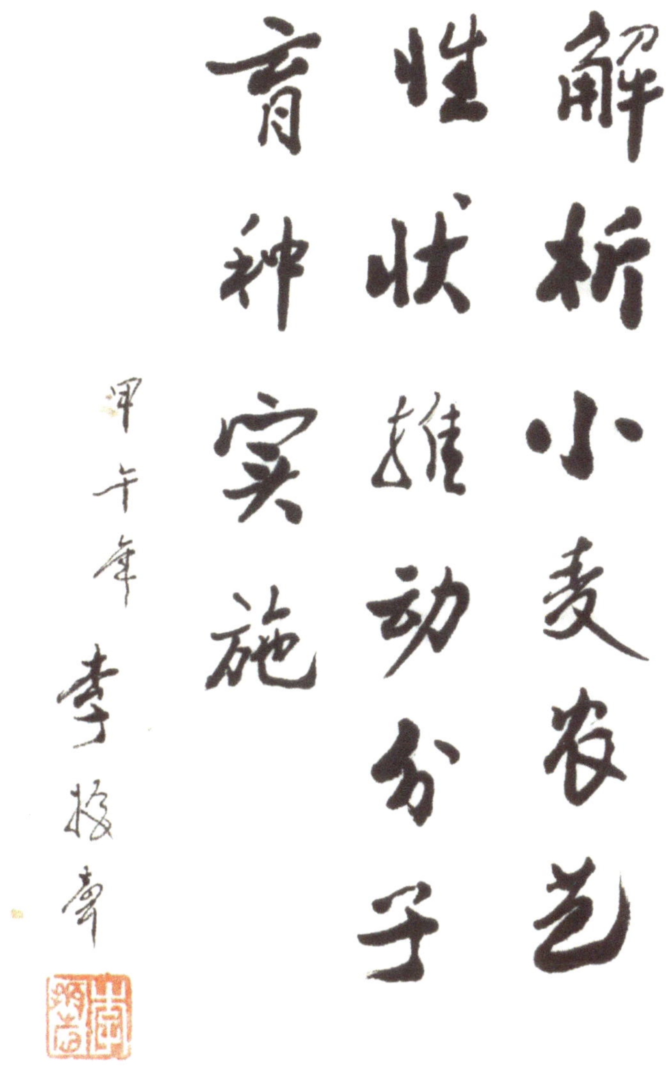

Translation: Dissecting Wheat Agronomic Traits and Promoting Molecular Breeding Implementation.

Author: Zhensheng Li is Academician of the Chinese Academy of Sciences and receiver of the National Supreme Scientific and Technological Award.

Foreword 1

Modern biotechnologies, prevalent in nearly all aspects of crop breeding programs, have been developing so rapidly that one could anticipate new breakthroughs on a regular basis. The invention of polymerase chain reaction (PCR) in 1985 by Kary Mullis (USA), for example, has allowed us to dissect various crops' quantitative trait loci (QTLs) at the level of a single gene. According to the statistics, more than 4200 genetic maps of various crops have been constructed by the end of 2012 using various molecular markers, many of which are SSR maps. These maps have made it possible to conduct the QTL mappings and efficacy analyses on traits associated with plant morphology, yield, quality, as well as stress resistance. Molecular markers developed by QTL mappings have been utilized in marker-assisted selections (MASs), resulting in enhanced tracking efficiency of major effect genes and QTLs, thereby accelerating germplasm development and the speed of variety development. With the advent of these breeding trends, Peleman and Vander Vort (Belgium Academy of Sciences) proposed a novel breeding concept in 2003, known as "breeding by design" which will undoubtedly become a mainstream technology in crop genetic improvement and enhance the in-depth crop breeding significantly in the near future. However, wheat "molecular breeding" is currently at the conceptual phase. As an allohexaploid, the wheat genome is much larger than that of rice, corn, and many other crops. Adding to the challenges, the genome sequencing is incomplete. The lack of progress in wheat molecular breeding is, in a similar fashion, due to the fact that most of the traits responsible for wheat yield, quality, and various others are controlled by multiple QTLs. Furthermore, some issues derived from MAS per se have yet to be resolved. This is evidenced by the fact that grain weight of a specific lineage containing large grain QTL gene is not necessarily high when selection of grain weight within the hybrid progenies is based on a single or a few QTL markers. This is especially true when selection is conducted within different selection populations with diverse genetic backgrounds, resulting in significant incertitude and perplexity for the application of MAS in wheat.

Fortunately, the author of this book has conducted MSA on the basis of his extensive experience in conventional breeding and he is also well versed in the advantages of molecular breeding. The author's unique background made it possible to create a synthesis of traditional breeding and MAS breeding. Noticeably, the focal points of this book, unlike many other molecular biology monographs (published domestically and abroad) majoring in introductions of basic concepts, research tools, and/or experimental techniques, are the results of an authentic summary of constructions of various genetic maps and applications of QTL analyses and molecular markers in wheat. The author, together with the rest of his team members, has contributed to identifications of a number of main QTLs and molecular markers that are associated with yield, quality, and stress resistance, which will provide a better foundation for the MAS breeding; creations of combinations of the breeding elements containing advantageous QTL genes; maintaining or eliminating any specific F1 hybrids based on the QTL gene gathering levels and their heterosis strengths; and applications of molecular markers to track QTL genes within the F2-F6 pedigree selections. Most importantly, these findings provide practical tools and techniques for MAS breeding of many other crops including wheat. It is for these reasons that I am honored to pen the preface for this book in order to express my support, recognition, and inspiration to the author and his team members.

Xu Liu
Academician of the Chinese Academy of Engineering
Beijing, China

Foreword 2

Tackling global food security depends on our ability to develop and deliver technologies that lead to increased food production. However, due to limitations in arable land, we must achieve this increase without expanding the area under production, and in future environments where the frequency and severity of climate shocks and extremes are expected to increase as a result of climate change. Over the past few decades, we have relied on access to water and energy to drive major yield increases through expansion of irrigation schemes and large-scale use of fertilizers, but this is not an option for the next phase of productivity gains. We will be increasingly dependent on the ability of our breeders to produce varieties that show improved yield under conditions where water and nutrients are likely to be limiting.

Agriculture has a long history of major production gains through the application of new technologies. The early farmers would have selected lines that supported their production systems and improved harvesting and storage. As a result, they developed lines that showed major changes in characteristics that suited them to farming compared to their wild relatives. A key change would have been selection for plants that grew well as a crop or community. Over time, farmers developed plants that suited their environment and this led to the generation of diverse landraces that supported human development.

A flow of genetic material over regions occurred as farmers exchanged grain with their neighburs and along trading routes. This communication, trade, and sharing would have slowly expanded the germplasm available to farmers and helped spread innovation. The process of practical germplasm exchange and selection underpinned modern agriculture, but it was not until the introduction of selective breeding that we saw a dramatic acceleration in the rates of genetic improvement. The discovery of the principles of genetics laid the foundations for the rapid improvements in crops that have occurred over the past century. As knowledge of genetics and genes expanded, the rates of genetic gain in crop improvement also accelerated.

Systematic breeding is based on the use of variation to develop new gene combinations. Breeders have selected the best performing plants from crosses to continually enhance the yield and quality of our crops. The opportunity for advances in breeding outcomes is closely tied to the diversity of the variation available, the size of the populations screened, and the intensity of selection. Therefore, selective breeding is essentially a numbers game since many important crop characteristics, such as yield and drought tolerance, controlled by a very large number of genes. Consequently, the chance of finding the best or improved combinations of genes can be difficult. However, new technologies have greatly assisted breeders; for example, mechanized sowing and harvesting has allowed breeders to grow and assess thousands of genetic combinations. In addition, improvements in computing power and sophisticated statistical methods have supported the accurate assessment of new plant lines. More recently, DNA markers have allowed breeders to follow individual genes, gene variants, and genomic regions as they are passed to the progeny of crosses. Based on the DNA fingerprint, the breeders can predict many of the key characteristics of the plants when grown as a crop, such as disease resistance, quality, and even yield. Marker technology has helped create new breeding strategies and reduced the costs of monitoring genes and controlling the frequency of useful alleles in breeding populations.

To tackle global food security and enhancing crop production, it is important to address factors limiting production for our major food crops. Wheat is the world's most widely grown crop and contributes around 20 % of calories and protein for the daily human diet. To meet the predicted food demands of a world population of over 9 billion people by 2050, wheat production will need to increase by 60 % relative to 2010. This means we need to increase the rates of yield improvement from the current level of 1 % per year to at least 1.6 %. The effective application of new technologies and increased investment in wheat research and breeding will be critical if these targets are to be met. This challenge was recognized in 2011 when the G20 group of countries agreed to establish a global Wheat Initiative with the "aims to encourage and support the development of a vibrant global public–private research community sharing resources, capabilities, data and game changing ideas, and technologies to improve wheat productivity, quality, and sustainable production around the world." The Wheat Initiative currently brings together 16 countries, nine private companies, and two international research centers and continues to grow. The Wheat Initiative recently launched a Strategic Research Agenda that identified key targets and priorities for global research. These included a major focus on developing our knowledge and understanding of the genetic control of major traits affecting wheat production, stress tolerance, disease resistance, and quality. The agenda also highlighted the importance of cooperation and exchange of information and knowledge about wheat.

Most work on wheat genetics over the past few decades has focussed on germplasm and traits of relevance to the European and North America production systems. These environments have been only poorly representative of the conditions facing breeders and farmers in the world's largest wheat producing countries, such as China and India. Of these two major wheat producing countries, India has been well connected to the mainstream of germplasm development through close links with the international wheat improvement program led by CIMMYT in

Mexico. The flow of germplasm has been both to and from India with a long history of use Indian germplasm in international breeding programs. In contrast, the Chinese programs have for a long period been overlooked by the international wheat breeding community. However, there have been several important examples where germplasm sourced from China has had a major impact on the international breeding community.

These two volumes represent a new milestone in international wheat research by providing an overview of modern wheat genetic research from the perspective of the Chinese wheat research community. Importantly, the work covers trait dissection based on Chinese germplasm and covering traits of relevance to wheat production in the world's biggest wheat producing country. Through this work, we gain an insight into both the success and challenges faced by Chinese researchers and breeders. The complexity of the challenges faced in China to deal with demands for improvements in the sustainability of production under pressures to reduce inputs and in the face of climate change represent a model for strategies to tackle the global challenges. These volumes will help develop a framework for wheat genetic research and open the Chinese experience to wheat researchers from around the world.

Peter Langridge
University of Adelaide

Foreword 3

It is projected that world population will reach 9.7 billion by 2050, and the global food need will increase 40 % by 2030 and 70 % by 2050 approximately (FAO). The challenge of meeting this demand is made still more difficult by climate change, global temperature warming, and more frequent extreme weather events. To increase cereal production is one of the most important ways to keep food security.

Total cereal yields are a function of land area devoted to cereals multiplied by the yield per unit of area. This means that if cereal production is to increase, either the agricultural land area or yield per unit area will need to increase, or a combination of both. In reality, the agricultural land area is not expected to increase in the future and may even decrease with currently productive land being lost to rising sea levels, desertification, etc. Therefore, if the challenge of increasing production is to be achieved, it will come from increasing yield per unit area. However, this achievement must be not adversely affecting the environment, just as John Beddington said "The challenge for global agriculture is to grow more food on not much more land, using less water, fertilizer and pesticides than we have historically done." Therefore, breeding new crop varieties to increase yield per acre is the best way of economic efficient and environment sustainable.

Wheat (*Triticumaestivum* L.) is one of the most important food crops around the world, which provides a fifth of human calories. High stable yield and good quality are the key objectives of breeding programs, but since 1980, the rate of increase in wheat yield has slowed. According to the statistics in China, the rate of increase in wheat yield per unit area was over 7 % in the 1980s and 1990s of the last century, but has slowed to less than 3 % in this millennium. However, higher yield cultivars are still being developed in China (e.g., Shannong 20 released 2010, 11.9 t/ha^{-1}; Lankao 198 released in 2012, 12.2 t/ha^{-1}; and Yannong 999 released 2011, 12.3 t/ha^{-1}), which have contributed greatly to increase total wheat production and emphasize the importance of continued genetic improvement.

In order to meet the fast-growing demand for wheat, researchers have presented different strategies to dramatically increase wheat productivity. In the UK, the Rothamsted research team developed the "20:20 Wheat" strategic program that seeks to provide the underpinning knowledge and tools to increase the yield potential of wheat in the UK to 20 $t.ha^{-1}$ in 20 years. This project identifies "maximizing yield potential and protecting yield potential" as central approaches. The identification of the genetic basis for relevant traits and the dissection of their interdependent relationships are critical to the realization of this goal. The rapid developing of genome sequencing, combined traditional breeding program, and molecular marker-assisted selection (MAS) increase the probability of successfully increasing yields.

However, the development of efficient molecular markers is important for molecular MAS or molecular design breeding. Therefore, quantitative trait loci (QTL) mapping for important wheat traits to capture major and stable QTLs is a key step. This book cohesively describes the developments in genetic mapping, QTL analysis, and molecular marker-assisted breeding that occurred over the past decades. In the first volume, this book introduces the core concepts and research methods of QTL; then, the authors illustrate six molecular genetics maps constructed by their group and QTL mapping for more than twenty important wheat traits, including quality, physiology, and various stress resistances. The second volume is mainly about conditional QTL mapping analyses and their applications to wheat breeding and cultivation. The authors introduced the concept and advantage of conditional QTL and illuminated their research results using the method for dissecting the temporal and spatial expressions and interrelations of some QTL. Molecular marker exploration methods and practical examples are also described in this part, which provided a good perspective on wheat breeding.

The book provides a great deal of novel information, in-depth knowledge of wheat genetics and molecular breeding, which will be extremely valuable to academics and to wheat breeders.

Prof. Martin Parry
Associate Director
Rothamsted Research, Harpenden,
Herts, AL5 2JQ, UK

Preface

Publishing a scientific research monograph not only requires extraordinary accumulation of data derived from technical endeavors that often spans a decade, but also requires the authors to invest many months or even years of writing. Writing and publishing a book offers the authors neither benefit nor satisfaction when taking into consideration the myriad of factors such as contemporary fast-paced research rhythm combined with professional title and salary promotions, research grant proposals and project evaluations, the high cost of book publications, limited number of readers, and the relatively small market. Nevertheless, there are multitudes of factors that motivated us in striving to compose and publish this book. First and foremost, there is a need for achieving breakthrough research in order to develop competitive wheat varieties. According to Li and Wan (2012), the demand for wheat production in China is projected to increase by at least 28 % by 2020. With the continuous depletion of arable land in China, the only way to meet this demand is to develop innovative varieties with high yield. It is true that the conventional breeding has made great contributions to the increased wheat production in China since 1949, and the techniques for field selections are still irreplaceable at present. However, this traditional breeding method has a number of disadvantages including selections being based only on phenotype, which inherently results in low efficiency and less superior varieties. For example, several major commercial varieties in China, such as Jimai 22, Aikang 58, Zhoumai 18, and Shanon 20, perform well within the boundaries of their plant habits and stress tolerances, but to reach another breakthrough presents entirely new sets of significant challenges. The development of super varieties with multiple beneficial traits controlled by collective elite alleles requires molecular markers to identify, track, and accumulate these superb genes, which needs the multidisciplinary knowledge (Peleman and Vander Vort 2003). Secondly, there is a need for combining molecular breeding and conventional breeding. Since the advent of modern molecular biology techniques represented by PCR, rapid developments in plant genetic diversity analysis and identification and cloning of elite genes have been made over the last three decades. The wealth of data in regard to genomics, proteomics, metabolomics, and phenotypes and numerous patents are too many to mention. It is my belief that "molecular breeding" and "molecular design breeding" are still at the stages of concept development and project applications. This is largely due to the poor combination of molecular breeding with traditional breeding. The current scientific research system is the cause of the "mismatch"—researchers on molecular breeding are mainly scattered in the confines of academic institutions and/or universities within which they do not fully understand or consider the needs of conventional breeding, while the conventional breeders who often work at local breeding stations and agricultural corporations have less interest in the "molecular design breeding" (it is "computer breeding" according to them). Furthermore, because wheat genome is characterized by its immense size and enormous complexity of QTLs, trait selections based on only one or a few molecular markers from populations with diverse backgrounds and environments are often not ideal. For example, genes with large grain gene/QTL and grain weight may not be necessarily high. Similarly, lines with disease resistant gene/QTL may be susceptible to diseases in the field. Having worked at Shandong Agricultural University for several decades, the author takes advantage of the unique situations experienced in both traditional breeding and molecular breeding and implemented the synthesis of the two breeding approaches with good results. This book publishes the summaries of my team research results and my past 16 years' research experience. Thirdly, we wish to express our gratitude for the monumental support from the national science and technology policy for many of our wheat breeding projects. Over the last decade, we have received research funds for a number of national research projects, including the State "973" program (No. 2009CB118301) for molecular improvement of high-yield wheat and development of molecular breeding elements aiming for creating super wheat high yield (supported by the Ministry of National Science and Technology); four projects (No. 30471082, 30671270, 30971764, and 31171554) supported by the Natural Science Foundation of China; two projects on wheat transformation supported by the National Development and Reform Commission; and the Mega Project on "Development and commercialization of super wheat varieties in Shandong Province." The success of these milestone projects and the wealth of research data presented in this volume are the results of the continuous support we received over the past ten years from the state and province, which allowed me and all of my team members (including all the graduated students) to focus and conduct these studies. By publishing this book, it is my intention to express my sincere thanks to the state and provincial leadership as well as all of the counterparts in China for their support and inspiration during this painstaking period of research.

Based on the foundation of the molecular biology and bioinformatics, Belgian scientist Peleman et al. (2003) recently proposed a novel breeding concept known as "breeding by design." This idea consists of three core concepts: mapping QTL-associated agronomic traits; evaluating the allelic variations at these loci; and implementing molecular design breeding. The premise of the research conducted by my laboratory over the past ten years was based on the concepts of molecular breeding and molecular design breeding. Constructive data (e.g., creations of molecular elements and molecular markers) derived from the research have been successfully applied to traditional breeding programs, enabling us to make the right cross combinations followed by good pedigree selections. This book compiles wheat molecular genetics map construction and genetic diagnosis of major wheat traits (QTL analysis). The book is divided into seven chapters. Chapters 1 and 2 mainly introduce "research progress of crop quantitative traits" and "the core concept and research methods of quantitative traits," which establish the necessary backgrounds for the contents of the subsequent chapters. Chapter 3 presents "six wheat genetic molecular maps" established by us with the details of map characteristics and their merits of applications. Chapters 4–7 discuss the following subjects: genetic analyses of QTLs associated primarily with wheat yield, quality, physiology, and stress resistance, respectively, have obtained more than 120 major QTLs of dozens of major traits and their molecular markers as well. In order to give readers a comprehensive understanding of the latest research progress, the volume presents not only the results of QTL mapping and efficacy analysis of each major QTL primarily based on our own research projects, but also, in addition, the summaries of similar projects at both home and abroad.

Introduction of the concepts and methods consists of only about 10 % of this volume, and the bulk of the content—more than 90 %—contains the summary of our research data, thereby indicating that this is not a biotechnological book with emphasis on the foundations of methodology and techniques. Rather, this book begins with establishment of molecular genetic maps, QTL analyses, followed by molecular marker-assisted breeding, thereby resulting in a science monograph with a comprehensive and in-depth research system. Ultimately, this publication is not only the collection of the findings of the emerging and ever-evolving wheat molecular marker breeding, but also the prerequisite for the implementations of the newly proposed "molecular design breeding."

The contents of this book are contributed by the members of my Wheat Quality Breeding Team stationed at the State Key Laboratory of Crop Biology, Shandong Agricultural University. Data presented in this volume are the results of several generations of wheat breeding efforts evidenced by development of a novel wheat variety (PH82-2-2) with high protein content and other superior qualities in the 1980s (awarded a 2nd Prize by the National Technology Invention); creations of seven new wheat varieties with high yield and superior quality over the past ten years, including Shannong Youmain #2 (evaluated at the provincial level in 2001 and at the state level in 2009), #3, Shannong #11 and #12 (evaluated at the provincial level in 2003, 2004, and 2005, respectively), #19 and #20 (evaluated at the state level in 2010 and 2011, respectively), and #26 (evaluated at the state level in 2014); and the comprehensive understanding of advantages and disadvantages of the conventional wheat breeding programs. The author has 36 years of career endeavors divided equally between teaching and research, with primary focus on plant physiology and biochemistry in addition to plant genetics and breeding. The fundamental knowledge of these two disciplines enabled me to successfully combine the traditional breeding with the modern molecular biology. For instance, the establishment of various genetic populations (RIL, DH, CIL, ad NL) began as early as 1998, which laid the foundation for the subsequent QTL mapping and molecular marker-assisted breeding. The rate of selected variety combinations versus cross combinations has increased from 1/1000 by traditional breeding to 1/500 by this strategy, whereas the selected lines for potential varieties from traditional breeding are only 1/1,000,000 compared to 1/10,000 using our selection system. Furthermore, land requirement for breeding studies is about 50 % less than that of a decade ago, and the cost of breeding has decreased significantly, while breeding efficiency has experienced remarkable improvement.

During my nearly 40 years of breeding experience, I have presided over a number of programs on molecular breeding and molecular design breeding at the state level. Participating in writing this book includes young faculty members, graduate students who have left the author's laboratory and are currently working across the country, and those who are currently still in their graduate programs at both Ph.D. and master levels, as well as the field technicians. Each of them provided his/her utmost effort to contribute to this publication. However, due to the rapid development of molecular biology and marker-assisted breeding technology, over time it is inevitable to identify insufficient information in this book. We hope that this volume would provide service and impart knowledge to the readers, but at the same time, we also welcome the readers to submit comments, feedbacks, or concerns.

Tai'an, China Jichun Tian
January 2015

Contents

Abstract

This book which cohesively encapsulates the developments in wheat genetics map and quantitative trait loci (QTL) analysis that occurred over the past sixteen years is comprised of seven chapters. Chapters 1 and 2, respectively, introduce the readers to the core concepts and research methods of wheat QTL. The third chapter illustrates the unique characteristics and breeding values of the six molecular genetic maps constructed by SSR, DarT, and SNP markers. Chapters 4–7 discuss the following subjects: genetic analyses of QTLs associated primarily with wheat yield, quality, physiology, and stress resistance, respectively, which have obtained more than 120 major QTLs of dozens of major traits and their molecular markers as well. In order to give readers a comprehensive understanding of the latest research progress, the volume presents not only the results of QTL mapping and efficacy analysis of each major QTL primarily based on our own research projects, but also in addition the summaries of similar projects both at home and abroad.

It is noteworthy that this is not a biotechnological book that serves to establish general methods. Rather, it is a scientific monograph with in-depth expositions of integral cohesive research system—stemming from the construction of molecular mappings to QTL analyses and followed by the marker-assisted breeding in the next book "Conditional QTL Analysis and Molecular Marker—Assisted in Wheat Breeding." This book provides a wealth of novel information, wide range of applications, in-depth knowledge of crop genetics, and molecular breeding, which should be valuable not only for plant breeders but also for academic faculties, senior researchers, and advanced graduate students who are involved in plant breeding and genetics.

Chapter 1
The Concept and Research Progress of Quantitative Traits

Abstract Quantitative traits are very common in nature; most agronomical important traits in crops are quantitative. In this chapter, the history and concept of molecular quantitative genetics, tools and methods to study quantitative traits, application of molecular markers, and progress and prospect of QTL mapping were introduced.

Keywords Quantitative traits · Molecular quantitative genetics · Genetic characters · Genetic markers · Applications of molecular markers · Progress of QTL mapping

1.1 History of Molecular Quantitative Genetics

In ancient times, our ancestors had established the methods of crop cultivation and animal domestication, which lead to the discoveries and applications of genetic variations and heredity. Since then, various hypotheses on the underlying mechanisms of genetic variations have been put forward. Based on his garden pea hybridization experiments, Gregor Mendel, an Austrian monk (1822–1884), established the Law of Segregation and the Law of Independent Assortment in 1864. However, these important discoveries were largely ignored and were not until thirty-five years later when Mendel's laws were rediscovered by Hugo de Vries (1848–1935), Carl Correns (1864–1933), and von Tschermak-Sysenegg (1871–1962) through experimentations with evening primrose (*Oenothera lamarckiana*), corn, and peas, respectively. These three European scientists further confirmed Mendel's conclusions and published their results in the Journal of German Proceedings of the Academy of Sciences (Journal Comptesrendus de l'Académie des sciences) in 1900, and hence the birth of genetics. During 1900–1952, genetic research entered into the cellular level, i.e., cytogenetics, thereby propelling subsequent research from the cellular level toward the current molecular level. Significant breakthroughs have been made since then: Watson and Crick discovered the DNA double helix in 1953, thus pioneering the modern molecular genetics and biology. In the early twenty-first century, the completion of the human genome

J. Tian et al., *Genetic Analyses of Wheat and Molecular Marker-Assisted Breeding, Volume 1*, DOI 10.1007/978-94-017-7390-4_1

sequencing project launched the post-genomic research era, and scientists are now ready for the next new biological challenges.

Throughout the history of the development of genetics, the main focuses have been on descriptions and analyses of quality traits. However, most agronomically important traits in crops, such as yield, quality, and disease resistance, are quantitative traits. In the initial stage, researchers had attempted to clarify the genetic mechanism of these traits and applied their findings to production practice. At the end of nineteenth century, Mendelian genetics and mathematics were combined to form the discipline of population genetics. In the 1920s, Fisher established quantitative genetics by combining population genetics and bio-statistics (Fisher 1918). Quantitative genetics is a branch of genetic research that places great emphasis on quantitative traits, and has been developed for nearly a century serving as the theoretical basis of breeding (Sun 2006).

It was Botstein et al. (1980) who first proposed the idea of using DNA restriction fragment length polymorphism (RFLP) as a genetic marker. Compared with cytological markers and biochemical markers, RFLP markers provide significant advantages such as higher level of polymorphism, larger numbers, better genomic coverage, fewer limiting factors, and easer to detect. Molecular markers are superb tools to study the genetic variations of biological traits and especially quantitative traits, and thus have galvanized great interest among geneticists and plant breeders, compelling continuous discoveries of new molecular genetic markers (e.g., RAPD, AFLP, SSCP, and SNP). These molecular markers brought great convenience and practicality. The merge of molecular genetics and quantitative genetics created the molecular quantitative genetics which will be the predominant impetus of the quantitative genetics in the twenty-first century. Revolutionary changes are expected to occur in the fields of QTL detection and localization, marker-assisted selection (MAS), marker-assisted introgression, and ultimately development of super quality new crop varieties. The most quintessential example is that the markers developed for QTL mapping have been used in MAS and breeding, which greatly improves the tracking and utilization efficacy of major QTLs, thereby accelerating germplasm innovation and variety selection. More recently, based on MAS breeding, two Belgian scientists (Peleman and van der Voort 2003) proposed a novel concept, known as "breeding by design." This new breeding idea is expected to take the breeding to a whole new level and may become the mainstream in the future crop genetic improvement technology.

1.2 Concept and Genetic Characters of Quantitative Traits

Unlike qualitative traits that are distinctively identifiable and can be grouped clearly within a population, quantitative traits show variations containing individuals with no clear-cut differences in a population (e.g., weight or height of animal and plant). Environments can have a great impact on quantitative traits and therefore the differences expressed by individuals within a group generally show a pattern of continuous normal distribution.

In order to characterize the genetic features of quantitative traits, a Swedish scientist (Eller 1909) put forward the "polygenic theory." He assumed that the information derived from qualitative character studies could still be used to explain the inheritance of quantitative traits. This theory is based on the assumption that the same trait is controlled by multiple genes and each of which contributes about equal, but minor effect. These multiple genes with minor effects were named as "minor genes," whereas those fewer in numbers with major effects as "major genes." The effects of minor genes controlling the same trait generally have additive effects, while alleles at these loci do not show dominant–recessive relationship. The key points of this "polygene hypothesis" of quantitative genetics are as follows: (1) quantitative trait controlled by multiple independent genes, each of which has minor effect and follows the Mendelian laws; (2) all genes have equal effect on the trait; (3) the alleles of each gene show an incomplete dominant or non-dominant relationship and contribute synergistic or antagonistic effect; and (4) express additive effects.

It has been now accepted that quantitative trait can be controlled by a few major genes with major effect or by a multiple minor genes with minor effect; genetic effect of each minor gene can be different; the mode of gene action can be additive or dominant within a pair of alleles, epistatic among the non-allelic genes as well as interaction effects between gene and environment.

Quantitative trait loci (QTLs) refer to their locations on the chromosomes in the genomes and their mode of actions and impact efficacy. QTLs are closely associated with the phenotypes derived from the continuous variable population. Currently, DNA molecular markers have been used in effective QTL mapping and efficacy evaluations.

1.3 Tools for Quantitative Trait Studies

Molecular markers are essentially genetic markers that are derived from DNA nucleotide changes occurring in individuals of a given population and represent the genetic polymorphisms at the DNA level. The rapid development of molecular biology techniques have generated a number of DNA molecular markers, and these markers have been successfully applied to plant breeding, genome mapping, gene mapping, identification of species evolutionary relationship, gene library construction, and gene cloning (Aneja et al. 2012).

Molecular markers, in a broad sense, refer to DNA sequences or biochemical markers (e.g., isozymes, allozymes, and some other protein-based markers) that are inheritable and detectable, while in a narrow sense, they are specific DNA sequences that can reflect the genetic differences among individuals or populations.

1.3.1 Types of Molecular Markers

Based on the techniques, molecular markers can be divided into hybridization-based markers, PCR-based markers, restriction enzyme-based markers, and DNA chip-based markers. The most commonly used markers today are PCR-based markers.

1. The hybridization-based markers include RFLP and Variable Number of Tandem Repeats (VNTR). The technique involves digestion of DNA with a restriction enzyme, separation of DNA fragments via gel electrophoresis, and transfer of DNA fragments to a filter for hybridization with radioactively labeled probes.
2. PCR-based molecular markers are commonly used. Among them, some are of arbitrary nucleotide sequence and according to PCR amplification of random segments of genomic DNA, for example, random amplified polymorphism DNA (RAPD) and DNA amplification fingerprinting (DAF).
   Also, PCR-based markers can be designed through the use of specific known nucleotide sequences as their primers (usually 18–24 bp long) and can be amplified through regular PCR programs. These primers can then be used for analyzing the polymorphisms present in a specific region of the genome. Sequence tagged sites (STS), simple sequence repeat (SSR), sequence-characterized amplified region (SCAR), single primer amplification reaction (SPAR), single-strand conformation polymorphism (SSCP), and dideoxy fingerprints (DDF) belong to this type of PCR-based markers.
3. Amplified fragment length polymorphism (AFLP) and cleaved amplified polymorphism sequences (CAPS) are examples of the restriction enzyme-based markers.
4. Single-nucleotide polymorphism (SNP) is an example of DNA chip-based molecular marker. SNP reveals a single-nucleotide change occurring within coding or non-coding regions of genes in a population. SNP can be used to identify the genetic difference between two or more individuals, and is known as the third generation of molecular marker. With the advancement of DNA microarray technology, SNP is expected to become the most important and effective molecular marker.

1.3.2 Applications of Molecular Markers

Molecular markers can be used in many areas of crop breeding programs, of which constructions of genomic and gene maps are the most important applications.

1.3.2.1 Germplasm Identification and Genetic Diversity Analysis

Genetic diversity evaluations provide a great scientific basis for studies on species evolution, variety identification, parent selection, variety protection, and ultimately better utilization of germplasm and more efficient breeding. Molecular markers are widely present in the genome. By comparing the polymorphisms of these randomly distributed genomic markers, one can evaluate the examined populations comprehensively, thereby revealing their genetic nature. These data can then be used in species/variety cluster analysis which will allow better understanding of their phylogenetic and genetic relationships. The high level of polymorphism of molecular markers will continue to play a very important role in kinship studies, species classification, germplasm identification, and protection.

1.3.2.2 Genetic Linkage Map Construction and Gene Mapping

A genetic map of a species or population provides information of the relative positions of genetic markers located on a chromosome. Such information is often based on the recombination frequency gained from crossovers occurring between homologous chromosomes and provides the basis for plant breeding and molecular cloning.

For a long time, morphological, physiological, or biochemical markers are commonly used for genetic map constructions of various species. However, these linkage maps are often associated with low resolution, low saturation, and too large distance between markers, resulting in limited application value. Discovery of abundant types of molecular markers allows the continuous additions of novel markers to the linkage maps and further increase in marker density, suggesting that construction of a genetic map with desirable density is highly achievable. High-density genetic maps make it possible to use relatively simple and effective molecular markers in gene mapping and cloning. Molecular marker has been proven to be a powerful tool for quick, reliable, and efficient gene mapping.

1.3.2.3 Map-Based Cloning

Map-based cloning, also known as positional cloning, was first proposed by Coulson in 1986. This method is based on the information of the targeted gene location on the chromosome without knowing its DNA sequence and its final expression product, but with the knowledge of the marker and the tightly linked gene of interest. Map-based cloning is commonly used in almost all gene identifications. The high-density genetic map, large physical map, broad sequences in the gene library, and whole-genome sequence have provided a good foundation for wide applications of map-based cloning.

1.3.2.4 Marker-Assisted Selection Breeding

Traditional plant breeding is mainly dependent on the phenotype selection. Many factors, such as environmental conditions, gene–gene interaction, and gene–environment interaction, can have significant impact on selection efficiency. MAS allows breeders to select the traits of interest at early generation or select the individuals showing the recurrent parental background based on the molecular markers that are closely linked with these genes.

Markers that are closely linked with those agronomically important traits are very valuable for marker-assisted breeding programs as they can further improve breeding efficiency and accelerate the variety development process. Therefore, screening of these types of markers is essential for MAS breeding.

1.4 Progress and Prospect of QTL Mapping

1.4.1 Methods of QTL Mapping

Classic quantitative genetics is established on the basis of multigene hypothesis. It considers the genes underlying the quantitative traits as an entirety, and focus on analyzing and estimating the decomposition of varied genetic effects and genetic variances. The advent of molecular linkage map made it possible to study the quantitative trait genes as the study quality trait genes. It allows the identification of a single QTL gene on a specific chromosome and evaluates the QTL genetic effect. Hence, this process is called QTL mapping.

Similar to mapping a single gene, all QTL genes can be mapped on the genetic map using the distance between the QTL and genetic marker (expressed with recombination rate). Based on the number of markers used, QTL mapping can be single-marker, double-marker, or multiple-marker mappings. Also, methods of statistical analysis for QTL mapping can be different: the mean and variance analysis, regression and correlation analysis, moments and maximum likelihood analysis, and marker intervals used in mapping can vary: single-interval mapping and multiple-interval mapping. In addition, comprehensive analysis derived from the combination of several of these analyses [e.g., QTL composite interval mapping (CIM), multi-interval mapping (MIM), multiple QTL mapping, and multitrait mapping (MTM)] has also been used for QTL mapping.

1.4.2 QTL Mapping Progress

Currently, molecular markers have been successfully applied to the breeding programs of corn, soybeans, chickens, pigs, and many others, and significant progress has been made in gene mapping, marker-assisted breeding, disease therapy, and other applications. Current plant QTL studies can be summarized as follows.

1.4.2.1 QTL Mapping in Crops

The repaid development of molecular genetics has rejuvenized the crop breeding programs: quantitative traits can be separated into a number of discrete components that are controlled by numerous Mendelian factors, and their locations on chromosomes and relationships with other genes can be determined.

Over the past few years, significant progress has been made in crop QTL mapping. Statistical data, by the end of 2012, showed that more than 4200 genetic maps had been constructed using various molecular markers, most of which being SSRs. These QTL traits are mainly associated with plant morphology, yield, grain quality, disease resistance, and many other agronomically important traits. Successful QTL mappings include food crops (e.g., rice, corn, bean, and wheat), economic crops (e.g., canola, hemp, and sunflower), vegetable crops (e.g., tomato, carrot, cucumber, cabbage, artichoke, bean, coriander, lettuce, pepper), fruit crops (e.g., apple, peach, apricot, walnut, plum, cherry, strawberry, wild pear, wild apricot, and mountain cherry), and some forage crops (e.g., *Astragalussmicus*). More than 60 % of these QTL mappings belong to food crops. Currently, there are about 180 genetic maps in wheat, most of which are SSR maps, involving morphology, yield, gain quality, disease resistance, as well as other important traits (Besnier et al. 2010; Wurschum 2012).

1.4.2.2 Progress of Molecular Markers Used in QTL Mapping

For the past 30 years, molecular marker technologies have developed very rapidly. To date, molecular markers comprised of four major categories which can be categorized into several dozens of specific kinds. Meanwhile, novel molecular marker systems emerge continuously. In addition to those that are commonly used in the past and present [e.g., RFLP, (Devos 1993), RAPD (Williams et al. 1990), AFLP (Vos 1995), and SSR (Torada 2006) markers], sequence-related amplified polymorphism (SRAP) has recently been used widely in crop quantitative genetics research due to its advantages such as higher polymorphism, more informative, easier to operate, higher reliability, repeatability and stability, and lower cost (Li et al. 2001; Anejaet et al. 2012).

Another promising marker system known as diversity arrays technology (DArT) is now available (Jaccoud et al. 2001). Advantages of this marker system include high-throughput, highly automatic, and known sequence is not required. For these reasons, it has been widely used in genetic map constructions and gene mapping in wheat. The SNP marker system (Brookes et al. 1999; Rafalski et al. 2002), on the other hand, is the most common variant in the genomes of all species and thus is more valuable than SSR markers in building high-density genetic map, fine mapping of targeted genes, and gene cloning. The rapid development of SNP detection and analysis, especially by combining DNA microarrays and microarray technology, makes SNP become the most promising third-generation molecular marker following RFLP and SSR. This suggests that the combination of a variety of techniques can generate novel marker systems that create more comprehensive and diversified DNA sequence fragments, thereby identifying gene expression differences, discovering new genes, and exploring molecular mechanism of stress resistance more efficiently.

1.4.2.3 Progress of QTL Mapping Method

Various populations can be used in QTL mapping and can be generally divided into two categories: temporary segregating population, including F_2 population, back cross population and permanent segregating population, including doubled haploid (DH) population and recombinant inbred lines (RIL) population. These populations usually have complex genetic background and have limited fine mapping accuracy (usually in 10–30 cm interval, Kearsey and Farquhar 1998). During the process of crop improvement, breeders need to select the best alleles for developing new cultivars, and yet the use of these two populations can only evaluate the two alleles contributed by two parents. The linkage disequilibrium (LD)-based association analysis developed recently can resolve this issue as the natural populations or inbred varieties can be used for QTL mapping studies, and can also identify genes controlling important agronomic traits and explore elite allelic variations. LD refers to the non-random association of alleles at two or more different loci, which descends from single, ancestral chromosomes. The level of LD is determined by many factors (e.g., genetic linkage, selection, and recombination rate). Therefore, QTL location can be determined by measuring the degree of LD between the marker and the QTL. Comparing with linkage mapping, association mapping has high mapping accuracy and saving time as construction of segregating populations are not required (Gupta et al. 2005).

1.4.2.4 Dynamic Studies of Plant QTL

Studies on quantitative traits remained mainly at the DNA level in the past. Recently, QTL mapping information combined with gene expression analysis has been applied to QTL studies. Gene expression level is considered to be a quantitative trait and generated genetic genomics or gene expression QTL (eQTL). eQTL is not simply an extension of QTL studies, but it puts more emphasis on gene interactions, gene regulation networks, and metabolic pathways. Therefore, eQT provides a novel approach to study the molecular mechanisms of complex quantitative traits and to establish genetic regulatory networks. Currently, applications of eQTL have been extended to *Zea mays*, *Arabidopsis,* and other crops (Sun 2012).

Quantitative or qualitative traits are usually the terminal expressions of these traits during plant ontogeny. QTL analysis can only explain accumulative effect, but cannot explain QTL expressions at different developmental stages, mode of action, and effect. Traits represent the results of continuous expressions of their underlying genes, and different QTLs usually generate different expression profiles. It is thus necessary to combine trait development with the quantitative genetics, so that QTL studies from a dynamic perspective is possible. The dynamic QTL analysis reveals the QTL effects under different developmental stages (Zheng et al. 2011).

1.4.3 Application Prospects of QTL Mapping

Although QTL studies in crops have made huge progress over the last ten years or so, the efficient use of the QTL information on novel gene discoveries and agronomically important crop breeding programs has not met the expectations. In order to fully understand and utilize the genes of quantitative traits and assist the breeding programs, theoretical and applicable researches of QTL mapping should be further strengthened.

1.4.3.1 QTL Applications to Identifying and Cloning Genes of Interest

As stated above, despite the fact that crop quantitative trait studies have been a popular subject for a long time, resulting in numerous reports, most studies still remain on the surface level. Understanding the genetics of QTL traits at the molecular level is essential. QTL cloning is considered to be one of the major challenges in life sciences in the twenty-first century. Until now, QTLs that have been cloned are those either with major effect or with relatively large effect on the traits of interest. This is largely because the genetic effects of minor QTLs are often overshadowed by environmental effects, resulting in unmatched phenotypes and genotypes and ultimately inaccurate QTL mappings. Nevertheless, the use of newly developed technologies such as high-throughput SNP will undoubtedly change this situation and more agronomically important major and minor QTLs will be mapped and cloned. For the crops with no reference genome or limited sequence information, their QTL mappings can be realized by using the comparative genomics approach. This approach allows identification of candidate genes based on the sequence information present in the homologous regions of the model plants (e.g., *Arabidopsis thaliana*) and makes the QTL mappings of these crops possible.

1.4.3.2 QTL Application in Gene Interaction and Regulatory Networks Studies

Quantitative traits are controlled by the cellular biochemical networks that are regulated by a series of genes. Better understanding of these networks will allow us to identify genes controlling the complex traits and elucidate their roles in metabolic pathways and plant development. The use of dynamic QTL analyses and mutants increased the number of detectable QTLs. Clearly, the complexity of quantitative traits is not only due to the QTLs being controlled by multi genes, but also due to gene interactions, epistatic and environmental effects. Currently, detections of QTL interactions mainly depend on the software-based statistical analyses. The authenticity of these results remains to be further tested, perhaps through the establishment of nearly isogenic lines or other means. The establishment of eQTL offers a novel approach for epistasis detections and regulatory network constructions.

1.4.3.3 QTL Application to the Efficient Breeding Programs

The goal of QTL studies is to conduct MAS to improve breeding efficiency. The reasons that molecular MAS has not been widely used in many crops are many, and these include the lack of fully understanding of QTL's genetic basis and labor intensive when working on multiple genes using molecular markers. Therefore, it is essential to develop more efficient QTL mapping techniques in order to fully understand and utilize QTL genes, and ultimately further improve the breeding efficiency. With the rapid developments of plant functional genomics and molecular biology techniques, in addition to the use of new markers and new ideas, the knowledge gained from QTL studies will undoubtedly be well applied to the more efficient breeding programs.

1.4.3.4 QTL Application in Improving Breeding Level

Modern plant breeding plays a great role in promoting both development of agriculture and the national economy. According to the statistics over the past 60 years, crop varieties in China have been substituted for 5–6 times, resulting in the average yield increase of seven times. Approximately 35–40 % increase in food production was contributed by successful crop breeding programs. Nevertheless, food supply has turned from relative excess to relative tight worldwide in the past ten years, while the situation in China has not been optimistic either, from a relative balanced status to currently structurally insufficient situation.

Therefore, to secure the global sustainable food and edible oil supply and meet the national strategic requirement, it is crucial to develop novel varieties that have higher yield, better quality, and more stress tolerant potential.

It has been generally accepted that conventional breeding is mainly dependent on phenotypic selection, resulting in low breeding efficiency, while MAS breeding has been difficult in assisting quantitative trait selection directly. Meanwhile, development of novel cultivars through the transgenic approach has been largely hampered by the fact that most of the agronomic traits are controlled by multiple genes, and transferring them all at once is a huge challenge at present. Recently, Belgian scientists Peleman and van der Voort (2003) proposed a novel concept: molecular design breeding (breeding by design). Molecular design breeding uses bioinformatics as its platform, genomics and proteomics database as its basis, and integrates the information on genetics, physiology, biochemistry, and biological statistics. Based on these data and breeding objectives and developmental environment required by any specific crops, one can design breeding program prior to conducting the real experiment. Specifically, molecular design breeding process includes screening polymorphic markers, constructing marker linkage maps, evaluating quantitative trait phenotypes, analyzing QTLs, obtaining and then assembling breeding elements of the specific QTL traits, and finally conducting field selections. Compared with other breeding methods, molecular design breeding simulates a model with a computer first and then considers a large number of factors. Therefore, this breeding approach is much more comprehensive, more efficient and accurate in selecting parental combinations, and ultimately better meeting the breeding requirement. It is expected that super crop verities with higher yields, better quality, and more resistance to various stresses will be developed through this breeding approach, which take the breeding program to a whole new level.

References

Aneja B, Yadav NR, Chawla V, Yadav RC. Sequence-related amplified polymorphism (SRAP) molecular marker system and its applications in crop improvement. Mol Breed. 2012;30:1635–48.

Besnier F, Le Rouzic A, Alvarez-Castro JM. Applying QTL analysis to conservation genetics. Conserv Genet. 2010;11:399–408.

Botstein D, White R, Skolnick M, Davis R. Construction of a genetic linkage map in man using restriction fragment length polymorphisms. Am J Human Genet. 1980;32:314–31.

Brookes AJ. The essence of SNP. Gene. 1999;234:177–86.

Coulson A, Sulston J, Brenner S, Karn J. Toward a physical map of the genome of the nematode C. elegans. Proc Natl Acad Sci USA. 1986;83:7821–5.

Devos KM, Millan T, Gale MD. Comparative RFLP maps of homoeologous group 2 chromosomes of wheat, rye, and barley. Theor Appl Genet. 1993;85:784–92.

Fisher RA. The correlation between relatives on the supposition of Mendelian inheritance. Trans R Soc Edinb. 1918;52:399–433.

Gupta PK, Rustgi S, Kulwal PL. Linkage disequilibrium and association studies in higher plants: present status and future prospects. Plant Mol Biol. 2005;57:461–85.

Jaccoud D, Peng K, Feinstein D, Kilian A. Diversity arrays: a solid state technology for sequence information independent genotyping. Nucleic Acids Res. 2001;29:25–31.

Kearsey MJ, Farquhar AG. QTL analysis in plants; where are we now? Heredity. 1998;80:137–42.

Li G, Quiros CF. Sequence-related amplified polymorphism (SRAP), a new marker system based on a simple PCR reaction: its application to mapping and gene tagging in Brassica. Theor Appl Genet. 2001;103:455–61.

Nilsson-Ehle, "KreuzungsuntersuchungenanHafer und Weizen," Lunds Univiversitets Årsskrift. N.F. Atd 2, 5, Nr. 2 (1909) 1–122, N.F. Afd. 2, 7 (1911) Nr. 6, 1–84.

Peleman JD, van der Voort JR. Breeding by design. Trends Plant Sci. 2003;8:330–4.

Rafalski JA. Application of single nucleotide polymorphisms in crop genetics. Curr Opin Plant Biol. 2002;5:94–100.

Sun YZ. Quantitative genetics. Beijing: China Agriculture Scientech Press; 2006. p. 1–8.

Sun W. A statistical framework for eQTL mapping using RNA-seq data. Biometrics. 2012;68:1–11.

Torada A, Koike M, Mochida K, Ogihara Y. SSR-based linkage map with new markers using an intraspecific population of common wheat. Theor Appl Genet. 2006;112:1042–51.
Vos P, Hogers R, Bleeker M, Reijans M, Van de Lee T, Hornes M, Frijters A, Pot J, Peleman J, Kuiper M, Zabeau M. AFLP: a new technique for DNA fingerprinting. Nucleic Acid Res. 1995;23:4407–14.
Williams JGK, Kubelik AR, Livak KJ, Rafalski JA, Tingey SV. DNA polymorphisms amplified by arbitrary primers are useful as genetic markers. Nucleic Acids Res. 1990;18:6531–5.
Wurschum T. Mapping QTL for agronomic traits in breeding populations. Theor Appl Genet. 2012;125:201–10.
Zheng LN, Zhai HQ, Zhang WW. Dynamic QTL analysis of rice protein content and protein index using recombinant inbred lines. J Plant Biol. 2011;54:321–8.

Chapter 2
Genetic Analysis Methods of Quantitative Traits in Wheat

Abstract It is very important of understanding how to construct the genetic population and genetic maps, what methods should be used, etc., to researchers; so in this chapter, the types and quality of genetic population, construction methods of genetic population, types of genetic markers, and statistical methods of QTL mapping were introduced; moreover some new methods and key notes from our study experience were especially provided.

Keywords Genetic populations · Genetic markers · Population construction · Populations Quality · QTL mapping · Genetic analysis methods

In order to conduct genetic analyses of wheat quantitative traits, it is essential to establish proper genetic populations, select appropriate DNA markers, construct genetic maps and conduct QTL mappings. As the old saying stated, "traditional breeding often depends on experience, while molecular breeding relies on materials." Here the word "materials" simply refers to the genetic populations for quantitative trait studies. Therefore, this chapter presents the type and quality of genetic populations followed by a brief introduction of the genetic analyses of wheat quantitative traits.

2.1 The Types and Quality of Genetic Populations

2.1.1 The Types of Genetic Populations

A genetic population, in a narrow sense, is often derived from a F_1 hybrid resulted from a cross between two pure breeding parents with distinct alleles, and it represents all genotypes of the family. In theory, such population should cover all loci, including those of homozygous and those of heterozygous. The basic guideline for creating an ideal genetic population is to avoid any man-made interruptions or selections during the population construction. However, to gain all genotypes is impossible due to infertility derived from some reproductive disorders, and mortalities or loss because of various environmental stresses or human factors.

J. Tian et al., *Genetic Analyses of Wheat and Molecular Marker-Assisted Breeding, Volume 1*, DOI 10.1007/978-94-017-7390-4_2

Currently, two types of genetic populations for genetic map construction are used: temporary population and permanent population. The former covers F_2 and its derivatives of F_3, F_4 and BC (backcross) lines, and the latter includes doubled haploid (DH), recombinant inbred line (RIL), immortalized F_2 (IF_2), and near-isogenic lines (NIL) populations. In recent years, association mapping based on (LD) linkage disequilibrium has been widely used to select natural genetic populations containing various variations, and such population can be a commercial variety, a newly bred variety candidate, or germplasm. This suggests that these natural populations belong to permanent populations.

Based on the accuracy of QTL mapping, there are two types of mapping populations: primary population and secondary population (also known as fine mapping population). The former includes F_2 and its derivatives (e.g., F_3, F_4 lines), BC_1, BC_2F_X, DH, RIL, and IF_2 (derived from DH or RIL). Due to the interference of the genetic background, the confidence intervals of the QTLs identified with these populations are often larger than 10 cm. According to the population sources, the secondary mapping populations can be further divided into two groups: those derived from primary mapping populations via further selections [(e.g., NILs, residual heterozygous lines (RHLs), and QTL isogenic recombinants lines (QIRs)]; those substitution populations that have no relationship with the primary populations, [e.g., introgressive lines (ILs), single-segment lines (SSLs), and chromosome segment substitution lines (CSSLs)]. The QTL fine mapping population can eliminate the interferences of various genetic backgrounds, thereby gain better accuracy in QTL mapping.

The following is a brief introduction of the characteristics of several commonly used mapping genetic populations and a few key points to consider during the constructions of QTL mapping populations.

2.1.1.1 F_2 and Its Derived F_3 Population

In theory, the F_2 population should represent all the possible recombinants due to the random fusions of male and female gametes produced by selfing F_1 individuals and thus should generate abundant genetic information. This unique population can be easily created as it is rather simple. However, using F_2 as mapping populations has several significant limitations, including the following: (1) since phenotypic identifications are based on the individual plants, detection of many QTL traits with low heritability can be difficult; (2) F_2 is a temporary unstable population and long-term maintenance is an issue as its genetic structure will change, leading to difficulties to conduct multiple experiments each year at multiple locations after its sexual reproduction; (3) the presence of heterozygous genotypes in F_2 makes identification of homozygous or heterozygous dominant loci difficult, leading to low mapping accuracy. Therefore, only QTLs with significant effects and stable expressions can be detected when F_2 population is used. An alternative to this issue is the use of "mixed F_3" families derived from F_2. This approach involves the use of mixed DNA of all F_3 individuals to analyze their F_2 individual genotypes. A genetic map can also be constructed when each F_3 individual is evaluated. For each locus, segregation ratio will be 3:2:3, instead of 1:2:1. This is because, for a F_2 heterozygous locus, there is only one chance, but two in F_3. The disadvantages, however, are labor intensive and the increase in sampling errors.

2.1.1.2 Backcross Population

BC_1 is the result of F_1 being backcrossed with one of its two parents and has been commonly used as mapping population. Since each segregating locus in BC_1 has only two genotypes, the resulted population represents the segregation ratio of F_1 gametes. Compared to the F_2 population, BC_1 provides the highest mapping efficiency. Yet, similar to F_2, the BC_1 population can be used for only one generation, cannot be maintained for a long term, and can provide limited information. For these reasons, direct use of BC_1 for QTL mapping is rare, unless there is a need for some special studies (e.g., cross-incompatibility). A permanent BC_1F_X population can be created via the repeated selfings of each individual line in $BC_{1.}$

2.1.1.3 DH Population

Doubled haploid (DH) lines are normally created through in vitro cultured anthers/immature microspores or ovaries/ovules (representing male or female gametes, respectively) followed by genome doublings (Fig. 2.1). Therefore, each DH

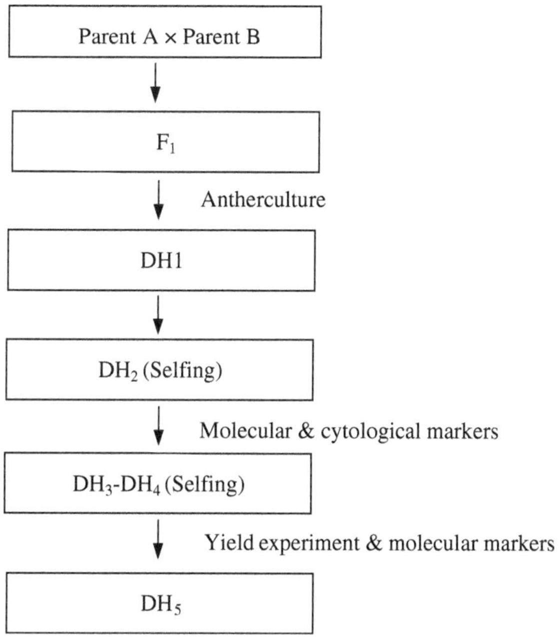

Fig. 2.1 Schematic diagram of DH generation and evaluation

line is completely homozygous and is a "permanent population." The advantages of such populations include the following: (1) they can be used for multiple experiments to minimize the experimental errors and (2) can be planted under various environments and in multiple years to study interactions between genotype and environment as DH lines are the most ideal materials. Moreover, the DH lines are also ideal for QTL fine mapping (Li et al. 1996) because their genetic structures represent the segregations and recombinations of alleles during F_1 gamete formation. The disadvantages of DH lines are as follows: (1) generation of DH lines often depends on the anther culture technology, resulting in interruptions of genetic structures and biased partitions due to possible differential responses of different gamete genotypes to the culture conditions; (2) limited rate of recombinations and lack of heterozygosities because there is only one meiotic division during gamete formation; and (3) DH lines can be used only for analyses of QTL additive effect, but not for dominant effect, which would affect the QTL mapping accuracy (Li et al. 2005).

2.1.1.4 RIL Population

A RIL is a permanent population and is developed by single-seed descent (select one individual plant each generation) of F_1, and this selfing process continues for multiple generations. Each individual genotype within the RIL population is a homozygous stable line, while different individuals may represent various genotypes. Similar to the DH populations, the RILs are permanent and can be experimented multiple times at various sites each year. However, compared with DH lines, RILs are unique that rates of recombinants within each linkage interval of the two homologous chromosomes increase significantly due to the repeated selfing process. Therefore, the use of RILs can break different QTLs located in the same chromosomal region, indicating that the RILs are ideal materials for QTL mapping (Burr 1988; Lin et al. 2008). The disadvantage of RIL population is that developing RILs are time-consuming and some lines may be eliminated during their continuous selfing process, leading to biased segregation.

2.1.1.5 NIL Population

Derived from a F_1 that was backcrossed repeatedly to one of its parents, the NIL population consists of lines with the same or similar genetic backgrounds, and yet with the variations only present in a few chromosomal regions. Hence, the NIL population is unique because the numbers of molecular markers required for mapping the targeted genes are fewer than other populations. Under the same genetic background, the NIL actually allow multiple QTLs that affect the same trait to break into individualized Mendel factors, convert the quantitative traits to qualitative traits, and thereby eliminate the interferences derived from genetic backgrounds and remove the masking effects of major QTLs over minor QTLs. NIL can be used to conduct map-based cloning of those targeted genes and fine gene mapping (Yano et al. 1997).

2.1.1.6 Segment Introgression Lines

A segment introgression line (also known as chromosome segment substitution line) is developed by crossing the F_1 with one of its parents repeatedly, by which a chromosome fragment of a variety is introgressed into the chromosome of the other. The desirable backcrossing generations can vary and often are based on the aims of the particular breeding and research programs. For breeding purpose, backcrossing two to three times is good enough when the transferred donor's DNA is about 12.5–6.25 % of the recipient's DNA. On the other hand, the backcrossing times should be increased so that a population with various DNA fragments can be created for QTL analyses.

2.1.1.7 Immortalized F_2 Population

Immortalized F_2 population (IF_2) was initially proposed by Hua et al. in (2003) and is unique as it combines the advantages of the segregating F_2 and the eternal RIL populations. The IF_2 is developed by the designed "2–2 crosses" between the homozygous lines of the eternal population. Importantly, the IF_2 population can not only provide rich information similar to the F_2 population so that the effects of both dominant and epistatic can be estimated effectively, but also, acted as the RIL or DH lines, can produce enough seeds to meet the demand for multiple trials at various sites each year, leading to gaining accurate phenotype data and identifying the closely linked QTL markers effectively.

Additionally, the IF_2 population can be used for heterosis QTL mappings since it can be trialed repeatedly at multiple sites within multiple years, which is impossible if F_2 or RIL eternal population is used alone. The main drawbacks to use IF_2 are (1) making massive cross combinations are both labor intensive and challenging, not all crosses can give rise to enough seeds, resulting in insufficient data and (2) completely random crosses are difficult to implement because heading time of RIL or DH populations can be different. These factors can be an issue as the established eternal IF_2 population may make the predicted data to be biased, and consequently lead to an abnormal population structure that has biased QTL locations and incorrect effect evaluations.

2.1.1.8 Fine Mapping Populations

Fine mapping populations include NILs, RHLs, QIRs, ILs (DNA segment introgression lines), single-segment substitution lines (SSSLs), and CSSLs.

The populations of NILs and ILs have been presented and shown below are discussions on the other remaining four mapping populations.

Residues of Alloplasmic Lines

Residues of alloplasmic lines (RHLs) are populations that consist of individuals with the presence of one or a few traits contributed by one parent, while other characters at other loci are derived from the other parent. Segregations of traits at the loci being examined occur constantly. RHLs have similar genetic background and can be used for marker-assisted selection, but cannot be used for epistatic effect evaluation.

QTL Isogenic Lines

QIRs are developed by a preliminary mapping initially using a small population followed by fine mapping using a large population. Each individual within the large population will have one recombination event occurring at the related QTL locus with no change at any remaining loci. QIR has both advantages (e.g., it is easy to construct and the molecular marker distance can be shorter than 1 cm) and drawbacks (e.g., background interference is present, and there are epistasis effect cannot be identified).

Single-Segment Substitution Lines

SSSL is very similar to NIL and is also developed through multiple generations of backcrossing (Fig. 2.2). An ideal SSSL should maintain all recipient genetic background with the exception of the targeted QTL DNA segment that comes from the donor's chromosome, strongly suggesting that SSSL can be used for fine mapping of single QTL. However, during the backcross process, the QTL gained via the preliminary mapping should be used for assistance selection, resulting in issues such as labor intensive and tedious.

Chromosome Segment Substitution Line

Different from SSSL, CSSLs are a series of NILs in which the substituted segments of the wide population contain the entire information of the donor, while each CSSL carries one or more donor chromosome segments in the genetic background of the recipient. The main characteristic of CSSLs is that the substituted segments of each CSSL are stable. As a result, CSSLs are useful for genetic studies in terms of the detection and fine mapping of QTLs for genome-wide target traits, and for studying the interactions between QTLs.

The introgression of fragments is mainly achieved through genetic recombination. Through backcross breeding, the lines carrying any genomic region can be produced. The selection mode adopted in the backcross process can be various; the final target should be that the lines have a single homozygous chromosome with the donor, while the other segments of the chromosome are from the acceptor parent.

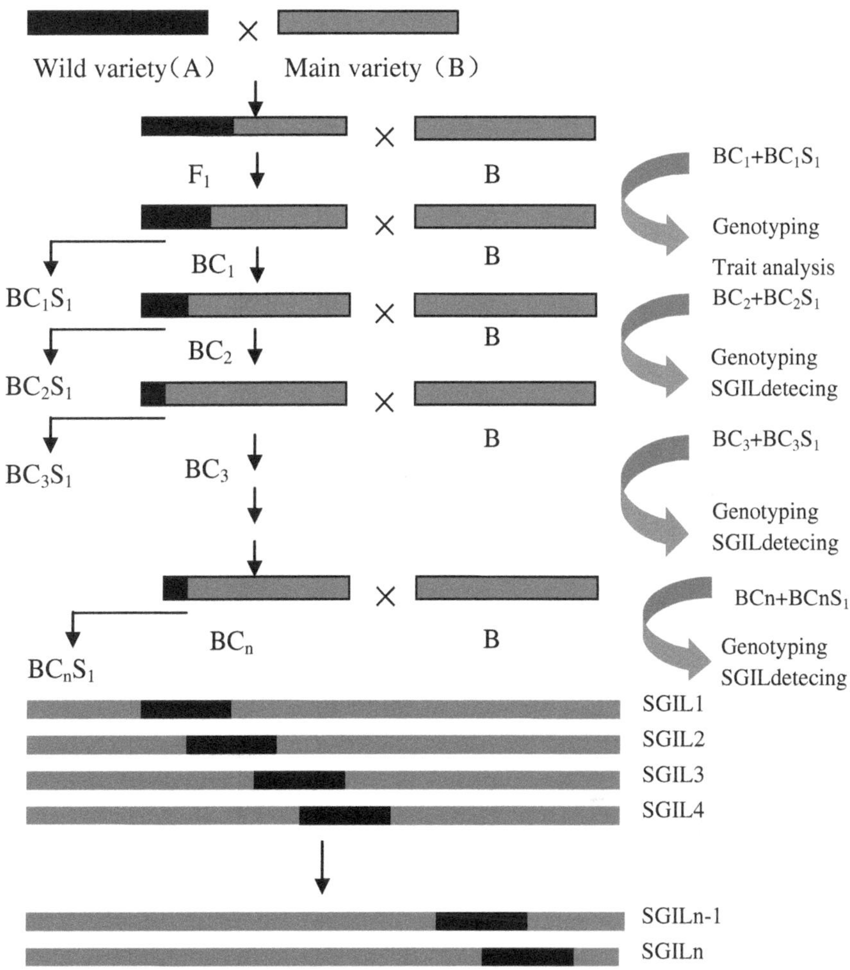

Fig. 2.2 Schematic diagram of SSSL construction (Reprinted from Zhang et al. 2004)

2.1.2 Genetic Population Construction and Some Key Notes to Consider

2.1.2.1 Construction Method

It is true that constructions of genetic populations should be based on each specific population even though most of which need to generate their F_1 as the initial step and share some common points.

F_2 Population Construction

A F_2 population is produced by selfing F_1 individuals. The key points include the following: their F_1 initial parents should show significant polymorphism the number of hybrid ears should be determined based on the F_2 population size, and each individual plant derived from each F_1 seed is a representative of the F_2 population.

IF_2 Population Construction

To establish an IF_2 population, first divide the DH or RIL population into two groups, each of which consists of a certain number of lines. Second, select one line from each group to make a cross match, followed by doing the same repeatedly from using the remaining lines of each group. After the 1st round crosses, a population with 1/2 of the DH or RIL lines can be obtained, and the 2nd round crosses will result in the population size equal to that of DH or RIL lines, thereby a whole set of IF_2 is created. This progress can be repeated yearly or one can make enough seeds per year to conduct trials at multiple sites per year.

BC_1 Population Construction

To create a BC_1 population, the F_1 is crossed back with one of its initial two parents. The backcross numbers are within 10–100 ears to develop a large population for the following desired studies.

NIL Population Construction

NIL is generated by crossing BC_1 with the same parent for at least four more generations (BC_5), resulting in the genetic compositions of the two isogenic lines are almost identical, with the exception of the targeted traits. The individuals selected for the backcrosses during the NIL developing process should be determined based on the NIL's targeted traits (e.g., ear size, plant height, grain weight, disease resistance, and quality). In addition to the field observation on the target traits, trait identification should be conducted by combining the biochemical markers with the DNA markers to speed up the NIL construction process (Fig. 2.3). Developing a NIL with 3–4 targeted traits using the same population is considered to be cost-effective.

RIL Population Construction

RIL populations are established via single-seed selections from individual plants of a F_2 population and the detailed process includes creating a F_2 population by selfing F_1, selecting 300 or more individuals (randomly select those showing significant

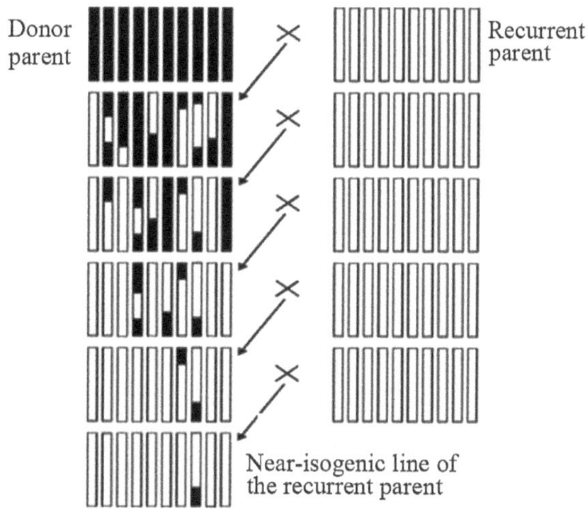

Fig. 2.3 Schematic diagram of NIL construction (Reprinted from Fang et al. 2001)

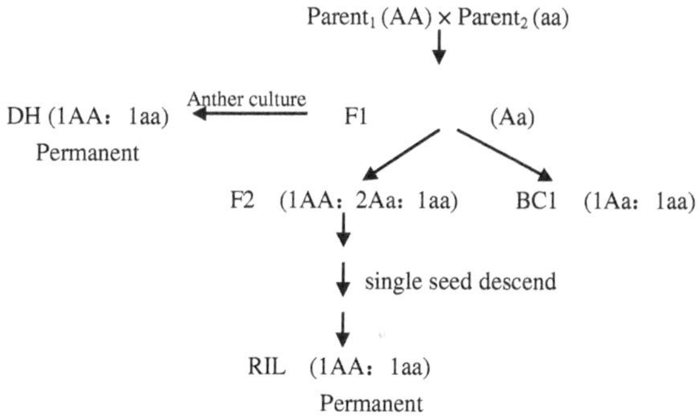

Fig. 2.4 Schematic diagram of genetic population constructions

polymorphisms) from the segregating F_2 population, and labeling them clearly and selfing the selected F_2 plants to obtain the F_3 seeds; this process is repeated until at least F_6 generations when all traits of each F_6 line are stabilized, which can be used for QTL mappings (Fig. 2.4). Noticeably, RIL construction is based on the "single-seed descent" principal, yet some genotypes can be eliminated due to poor sowing conditions if only one seed of each line is sown per generation, resulting in the number of lines in the population does not meet the QTL mapping requirements. To solve this issue, we select one ear/line/generation and sow all seeds in a single row starting from the F_2 generation until the F_6 generation when all traits are stabilized and are ready for QTL mapping.

DH Population Construction

DH populations are normally derived from the in vitro-cultured gametophytic cells of F_1 hybrids (known as anther/ovule culture) followed by genome doubling, although haploids can also generated via cultured zygotes of some rare interspecific hybrids by eliminating the whole genome of one parent. Recently, a novel approach to inducing haploids is developed: when a centromere mutant (cenh3) is crossed to a wild type, the mutant genome is eliminated, and hence DH lines of the paternal or maternal wild parent are generated after genome doubling. In our laboratory, we develop wheat DH lines through anther culture. One of the key factors for successful DH production is to culture the anthers containing embryogenic microspores at mid-uninucleate to late uninucleate stages. Specifically, F_1 hybrid's anthers are harvested, in the field prior to heading, when the spike length reaches 2/3 of leaf sheath in the northern winter wheat region, and young panicle length is at 1 cm of the auricles or spikes and auricles are equal in length in the south of Huanghuai district, respectively. Importantly, anther length is about 2/3 of its total length and anther should look green and opaque.

Prior to culturing, ear with the sheath (after other leaves are removed) is disinfected in 70 % alcohol for 10 s after which sheath is removed. Anthers are taken out from the outer and inner glumes and placed onto the culture medium for cell de-differentiation or callus induction. The whole process is conduced under strict sterile condition in a laminar flow hood. The initial culture should be in the darkness and the tissue culture room temperature should maintain at 28–30 °C.

Cultures are transferred onto the plantlet initiation medium when callus size is about 1–1.5 mm in diameter, which requires about 30 days de-differentiation. This process takes about 7–15 days at conditions of 23–25 °C and daylight of 10 h. Plantlets of 2–3 cm long are then transferred onto a new medium for their further development for 7 days after which being placed in a growth chamber with temperature of 6–10 °C and light until mid-October. Plantlets are finally taken out and planted in a plot where transparent plastic film is provided on the top for acclimatization. Ten days later, the plastic cover can be removed and plants should be ready to go through the winter.

To double the genomes of these haploids, apical shoots of all tillers are exposed in an antimicrotubule agent such as colchicine or several types of herbicides. We normally take the young plants from the field, clean them and expose the whole plants in 0.04 % colchicine containing 1.5 % dimethyl sulfoxide (DMSO) solution for 8–24 h at 9–10 °C. After treatment, the plants are washed and transplanted to the field for fertile DH development as shown in Fig. 2.4.

Construction of Chromosome Segment Substitution Lines

The chromosome segment substitution lines are initially derived from F_1 hybrids. Specifically, to obtain more than 100 BC_1F_1 spikes, F_1 is backcrossed to one of its two parents (ideally use 1 ear per F_1 individual for the cross or 2–3 ears if F_1 plants are insufficient) and randomly selected 2–3 BC_1F_1 ears per BC_1F_1 row to produce BC_2F_1 populations next year. In general, the number of backcross generations depends on the specific projects. For example, BC_2F_1 is normally good enough (some projects require BC_3F_1 or BC_4F_1). At BC_2F_1, chromosome component contributed from the donor parent is about 10 % of the recipient at which time the recombination rate is high and ideal phenotypes or variety can also be selected. The BC_2F_2 lines contain all donor parental genes following a single BC_2F_1 selfing. Generally, BC_2F_4 can be used as a stable population for any research projects related to QTL mapping.

Population Constructions for Fine Mapping

1. Similar to the RIL population, the RHLs are resulted from the F_2's continuous selfing. The uniqueness of RHLs is that one or a few traits from one parent can be maintained, while other loci may be the same as the other parent. Moreover, segregations at all loci examined are always present, resulting in a unique population with the uniform genetic background. The population size can be determined based on the selected traits and data of the observations. For example, evaluation and selection can be initiated at F_3, F_4 or F_5 according to the targeted trait.
2. On the bases of the primary QTL mapping, the QIR population is derived from the backcross to one of parents, and foreground and background selections of different molecular markers based on the major QTLs mapped and the markers located at their flanking sides. For example, we identified the major QTLs for heading and their flanking markers of *Xbarc*320 and *Xwmc*215 and are used for evaluating each backcross population. Meanwhile, genetic background selections are conducted using 200 molecular markers and QTL heterozygotes have now been identified. Noticeably, at least one marker on each chromosome arm should be used during genetic background selection.
3. Similar to QIR construction, SSSLs are the same population as the recipient parent with the exception of the targeted QTL traits contributed by the donor's chromosome segment. This process involves the tracking of the assisted selections on the targeted QTL traits derived from the primary mapping.
4. CSSLs are substitution lines consisting of overlapping chromosome segments which cover the entire genome. They are produced by crossing the multiple donor parents and recipient followed by repeated backcrossing to the latter.

2.1.2.2 Key Notes to Consider During Genetic Population Constructions

Population construction approaches may be different based on different research purposes, and thus some key notes to consider may be different even though they share some in common. Several comments are presented in this section.

1. The two parents of F_1 must be in accordance with the project purpose. In addition, the donor parent (DP) is often from the core germplasm or unique

resource that cannot be used directly, while the recipient parent (RP) represents the best commercial native variety (line).

2. The two parents of F_1 must be highly homozygous; in addition to keep the selected ear, seeds from the remaining ears of the F_1 should be stored as a backup at low temperature, so that it can be used for later propagation after the population is constructed; seeds for recurrent crosses must be from strict selfing (ears must be bagged) of the recurrent parent to avoid false hybrids.
3. While a larger number of F_1 ears are required for F_2 population construction, only 1–2 ears during the initial crosses are needed for other population construction. During this process, strict emasculation of the selected plants must be practiced to avoid self-pollination. Backcross of F_1 generally needs 3–6 ears (using different individuals), whereas BC_1F_1 requires 20 ears (plants). One kg seeds of BC_2F_2 from BC_2F_1 selfing are ideal.
4. For phenotypic evaluation, fertile land should be used for F_2 (BC_2F_2) and F_6 generations, and land for other generations is not critical. Care must be taken when more than one generation/year is practiced using either greenhouse or areas with different climates. Special attention should be taken to prevent loss of some, most, or all of the lines due to undesirable climates or any other bad conditions.
5. To construct near-isogenic lines, backcross lines, or chromosome segment substitution lines, it is the best to use SSRs or biochemical markers for identifying specific genes contributed by the donor parent, thereby speeding up the construction process and improving population qualities.

2.1.3 *Quality of Genetic Populations*

For QTL mapping, the quality of genetic population includes the population adaptability, population size, homozygosity, and genetic diversity.

2.1.3.1 Parental Selection

Parental selection is closely associated with the adaptabilities of the genetic populations. For example, for grain weight QTL mapping, the two parents must show significant difference in grain size; for Fusarium head blight (FHB) resistance QTL mapping, the two parents must have obvious different resistance to FHB, whereas for quantity trait QTL mappings, it is the best that one parent has the strong gluten, the other has the weak gluten character. It is expected that the larger the differences between the two parents are, the more diversified populations and the better for mapping the traits of interest. Of course, when the two parents show obvious differences in several QTL traits, the resulting populations can be used for their related genetic studies. In general, three principles should be considered for parental selection. (1) Genetic difference between the parents: the genetic difference between the parents should be neither too large nor too small. Too large difference might affect the hybrid's homologous chromosome pairing, leading to severe reduction of crossovers, thus low recombination rates for loci in the linked segments, biased partitions, and possible hybrid sterility. As a result, the degree of confidence would be dropped dramatically. Too small difference, on the other hand, means low DNA polymorphism and fewer polymorphism markers, resulting in low mapping accuracy. (2) Purity of parents: selfing can guarantee the homozygosity of the parents. (3) Cytological analysis of the parents and their F_1 hybrids: If the parents have been involved in chromosome translocation, chromosome deletion, or being monosomics, they are not ideal for population constructions.

2.1.3.2 Population Size

The mapping population size has a great impact on the accuracy and effectiveness of genetic analysis and QTL detection, especially on QTL detection number, QTL efficacy estimation, and QTL detecting sensitivity (Buckler et al. 2009; Schön et al. 2004; Zou et al. 2005). Xu (1994) found that, with an increase in mapping population size, the estimated likelihood ratio test (LOD) value went up, while both estimated bias of recombination rate and the biased estimation of the QTL genotypic mean and variance decreased. In addition, Beavis (1998) indicated that false positive QTL could still be detected with a population size containing as many as 200 individuals. These studies suggest that constructing a large genetic segregating population is challenging due to the heavy workload and high cost in the process of establishing a genetic linkage map, trait genetic analyses, and QTL detections. Therefore, it is suggested that the size of the population should be determined based on the study purpose. Importantly, the population size should be determined based on the aim of the research. For example, it is estimated that about 200 lines are needed for primary QTL mapping, whereas the population size for fine mapping studies should be as large as possible as the lager the population size, the higher the mapping precision. Additionally, the population size should also be determined by the types of the population being constructed. We believe that F_2 followed by RIL requires a large population in order to allow expressions of all possible genotypes in the population. In general, to achieve high mapping accuracy, the order of population sizes are F_2 > RIL > BC_1 > DH.

2.1.3.3 Homozygosity and Genetic Diversity

As already discussed, parental homozygosity is critical for constructing highly homozygous populations and elimination of false hybrids is equally important. For populations that need multiple selfing generations, they cannot be used until each line in the population is highly homozygous. Population diversity is closely associated with parental selection.

2.2 Types and Applications of Genetic Markers

Having established various genetic populations, we can now conduct identifications of the genotypes and construct genetic linkage maps using the molecular markers. The advantages of the molecular marker-based detections include the following: (1) the map is presented by DNA sequences; (2) there are a large number of markers distributed in the genome; (3) some markers are highly polymorphic; and (4) some markers are codominant, resulting in intact genetic information. Therefore, molecular markers play important roles in genetic mappings.

Locating QTLs/genes on chromosome(s) can be realized using the four types of genetic markers (morphological markers, cytological markers, biochemical markers, and molecular markers) as discussed below.

2.2.1 Morphological Markers

Morphological markers represent the expressions of plant phenotypic traits that are generated at a specific developmental stage or under a specific environmental condition. Currently, these markers have been widely used in the studies of rice, corn, soybean, wheat, and many others. In wheat studies, for example, researchers have used various aneuploids (e.g., monosomics, nullisomics, trisomics, tetrasomics, and ditelosomics) in mapping some targeted genes on a specific chromosome even on a specific chromosome arm. However, because the use of these types of markers is largely limited by numbers and polymorphisms as well as some of their expressions are affected by gene expression regulation, developmental stage, and environmental factors, this approach has several drawbacks, including that phenotypic differences often do not represent genotypic variations and time-consuming.

2.2.2 Cytological Markers

Cytological markers mainly refer to the chromosome karyotype (i.e., chromosome number, size, centromere position) and bending patterns. These markers have been applied to the mappings of some exogenous genes. To improve wheat agronomic traits, researchers have successfully introduced a number of elite genes into commercial varieties via wide hybridizations and chromosome engineering. The locations of many of these genes/chromosome segments (e.g., C belt, N belt type) have been determined through karyotype and banding pattern analyses. However, the use of these markers has also some limitations, including the following: (1) generation of such resource is difficult, (2) lines with changes in chromosome structures and numbers often show high rate of mortalities due to their poor tolerance to environmental stresses, and (3) fewer number of cytological markers are available.

2.2.3 *Biochemical Markers*

Biochemical markers refer to the products of gene expressions such as enzymes and proteins (e.g., isozymes and seed storage proteins). To date, there are about more than 180 biochemical markers that have been located on wheat chromosomes, and most of which have been very useful for identifications of exogenous genes in wheat. The use of biochemical marker for identification of wheat high molecular weight glutenin subunits is a great specific example.

2.2.4 *DNA Molecular Markers*

Rapid development of molecular biotechnologies has made it possible to use the nucleotide sequence variations as genetic markers. The discovery of the restriction fragment length polymorphism (RFLP) reported by Botstein et al. (1980) was considered to be the onset of the use of DNA markers. With the development of (PCR) polymerase chain reaction technology, the second-generation marker systems of random amplified polymorphic DNA (RAPD), amplified fragment length polymorphism (AFLP), microsatellite markers (SSR), and sequence characterized amplified region (SCAR) were also developed. Currently, the third-generation marker systems are available, including single-nucleotide polymorphism (SNP) and expressed sequence tag (EST) markers. In recent years, SSR, AFLP, EST, and SNP markers have been widely applied to genetic linkage map constructions, QTL mapping, genetic diversity analyses, and molecular marker-assisted breeding.

2.2.4.1 Restriction Fragment Length Polymorphism (RFLP)

RFLP is useful in revealing the allele variations through autoradiography or non-radioactive graphic techniques. The process involves the digestions of genomic DNA with specific restriction endonucleases, analyses of various DNA fragments separated by agarose gel electrophoresis, and hybridizations with radio- or non-radioactively labeled probes. This system can generate a large number of variants, show codominance and produce intact information, and has a good repeatability and stability. The disadvantages, however, include: it is only useful to analyze the gene rich region; the process is long, tedious, and expensive; and produces lower polymorphism in wheat because this technique is based on southern blot hybridization.

2.2.4.2 RAPD Markers

RAPD is based on PCR (Williams et al. 1990). DNA amplifications are random and polymorphism is generated. Compared with other marker systems, RAPD needs less DNA, it is easy to operate, and one set of primers can be used in analyzing genomes of different species, it can produce higher polymorphism than RFLP, and ultimately can detect large quantities of genetic polymorphisms rapidly. The issues with this system are as follows: (1) because of its dominant marker nature, it can only be used for dominant analysis; (2) cannot distinguish the homozygous from heterozygous genotypes; (3) cannot provide complete genetic information; and (4) has poor stability and reproducibility. This suggests that the RAPD system is useful for gene mapping, alien chromosome fragment detection, and variety/species diversity studies, but has less value for constructing genetic linkage map in wheat.

2.2.4.3 AFLP Markers

AFLP was first developed by Zabeau and Vos (1993). It combines the advantages of both RFLP and RAPD markers. The digested genomic DNA fragments with the two types of restriction nucleases produce sticky ends which are ligated with the synthetic double-stranded DNA. A subset of the restriction fragments is then selected to be amplified. The amplified fragments are separated and visualized on denaturing polyacrylamide gels, either through autoradiography or fluorescence methodologies, or via automated capillary sequencing instruments. The marker has the advantages of simple and fast, high polymorphism and good stability and therefore can be used to study the genetic diversity of germplasm resources, genetic map construction, and gene mapping. This mark also has dominant nature and shows the DNA fragment length polymorphism. But its wide application has been hampered because of its inability to distinguish DNA of different sequences with the same length.

2.2.4.4 SSR Markers

Simple sequence repeat (SSR) marker is also called the microsatellite marker and has DNA repeat sequences of 1–6 base pairs. The same satellite DNA sequence can be distributed in different genome locations. The polymorphism of SSR comes from random repeat variables in number. Wheat is allohexaploid, and thus its genome size is huge in which approximately 80 % DNA is repetitive. Therefore, the use of SSR markers is more effective in wheat QTL mappings and related studies. Indeed, they have been used in linkage map construction, QTL mapping, and other important QTL locus/gene identifications in various cereal crops. Moreover, because SSR is codominant, it has the ability to detect heterozygote and homozygote. It also provides relative complete genetic information, has good stability and reproducibility, and is easy to use.

2.2.4.5 EST-SSR Markers

Expressed sequence tag (EST) is a gene expression fragment with 300–500 bp. This technology involves in cDNA library construction via mRNA reverse transcriptions, followed by cloning into vectors. The randomly selected clones are sequenced at 5′ or 3′ end, and are compared with the known sequences. Information on species evolution, variation, and senescence can be obtained using this marker system. Therefore, EST-SSR can show the specific gene expression of a specific organ/tissue at a specific developmental stage. The most remarkable feature of EST-SSR is that it can identify functional genes directly as reported in wheat, rice, corn, and other crops.

2.2.4.6 Inter Simple Sequence Repeat (ISSR) Markers

ISSR marker is developed by anchoring 1–4 purine or pyrimidine bases at the 5′ or 3′ end of a SSR sequence and using the anchored bases as primer to amplify a DNA sequence with both ends being reverse complements to the SSR sequence. The polymorphisms of different strains are determined based on the presence or absence of electrophoretic bands. This method is simple, convenient for detection, and repeatable. Therefore, it has been widely applied to gene mapping, species diversity, and systematic developmental studies.

2.2.4.7 Sequence-Character Amplified Regions (SCAR)

SCAR marker is proposed by Paranin et al. in 1993 based on the RAPD technology. Technically, the target DNA sequence can be recovered from the gel after RAPD analysis. The specific primers with 18–24 bases are designed and used to amplify the genome DNA through PCR. Alternatively, inserting about 14 bases at the end of the original primer based on the terminal sequence of the RAPD marker will produce a specific primer that is complementary to the original RAPD marker terminal sequence. Because the SCAR marker is transformed from the RAPD marker, it has a longer primer sequence and thus has a better stability and repeatability than RAPD.

2.2.4.8 Sequence Tagged Sites (STS)

STS primers are RFLP single probe copy probes and microsatellite sequences. Specifically, these primers are developed based on the known probe sequence of both ends of RFLPs. The amplified product is about 200–500 bp DNA sequence occurring only once in the genome and can be used for locating specific gene sites (Olson et al. 1989). Compared with RFLP, STS marker is most advantageous as it does not need to maintain the probe clones, and obtaining the sequence information from the database is simple and easy. Moreover, the STS marker is codominant and marker information can be shared within genetic profiles of various crops. It is considered to be an intermediary in integrating plant genetic map and physical map. The STS, as a new promising molecular marker, is also codominant, can provide high polymorphism, and generate a great deal of information and therefore has very high application value even though the cost to develop such marker is relatively high.

2.2.4.9 Diversity Arrays Technology (DArT)

Developed by Jaccoud et al. (2001), DArT is a new molecular marker derived from microarray hybridizations that detect the presence versus absence of individual fragments in genomic representations. DArT detects the polymorphism of the DNA fragment harvested from genomic DNA digested by restriction enzymes. The basic principle of DArT is that genomic DNA from different samples is equivalent mixed first, followed by digestion of restriction enzymes, retrieving DNA sections of varying sizes based on electrophoresis, then processing through several steps to produce multiple copies of the smaller fragments, which is called a "representation" and will reduce the complexity of genetic material, and eventually, these fragments are placed as tiny spots onto a batch of identical glass slides using a microarray machine. Each point represents a DNA fragments from a different sample genome, and there are also specific fragment presented only in a few samples. In order to detect the genetic differences among different samples, DArT requires that the "representations" harvested from different samples but digested by the same enzyme are used as probes, to produce corresponding probe combinations which will be used to hybridize with the chip. Since the DNA sequences from different sample are different, the hybrid result for each sample will be different, depending on which spots the DNA binds to, and these differences show the degree of diversity among the samples. These differences can be harvested by machine scanning for signals generated from each DNA spot. In the analysis of polymorphism, the different signals generated from each DNA spot are DArT markers, which can be a representative of a genome polymorphism fragments and can be used as markers for following studies (Jaccoud et al. 2001; Hong et al. 2009).

DArT includes the following steps: the producing of genome fragments using specific methods can reduce complexity of the genome, DArT library construction, chip preparation, sample preparation, chip hybridization, and signal scanning and data processing. The method does not have to know the DNA sequence previously and has the advantage of high quality, high degree of automation, high throughput, and stable results. Although requires a relatively high purity of DNA, DArT has many advantages which enable it applied in genetic linkage map, QTL mapping, identification of germplasm resources, and evolution analysis (Wenzl et al. 2004; Wittenberg et al. 2005; Xia et al. 2005; Akabari et al. 2006).

2.2.4.10 Single-Nucleotide Polymorphism (SNP) Markers

SNP is considered to be one of the third-generation genetic markers and its polymorphism is derived from the change of a single nucleotide in the genome, including single-nucleotide transition, deletion, transversion, and insertion. SNP differences can be identified by analyzing their DNA sequences or alignments of known DNA sequences. The simplicity way is via amplifying a genome segment with specific primers through PCR, followed by PCR product sequencing and alignment. Massive SNP identification can also be achieved by DNA chip technology.

2.3 Statistical and Mapping Method of Quantitative Traits

It is well known that most of the wheat important agronomic and quality traits (e.g., yield, growth cycle, disease/stress resistance, and flour quality) are quantitative traits controlled by polygenic genes (Michelmore et al. 1988). Traditional quantitative genetics neither can determine the number of QTLs controlling these characters, nor can it determine the genetic effects of a single QTL or its location in the chromosome. Researchers often treat the polygenic traits as a whole entity to control a whole group of quantitative traits, and estimate the overall genetic and environmental effects. Although the concept of using genetic markers to detect QTLs had been proposed in 1923 by Sax, few genetic markers for QTL studies were available prior to the 1980s. Due to the rapid development of molecular biology and computer technology since the 1990s, new methods and statistical tools have been developed. The use of these methods and tools, studies on QTL nature, QTL location, and relationship between QTLs and environments have been possible (Paterson 1988; Lander et al. 1989; Dudley et al. 1996; Darvasi et al. 1994).

2.3.1 The Principle of QTL Mapping

QTL mapping is essentially a set of procedure aimed at first detecting and then locating a QTL. It aims to locate the potential genes controlling a quantitative trait using the expected association between the putative genes and the known genetic markers first, and then to estimate their effects.

It is a combination of linkage mapping and quantitative genetics approaches to find an association between genetic marker and a phenotype that one can measure or that can be measured. Therefore, QTL mapping is based on a hypothetical genetic model and a concept of statistics. There are three steps to QTL mapping: (1) linkage map construction; (2) phenotypic evaluation and identification of polymorphism of genetic markers; and (3) statistical analyses to identify and estimate the effect of the loci hat affect the trait(s) of interest.

The essential conditions of the steps are (1) high density linkage map (the average distance between markers should be less than 10 cm) and corresponding statistical analysis method; (2) the target traits can separate in the population and show a continuous pattern (the selected parents should have significant differences and distinct genetic relationship).

2.3.2 *Methods of QTL Mapping*

With the rapid development of molecular markers, the studies on quantitative traits are now at the QTL mapping stage. QTL mapping and the estimated efficacy value can be realized based on the statistical models used in the QTL mapping. To date, a number of QTL mapping approaches are available, and these include single-interval mapping, composite interval mapping, mixed linear model-based composite interval mapping, inclusive composite interval mapping, as well as the Bayesian analysis.

2.3.2.1 Single-Marker Analysis

Single-marker analysis is used to compare the variations of different QTL genotypic means and identify the relationship between each molecular marker and phenotype based on additive–dominant model, and a variety of statistical analyses, including t-tests, ANOVA, regression, maximum likelihood estimations, and log likelihood ratios. However, the disadvantages of this approach are as follows: (1) it cannot distinguish a single QTL from multiple linked QTLs; (2) cannot estimate the QTL's possible position; (3) QTL effects can be declined due to recombination between the marker and QTL; (4) false positive may be an issue; and (5) requires a large population and has low detection efficiency. Therefore, new model and method are needed for QTL mapping researches.

2.3.2.2 Single-Interval Mapping

The interval (between two adjacent markers) mapping method was then put forward due to the drawbacks of the single-marker analysis. The first model proposed by Jensen was suitable for analyzing DH population and could be used for estimating biased partition. Later, a model suitable for more broad populations was also developed by Knapp et al. (1990). Lander and Botstein (1989) proposed an improved model that based on a maximum likelihood equation with normal mixture distribution and simple regression, which can calculate the LOD score, which indicates the probability that a QTL is present at that position, of any position between two adjusted markers based on genetic linkage maps. LOD scores are plotted along the chromosome map, and those that exceed a threshold significance level suggest the presence of a QTL in that chromosome region. The most likely QTL position is interpreted to be the point where the peak LOD score occurs. This method can not only determine the possible position QTL in the interval, but also can reduce the size of population required in QTL mapping. In 1992, Haley & Knott and Martinez & Curnow proposed a simple regression interval mapping analysis method. Given the error is independent and followed a normal distribution, the regression analysis and the maximum likelihood ratio test have following relationship, likelihood ratio test UpMSR/MSE Up$F_{regression}$. Linear regression method can obtain the approximate the same result as maximum likelihood method, but has greatly reduced calculation amount, and thus the simple regression method is widely accepted.

There are also some limitations for simple interval mapping. The indicated positions of QTLs are sometimes ambiguous or influenced by other QTLs, and it is difficult to separate effects of linked QTLs. Further, it only uses the information of two markers in each test, while the information of other markers is not fully utilized, and the final mapping results may not be correctly represented.

2.3.2.3 Composite Interval Mapping

In order to overcome the defects of simple interval mapping, Zeng (1994) proposed composite interval mapping (CIM) which combines the multiple regression analysis with interval mapping, and detect QTL in multiple intervals using multiple molecular marker information. The main difference with interval mapping is that the CIM applies multiple regression model in the maximum likelihood analysis which filters out the influence of QTL lying out the interval in the detection of markers at any given point. The basis of this method is an interval test that attempts to separate and isolate individual QTL effects by combining interval mapping with multiple regressions.

The method has the following main advantages: (1) it uses the QTL likelihood map to display the QTL possible positions and significance, thus maintaining the advantages of interval mapping; (2) each time only one interval is examined; (3) estimating of QTL location and effect is asymptotically unbiased if there is no epistasis and QTL–environment interaction; (4) the whole genomic marker information is fully utilized; and (5) with the multiple markers selected, genetic variations generated from other genome regions can be controlled, and mapping accuracy and efficacy can be increased (Zeng 1994).

Disadvantages of this method include the following: (1) it cannot analyze some complex genetic issues, such as epistasis and QTL–environment interaction, and (2) because the intensive computations are used, CIM can be a slow process. Wu et al. (1997) proposed a composite interval mapping method based on the least-square estimation, which is simpler and faster than CIM based on maximum likelihood estimation.

2.3.2.4 Mixed Composite Interval Mapping

Due to the disadvantages of the composite interval mapping, Zhu (1996) proposed the mixed model composite interval mapping which integrates additive, dominant, epistatic and environmental interaction effect into one model. This model assumes (1) a quantitative trait is controlled by multiple QTL genes; (2) the population mean and every major QTL effect (including additive, dominant and epistatic effects) are fixed effects; and (3) the effects of environment, QTL–environment interaction, molecular marker, marker–environment interaction, and the residual error are considered to be random effects. The method combines the effect estimation and QTL location, and is a joint QTL mapping under multiple environments, resulting in a significant improvement in mapping precision and efficiency. The use of the mixed linear model approach in QTL mapping can result in unbiased estimation of QTL–environment effect and has a great deal of flexibility and extensibility. Because this model is based on the composite interval mapping, it can be extended to the analysis of main QTL effects of additive–additive, additive–dominant, dominant–dominant, dominant–epistasis and genetic–environmental interactions. The estimated values of these effects can be used to predict the heterosis of major QTL effect and QTL–environment effect and plant individual breeding value. On the basis of breeding value, individuals with great potential can be selected, and thus the breeding efficiency can be improved. Therefore, the mixed linear model-based composite interval mapping has a greater application prospect. At present, it has been used successfully in the analyses of many important QTL traits that are associated with crop yield and quality.

2.3.2.5 Inclusive Composite Interval Mapping

Composite interval mapping has been widely used in QTL mappings over the past 10 years, yet has some defects in the algorithm. Under the assumption of additivity of QTL effects on the phenotype of a trait of interest, the additive effect of a QTL can be completely absorbed by the two flanking marker variables, and the epistatic effect between two QTL can be completely absorbed by the four marker-pair multiplication variables between the two pairs of flanking markers. For these reasons, Wang et al. (2009) proposed inclusive composite interval mapping (ICIM) method. Two steps are involved in ICIM. In the first step, stepwise regression was applied to identify the most significant regression variables in both cases but with different probability levels of entering and removing variables. In the second step, a one-dimensional scanning or interval mapping was conducted for mapping additive and a two-dimensional scanning was conducted for mapping digenic epistasis. The mapping strategy simplifies the process of background genetic variation controlling in composite interval mapping. The method has low sampling error but high mapping efficiency. When QTL exists, the LOD value is higher; when no QTL exists, the LOD values are close to 0. The interaction of additive QTL of epistasis mapping can be analyzed with this method.

The main advantages of ICIM are as follows: (1) it has a low sampling error, but high mapping efficiency; (2) it has a good robustness for the mapping parameters; (3) epistasis mapping is easy; and in epistasis mapping, it can not only detect the interaction of additive QTLs, but can also detect the interaction of QTLs with no significant additive effects. The disadvantage of this method is that it usually generates too many epistatic loci, resulting in difficulties in the process of selection.

2.4 New Methods of QTL Mapping

As stated above, the development of molecular marker technology since 1980s in the twentieth century has made it possible in the close conjunctions of quantitative traits with the DNA sequence information so that complex QTLs can be mapped at the whole-genome level with the statistical assistance. It provides a great deal to better understand the genetic basis of complex traits and conduct DNA isolation, gene cloning, and marker-assisted breeding. Future research should focus on how the QTL mapping can help reveal the gene dynamic changes during plant development and understand how differences in gene expressions can affect trait development. In recent years, novel QTL analysis approaches and software have emerged (e.g., Fig. 2.5) and is presented below.

2.4.1 *Conditional QTL*

The recent rapid development of biotechnologies, especially the molecular biology and information technology, brought about a novel concept of studies known as

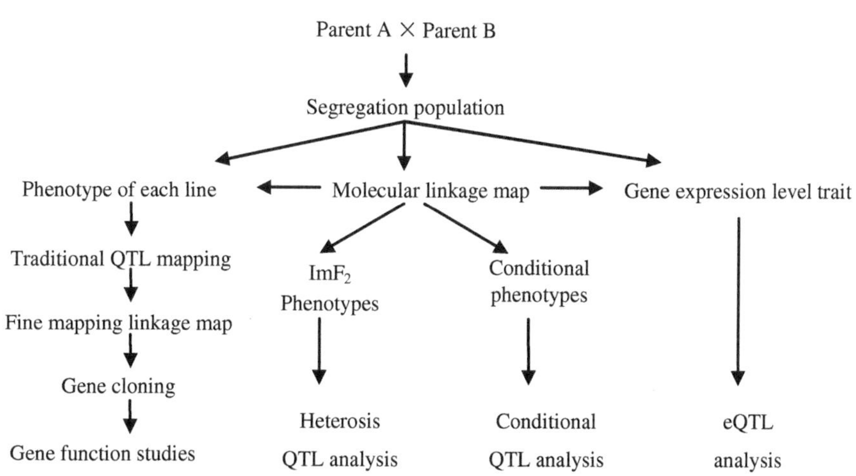

Fig. 2.5 Schematic diagram of several QTL analysis methods

"Conditional QTL." Zhu (1996) published the first paper with such concept in the scientific journal of "Genetics." Soon, Wu et al. (1997) also proposed a method of QTL dynamic mapping and developed a software program to analyze, using the net genetic effect, the QTLs being expressed or silenced during the time period of $t-1$ to t. Since then, a number of traits controlled by conditional QTLs have been analyzed in various organisms, including mice (Atchley et al. 1997), cotton (Ye et al. 2003), and rice (Shi et al. 2001; Guo et al. 2005) using these programs. However, there were no descriptions on the difference between conditional and traditional QTLs previously. Basically, conditional QTL refers to the net inheritable effects resulted from plant growth and development at a specific developmental stage and/or specific agricultural practice under a given condition. It is a method of genetic analysis for analyzing the temporal and spatial expressions and interactive effects of a certain QTLs under various conditions. It can be used to dissect the comprehensive expressions of multiple QTLs during a given organism's (or a specific trait) whole life cycle. Therefore, conditional QTL can be used to study the dynamic development of agronomic traits, QTL response under different agricultural treatments and the cause–effect relationship (see Chap. 1, Volume 2).

For the last ten years, our group focused on the genetic analysis of quality traits with conditional QTL, such as conditional QTL mapping for the dynamic accumulation behavior of grain protein content, for developmental behavior of total starch and its components content, for protein and starch interaction, and for sedimentation volume on seven quality traits (see Chap. 2, Volume 2). In addition, the genetic studies of yield traits with conditional QTL, which include conditional QTL mapping for wheat canopy traits under two different nitrogen application levels, for plant height at different growth stages, for wheat spike dry weight and thousand-kernel weight during different development stages, and for **grain yield and its three components**, etc. **were studied** (see Chap. 3, Volume 2).

2.4.2 eQTL Mapping Method

In 2001, Jansen and Nap proposed the expression QTL (eQTL) mapping method, which uses phenotypic observations, molecular markers, and expression profile data in identification of quantitative trait loci. In the method, the expression level of each gene is treated as a trait, and all the expression information of genes in an individual forms a profile, the linkage of the profile and molecular marker is analyzed. The QTL detected in this method is eQTL (Jansen and Nap 2001). When a gene position is consistent with eQTL-linked marker, the gene underlying the quantitative trait can be determined. The eQTL can be divided into CIS eQTL (Cis-acting eQTL) and trans-eQTL (trans-acting eQTL). CIS eQTL refers to that the eQTL is located to a genome region containing the target gene and indicates that the expression level polymorphism may determine the gene expression difference. An anti-eQTL refers that the eQTL polymorphism underlies the expression difference of a gene located in difference genomic region. A CIS eQTL can directly provide the information of candidate genes, while a trans-eQTL not only can be combined with other methods to obtain the control network, but also reduce the candidates of the nodes (Rosa et al. 2006). At present, the main method used to study eQTL includes cDNA-AFLP (Vuylsteke et al. 2006), qRT-PCR, and gene chip technology (Potokina et al. 2008). Of these, the gene chip technology has the advantages of high flux, high sensitivity, and is the main approach to study eQTL. The mainly used methods include the following: (1) In transcript-based mapping method, each expression trait (e-trait) is analyzed independently first and then all eQTL of the expression traits are obtained; (2) In marker-based method, the genotype of every marker is identified first and then the expression difference among different marker genotypes was analyzed, testing whether it is associated with the eQTL, and whole-genome-wide scan was performed finally.

eQTL study not only helps us to estimate the heritability of gene expression levels, construct gene regulatory network and mine candidate gene, but also provides conditions for our understanding of gene–gene and gene–environment interactions, so that we can understand the molecular mechanism of biological and genetic basis of complex traits more deeply. eQTL has some shortcomings: (1) the cost is relatively high; (2) improper selection of individuals may lead to partial separation and biased results of QTL mapping; and (3) the current methods for QTL mapping restrict the eQTL study (Liu et al. 2008).

2.4.3 QTL Mapping Methods of New Gene Mining Germplasm

QTL mapping population in crop is generally developed by crossing two homozygous inbreeding lines that often show significant differences. When the two parents carry the same allele, the gene effect cannot be detected even though their effects are large. Therefore, Rao and Xu (1998) proposed a four- even eight-crossway hybridization in order to increase the number of parents. Nevertheless, the parent number is still small. Using statistical techniques, scientists have been able to find novel genes from the germplasm resources, instead of the existing commercial varieties. The QTL statistical method for discovering new genetic resources mainly includes the hitchhiking effect based on association analysis method and a mixed model method based on IBD (Zhang 2006).

Association mapping, also known as "linkage disequilibrium mapping," is a method of mapping QTLs that takes advantage of historic linkage disequilibrium to link phenotypes to genotypes. The primary idea of association mapping is that performing whole-genome scanning and searching for genome segment (or loci) containing selection signal, and then scanning further for important segment and elite alleles (You and Zhang 2007). The main idea of mixed model method based on IBD is that the pedigree is used to compute identical value of the offspring first, and IBD value is embedded in variance component to determine the position and effect of QTL, and then the best unbiased linear model is used for estimating QTL effect. According to the estimated QTL effect of the varieties, parental selection and molecular design breeding can be carried out, and transferring of genes in varieties can also be studied (Zhang et al. 2005). China is rich in germplasm, and if enough information is obtained, a large number of useful genes can be mined and their effect can be predicted; and the consequent results can be used in molecular design breeding and breeding efficiency improvement (Zhang 2006).

The recent rapid development of crop QTL research progresses has played a significant role in mining new crop gene resources which has benefited the crop genetics and breeding programs. It is imperative to establish efficient QTL mapping methods, including the constructions of new mapping populations, new methods of statistical analysis, and molecular marker-assisted methods for QTL fine mapping, cloning, and selection. This will allow us to better understand crop QTLs and ultimately apply the QTL knowledge to the efficient crop breeding programs. Also, QTL studies will undoubtedly help for the better understanding of gene functions at the whole-genome level and better interpretations of molecular network associated with plant development, interactions between QTLs and environmental changes and biological basis.

References

Akbari M, Wenzl P, Caig V, Carling J, Xia L, Yang SY, Uszynski G, Mohler V, Lehmensiek A, Kuchel H, Hayden MJ, Howes N, Sharp P, Vaughan P, Rathmell B, Huttner E, Kilian A. Diversity arrays technology (DArT) for high-throughput profiling of the hexaploid wheat genome. Theor Appl Genet. 2006;113:1409–20.

Atchley WR, Zhu J. Developmental quantitative genetics, conditional epigenetic variability and growth in mice. Genetics. 1997;147:765–76.

Beavis WB. QTL analyses: power, precision, and accuracy. In: Patterson AH (ed) Molecular dissection of complex traits. Boca Raton: CRC Press; 1998.

Botstein D, White RL, Skolnik M, Davis RW. Construction of a genetic linkage map in man using length polymorphism. Am J Hum Genet. 1980;32:314–31.

Buckler ES, Holland JB, Acharya CB. The genetic architecture of maize flowering time. Science. 2009;325:714–8.

Burr B, Burr EA, Thompson KH, Albertson MC, Stuber CW. Gene mapping with recombinant inbreds in maize. Genetics. 1988;118:519–26.

Darvasi A, Solier M. Optimum spacing of genetic markers for determining linkage between marker loci and quantitative trait Ioci. Theor Appl Genet. 1994;89:351–68.

Dudley JW, Lamkey KR, Geadelmann JL. Evaluation of populations for their potential to improve three maize hybrids. Crop Sci. 1996;36:1553–9.

Fang XJ, Wu WR, Tang JL. DNA makers assisted breeding in crops. Australia: Science Press; 2001, 2.

Guo LB, Xing YZ, Mei HW, Xu CG, Shi CH, Wu P, Luo LJ. Dissection of component QTL expression in yield formation in rice. Plant Breed. 2005;124:127–32.

Hong Y, Xiao N, Zhang C, Su Y, Chen J. Principle of diversity arrays technology (DArT) and its applications in genetic research of plants. HEREDITAS (Beijing). 2009;31:359–64.

Hua JP, Xing YZ, Wu WR, Xu CG, Sun XL, Yu SB, Zhang QF. Single-locus heterotic effects and dominance by dominance interaction can adequately explain the genetic basis of heterosis in an elite hybrid. Proc Natl Acad Sci USA. 2003;100:2574–9.

Jaccoud D, Peng K, Feinstein D, Kilian A. Diversity arrays: a solid state technology for sequence information independent genotyping. Nucleic Acids Res. 2001;29:e25.

Jansen RC, Nap JP. Genetical genomics: the added value from segregation. Trends Genet. 2001;17 (7):388–391.

Knapp SJ, Bridges WC, Birkes D. Mapping quantitative trait loci using molecular marker linkage maps. Theor. Appl. Genet. 1990;79:583–592.

Lander ES, Botstein S. Mapping mendelian factors underlying quantitative traits using RFLP linkage maps. Genetics. 1989;121:185–99.

Li W, Wu W, Lu H. Worland A J, Law C N. QTL mapping and effect estimating on chromosome 7D of wheat. Acta Agro Sinica. 1996;22:641–645.

Lin F, Xue SL, Tian DG, Li CJ, Cao Y, Zhang ZZ, Zhang CQ, Ma ZQ. Mapping chromosomal regions affecting flowering time in a spring wheat RIL population. Euphytica. 2008;164:769–77.

Liu G, Peng H, Ni Z, Qin D, Song F, Song G, Sun Q. Integrating genetic and gene expression data: methods and applications of eQTL mapping. Heredity (Beijing). 2008;30(9):1228–36.

Liu J, Liu Y, He N, Cui D. Genetics analysis of several quantitative traits of doubled haploid population in wheat. J Triticeae Crops. 2005;25:16–9.

Michelmore RW, Shaw DV. Quantitative genetics: character dissection. Nature. 1988;335:672–3.

Olson M, Flood L, Cantor D, Boston D. A common language for physical mapping of the human genome. Science. 1989;245:1434–5.

Paran I, Michelmore RW. Development of reliable PCR-based markers linked to downy mildew resistance genes in lettuce [J]. Theor Appl Genet, 1993;85:985–993.

Paterson AH. Resolution of quantitative traits into Mendelian factors by using a complete linkage map of restriction fragment length polymorphysiums. Nature. 1988;335:721–6.

Potokina E, Druka A, Luo ZW, Wise R, Waugh R, Mike K. Gene expression quantitative trait locus analysis of 16000 barley genes reveals a complex pattern of genome-wide transcriptional regulation. Plant J. 2008;53:90–101.

Rao SQ, Xu S. Mapping quantitative trait loci for ordered categorical traits in four-way crosses. Heredity. 1998;81:214–24.

Rosa GJM, Leon N, Rosa AJM. Review of microarray experimental design strategies for genetical genomics studies. Physiol Genomics. 2006;28:15–23.

Schön CC, Utz HF, Groh S, Truberg B, Openshaw S, Melchinger AE. Quantitative trait locus mapping based on resampling in a vast maize testcross experiment and its relevance to quantitative genetics for complex traits. Genetics. 2004;167:485–98.

Shi CH, Wu JG, Fan LJ, Zhu J, Wu P. Developmental genetic analysis of brown rice weight under different environmental conditions in indica rice. Acta Bot Sin. 2001;43:603–9.

Vuylsteke M, Daele HVD, Vercauteren A, Zabeau M, Kuiper M. Genetic dissection of transcriptional regulation by cDNA-AFLP. Plant J. 2006;45:439–46.

Wang J. Inclusive composite interval mapping of quantitative traits genes. Acta Agro Sinca. 2009;35:239–45.

Wenzl P, Carling J, Kudrna D, Jaccoud D, Huttner E, Kleinhofs A, Kilian A. Diversity arrays technology (DArT) for whole-genome profiling of barley. Proc Natl Acad Sci USA. 2004;101:9915–20.

Wu W, Li W, Lu H. Dynamic mapping strategy of quantitative trait locus. J Biomathematics. 1997;12:490–8.

Williams J, Kubelik A, Livak K, Rafalski J, Tingey S. DNA polymorphisms amplified by arbitrary primers is useful as genetic markers. Nucl Acid Res. 1990;18:6531–5.

Wittenberg A H J, van der L, Cayla C, Kilian A, Visser R G F, Schouten H J. Validation of the high-throughput marker technology DArT using the model plant a rabidops is thaliana. Mol Genet Genomics. 2005; 274:30–39.

Xia L, Peng K, Yang SY, Wenzl P, Vicente MCD, Fregene M, Kilian A. DArT for high-throughput genotyping of Cassava (Manihotesculenta) and its wild relatives. Theor Appl Genet. 2005;110:1092–8.
Xu Y. Factors influencing the power of QTL mapping: population size. J Zhejiang Agric Univ. 1994;20:573–8.
Yano M, Harushima Y, Nagamura Y, Kurata N, Minobe Y, Sasaki T. Identification of quantitative trait loci controlling heading date in rice using a high-density linkage map. Theor Appl Genet. 1997;95:1025–32.
Ye ZH, Lu ZZ, Zhu J. Genetic analysis for developmental behavior of some seed quality traits in upland cotton (*Gossypumhirsutum* L.). Euphytica. 2003;129:183–91.
You G, Zhang X. Identification of important genes by marker-trait associationg analysis based on hitchhiking mapping. Heredity (Beijing). 2007;29(7):881–8.
Zeng ZB. Precision mapping of quantitative trait loci. Genetics. 1994;136:1457–68.
Zhang L, Liu P, Liu X. Construction of chromosome single segment substitution lines and QTL fine mapping. Mol Plant Breed. 2004;2(3):743–6.
Zhang Y. Research progress on crop QTL mapping methods. Chin Sci Bull. 2006;51:2223–31.
Zhang YM, Mao YC, Xie CQ, Smith H, Luo L, Xu SZ. Mapping QTL using naturally occurring genetic variance among commercial inbred lines of maize (*Zea mays* L). Genetics. 2005;169:2267–75.
Zhu J. Analytic methods for seed models with genotype × ′environment interaction. Chin J Genet. 1996;23:11–22.
Zabeau M, Vos P. Selective restriction fragment amplification: a general method for DNA fingerprints. European Patent Application. Pub, 1993.
Zou F, Gelfond JAL, Airey DC, Lu L, Manly KF, Williams RW, Threadgill DW. Quantitative trait locus analysis using ecombinant inbred intercross (RIX): theoretical and empirical onsiderations. Genetics. 2005;170:1299–310.

Chapter 3
Construction of Molecular Genetic Map of Wheat

Abstract Molecular genetic map not only provides a powerful tool for the analysis of quantitative trait loci (QTLs) and marker-assisted selection (MAS) at the genomic level, but also lays a foundation for fine mapping and cloning important genes. In this chapter, the unique characteristics and breeding values of the six molecular genetic maps (1 DH, 3 RIL, and 2 natural populations) constructed by SSR, DarT, and SNP markers were illustrated. The parents for each genetic map have some distinguishing features, such as agronomic traits, yield, and/or quality traits. The average distances between adjacent markers in the wheat maps were appropriate (0.44–9.77 cM), thus meeting the recommended requirement for genome-wide QTL scanning. The molecular genetic maps have been used to QTL mapping for some agronomic traits, yield and quality traits, and the good results have been achieved.

Keywords Wheat · Molecular genetics map · High-density genetic linkage map · RIL population · DH population · SNP markers · DArT markers

Molecular genetic map not only provides a powerful tool for the analysis of quantitative trait loci (QTLs), marker-assisted selection (MAS) at the genomic level, but also lays a foundation for fine mapping and clone important genes. The chromosome complement of common wheat (2n = 6x = 42) consists of three genomes (A, B, and D). Genetic analysis and gene discovery in hexaploid wheat have been arduous and slower than rice, maize, and other crop because of its large genome size, abundance of repetitive DNA sequences, and limited polymorphism. The first intervarietal map for bread wheat, based on restriction fragment length polymorphism (RFLP) markers, was published in 1997 (Cadalen et al. 1997). The mapping population was composed of 106 doubled haploid (DH) lines derived from a cross between Chinese Spring and the French semi-dwarf winter wheat cultivar, Courtot. This map contained 266 loci covering 1772 cM. It had a poor representation of the D genome chromosomes, and no markers were assigned to chromosomes 2D, 4D, and 5D. The first microsatellite map in wheat possessed 279

J. Tian et al., *Genetic Analyses of Wheat and Molecular Marker-Assisted Breeding, Volume 1*, DOI 10.1007/978-94-017-7390-4_3

microsatellites with an International Triticeae Mapping Initiatives (ITMI) population from an interspecific cross between a common wheat variety and synthetic wheat (Röder et al. 1998). A microsatellite consensus map was constructed by joining four independent genetic maps of bread wheat (Somers et al. 2004). Three of the maps were F_1-derived, doubled haploid line populations, and the fourth population was "Synthetic" × "Opata," an F_6-derived, recombinant inbred line population. A total of 1235 microsatellite loci were mapped, covering 2569 cM, giving an average interval distance of 2.2 cM. This consensus represented the highest density public microsatellite map of wheat. A genetic linkage map with 464 loci developed from a DH population of 96 lines, which was generated from the cross between two common wheat varieties, Kitamoe and Munstertaler (Torada et al. 2006). Similarly, significant progress in constructing wheat genetic maps using different populations has also been made over the past decade or so (Zhang et al. 2008).

Crop improvement relies on the effective utilization of genetic diversity. Molecular marker technologies promised to increase the efficiency of managing genetic diversity in breeding programmes (Akabari et al. 2006). Numerous marker technologies have been developed over the last 25 years. RFLP, randomly amplified polymorphic DNA (RAPD), amplified fragment length polymorphism (AFLP), and microsatellites or simple sequence repeats (SSRs) are the most commonly used markers for wheat map construction. Of these, the SSR markers are widely used in map construction because they are highly polymorphic, widely distributed in the genome, required a small amount of DNA, and can be easily automated. Recently, diversity arrays technology (DArT) was developed as a hybridization-based alternative, which captures the value of the parallel nature of the microarray platform (Jaccoud et al. 2001). DArT simultaneously types several thousand loci in a single assay. DArT generates whole-genome fingerprints by scoring the presence versus the absence of DNA fragments in genomic representations generated from samples of genomic DNA. DArT has been used in genetic mapping (Akabari et al. 2006; Mantovani et al. 2008; Peleg et al. 2008; Yao et al. 2010). Single-nucleotide polymorphisms (SNPs) are found throughout genomes, which are an abundant form of genome variation (Brookes 1999; Rafalski 2002). Today, SNPs have become the marker of choice in most species for genome-wide association studies (GWAS), phylogenetic analyses, MAS, bulked segregant analysis, and genomic selection. Generally, SNPs are biallelic, whereas "satellite" polymorphisms are multiallelic. SNPs are more valuable than SSR in high-density genetic map, fine mapping, and gene clone, etc.

Since 2000 year, the team has built more than 10 various types of genetic population according to the breeding requirements in common wheat. We also have constructed 6 pieces of molecular genetic maps (1 DH, 3 RIL, and 2 natural populations) using SSR, DArT, and/or SNP markers. The parents of mapping population had more differences in yield, quality, resistance, and agronomic traits. The average distances between adjacent markers in the wheat maps were appropriate (0.44–9.77 cM), thus meeting the recommended requirement for genome-wide QTL scanning.

3.1 Genetic Map and Construction Methods

3.1.1 Concept of Genetic Map

Genetic map refers to chromosome reorganization of the exchange rate for the length of the genome, which is the basis of systematic study of the genome. Genetic map is divided into classical genetic map and molecular genetic map.

Classic genetic map is according to the laws of the chain of exchange and genetic loci exchange theory to construct linkage map. Due to the limitation of genetic markers, classic genetic map tag number is less, which is difficult to establish a saturated genetic map. It is very difficult to integrate different genetic maps. In addition, marker-associated traits cannot be a full performance due to the epistatic and interaction of genes. Due to lower efficiency and poor application value, classical genetic map has been slow in the past half century, and its application is largely restricted.

Molecular genetic map is constructed by the DNA markers, which is the basis for QTLs location, map-based gene cloning, comparative genomic research, molecular marker-assisted breeding, etc. Since the 1980s, the discovery of DNA molecular marker provides technical support for the construction of molecular genetic map. Molecular genetic map construction and QTL mapping have been becoming the current hot focus in the field of genetic breeding.

3.1.2 Construction Methods of Genetic Map

The theory basis of construction genetic map is chromosome exchange and restructuring. When the cells are meiosising, the genes on the non-homologous chromosomes are independent and free combination, while the linkage genes on the homologous chromosomes occur exchange and restructuring. The genes' exchange frequency has been closely related to the distance on the homologous chromosomes. So frequency of restructuring can show the genetic distance between genes.

Construction genetic map mainly includes the following: (1) analyze the genetic variance of parents; (2) establish proper mapping population; (3) test the polymorphism markers between parents, and scanning separation population; (4) construct a linkage group; and (5) determine the order of genetic markers and distances between the markers.

Common softwares for genetic mapping: MAPMAKER/EXP 3.0, JoinMAP 4.0, IciMapping 3.0, Mapchart 2.1, etc.

3.2 Genetics Map Construction

Since 2000 year, the team has built various types of mapping population, including RIL and DH populations, and constructed several genetic maps, so as to detect the QTLs of controlling yield, quality, and resistance in common wheat. The genetic maps were introduced as follows.

3.2.1 Genetic Map Construction Using DH Population Derived from Huapei 3 × Yumai 57

3.2.1.1 Materials and Methods

3.2.1.1.1 Plant Materials

A population of 168 DH lines was produced by hybridization with maize pollen grains of wheat F1 hybrid plants from the cross between two Chinese common wheat varieties, Huapei 3 (female parent) × Yumai 57 (male parent), and was used for the construction of the genetic map reported here. Huapei 3 is an elite variety with desirable agronomic characteristics for early maturity, high yield, and high resistance to several diseases (Hai et al. 2007), whereas Yumai 57 is a widely cultivated variety for its yield stability under different ecological conditions (Guo et al. 2004). Huapei 3 and Yumai 57 were registered in Henan province in 2006 (Hai et al. 2007) and by the State (China) in 2003 (Guo et al. 2004), respectively. The two parental varieties differ in many agronomically important traits including the baking quality (Guo et al. 2004; Hai et al. 2007).

The genetic mapping population differs in plant height, heading date, yield, and quality, which was mainly used for QTL mapping of above-related traits.

3.2.1.1.2 DNA Extraction and Isolation of SSRs from Public Databases

DNA was extracted from frozen-dried leaves as described by Ellis et al. (2005). A total of 2002 SSR and EST-SSR markers were tested for polymorphism between the two parents. Primer sequences for 260 *Xgwm* and 82 *Xgdm* SSR markers were available from Röder et al. (1998) and Pestsova et al. (2000), whereas those of 621 *Xwmc* and 480 *Xbarc* markers were described in the Graingenes Web site (http://www.wheat.pw.usda.gov/ggpages/SSR/WMC/) and the US wheat and barley scab initiative Web site (http://www.scabusa.org/pdfs/BARC_SSRs_011101.html), respectively. Thirty BE markers were available from Chen et al. (2005). Fifty *Xcfa*, 130 *Xcfd*, 48 *Xcwem,* and 301 *Xcfe* markers were kindly provided by Dr. Xianchun Xia, Chinese Academy of Agricultural Sciences, Beijing, China.

3.2.1.1.3 PCR Conditions

Amplification reactions were carried out in a 20 μL reactionmixture, containing 3.6 μL of genomic DNA (20 ng/μL), 1.5 μL $MgCl_2$ (25 mmol/L), 1.2 μL deoxyribonucleotide triphosphate mixtures (2.5 mmol/L), 2.0 μL of 10 × PCR buffer, 0.5 μL primer (10 μmol/L), 0.2 μL Taq polymerase enzyme (5 units/μL), and 11.0 μL of double-distilled H_2O. The PCRs were performed in 96-well microtiter plates using the Eppendorf AG 22331 Hamburg thermal cycler (Eppendorf, Hamburg, Germany). DNA amplification was programmed at 95 °C for 5 min, followed by 36 cycles of 95 °C for 1 min, 50–65 °C (depending on the primer combinations) for 50 s, 72 °C for 1 min, and a final extension of 10 min at 72 °C before cooling to 4 °C. After amplification, the PCR products were mixed with 5.6 μL loading buffer (2.5 mg/mL bromophenol blue, 2.5 mg/mL diphenylamine blue, 10 mmol/L ethylene diamine tetraacetic acids, 95 % (v/v) formamide), denatured for 5 min at 95 °C, and chilled on ice for 5 min. The PCR products were separated using 6 % (w/v) denatured polyacrylamide gel and were detected by silver staining (Karakousis et al. 2003).

3.2.1.1.4 Method of Genetic Map Construction

The genetic map was constructed with MAPMAKER/Exp ver. 3.0 b (Lincoln et al. 1993). The commands, "group" with a logarithm of the odds ratio scores 3.0 or more, "try," "compare," and "ripple," were used to develop the linkage map. The resulting groups were oriented and placed to the chromosomes based on the microsatellite consensus map (Somers et al. 2004) and the composite wheat map (http://wheat.pw.usda.gov). The Kosambi mapping function was used to convert recombination fractions into cM values as map distances. The linkage map was finally drawn using the software Mapchart ver. 2.1 (Voorrips 2002).

3.2.1.2 Result and Analysis

3.2.1.2.1 SSR and EST-SSR Marker Analysis

As the first step toward constructing the genetic map, we tested the polymorphism of 2002 available markers (1623 SSRs and 379 EST-SSRs) between the two parental varieties. Of these, 270 SSR and 17 EST-SSR markers revealed polymorphism between the parents of the mapping population. The great majority of the SSR primer sets (250 pairs in total) amplified single polymorphic loci (Fig. 3.1). However, multispecificities were encountered for two groups of SSR primer sets. The first group contained 17 sets, each of which amplified two loci (Fig. 3.2), whereas the second group included three sets, each of which amplified three loci (Fig. 3.3). Similarly, 12 EST-SSR primer sets exhibited single specificity, while five displayed double specificities. Consequently, the 270 SSR and 17 EST-SSR primer sets detected a total of 315 loci (293 SSR and 22 EST-SSR locations).

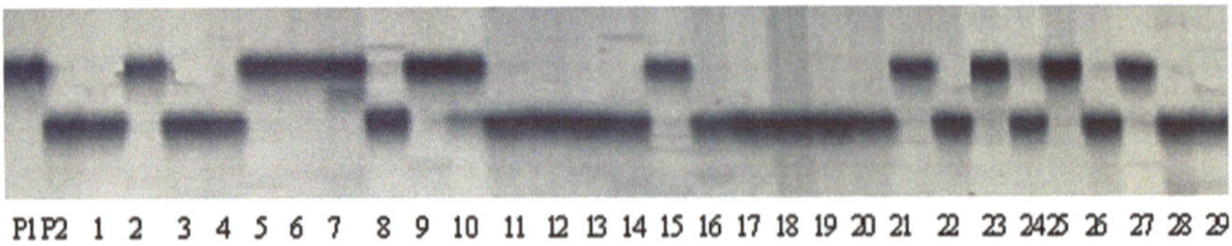

Fig. 3.1 SSR amplified profile of primer Xgwm186 in the DH population P1: Huapei 3; P2: Yumai 57

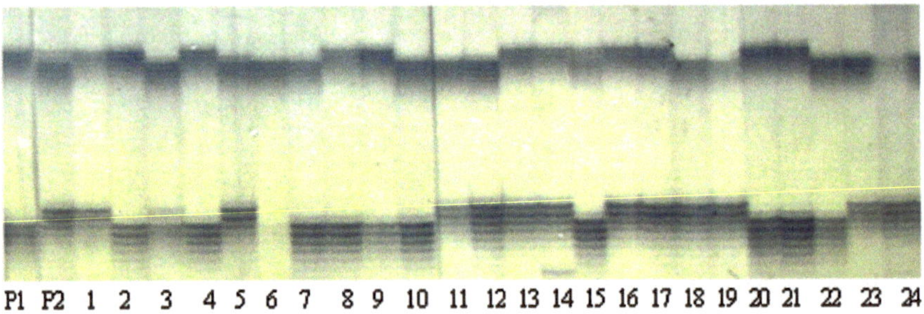

Fig. 3.2 SSR amplified profile of primer Xgwm133 in the DH population P1: Huapei 3; P2: Yumai 57

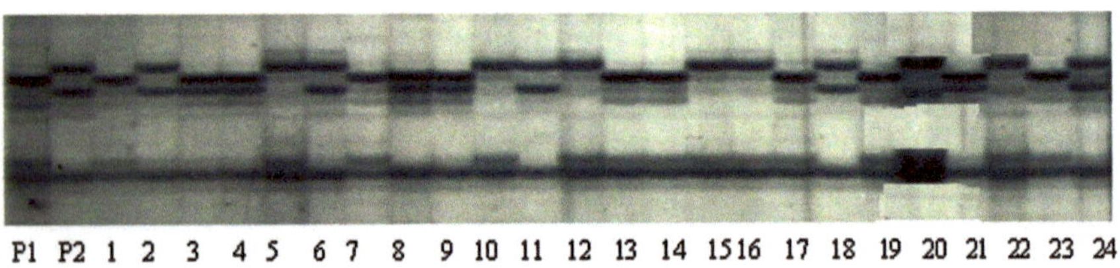

Fig. 3.3 SSR amplified profile of primer Xcwem32 in the DH population P1: Huapei 3; P2: Yumai 57

3.2.1.2.2 Map Characteristics

Using the polymorphic markers described above and the map construction program, a genetic linkage map containing 305 SSR markers, including 283 SSR and 22 EST-SSR loci, was finally developed. Ten SSR loci remained unlinked. The map covered a total length of 2141.7 cM with an average distance of 7.02 cM between adjacent markers on the map, which resulted in 24 linkage groups comprising 3-24 loci (Table 3.1; Fig. 3.4). Each of the linkage groups could be assigned to one of the 21 chromosomes based on the information from previous mapping studies (Somers et al. 2004; http://wheat.pw.usda.gov). The A genome map was comprised of 102 loci and spanned 666.2 cM. The corresponding values for the B genome map were 116 loci and 551.7 cM, and those for the D genome were 87 loci and 923.7 cM, respectively.

The linkage group size ranged from 13.6 cM for linkage group 5A-2 to 167.8 cM for linkage group 2D, with a mean of 89.23 cM per linkage group. The number of loci per linkage group ranged from 3 (5A-2) to 24 (3B) with a mean of 12.71 loci per linkage group (Table 3.1). Less loci were mapped on the D genome (28.6 %) compared to those on the A (33.4 %) or B (38.0 %) genomes. The global map density was one locus/7, with one locus/6.53 cM for the A genome, one locus/4.76 cM for the B genome, and one locus/10.62 cM for the D genome. Gaps were found for the linkage maps of three different chromosomes (5A, 5B, 7B) (Fig. 3.4). There were chromosomal regions (such as 1D) that harbored few SSR markers compared with previously published maps (Somers et al. 2004). As found previously (Torada et al. 2006), the clustering of microsatellites near the centromere was observed in several chromosomes (i.e., 1A, 1B, 2A, 2B, 3A, 3B) (Fig. 3.4).

Table 3.1 Genetic distance and marker distribution as well as distorted locus among different linkage groups

Linkage group (LG)	Number of markers			Length of LG (cM)	Average distance (cM)	Distorted locus		
	SSR	EST-SSR	Total			SSR	EST-SSR	Total
1A	17	2	19	75.00	3.95	4	1	5
1B	18	6	24	83.00	3.46	8	4	12
1D	6	0	6	57.50	9.58	1	0	1
2A	18	1	19	105.40	5.55	2	0	2
2B	20	2	22	106.30	4.83	1	0	1
2D	18	0	18	167.80	9.32	3	0	3
3A	19	0	19	160.40	8.44	2	0	2
3B	22	2	24	131.70	5.49	19	2	21
3D	13	0	13	94.40	7.26	2	0	2
4A	12	0	12	43.50	3.63	2	0	2
4B	8	0	8	18.40	2.30	0	0	0
4D	9	3	12	155.20	12.93	0	0	0
5A1	4	2	6	79.90	13.32	1	0	1
5A2	1	2	3	13.60	4.53	0	0	0
5B1	5	0	5	13.70	4.03	1	0	1
5B2	7	0	7	28.20	2.74	0	0	0
5D	15	0	15	122.00	8.13	0	0	0
6A	10	2	12	87.40	7.28	0	0	0
6B	13	0	13	107.50	8.27	13	0	13
6D	10	0	10	161.80	16.18	4	0	4
7A	12	0	12	101.00	8.42	0	0	0
7B1	7	0	7	49.90	7.13	4	0	4
7B2	6	0	6	13.00	1.19	0	0	0
7D	13	0	13	165.00	13.75	3	0	3
Total	283	22	305	2141.70	NA	70	7	77
Average	11.79	0.92	12.71	89.23	7.02	2.91	0.29	3.21

NA not applicable

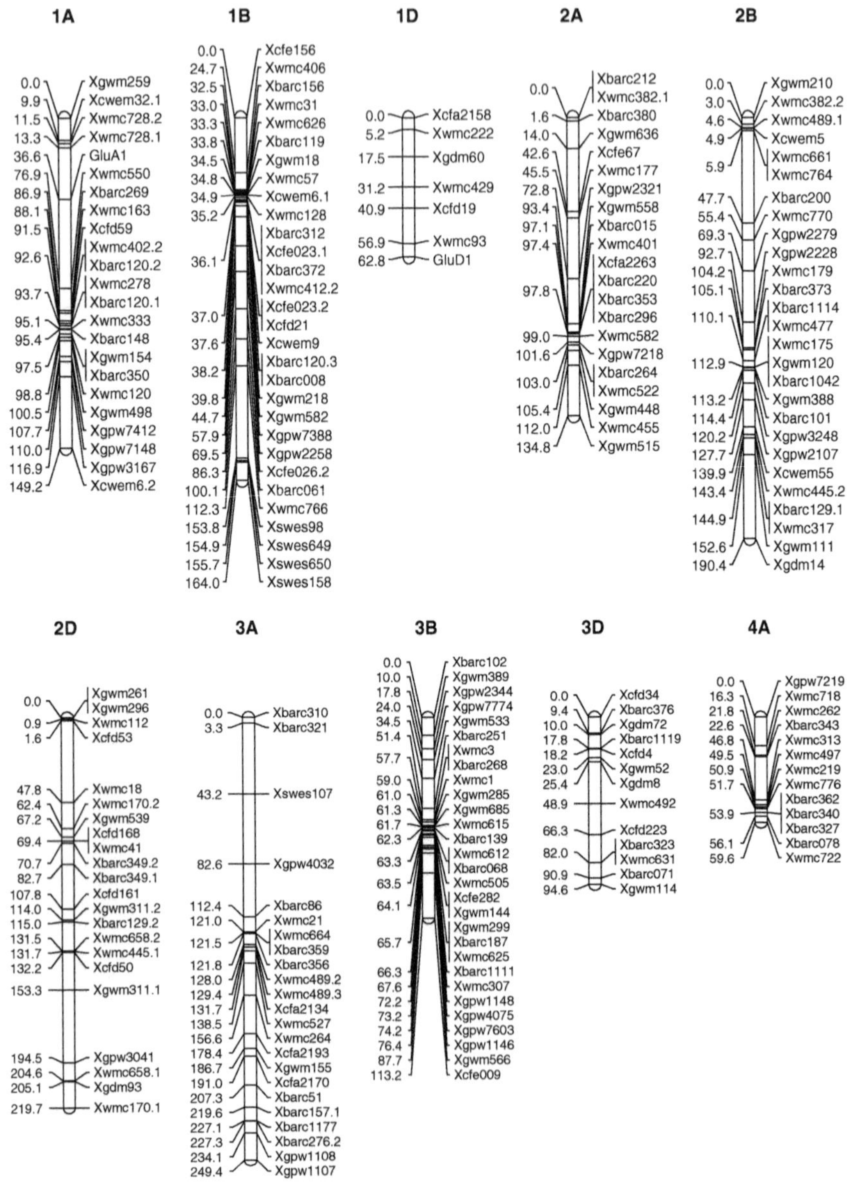

Fig. 3.4 Genetic map developed by using a double haploid population from Huapei 3 × Yumai 57

The chromosomal locations and orders of the mapped markers in this work were similar to the ones reported previously (Somers et al. 2004) and the composite wheat linkage map (http://wheat.pw.usda.gov). However, for 20 markers (19 SSRs and one EST-SSR, Table 3.2), their chromosomal locations revealed by this work

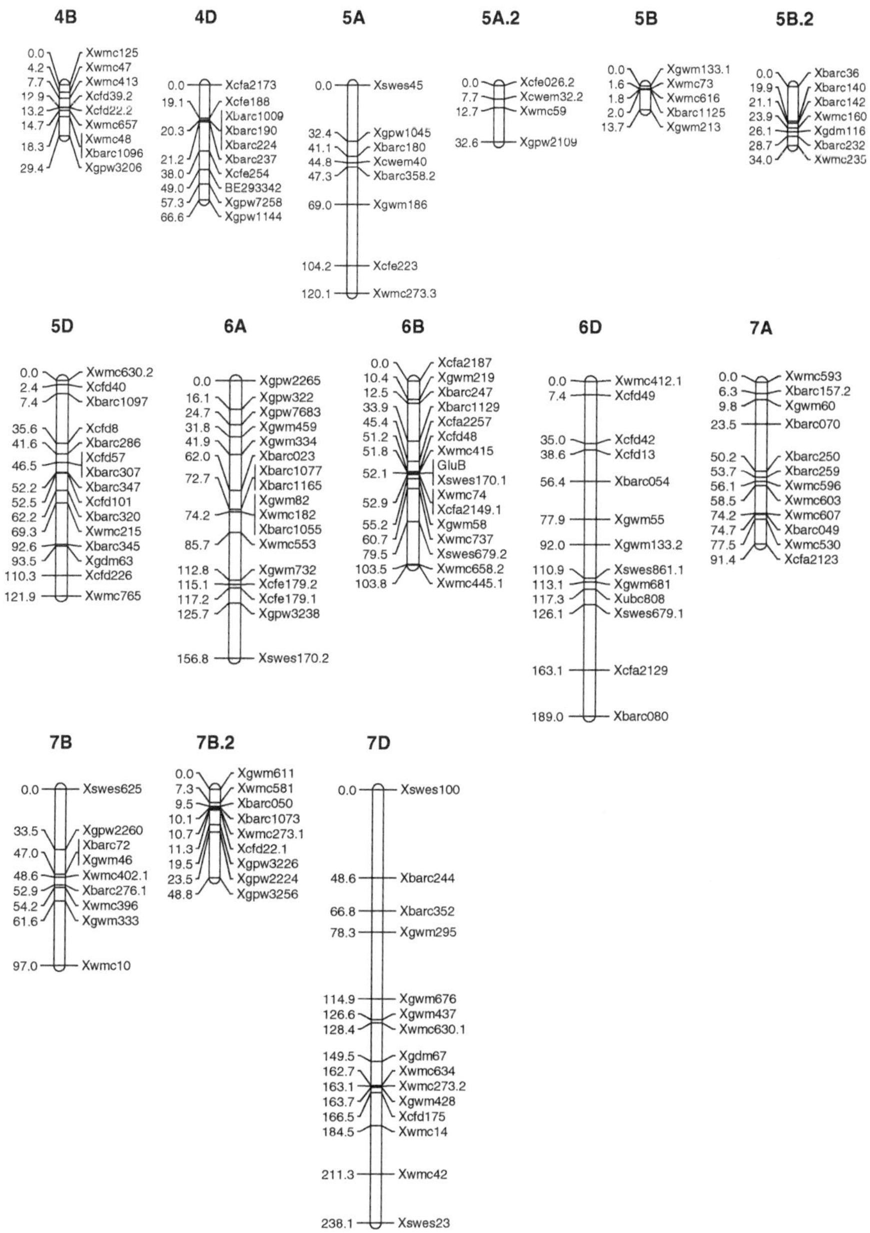

Fig. 3.4 (continued)

differed from the ones in previously published maps (Röder et al. 1998; Elouafi and Nachit 2004; Somers et al. 2004). For example, *Xgwm154* was located on the 1A chromosome in this map, but it had been reported to be located on 5A, 7A, and 3B

Table 3.2 Loci (markers) that showed different map positions between our study and those published previously

Previous map position	Map position found by this study	Number of markers
Xgwm259 (1B)a, Xgwm154 (5A/7A/3B), Xwmc728 (1B/2A/5B)	1A	3
Xcfd21 (7D/1D/1A), Xgwm218 (3A)	1B	2
Xcfa2158 (1A/1B)	1D	1
Xgwm111 (7D/7B/4A/6D)	2B	1
Xbarc187 (1B)	3B	1
Xbarc376 (3B)	3D	1
Xbarc340 (5B/7B)	4A	1
BE293342 (4A), Xbarc190 (4A), Xbarc224 (4A/5B)	4D	3
Xbarc1077 (3B)	6A	1
Xwmc415 (5A/5B), Xcfd48 (1B/1D), Xbarc1129 (1B)	6B	3
Xcfa2129 (1A/1D/1B), Xbarc080 (1B), Xgwm681 (6B/7A)	6D	3
Total		20

Note The wheat chromosomes in which the listed markers have been mapped onto by previous studies are in the parentheses

Table 3.3 New loci that were mapped on the linkage groups in the DH population

Linkage group	Locus (marker)
1A	*Xbarc269, Xbarc350*
1B	*Xbarc312, Xbarc372*
2A	*Xbarc380, Xbarc264*
2B	*Xbarc373*
3A	*Xbarc276, Xbarc359*
3B	*Xbarc251*
4A	*Xbarc362*
4D	*Xbarc237, Xbarc1009*
5B	*Xbarc36, Xbarc1125*
5D	*Xbarc307, Xgdm63*
6A	*Xgwm732*
6B	*Xbarc247*
7A	*Xbarc259, Xbarc250*
7D	*Xbarc244*

chromosomes (Röder et al. 1998; Elouafi and Nachit 2004). Importantly, this work mapped the chromosomal locations of 22 new SSR markers (Table 3.3). They were found to distribute on 14 linkage groups (Table 3.3; Fig. 3.4).

3.2.1.2.3 Segregation Distortion

From the 315 loci analyzed using 168 DH lines, 77 loci (24.4 %) showed segregation distortion. Of these loci, 44 markers (57.1 %) showed distortion in favor of the female (Huapei 3) alleles, whereas 33 (42.9 %) were in favor of the male (Yumai 57) alleles. The distortion loci were not evenly distributed among the A, B, and D genomes (Table 3.1; Fig. 3.4), with 12, 51, and 14 loci mapped on the A, B, and D genomes, respectively.

The segregation distortion loci were mainly clustered on chromosomes 1A (5), 1B (12), 3B (21), and 6B (13), respectively. For example, the 6B chromosome possessed 13 distortion loci, which were in favor of the Huapei 3 alleles. Twenty-one distortion loci on 3B chromosome inclined toward the Yumai 57 alleles. There were no segregation distortion loci on chromosomes 4B, 4D, 5D, 6A, and 7A (Fig. 3.4).

3.2.1.2.4 Increased Density of the Genetic Map

The molecular genetic map was increased density to 357 loci in 2011. The map covered a total length of 2780.9 cM with an average distance of 7.79 cM between adjacent markers on the map (Fig. 3.4).

3.2.1.2.4 Comparison of the Present Map with Others

In the present study, a new genetic map was constructed using SSR and EST-SSR markers. Compared to the previously published intervarietal linkage maps (discussed above), our map has several unique features. First, the parental lines are elite wheat varieties and widely cultivated in Henan province, which, together with the neighboring Shandong province, form the largest winter wheat cultivation region in China. The two parental varieties and the germ plasma derived from them will be important genetic resources for further improvement of winter wheat varieties in China. Second, our mapping population is comparatively large (being composed of 168 DH lines) and genetically stable, thus suitable for QTL analysis based on multisite and multiyear trials. Third, the map provides fairly adequate genome coverage for whole-genome QTL analysis and a reasonable amount of markers to tag chromosomal regions of interest. There commended map distance for genome-wide QTL detection is 10 recombinations/100 meiotic events, or an interval length smaller than 10 cM (Doerge et al. 2002). In our map, the average distance between adjacent markers is 7.79 cM, thus meeting the recommended requirement for genome-wide QTL scanning. Finally, there are relatively less linkage gaps in our map, which is advantageous for high-resolution mapping and tagging of QTLs.

3.2.2 Genetic Map Constructed Using RIL Population Derived from Nuomai 1 × Gaocheng 8901

3.2.2.1 Materials and Methods

3.2.2.1.1 Plant Materials

A population of 256 RILs was generated by single seed 10 descent from the cross of Nuomai 1 and strong gluten Gaocheng 8901 (Zhai et al. 2007). Nuomai 1, soft wheat, was produced from Jiangsu Baihuomai × Guandong 107 by China Agricultural University. It has deletions of three Wx protein subunits. Gaocheng 8901, hard wheat, was produced from 77546-2 × Lingzhang by Gaocheng Academy of Agricultural Sciences. Nuomai 1 and Gaocheng 8901 were registered in Beijing in 2005 and in Hebei Province in 1998, respectively. The rapid viscosity analysis (RVA) indicated that the pasting temperature and setback value of Nuomai 1 are lower than Gaocheng 8901, but peak viscosity value is higher than Gaocheng 8901. These two parents have different HMW-GS compositions. Nuomai 1 has null, 7 + 8, 2.2 + 12 at *Glu-A1*, *Glu-B1,* and *Glu-D1* loci, respectively, whereas Gaocheng 8901 has 1, 7 + 8, and 5 + 10 at these corresponding loci.

This genetic map construction is mainly used for wheat starch, protein, and processing quality traits of QTL mapping.

3.2.2.1.2 DNA Extraction and Molecular Marker Detection

DNA was extracted by approved method of CTAB as described by https://www.triticarte.com.au. Agar gel (0.8 %) electrophoresis was used to detect the concentration and purity of extracted DNA.

DArT markers were provided and detected by Triticarte Pty. Ltd. (https://www.triticarte.com.au) in Australia. Primer sequences for 33 GWM SSR markers were available from Röder et al. (1998) and Pestsova et al. (2000), whereas 62 GPW SSR markers were described in GrainGene 2.0 (http://wheat.pw.usda.gov/GG2/index.shtml), and 36 CFD SSR markers were kindly provided by Dr. Xia Xianchun, Chinese Academy of Agricultural Sciences, Beijing, China.

3.2.2.2 Result and Analysis

3.2.2.2.1 Molecular Marker Distribution

Among the 916 DArT marker provided by Triticarte Pty. Ltd., Australia, 479 primers showed polymorphism in the RIL population. According to provisional construction of linkage map, we found few markers located on 4D, 5D, and 6D chromosomes. Therefore, 131 SSR markers were found from GrainGene 2.0 to test the polymorphism between the two parental varieties. Of these, 31 SSR markers revealed polymorphism between the parents of the mapping population, and 14 markers were used to construct the genetic map. The markers were not evenly distributed (Table 3.4, Fig. 3.5.) on chromosomes 1A, 2A, 3B, 6A, 4D, 5A, 5D, and 6D, respectively. SSR markers were mainly located on 1D, 3A, 3B, 3D, 4D, 5D, 6A, 6B, and 6D, respectively. Two HMW-GS loci were mapped on chromosomes 1A and 1D, and three Waxy protein subunit loci were located on chromosomes 7A, 4A, and 7D, respectively.

3.2.2.2.2 The Characteristics of the Genetic Map

Four hundred and ninety-eight markers (479 DArTs, 14 SSRs, two HMW-GS loci, and three Waxy protein subunit loci) were mapped on the 24 linkage groups (Fig. 3.5). The genetic map spanned about 4229.7 cM in wheat genome, and the average distance of markers was 9.77 cM (including 65 overlapping sites). The A genome contained 211 markers (including 29 overlapping sites), with total length of 1617.3 cM, and the average distance of markers was 8.89 cM, while the B genome had 166 markers (including 15 overlapping sites), covering 1682.2 cM with an average distance of 11.14 cM. The D genome contained 121 markers (including 21 overlapping sites), and the average distance of markers was 9.30 cM and spanning about 930.2 cM. The average length of each linkage group is 176.2 cM. The 3B is the longest chromosome (411.2 cM), and 5D is the shortest, with only 34.4 cM. Each linkage group possesses three (e.g., 5D and 6D chromosomes) to 42 markers (e.g., 3B chromosome). However, large gaps were found on chromosomes 1A, 6A, and 7D, respectively.

In addition, 55 new markers were mapped on the 18 linkage groups without the involvement of chromosomes 1B, 2A, 2B, 3A, 4B, 5B, and 6A, respectively (Table 3.5). The genomes of A, B, and D consisted of 10, 13, and 32 markers, respectively.

3.2.2.2.3 Segregation Distortion

Chi-squared test indicated that, among the 498 markers, 168 markers showed the genetics distortion segregation ($P < 0.05$), accounting for 33.7 % (Table 3.4). Sixty-two DArT markers and one SSR marker (37.5 %) exhibited distortion in favor of the female parent Nuomai 1 alleles, whereas 104 DArT markers and one *HMW-GS* marker (62.5 %) were in favor of male parent Gaocheng 8901 alleles. The distortion loci were not evenly distributed among the A, B, and D genomes (Table 3.4), with 57, 70, and 42 loci mapped on the A, B, and D genomes, respectively. The segregation distortion loci were mainly clustered on chromosomes 2A, 2B, 4D, 5A, 5B, and 7D2, respectively. All loci on the 5A and 7D2

Table 3.4 Marker distribution as well as distorted locus among different linkage groups

Linkage group	Number of markers				Length of linkage group (cM)	Average distance (cM)	Distorted locus			
	DArT	SSR	HMW-GS/ *Wx* loci	Total			DArT	SSR	HMW-GS/*Wx* locus	Total
1A1	23	0	*Glu-A1*	24	155.9	7.80	2	0	0	2
1A2	33	0	0	33	262.3	11.40	15	0	0	15
1B	18	0	0	18	234.8	16.77	6	0	0	6
1D	20	1	*Glu-D1*	22	240.5	13.36	3	0	0	3
2A	28	0	0	28	316.5	12.17	20	0	0	20
2B	15	0	0	15	90.3	6.45	11	0	0	11
2D	30	0	0	30	221.3	8.85	7	0	0	7
3A	18	1	0	19	195.9	10.88	5	0	0	5
3B	41	1	0	42	411.2	10.54	14	0	0	14
3D	18	2	0	20	169.9	12.13	5	0	0	5
4A	35	0	*Wx-B1*	36	156.6	4.89	1	0	0	1
4B	11	0	0	11	121.8	12.18	3	0	0	3
4D	5	1	0	6	88.5	14.75	4	0	0	4
5A	5	0	0	5	70.1	17.52	5	0	0	5
5B	24	0	0	24	239.6	10.41	15	0	0	15
5D	2	1	0	3	34.4	11.47	0	0	0	0
6A1	35	1	0	36	137.3	4.29	5	0	0	5
6A2	17	2	0	19	160.3	9.42	1	0	0	1
6B	34	2	0	36	339.1	10.60	19	0	0	19
6D	1	2	0	3	35.0	11.67	1	1	0	1
7A	10	0	*Wx-A1*	11	162	16.2	3	0	0	3
7B	20	0	0	20	245.4	12.91	2	0	0	2

(continued)

Table 3.4 (continued)

Linkage group	Number of markers				Length of linkage group (cM)	Average distance (cM)	Distorted locus			
	DArT	SSR	HMW-GS/*Wx* loci	Total			DArT	SSR	HMW-GS/*Wx* locus	Total
7D1	20	0	0	20	97.8	6.11	4	0	0	4
7D2	16	0	*Wx-D1*	17	42.8	2.85	16	0	1 *Wx*	17
Total	479	14	5	498	4229.7		166	1	1	168
Average	20.75				176.2	9.77	7			

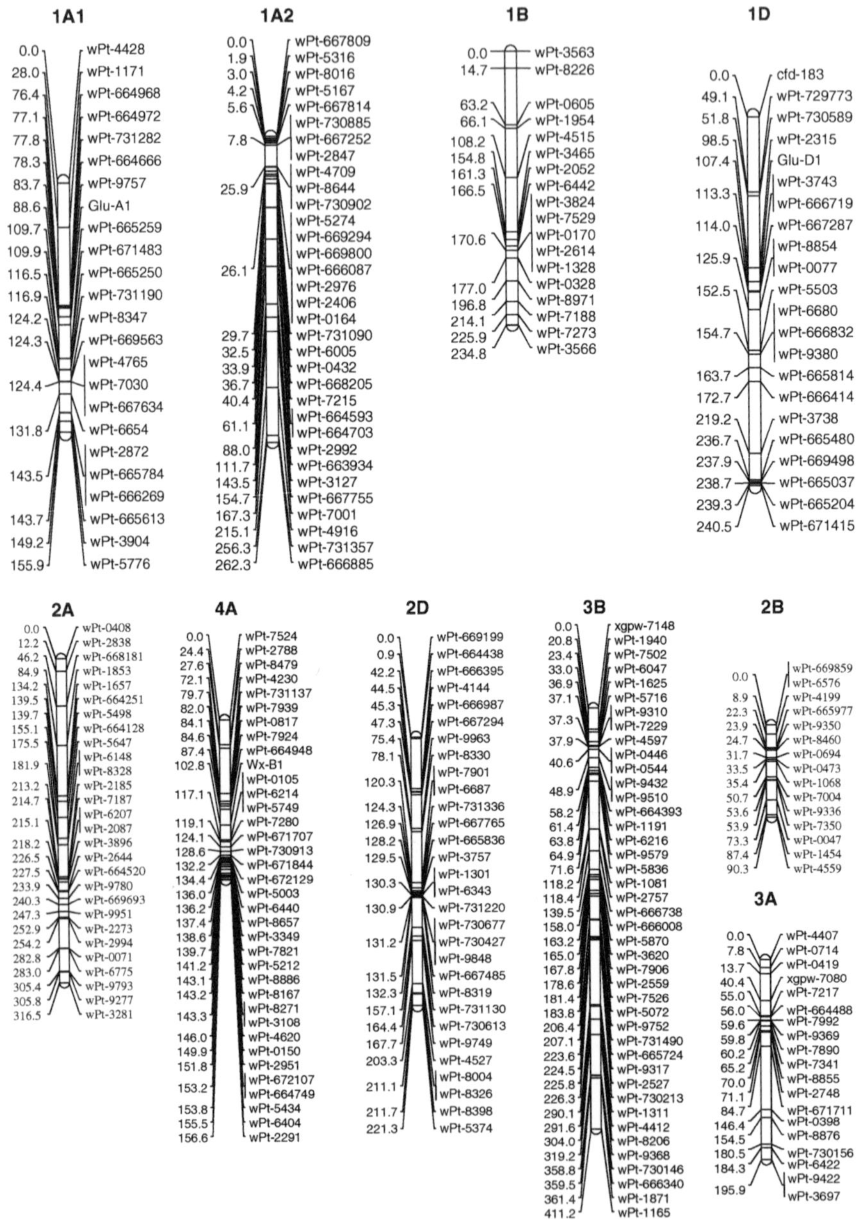

Fig. 3.5 Genetic map developed by using recombinant inbred line population from Nuomai 1 × Gaocheng 8901

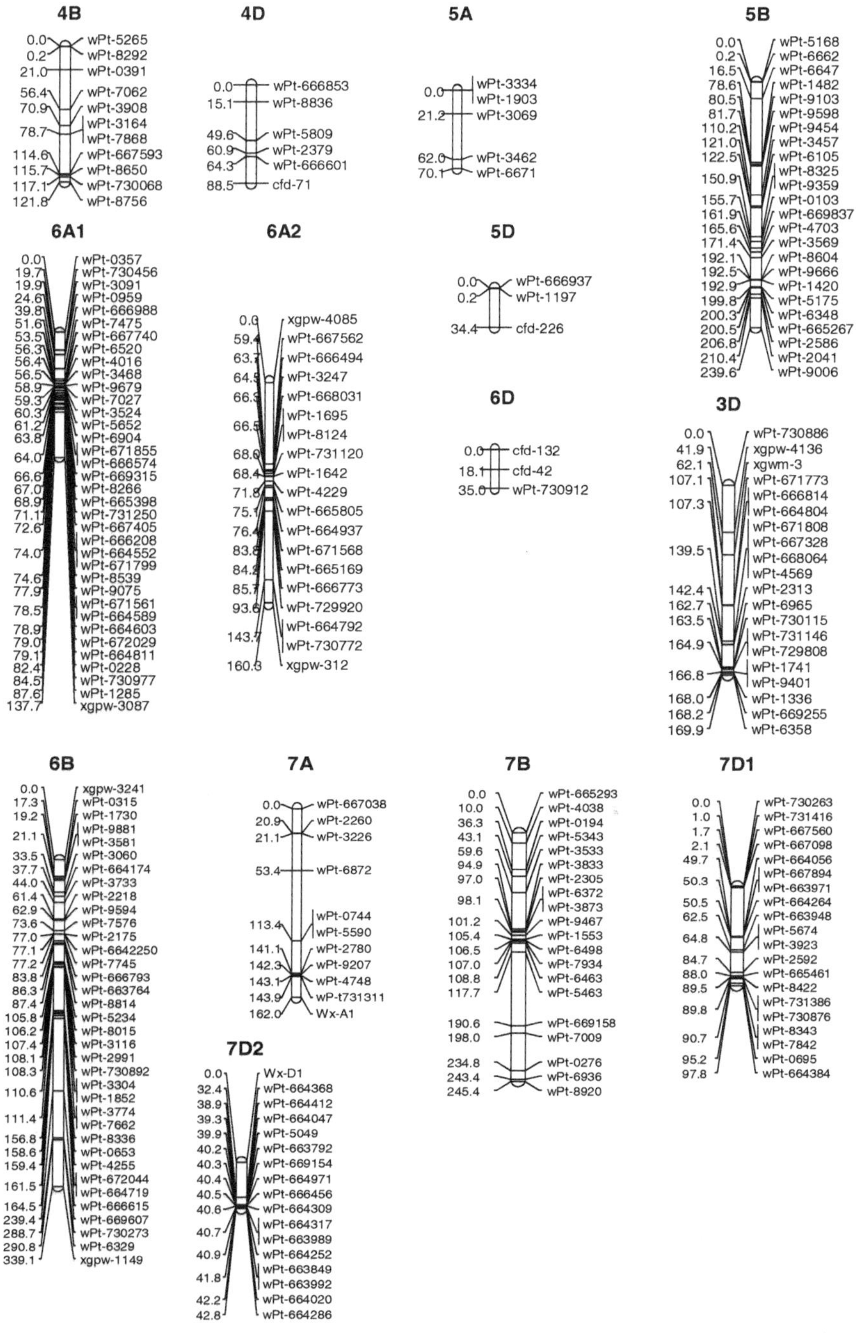

Fig. 3.5 (continued)

Table 3.5 Distribution of new loci that were mapped on the linkage groups

Chromosome	Marker	Number
1A1	*wPt-665,250, wPt-731,190*	2
1A2	*wPt-731,357, wPt-666,885*	2
1D	*wPt-729,773, wPt-730,589, wPt-666,414, wPt-665,480, wPt-669,498, wPt-665,037, wPt-665,204, wPt-671,415*	8
1D	*wPt-729,773, wPt-730,589, wPt-666,414, wPt-665,480, wPt-669,498, wPt-665,037, wPt-665,204, wPt-671,415*	8
2D	*wPt-669,199, wPt-664,438, wPt-666,395, wPt-666,987, wPt-667,294, wPt-730,613, wPt-666,008*	7
3B	*wPt-664,393, wPt-731,490, wPt-730,146, wPt-666,340, wPt-1871*	5
3D	*wPt-730,146, wPt-666,340, wPt-1871, wPt-730,115, wPt-731,146, wPt-729,808, wPt-669,255*	7
4A	*wPt-664,948*	1
4D	*wPt-666,853, wPt-666,601*	2
5A	*wPt-3069*	1
5D	*wPt-666,937*	1
6A2	*wPt-664,937, wPt-729,920*	2
6B	*wPt-0315, wPt-664,174, wPt-672,044, wPt-664,719, wPt-666,615, wPt730273*	6
6D	*wPt-730,912*	1
7A	*wPt-667,038, wPt-731,311*	2
7B	*wPt-669,158, wPt-7009*	2
7D1	*wPt-730,263, wPt-731,416, wPt-667,560, wPt-667,098, wPt-664,056*	5
7D2	*wPt-664,252*	1

chromosomes are distortion loci, which were in favor of the male parent GC 8901 alleles. No markers showed the genetics distortion segregation on the chromosome 5D (Fig. 3.5).

3.2.2.2.4 Comparison of the Present Map with Previous Others

The unique feature of this study, compared with many others, is that there are some clear differences in grain quality between the two parents. Nuomai 1 has special starch quality due to the deletions of its three Wx loci, and Gaocheng 8901 shows good protein quality because of its strong gluten. This means the Nuomai 1's starch quality (e.g., starch content and starch pasting properties) and the Gaocheng 8901's dough rheological characteristics (e.g., dough stability time, mixograph, and extensograph) can be analyzed successfully based on this genetic map. It is true that there is a large gap between makers on some chromosomes due to fewer markers available at this point. We believe that more markers needed be added in this map in the near future.

3.2.3 Genetic Map Constructed Using RIL Population Derived from Shannong 01-35 × Gaocheng 9411

3.2.3.1 Materials and Methods

3.2.3.1.1 Plant Materials

A population of 182 RILs was generated by single seed 8 descent from the cross of Shannong 01-35 (39-1/Hesheng 2) and Gaocheng 9411 (777546/Linzhangmai). The female parent Shannong 01-35 was middle gluten wheat, with low percentage of ear bearing tiller (about 3 million tillers per hectare), long spike, and larger grain with more than 60 g of thousand-grain weight, while for the male parent was strong gluten wheat with high percentage of ear bearing tiller (about 7.5 million tillers per hectare), short spike, more grains per spike, and about 38 g of thousand-grain weight. Both of the parents had more distant relatives and different in many agronomic traits, yield and quality traits.

This genetic map was mainly used for QTL mapping of grain weight, starch, protein, and processing quality traits.

3.2.3.1.2 DNA Extraction and Molecular Marker Detection

DNA was extracted by approved method of CTAB as described by https://www.triticarte.com.au. Agar gel (0.8 %) electrophoresis was used to detect the concentration and purity of extracted DNA. DArT and SSR markers and detection were same as "Nuomai 1 × Gaocheng 8901" in Sect. 3.2 of this chapter.

3.2.3.2 Result and Analysis

3.2.3.2.1 Molecular Marker Distribution

A total of 502 primers showed polymorphism in the RIL population (Table 3.6), including 442 DArT, 59 SSR markers, and one *TaGW2*-CAPS marker of grain weight. Among 442 DArT markers, 398 were located on 21 wheat chromosomes. Chromosome 1B had the most markers with 37. Only one marker was located on chromosome 4D. A total of 49 markers had no location information. Forty-nine SSR markers were located on 18 wheat chromosomes except 2B, 1D, and 2D.

3.2.3.2.2 Characteristics of the Genetic Map

Five hundred and two markers (442 DArTs, 59 SSRs, 1 *TaGW2*-CAPS marker) were mapped on the 24 linkage groups (Table 3.6, Fig. 3.6). The genetic map was covered

Table 3.6 Markers and distorted locus distribution among different linkage groups

Linkage group	Number of markers			Length (cM)	Average distance (cM)	Distorted loci			Favorable to Shannong 01-35	Favorable to Gaocheng 9411
	DArT	SSR	Total			DArT	SSR	Total		
1A.1	20	0	20	81.1	4.1	12	0	12	12	0
1A.2	29	3	32	143.9	4.5	1	0	1	0	1
1B.1	45	3	48	113.5	2.4	45	3	48	0	48
1B.2	6	1	7	44.8	6.4	0	0	0	0	0
1D	16	0	16	108.9	6.8	1	0	1	1	0
2A	11	6	17	328.1	19.3	4	0	4	4	0
2B	16	1	17	210.7	12.4	10	0	10	0	10
2D	30	2	32	184.3	5.8	6	2	8	4	4
3A	32	6	38	232.6	6.1	14	0	14	14	0
3B.1	27	1	28	289.8	10.4	6	0	6	4	2
3B.2	5	0	5	2.0	0.4	5	0	5	5	0
3D	8	6	14	285.1	20.4	2	2	4	0	4
4A	17	2	19	173.3	9.1	4	1	5	4	1
4B.1	6	2	8	99.2	12.4	3	0	3	0	3
4B.2	4	1	5	36.8	7.4	1	1	2	2	0
4D	1	2	3	80.6	26.9	0	0	0	0	0
5A	5	2	7	184.9	26.4	0	1	1	1	0
5B.1	17	4	21	126.8	6.0	5	1	6	5	1
5B.2	18	0	18	159.6	8.9	3	0	3	3	0
5D	6	4	10	210.9	21.1	2	0	2	2	0
6A.1	29	3	33[a]	152.6	4.6	3	0	4[a]	0	4
6A.2	3	0	3	0.9	0.3	0	0	0	0	0
6B.1	6	0	6	30.3	5.1	1	0	1	0	1

(continued)

Table 3.6 (continued)

Linkage group	Number of markers			Length (cM)	Average distance (cM)	Distorted loci			Favorable to Shannong 01-35	Favorable to Gaocheng 9411
	DArT	SSR	Total			DArT	SSR	Total		
6B.2	5	0	5	62.5	12.5	2	0	2	0	2
6B.3	3	1	4	34.3	8.6	2	1	3	3	0
6D	8	3	11	178.3	16.2	1	1	2	1	1
7A	26	4	30	258.3	8.6	2	0	2	0	2
7B	23	2	25	217.8	8.7	3	0	3	2	1
7D	20	0	20	52.6	2.6	16	0	16	16	0
Total	442	59	502[a]	4084.5	8.1	154	13	168[a]	83	85

[a]Including one *TaGW2*-CAPS marker

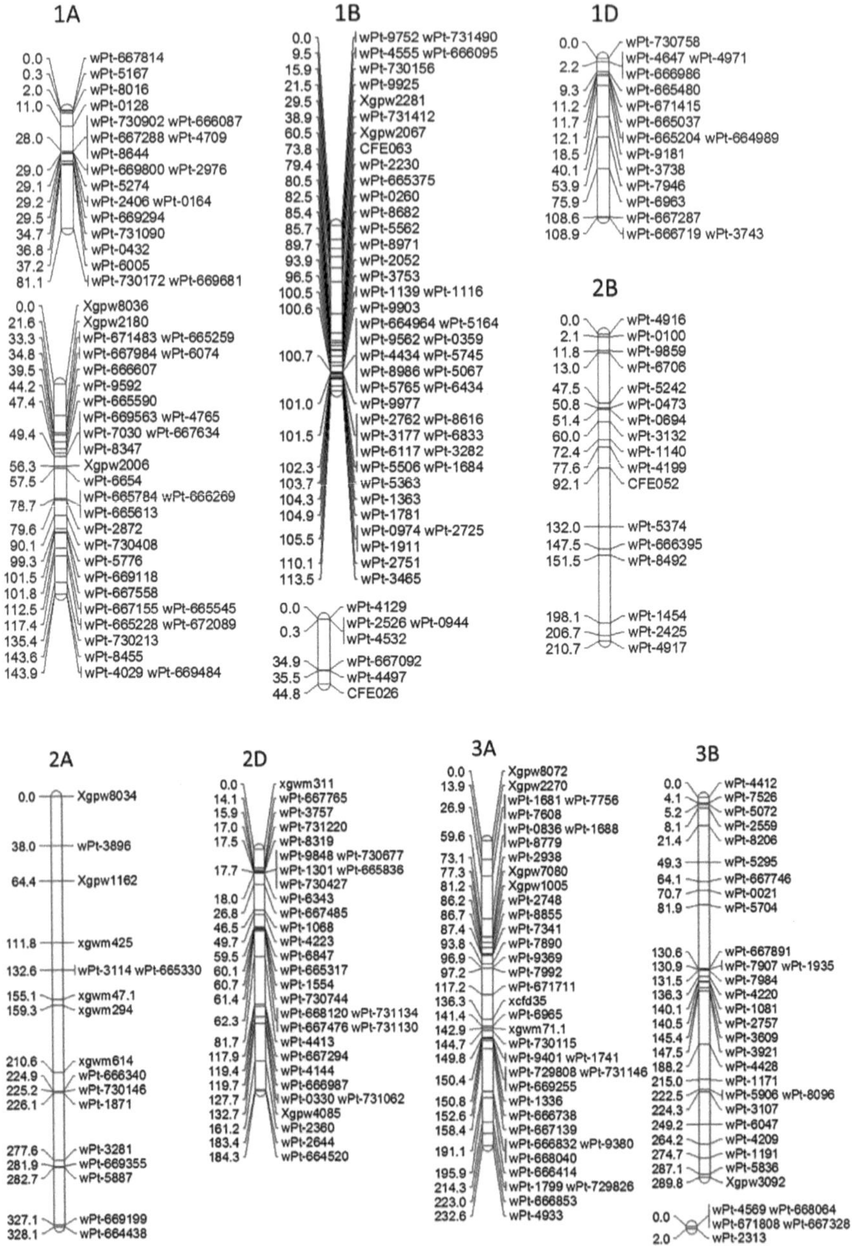

Fig. 3.6 Genetic map developed by using RIL population from Shannong 01-35 × Gaocheng 9411

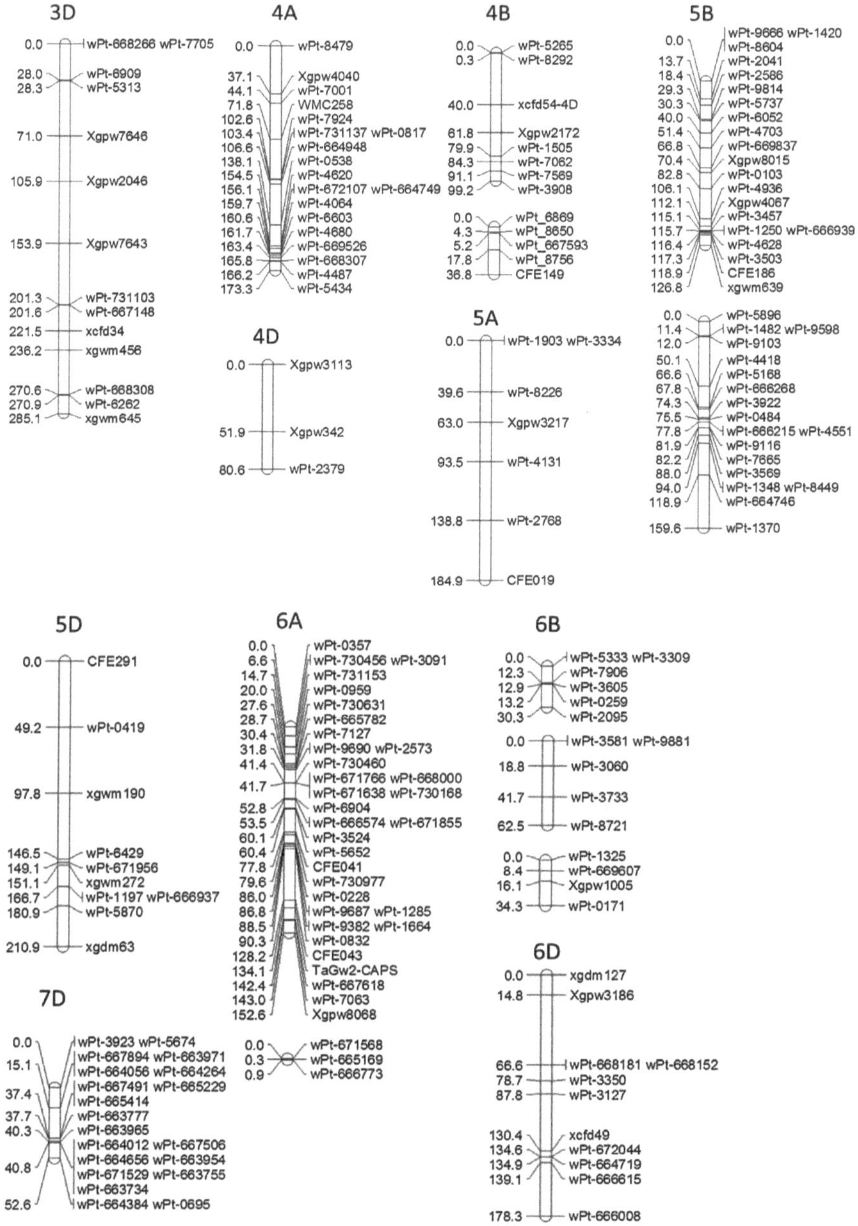

Fig. 3.6 (continued)

4084.5 cM in wheat genome, and the average distance of markers was 8.13 cM. Twenty-nine linkage groups were involved in 21 wheat chromosomes. Genomes A and B had similarly markers, 199 and 197, with the covering length is 1555.7 cM and

Fig. 3.6 (continued)

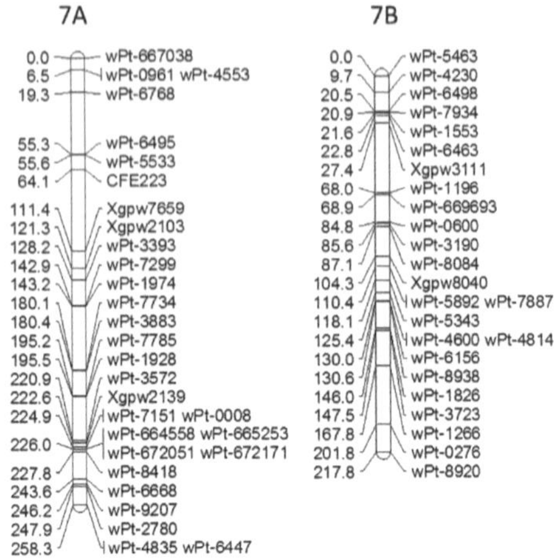

1428.1 cM, respectively. The D genome contained 106 markers, and spanning about 1100.7 cM. The *TaGW2*-CAPS marker was located on the chromosome 6A. In addition, 54 new markers (including 44 DArTs and SSRs)were mapped on the 18 linkage groups without the involvement of chromosomes 4D, 6B, and 7B (Table 3.7).

3.2.3.2.3 Segregation Distortion

Chi-squared test indicated that most polymorphism markers in the RIL population had the separation of 1:1 ratio ($\chi 2 < \chi\, 2_{0.05,1} = 3.84$). A total of 168 markers showed the genetics distortion segregation ($P < 0.01$), accounting for 33.5 %. Among them, 83 markers exhibited distortion in favor of the female parent alleles, whereas 85 markers were inclined to the male parent alleles. The distortion loci were not evenly distributed among the A, B, and D genomes. The B genome distribution frequency was the highest (55.4 %), while D genome distribution frequency was the lowest (19.6 %). The segregation distortion loci were mainly clustered on chromosomes 1A.1, 1B.1, 2B, 3A, and 7D, respectively. Forty-eight loci on the chromosome 1B.1 were in favor of the male parent, whereas 16 segregation distortion loci were in favor of the female parent alleles (Table 3.6).

3.2.3.2.4 Comparison of the Present Map with Previous Others

There are some differences between the two parents comparing with others in this study. First, the two parents had some differences in agronomic traits, yield and quality traits. Abundant polymorphism of the markers was tested between the parents. The female parent Shandong 01-35 was bred in Shandong Agricultural

Table 3.7 New loci that were mapped on the linkage groups in the RIL population

Linkage group	Number	Marker name
1A.1	2	*wPt-730,172, wPt-669,681*
1B.1	6	*wPt-666,095, wPt-731,490, wPt-667,092, wPt-665,375, CFE063, CFE026*
1D	7	*wPt-666,986, wPt-665,037, wPt-664,989, wPt-671,415, wPt-665,204, wPt-665,480, wPt-730,758*
2A	5	*wPt-730,146, wPt-664,438, wPt-666,340, wPt-1871, wPt-669,199*
2B	2	*wPt-666,395, CFE052*
2D	2	*wPt-667,294, wPt-666,987*
3A	7	*wPt-729,826, wPt-730,115, wPt-731,146, wPt-729,808,wPt-666,414, wPt-666,853, wPt-669,255*
3B	3	*wPt-667,328, wPt-668,064, wPt-671,808*
3D	2	*wPt-668,266, wPt-668,308*
4A	1	*wPt-664,948*
4B.2	1	*CFE149*
5A	1	*CFE019*
5B.1	1	*CFE186*
5D	3	*CFE291, wPt-666,937, wPt-671,956*
6A	3	*wPt-731,153, CFE043, CFE041*
6D	5	*wPt-668,152, wPt-666,615, wPt-666,008, wPt-664,719, wPt-672,044*
7A	2	*CFE223, wPt-667,038*
7D	1	*wPt-664,056*

University, with large grain (more than 60 g of thousand-grain weight), high yield, and middle gluten wheat (dough stability time 1.9 min, sedimentation value 30 mL). But it had low percentage of ear bearing tiller (about 3 million tillers per hectare). While the male parent was strong gluten wheat (dough stability time 14.4 min, sedimentation value more than 40 mL) with high percentage of ear bearing tiller (about 7.5 million tillers per hectare), small spike and more grains, and about 38 g of thousand-grain weight every year. Second, Nuomai 1 and Gaocheng 9411 are elite Chinese wheat cultivars. So the map had higher utilization value in breeding. Third, the genetic map contained more molecular markers (502 loci). It spanned about 4084.5 cM in wheat genome and the average distance of markers was 8.13 cM. Fourth, the genetic map is given priority to with the DArT markers, which were more inclined to gene enrichment region distribution away from the centromere comparing to RFLP or SSR markers (Semagn 2006). DArT was developed as a hybridization-based alternative, which captures the value of the parallel nature of the microarray platform (Jaccoud et al. 2001). Fifth, one CAPS marker associated with grain weight (*TaGW2*) was located on the map of chromosome 6A, which had the same position as Su et al. (2011) reported.

3.2.4 High-Density Genetic Linkage Map Constructed Using SNP Markers and Others Markers

SNP is the most in any biological genome and the most common form of polymorphism (Brookes, 1999). SNP is more valuable than SSR and other more repeat sequences in high-density genetic map construction, fine mapping target genes, and gene cloning. Advances in next-generation sequencing have significantly facilitated the discovery of SNPs by whole-genome transcriptome or reduced-representation sequencing in diverse populations of individuals (Wang et al. 2014). High-density SNP arrays have been developed for a number of economically important crops and animals (Ganal et al. 2011; Song et al. 2013; Wiedmann et al. 2008) and successfully used for genetic studies. The recently developed 9 K SNP wheat chip was used to detect genomic regions targeted by breeding and improvement selection in wheat (Cavanagh et al. 2013). In this study, a high-density genetic map was constructed by integration of SNP, DArT, and SSR markers derived from the RIL population of Shannong 01-35 × Gaocheng 9411, which could be more precise to QTL for grain weight and quality traits in common wheat.

3.2.4.1 Materials and Methods

3.2.4.1.1 Plant Materials

A population of 182 RILs was generated by single seed 8 descent from the cross of Shannong 01-35 (39-1/Hesheng 2) and Gaocheng 9411 (777546/Linzhangmai).

3.2.4.1.2 Molecular Marker Detection

DNA was extracted by approved method of CTAB as described by https://www.triticarte.com.au. Agar gel (0.8 %) electrophoresis was used to detect the concentration and purity of extracted DNA. DArT and SSR markers and detection were same as "Nuomai 1 × Gaocheng 8901" in Sect. 3.2 of this chapter.

Infinium iSelect SNP genotyping was performed on the BeadStation and iScan instruments according to the manufacturer's protocols (Illumina). SNP clustering and genotype calling were performed using GenomeStudio v2011.1 software (Illumina).

3.2.4.1.3 Data Analysis

The genetic map was constructed with MAPMAKER/Exp ver.3.0b (Lincoln et al. 1993) and Mapchart 2.1 (Voorrips 2002). The segregation distortion loci were analyzed by IciMAPING software.

3.2.4.2 Result and Analysis

3.2.4.2.1 Molecular Marker Distribution

A total of 9576 markers revealed polymorphism between the parents of the mapping population, including 9072 SNPs, 59 SSRs, 442 DArTs, 2 *HMW-GS,* and one *TaGW2*-CAPS markers.

3.2.4.2.2 The Characteristics of the Genetic Map

A total of 6241of polymorphism markers (6000 SNPs, 216 DArTs, 25 SSRs, and one *TaGW2*-CAPS) were mapped on the 60 linkage groups, which were involved in 20 wheat chromosomes except 4D (Table 3.8). The genetic map was covered 4825.29 cM in wheat genome, and the average distance of markers was 0.77 cM.

The A genome contained 2390 markers (38.3 %), with total length of 1913.4 cM, and the average distance of markers was 0.801 cM. The B genome had the most markers, including 3386 markers, covering 2540.95 cM with an average distance of 0.75 cM. Among them, the D genome was the shortest length (370.94 cM), containing 465 markers (7.5 %) with an average distance of 0.798 cM. The 7A was the longest chromosome (697.78 cM), and 5D was the shortest, with only 34.4 cM. The chromosome 2D had the smallest average distance of markers with 0.3 cM. However, no linkage groups were mapped on chromosome 4D.

3.2.4.2.3 Segregation Distortion

Chi-squared test indicated that, among the 6241 markers, 1307 markers (20.9 %) showed the genetics distortion segregation ($P < 0.01$). Of these, 421 markers (34.5 %) exhibited distortion in favor of the female parent Shannong 01-35 alleles, whereas 895 markers (65.5 %) were in favor of male parent Gaocheng 9411 alleles. The distortion loci were not evenly distributed on the 3 genomes (Table 3.8), with 337, 934, and 36 loci mapping on the A, B, and D genomes, respectively. The segregation distortion loci were mainly clustered on chromosomes. The loci on chromosomes 1B, 2B, 6A, and 6B were in favor of the female parent Shannong 01-35 alleles. The distortion loci on chromosomes 1A and 5B inclined toward the male parent alleles.

Distortion separation is common phenomenon in the nature. Most of the polymorphism loci of the genetic map were separated according to 1:1. The distortion loci accounted for 20.9 % of the total number of polymorphic markers, which was consistent with previous reports.

3.2.4.2.4 Comparison of the Present Map with Previous Others

First, the genetic map had high-density markers. The genetic linkage map was integrated three kinds of markers, including SNP, SSR, and DArT markers.

Table 3.8 Markers and distorted locus distribution among different linkage groups

Linkage group	Markers			Marker number	Length (cM)	Linkage	Average distance (cM)	Markers		Distorted locus	Distorted to parent	
	SNP	DArT	SSR					SNP	DArt		Shannong 01-35	Gaocheng 9411
1A	497	40	2	539	605.36	5	1.123	52	0	66	62	4
1B	355	9	0	364	280.62	8	0.771	27	1	139	0	139
1D	27	11	0	38	60.82	3	1.601	4	2	5	5	0
2A	155	7	2	164	230.63	6	1.406	19	2	34	12	22
2B	441	15	0	456	541.6	8	1.188	114	1	213	0	213
2D	166	21	0	187	55.98	3	0.299	12	2	3	0	3
3A	429	7	5	441	323.13	5	0.733	55	0	72	28	44
3B	664	21	1	686	655.11	8	0.955	113	0	86	52	34
3D	71	8	3	82	90.79	2	1.107	5	5	12	11	1
4A	348	15	1	364	357.98	6	0.983	33	1	39	16	23
4B	330	19	1	350	390.68	8	1.116	53	0	59	5	54
4D												
5A	229	5	1	235	302.32	10	1.286	66	0	56	7	49
5B	987	31	3	1021	628.14	11	0.615	74		143	141	2
5D	71	4	1	76	42.89	2	0.564	24	3	10	10	0
6A	528	20	3	551	319.72	3	0.580	83	0	110	5	105
6B	875	13	1	889	423.31	8	0.476	35	1	341	96	245
6D	69	5	1	75	82.91	3	1.105	3	1	4	3	1
7A	647	24	2	673	697.78	13	1.037	53	1	58	44	14
7B	462	20	2	484	457.61	7	0.945	56	0	5	4	1
7D	5	2	0	7	37.55	1	5.364	2	2	2	2	0
A	2833	118	16	2967	2836.92	45	0.956	361	4	435	174	261

(continued)

Table 3.8 (continued)

Linkage group	Markers			Marker number	Length (cM)	Linkage	Average distance (cM)	Markers		Distorted locus	Distorted to parent	
	SNP	DArT	SSR					SNP	DArt		Shannong 01-35	Gaocheng 9411
B	4114	128	8	4250	3377.07	45	0.795	472	3	986	298	688
D	409	51	5	465	370.94	12	0.798	50	15	36	31	5
Total	7356[a]	297	29	7682	6584.93	120	0.857	883	22	1457	503	954

[a]Including one *TaGW2*-CAPS

The integrated map holds some prominent features, such as more molecular markers (6241 loci), covering longer genetic distance (4825.29 cM), and small average distance between markers (0.77 cM). Secondly, some marker loci were newly mapped on chromosomes. New loci (905 markers) were located on 20 chromosomes except 4D. Among them, there were 883 SNPs and 22 DArTs. Thirdly, there were several defections in the genetic map. D genome had fewer loci. In particular, no linkage groups were located on chromosome 4D because of long genetic distance, though dozens of polymorphism loci on chromosome 4D were detected in the RIL population. This is largely due to the following reasons. D genome in common had higher conservative than others. There were some similar previous reports in the genetic map.

3.2.5 High-Density Genetic Linkage Map Constructed Using SNP Markers in Natural Population

3.2.5.1 Materials and Methods

3.2.5.1.1 Plant Materials

A set of 205 common wheat varieties, released largely from the 1980s to 2000s, including bred varieties, founder parents and higher lines. Among them, there were 132 elite bred varieties, which came from the provinces of Shandong (65), Henan (24), Hebei (14), Anhui (8), Jiangsu (6), Beijing (5), Shanxi (4), Gansu (2), Guizhou (1), Ningxia (1), and two foreign varieties. Seventy-three higher lines were bred in Shandong province (Table 3.11). The natural population had significant differences in grain weight and other agronomic traits, dough stability time, and other flour quality traits.

3.2.5.1.2 DNA Extraction

DNA was extracted from young leaf tissues of each variety using the protocol recommended by Triticarte Pty. Ltd (http://www.triticarte.com.au). DNA quality was checked by electrophoresis on 0.8 % agarose gels, and DNA concentration was determined with a NanoDropND-1000 UV-Vis spectrophotometer (NanoDrop Technologies, Wilmington, USA).

3.2.5.1.3 SNP Genotyping

Two hundred and five DNA samples were genotyped by SNP markers. SNP fingerprinting was performed at the University of California at Davis biotechnology inspection center using the Illumina and Kansas state university to jointly develop wheat 90 k gene chip analysis of genome-wide scan. Gene chip included 81,587 SNPS loci, covering the whole genomes of wheat. For the detection of SNP loci classification data, according to genotype detection rate is less than 80 % and the minimum gene frequency is less than 0.03 as a standard for quality control.

3.2.5.1.4 Genetic Diversity

Powermarker 3.25 (Liu et al. 2005, http://statgen.ncsu.edu/powermarker/) was used to calculate gene diversity, polymorphic information content (PIC), classical Fst values, and Nei's genetic distance (1983).

3.2.5.1.5 Population Structure

Population structure analysis for the 205 wheat accessions was performed using the Structure 2.3.1 software based on the genotyping data obtained with 3297 SNP markers distributed on 21 common wheat chromosomes.

3.2.5.1.6 High-Density SNP Genetic Linkage Map

The SNP markers were integrated into a composite map based on previously reported maps from diverse genetic populations by MSTmap software. They included (1) BT-Schomburgk × AUS33384, (2) Young × AUS33414, (3) Chara × Glenlea, (4) W7984 × Opata M85, (5) Sundor × AUS30604, and (6) Westonia × Kauz (Wang et al. 2014).

3.2.5.2 Result and Analysis

3.2.5.2.1 SNP Genotyping and Quality Control

Using gene chip 90 k (contain 81587 SNPS loci) technology, a total of 38381 SNPS of polymorphism loci were detected in the 205 varieties. Among these markers, 8086 SNPS loci of polymorphism loci had no information of corresponding chromosomal location. In all, 24355 SNP markers were selected for construction of genetic map and association mapping.

3.2.5.2.2 Genetic Diversity Analysis

A total of 48,710 allelic variations were detected, with two alleles per locus of SNP. About 59.9 % (14597/24355) of the SNPs with low frequency allele were greater than 0.20, whereas 9.60 % (2337/24355) of the SNPs were close to 0.5.

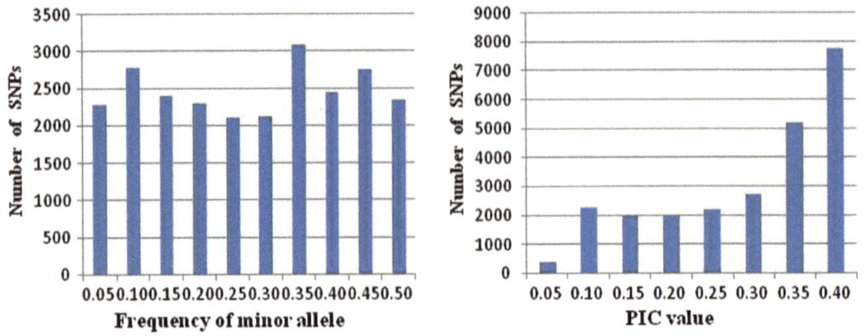

Fig. 3.7 Frequency distribution of minor allele and polymorphic information content (PIC) in the panel of 205 elite wheat line based on 24,355 SNP

So the two alleles of the same seat had similar frequency. The PIC values estimated for the 24355 SNP markers ranged from 0.05 to 0.38, and the mean was 0.27, which clustered between 0.35 and 0.38 (Fig. 3.7).

3.2.5.2.3 Detection of Population Structure

The potential existence of subpopulations among the 205 varieties was detected using STRUCTURE program. According to the method of Evanno et al. (2005), the ΔK value was plotted against the number of hypothetical subgroups K, with the highest ΔK observed with $K = 4$ (Fig. 3.8). The 205 varieties were segregated into four groups differing mainly in geographic origin (Fig. 3.9). Group 1 was dominated by the varieties from Henan province, group 2 by those from Hebei province, group 3 by those from Shandong province, and group 4 by those from three provinces (Shanxi, Jiangsu, and Ningxia provinces).

3.2.5.2.4 Neighbor-Joint Cluster Analysis

A neighbor-joining tree of 205 was constructed based on Nei's genetic distance (Fig. 3.10). The genetic distance of the 205 wheat materials ranged from 0 to 0.35.

Fig. 3.8 Rate of change in the log probability of data between successive K values (ΔK)

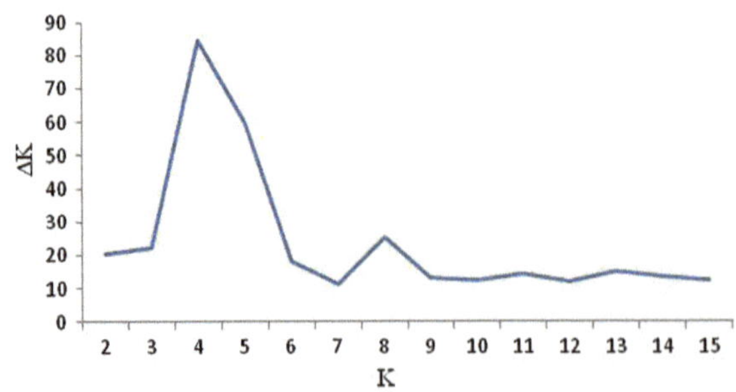

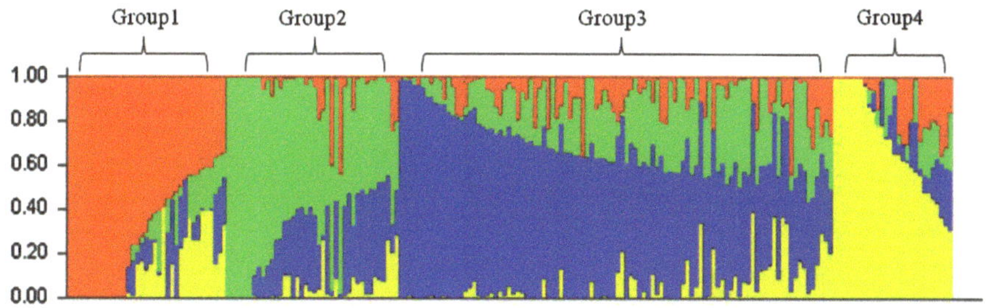

Fig. 3.9 Population structure of 205 accessions based on SNP markers

Fig. 3.10 Neighbor-joining (NJ) tree for the 205 wheat lines based on SNP analysis data

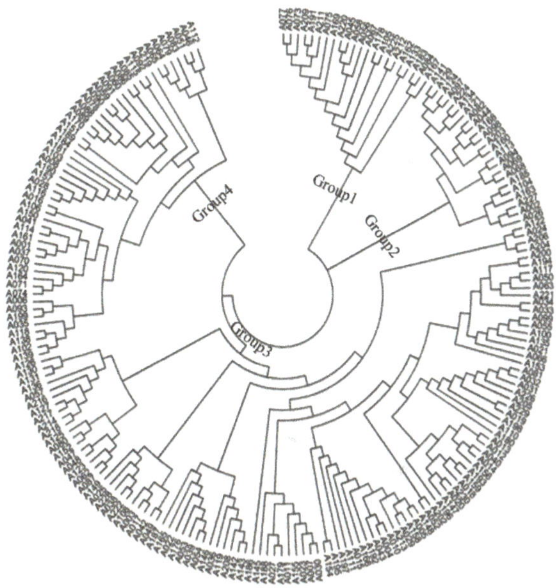

Four subgroups in the genetic population were divided in this study, consistent with the STRUCTURE analysis outcome.

3.2.5.2.5 Genetic Relationships Among Subpopulations

In order to further analyze population substructure among 205 lines, pairwise Fst among these subpopulations was calculated. Four subgroups in the genetic population were divided in this study (Table 3.9). The Group 1 (Hebei) and Group 3 (Shandong) had the biggest genetic distance, which was 0.1318. While the Group 3 (Shandong) and Group 4 (Henan) had the shortest genetic distance (0.1131). Above information indicated that the genetic differences of varieties held between Shandong and Hebei. However, more breeding resource materials were more communicative between Shandong and Henan with shorter genetic distance.

Table 3.9 Genetic distances between different germplasm groups

Group	Group 1	Group 2	Group 3
Group 1	0		
Group 2	0.1316	0	
Group 3	0.1318	0.1214	0
Group 4	0.1290	0.1200	0.1131

3.2.5.2.6 Construction of High-Density SNP Genetic Linkage Map

A composite SNP genetic linkage map was constructed carrying 24,355 loci (Table 3.10). The total genetic distance covered was approximately 3674.16 cM, with a mean genetic distance of 0.15 cM between adjacent loci. Each chromosome was 118.91 ~ 241.38 cM. The B genome contained the most loci (n = 12,321) with the length of 1150.47 cM, followed by A (n = 9523) with the length of 1252.51 cM, and D genomes (n = 2511) with the length of 1271.18 cM. Group 1 chromosomes carried the largest number of loci (n = 4525), followed by group 2 (n = 4208). Group 4 chromosomes had the smallest number of loci (n = 2105). Among the 21 chromosomes, chromosome 1B had the most loci (n = 2390), followed by 5B (n = 2187), whereas chromosome 4D had the fewest loci (n = 78).

3.2.5.2.7 Comparison of the Present Map with Previous Others

Natural population was used in this study. Natural population held rich genetic diversity, which had become one of the hot focuses of current international plant genomic research to discover quantitative trait genes. In contrast to QTL mapping using biparental populations, genome-wide association permits a survey of a wide range of alleles at each locus, detection of marker–trait associations at the whole-genome level, and identification of elite alleles for significantly associated loci. Using gene chip 90 k (contain 81,587 SNPS loci) technology, a total of 38381 SNPS of polymorphism loci were detected in the 205 varieties from Yellow and Huai river valleys. A high-density SNP composite genetic map was constructed, which aimed at association mapping important trait in common wheat, identifying elite allelic variation for molecular marker-assisted breeding (Table 3.11).

SNP markers were used to construct the genetic linkage map in this study. A total of 48,710 allelic variations were detected, with two alleles per locus of SNP. However, SSR markers generally produced 2 or more allelic variation (Lu et al. 2009). SNP markers had lower mutation frequency and higher genetic stability than SSR loci in genome. PIC values' mean was 0.27 in this study, which was higher than American varieties (PIC = 0.23), whereas lower than Europe varieties (PIC = 0.33) (Tobias et al. 2013). This shows that the genetic diversity of wheat varieties in our country was higher or close to the United States' ones, less than European varieties.

The composite genetic linkage map was high density. Comparing of the previous genetic map by traditional molecular markers (e.g., ALFP, SSR), SNP markers used

Table 3.10 The information of integrated linkage maps

Chr.	Number of marker	Length of LG (cM)	Average distance (cM)
Chr1(1A)	1506	161.35	0.11
Chr2(1B)	2390	174.10	0.07
Chr3(1D)	629	196.97	0.31
Chr4(2A)	1462	185.46	0.17
Chr5(2B)	1977	180.33	0.09
Chr6(2D)	769	151.92	0.20
Chr7(3A)	1154	184.56	0.16
Chr8(3B)	1628	150.97	0.09
Chr9(3D)	331	156.06	0.47
Chr10(4A)	1145	164.12	0.14
Chr11(4B)	882	118.91	0.13
Chr12(4D)	78	161.10	2.07
Chr13(5A)	1243	144.15	0.12
Chr14(5B)	2187	219.77	0.10
Chr15(5D)	240	207.32	0.86
Chr16(6A)	1463	180.74	0.12
Chr17(6B)	1786	127.54	0.07
Chr18(6D)	234	156.53	0.67
Chr19(7A)	1550	232.13	0.15
Chr20(7B)	1471	178.85	0.12
Chr21(7D)	230	241.28	1.05
Genome A	9523	1252.51	0.13
Genome B	12,321	1150.47	0.09
Genome D	2511	1271.18	0.51
Group1	4525	532.48	0.12
Group2	4208	517.71	0.12
Group3	3113	491.59	0.16
Group4	2105	444.13	0.21
Group5	3670	571.24	0.16
Group6	3483	464.81	0.13
Group7	3251	652.26	0.20
Total	24,355	3674.16	0.15

in this study were more stable, wider distribution in genomes and compatibility with high, multiplex detection systems. Advances in SNP marker development in wheat and the availability of various SNP genotyping platforms now permit high-throughput and cost-effective genotyping (Akhunov et al. 2009). SNP may be the best DNA markers to construct high-density linkage map. A composite SNP genetic linkage map was constructed carrying 24,355 loci with a mean genetic distance of 0.15 cM between adjacent loci, which was higher density than previous reported composite maps (2.2 cM, Somers et al. 2004; 3.4 cM, Zhang et al. 2013).

Table 3.11 Wheat varieties used in this study

Number	Variety name	Local	Number	Variety name	Local
1	Shannong 15	Shandong	68	Gan 5092	Hebei
2	Shannong 17	Shandong	69	Gan 05-093	Hebei
3	Shannong 19	Shandong	70	Heng 5229	Hebei
4	Shannong 20	Shandong	71	Heng 5346	Hebei
5	Shannong 10-2	Shandong	72	Heng 4371	Hebei
6	Shannong 11	Shandong	73	Hengguan 76	Hebei
7	Shannong 12	Shandong	74	Hengguan 35	Hebei
8	Shannong 06-278	Shandong	75	He 0927	Henan
9	Shannongyoumai 3	Shandong	76	Jifu 8512	Hebei
10	Shannong 0919	Shandong	77	Xuzhou 24	Jiangsu
11	Shannong 55843	Shandong	78	Yumai 34	Henan
12	Shangnong 22	Shandong	79	Yunong 416	Henan
13	Shannong 23	Shandong	80	Yunong 949	Henan
14	Shannong 055849	Shandong	81	Yu 70-36	Henan
15	Xinshangong 11	Shandong	82	Zhengmai 7698	Henan
16	Taishan 21	Shandong	83	Zhengmai 0856	Henan
17	Tainong 18	Shandong	84	Zhengzi 8780-2	Henan
18	Tainong 9236	Shandong	85	Zhoumai 16	Henan
19	Tainong 19	Shandong	86	Zhoumai 22	Henan
20	Lumai 14	Shandong	87	Zhoumai 24	Henan
21	Lumai 23	Shandong	88	Zhoumai 26	Henan
22	Luyuan 502	Shandong	89	Zhoumai 18	Henan
23	Luyuan 205	Shandong	90	Luo 86036	Henan
24	Jinin 16	Shandong	91	Luo 88079	Henan
25	Jining 6058	Shandong	92	Xinong 157	Shanxi
26	Jimai 19	Shandong	93	Xinong 979	Shanxi
27	Jimai 21	Shandong	94	Lian 0808	Jiangsu
28	Jimai 22	Shandong	95	Lian 0756	Jiangsu
29	Jinan 17	Shandong	96	Lianfeng 85	Henan
30	Yannong 21	Shandong	97	Fengyou 04	Guizhou
31	Yannong 19	Shandong	98	Ci-5	Mexico
32	Yannong 999	Shandong	99	Huacheng 3366	Anhui
33	Yan 99102	Shandong	100	Xinmai 296	Shandong
34	Weiyin 84137	Shandong	101	Shannong 33	Shanxi
35	Wei 60182	Shandong	102	Shiluan 02	Hebei
36	Weimai 8	Shandong	103	Zhu 0263-541	Henan
37	Zimai 12	Shandong	104	Fan 7030	Henan
38	Lainong 8621	Shandong	105	Luo 22	Henan
39	Qifeng 2	Shandong	106	Yuanjin 97-28	Shandong
40	Wenhang 1	Shandong	107	DH-1	Shandong
41	Wennong 17	Shandong	108	Ningdong 11	Ningxia

(continued)

Table 3.11 (continued)

Number	Variety name	Local	Number	Variety name	Local
42	Bo 8	Shandong	109	Xingmai 11	Hebei
43	Bonong 6	Shandong	110	Zhen 8906	Jiangsu
44	Liangxing 00	Shandong	111	Wo 85	Anhui
45	Xinmai 18	Shandong	112	Lunzao 3	Shandong
46	Ningmaizi 22	Shandong	113	Xiaoyan 22	Shanxi
47	Ningmaizi 28	Shandong	114	Aifeng 3	Gansu
48	Hemai 13	Shandong	115	Mengxian 201	Henan
49	Hemai 17	Shandong	116	Gaocheng 8901	Hebei
50	He 9946	Shandong	117	Xizhi 8222	Shandong
51	Laizhou 9361	Shandong	118	Yuejin 5	Shandong
52	Lin 4	Shandong	119	Bima 6	Gansu
53	Linmai 2	Shandong	120	Aikang 58	Henan
54	Lianmai 2	Jiangsu	121	Zhenmai 18	Jiangsu
55	Kenong 199	Beijing	122	Fanmai 5	Henan
56	Kenong 2009	Beijing	123	Fa 0356	France
57	Kenong 3106	Beijing	124	Huapei 3	Henan
58	Zhongyu 01089	Beijing	125	Yumai 57	Henan
59	Zhongyu 01095	Beijing	126	Shannongyoumai 2	Shandong
60	Jinghe91-P 39	Anhui	127	D 209	Shandong
61	Wan 38	Anhui	128	D 131	Shandong
62	Wan 50	Anhui	129	D 180	Shandong
63	Wanmai 52	Anhui	130	D 45	Shandong
64	Wanmai 53	Anhui	131	He 0302	Shandong
65	Bu 84111	Anhui	132	Shi 4185	Hebei
66	ShiB 07-4056	Hebei	133-205	Higher breeding lines	Shandong
67	Shi 08-534	Hebei			

3.2.6 Genetics Map Constructed Using a Wheat Backbone Parent "Aimengniu" and Derived Lines

3.2.6.1 Materials and Methods

3.2.6.1.1 Plant Materials

This study was based on a collection of 109 wheat lines, including three parents of Aimengniu, seven sister lines, and their derived varieties (or lines) (Table 3.12). Of which, 101 accessions were from different geographic areas (provinces of Shandong, Jiangsu, Anhui, Hebei, Henan, Shanxi, and so on) in China. There were only 8 accessions from elsewhere in the world. The varieties differed in many agronomic traits, yield and quality traits.

Table 3.12 Core parental cross Aifeng3//Mengxian201/Neuzucht and its derivatives

Line	Name	Line	Name	Line	Name	Line	Name
1	Shannong 60182	29	Zhengzi 8778-1-1	57	Tai 918933	85	Mengxian 201
2	Luami 23	30	Shannong 114653	58	Tai 918954	86	Neuzucht
3	Lumai 11	31	Shannong 114753	59	Tai 910989	87	Aimengniu I
4	Shannong 742006	32	Shannog 113047	60	Lunzao 1	88	Aimengniu II
5	Shannong 3373	33	Shannong 114427	61	Lunzao 3	89	Aimengniu III
6	Shannong 1881	34	Shannong 215953	62	Jining 844437	90	Aimengniu IV
7	Yumai 34	35	Shannong 311334	63	Jining 864872	91	Aimengniu V
8	Shannong 414680	36	Shannong 435001	64	Jining 87chu20	92	Aimengniu VII
9	Shannong 411843	37	Shannong 521248	65	Jining 876046	93	Zhoumai 18
10	Shannong 514095	38	Shannong 721511	66	Jining 844433	94	Xiaonong 8506-1
11	Shannong 560472	39	Shannong 722063	67	Lunong 85(5)135	95	Zhengzi 8780-2-5
12	Shannong 514038	40	Shannong 722309	68	Lunong 85(5)066	96	Xiaonong 4248
13	Shannong 513714	41	Shannong 863410	69	Lunong 54368	97	Linyuan 84-6013
14	Shannong 561149	42	Shannong 113367	70	Lunong 54369	98	Linyuan 86-50240
15	Shannong 560193	43	Zhen 8903	71	Lainong 8621	99	Xizhi 8222-2-82
16	Shannong 512946	44	Zheng 8906	72	Luyuan9	100	Jiyuanxiaofoshou
17	Shannong 560281	45	Weiyin 84137	73	Zhu 84053	101	Yuejin 5
18	Shannong 513250	46	Wei 60182	74	Tai 834497	102	Bima 6
19	Shannong 560244	47	Wei 60366	75	Shannongyoumai 2	103	Fife
20	Shannong 560648	48	Zhoumai 9	76	Shannongyoumai 3	104	Frassineto
21	Shannong 664	49	Bu 84,111	77	Taishan 23	105	Gentil rosso
22	Shannong 1135	50	Wo 85	78	Jinan 02	106	Marquis
23	Shannong 3295	51	Jinghe 91-P39	79	Zhoumai 13	107	Biyumai
24	Ningzimai 22	52	Jinghe 91-7086	80	Hemai 13	108	Wilhelmina

(continued)

Table 3.12 (continued)

Line	Name	Line	Name	Line	Name	Line	Name
25	Ningmaizi 27	53	Jifu 8503	81	Xuzhou 24	109	Squarehead extra
26	Ningmaizi 28	54	Jifu 8506	82	Jimduo 1		
27	Zhengzi 8778-0-2	55	Jifushe 85,012	83	Xiaoyan 22		
28	Zhengzi 8778-0-3	56	Tai 836,214	84	Aifeng 3		

Aimengniu germplasm combined several genes of disease resistance, high yield, and quality, coming from dozens of materials at home and abroad. More than 10 science research facilities had applied Aimengniu germplasm in wheat breeding programmes. Dozens of national or provincial varieties were registered successively. Among these, Lumai 1, Lumai 15, Yumai 57, Linmai 2, Zhongyu 9, and Zhoumai 18 had planted over 10 million acres one year, which won national awards of high-yield varieties. It was the idea population for locating and mining high-yield genes by association mapping method, which might useful for marker/trait association studies.

3.2.6.1.2 DArT Genotying

The 109 accessions of the collection were genotyped by DArT markers at the DArT Pty Limited (Canberra, Australia; http://www.triticarte.com.au).

3.2.6.1.3 Genetic Linkage Map Construction

The DArT markers employed for the association mapping analysis were integrated into a composite map based on previously reported maps from diverse genetic populations. They included (1) Cranbrook × Halberd, 339 DArT (Wenzl et al. 2006); (2) Arina × NK93604, 189 DArT (Semagn et al. 2006); (3) Avocet × Saar, 112 DArT (Lillemo et al. 2008); (4) Colosseo × Lloyd, 392 DArT (Mantovani et al. 2008); (5) 779 DArT (Wenzl et al. 2006); (6) 3B physics map (Paux et al. 2008); and (7) the selected linkage groups of markers from nine different populations archived on the Triticarte Web site (http://www.triticarte.com.au). The linkage map was finally drawn using the software Mapchart ver.2.1 (Voorrips 2002).

3.2.6.1.4 Genetic Diversity, Population Structure, and Linkage Disequilibrium

Powermarker 3.25 (Liu et al. 2005, http://statgen.ncsu.edu/powermarker/) was used to calculate gene diversity, PIC, classical Fst values, and Nei's genetic distance (1983). Analysis of population structure among wheat accessions was implemented using the model-based software STRUCTURE 2.2 (Pritchard et al. 1983; Falush et al. 2003). Five independent runs were performed setting the number of populations (K) from 1 to 10, burn in time and MCMC (Markov Chain Monte Carlo) replication number both to 500,000. The K value was determined by ln P(D) in STRUCTURE output and an ad hoc statistic ΔK based on the rate of change in ln P (D) between successive *K* (Evanno et al. 2005). Lines with membership probabilities ≥0.5 were assigned to corresponding clusters; lines with membership probabilities <0.5 were assigned to a mixed group.

LD between mapped DArT loci was calculated by the squared allele frequency correlation coefficient (r^2) implemented in TASSEL 2.0.1 (http://www.maizegenetics.net). The comparison-wise significance was computed by 1000 permutations after removal of loci with rare alleles ($F < 0.10$). LD was calculated separately for loci on the same chromosome (intra-chromsomal pairs) and unlinked loci on different chromosomes (inter-chromsomal pairs). The LD analysis for the mapped markers was performed for the whole wheat collection and separately for the subpopulations from STRUCTURE. A critical value for r^2, as an evidence of linkage, was derived using the 95 % percentile of unlinked loci according to Breseghello and Sorrells (2006). If within a chromosome region, all pairs of adjacent loci were in LD, this region was referred to as LD blocks.

3.2.6.2 Result and Analysis

3.2.6.2.1 DArT Diversity

DArT analysis produced 971 biallelic markers with corresponding PIC value ranging from 0.054 to 0.5, with the mean of 0.408. However, due to some markers with unknown position and the elimination of rare alleles and markers with more than 10 % of missing values, only 446 DArT markers were kept for further analysis. All of 446 DArT markers, randomly distributed across the genome, were polymorphic across the 109 wheat accessions (Fig. 3.11). The mean PIC value was 0.411, ranging from 0.088 to 0.5.

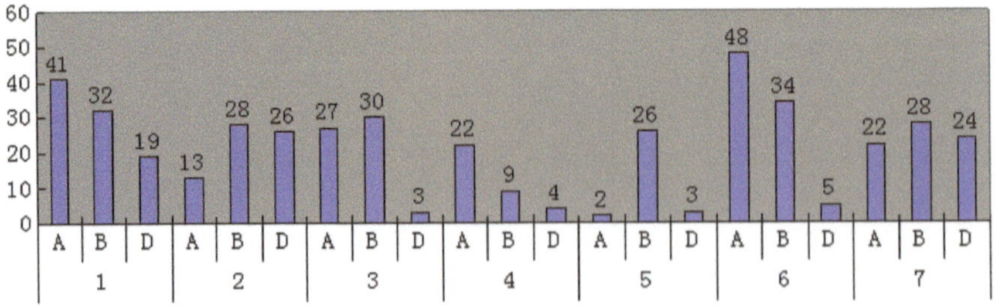

Fig. 3.11 The number of DArT markers on 21 chromosomes of wheat

3.2.6.2.2 Molecular Marker Distribution and Genetic Map

A composite genetic linkage map was constructed carrying 446 loci (Fig. 3.11). The B genome contained the most loci (n = 187), followed by A (n = 175) and D genomes (n = 84). Group 1 chromosomes carried the largest number of loci (n = 92), followed by group 6 (n = 87). Group 5 had the smallest number of loci (n = 31).

3.2.6.2.3 Population Structure

The analysis of population structure was inferred with the STRUCTURE 2.1 software. According to the method of Evanno et al. (2005), ΔK was plotted against the number of subpopulations K. The maximum value of ΔK occurred at K = 4, such that k = 4 (four subpopulations) was defined to provide the optimal structure (Figs. 3.12 and 3.13).

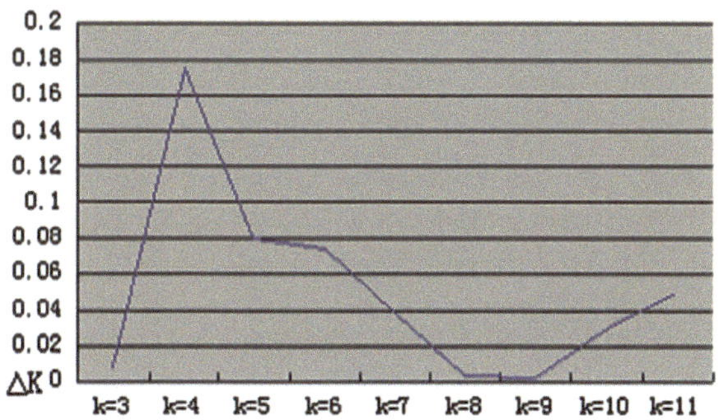

Fig. 3.12 Analysis of population structure based on ΔK parameter

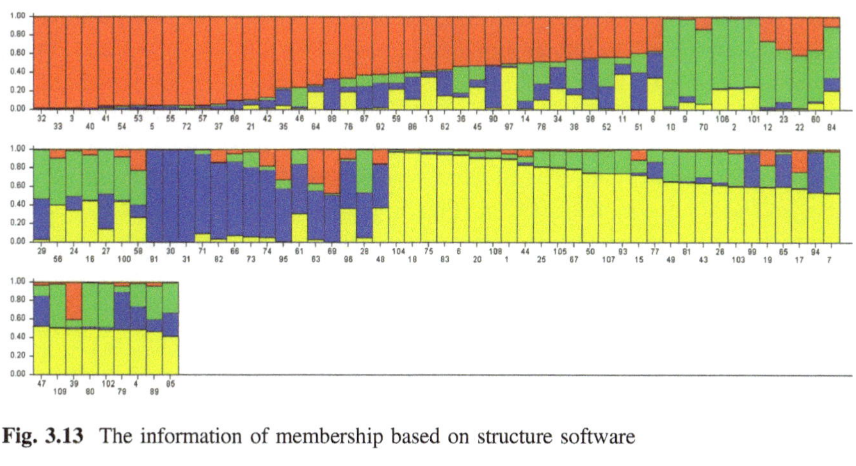

Fig. 3.13 The information of membership based on structure software

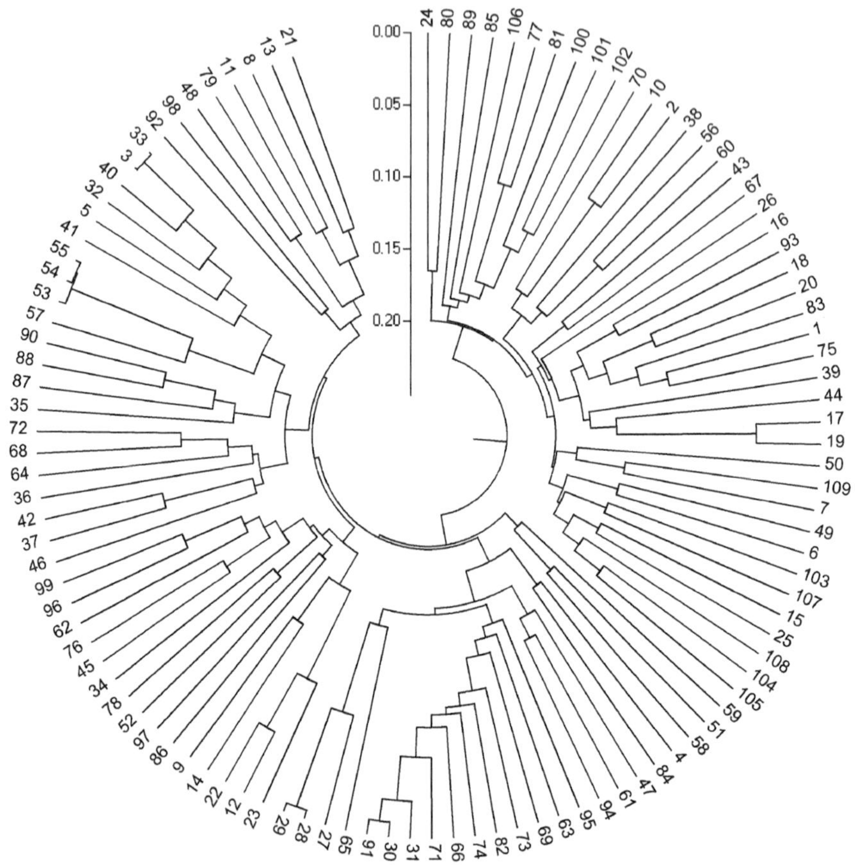

Fig. 3.14 Dendrogram representing the relationships among the 109 wheat accessions as revealed by UPGMA cluster analysis based on Nei's genetic distance

One hundred and nine lines were assigned to the four subpopulations with probability 50 % or higher. The four groups, designated as POP1, POP2, POP3, and POP4, consisted of 30, 13, 13, and 30 cultivars, respectively (Table 3.14). The remaining 23 lines failed to group with a probability higher than 50 %. These 23 lines with mixed ancestral genetic backgrounds were artificially assigned to the "mixed subpopulation" (POP5) for further analysis. This structure results accord with respect to the known origin and pedigrees of the material. Figure 3.14 presents a visualization of the structure results of each line with corresponding membership coefficients.

3.2.6.2.4 Genetic Relationships Among Subpopulations

In order to further analyze population substructure among 109 lines, pairwise Fst among these subpopulations was calculated (Table 3.13). The pairwise Fst value in

Table 3.13 Pairwise estimating of Fst and Nei's genetic distance based on DArT markers among the model-based subpopulations

Cluster	POP1	POP2	POP3	POP4	POP5
POP1		0.0856	0.0901	0.0928	0.0504
POP2	0.1961		0.121	0.0562	0.0401
POP3	0.2393	0.3142		0.1257	0.0698
POP4	0.2284	0.1193	0.3088		0.0336
POP5	0.1142	0.609	0.1517	0.0753	

Genetic distance estimates appear above the diagonal and pairwise Fst appears below the diagonal. All Fst values are significant ($P < 0.001$)

this study ranged from 0.0753 (between POP4 and MIXED) to 0.609 (between POP2 and MIXED), with the mean value of 0.2356, indicating moderate differentiation. The genetic distance data agreed with the Fst estimates. POP4 showed the smallest genetic distance with MIXED, while showing the greatest genetic distance with POP3. These results indicated that the collection had moderate structure.

3.2.6.2.5 Linkage Disequilibrium

Of the 446 polymorphic DArT markers had known map position, only 18 markers were filtered out account of a minimum allele frequencies (MAF) lower than 0.01 in order to prevent the detection of spurious LD. Across the entire collection, as many as 7.03 % of the total markers pairs were in LD ($P < 0.01$), with the marker pairs in LD reduced ranging from 0.501 % (POP3) to 2.64 % (POP1) within each subpopulation (Table 3.14). In the entire collection under investigation, 24.7 % of the intra-chromosomal marker pairs showed a significant level of LD ($P < 0.01$) higher than 5.54 % of the inter-chromosomal marker pairs. The average of r^2 for all pairs

Table 3.14 Percentage of DArT loci pairs in significant ($P < 0.01$) linkage disequilibrium (LD)

	All	Pop1	Pop2	Pop3	Pop4	Pop5
Intra-chromosomal						
% Significant pairs in LD ($P < 0.01$)	24.7	13.63	1.04	5.13	6.54	8.74
Mean r^2	0.111	0.144	0.125	0.239	0.107	0.135
Inter-chromosomal						
% Significant pairs in LD ($P < 0.01$)	5.54	1.92	0.286	0.178	0.113	0.55
Mean r^2	0.02	0.0548	0.108	0.109	0.043	0.05
Mean distance at r^2 > critical (cM)	23.6	11.7	6	6.7	15	13.3
Total % Significant pairs in LD ($P < 0.01$)	7.03	2.64	0.578	0.501	1.14	1.2

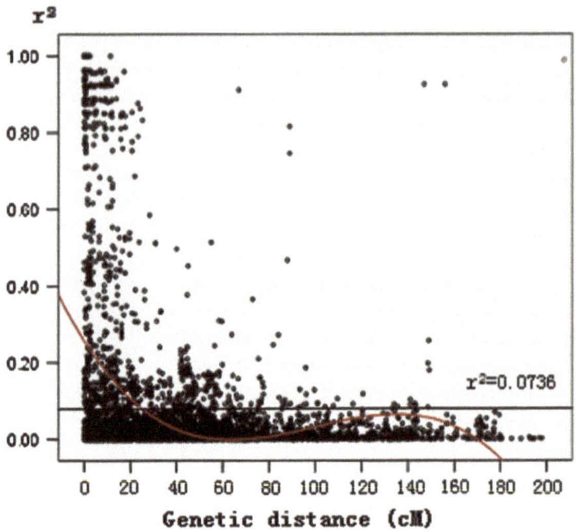

Fig. 3.15 Scatter plot of LD (r^2) of marker pairs versus genetic distance in cM. *Note* A *horizontal line* indicates the 95 % from the distribution of unlinked DArT data set

was 0.111, ranged from 0 to 1. Within the subpopulations, r^2 values of inter-chromosomal pairs ranged from 0.043 to 0.109, while from 0.107 to 0.239 for intra-chromosomal pairs among different subpopulations.

As shown in Fig. 3.16, within the subpopulations, the 95th percentile r^2 values of inter-chromosomal pairs ranged from 0.156 (Pop4) to 0.409 (Pop3). The intra-chromosomal LD decaying pattern varied among different subpopulations, LD decayed within 6–15 cM.

In the collection of 109 wheat accessions, the 95th percentile of the distribution of unlinked r^2 pairs estimates is 0.0736. It was estimated that the loci in LD of the values $r^2 > 0.0736$ were probably due to genetic linkage according to the research of Breseghello and Sorrells (2006). The result of LD decay curve showed that LD extended 23.6 cM (Fig. 3.15).

In the map, several genome-wide LD blocks were identified ($P = 0$, $r^2 = 1$) and highlighted in Fig. 3.16. Most LD blocks were <10 cM in genetic distance. Long-distance LD blocks were identified on chromosomes 1B, 1D, and 3B, respectively. The longest LD blocks (about 20 cM) was located on chromosomes 1D. Combined analysis of all pairwise r^2 from 21 chromosomes indicated that the r^2 values declined within 6–15 cM.

3.2.6.2.6 Comparing of the Genetic Map from the Founder Parent Aimengniu and Its Progenies and Others

First, the population came from the founder parent Aimengniu and its progenies with rich genetic base, which came from "Aifeng 3//Mengxian 201/Niuzhut." Aifeng 3 held complex blood relationships, such as dwarf source of Shuiyuan 86 (including

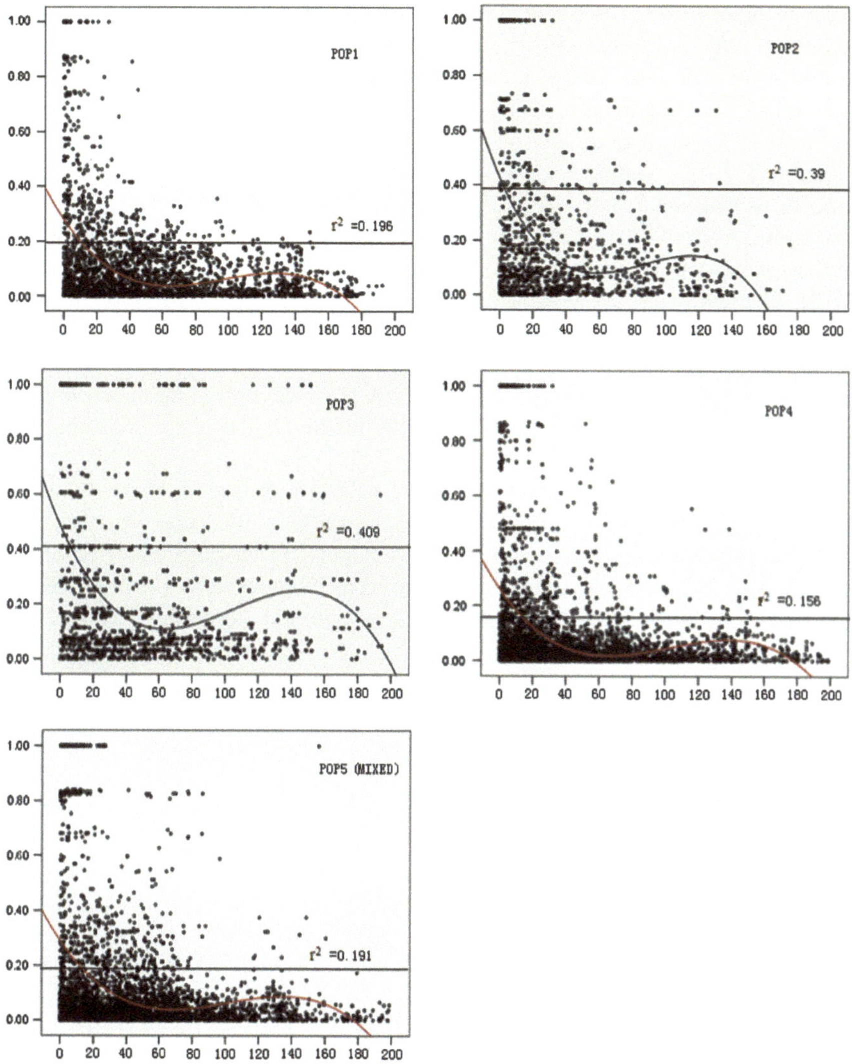

Fig. 3.16 *Scatter plot* of linkage disequilibrium (LD, r^2) between DArT pairs versus genetic distance in cM for each of the five subpopulations. *Note* A *horizontal line* indicates the 95 % from the distribution of unlinked DArT data set

Rht1and Rht2), Chi wheat (including Rht8 and Rht9), Zhongnong 28 with rich resistances, Biyumai, Danmai 1, Mazhamai, Guanzhonglaomai, and Xiaofushou. Mengxian 201 had high yield and resistance genes coming from Abo and Cicheng1 of Japan. Niuzhut was a 1RS/1BL translocation with genes of disease resistance (Yr9 and Pm8) and high yield. So the germplasm of Aimengniu carried good genes restructuring from four continents, with different genes of resistance and high yield. The varieties coming from Aimengniu had not only high yield but also outstanding performance in terms of quality, which created huge economic benefits.

Secondly, in contrast to QTL mapping using biparental populations, genome-wide association analysis is a recently developed, high-resolution method for genetic mapping using existing germplasm (such as landraces, elite cultivars, and advanced breeding lines) based on linkage disequilibrium (LD) (Zhang et al. 2013). Genome-wide association analysis permits a survey of a wide range of alleles at each locus, detection of marker–trait associations at the whole-genome level, and identification of elite alleles for significantly associated loci. In this study, association mapping was performed to analysis the loci controlling high yield and related traits. Such analysis may result in the identification of chromosomal loci linked to the target traits.

Thirdly, DArT developed by Jaccoud et al. (2001) was used in this study, which was the best platform to generate thousands of markers in species like yield and related traits with meager genomic resources. The DArT markers clustered on the chromosome distribution, which was consistent with several previous results (Semagn et al. 2006; Akbari et al. 2006; Mantovani et al. 2008; Peleg et al. 2008; Francki et al. 2009).

Fourthly, LD distribution across chromosomes may significantly affect the power of association mapping and effectiveness of marker-assisted breeding. In the entire collection under investigation, 24.7 % of the intra-chromosomal marker pairs showed a significant level of LD ($P < 0.01$) higher than 5.54 % of the inter-chromosomal marker pairs. The result of LD decay curve showed that LD extended 23.6 cM, which suited for association mapping.

3.3 Research Progress of Genetic Map Construction

3.3.1 Comparsion of the Genetic Map with that of Previous Studies

Genetic map previously mainly constructed by interspecifc populations or distant hybridization materials due to various markers with low polymorphism in common wheat (Gale et al. 1995; Röder et al. 1998; Messmer et al. 1999; Pestsova et al. 2000; Gupta 1994; Somers et al. 2004). However, variety maps are not only necessary but also close to the goals for applying practice. Especially some important economic and agronomic traits may not be performed completely between subspecies, so the whole genetics information was difficult to be obtained. Since 2000 year, the team has constructed 6 pieces of genetic maps with elite cultivated wheat varieties with some characteristics. The genetic map derived from DH population was mainly constructed by SSR marker, which also intergrated several EST-SSR and one HMW-GS. The genetic map from the RIL population of Nuomai 1 × Gaocheng 8901 carried DArT markers, a new type markers, which had located 2 *HMW-GS* loci (*Glu-A1* and *Glu-D1*) and 3 *Wx* protein markers (*Wx-A1*, *Wx-B1*, *Wx-D*). The high-density map of Shannong 01-35 × Gaocheng 9411 was obtained by using 90 K chipSNP and integrated some DArT and SSR markers, which was mainly used for QTL mapping of grain weight, starch, protein, and processing quality traits. The genetic map from the founder parent Aimengniu and its progenies was revealed by DArT markers, which intended to reveal the genetic structure of Aimengniu and the transmission of yield and quality traits in derivatives. A composite high-density SNP genetic map was constructed with 205 nature population, which mostly came from Yellow and Huai river valleys. The map carried 24,355 loci with a mean genetic distance of 0.15 cM between adjacent loci, which was higher density than previous reported composite maps (2.2 cM, Somers et al. 2004; 3.4 cM, Zhang et al. 2013).

In short, parental selection for each genetic map has some distinguishing features, such as agronomic traits, yield, and/or quality traits. Genetic map markers were improving with advances development in wheat and increasing marker density. The molecular genetic maps have been carried out QTL mapping for some agronomic traits, yield and quality traits, and have achieved good results (Chaps. 4, 5, 6, 7, and 8). Furthermore, some main-effect QTLs have been used for molecular marker- assisted breeding in common wheat.

3.3.2 Summary of Wheat Genetics Maps

About 50 pieces of genetic maps in wheat have been published (Table 3.15), which were mainly constructed by SSR markers. Most of the genetic maps derived from single population, such as DH, RIL, and F_2 mapping populations. Several composite maps were integrated two or more mapping population (Somers et al. 2004; Francki et al. 2009).

Table 3.15 Summary of genetic linkage maps published in wheat

Type of population	Parents	No. of loci	Type of marker	Length (cM)	Reference
F_2		204	RFLP	1800	Liu et al. (1991)
F_2		1015	RFLP	2828	Gale et al. (1995)
RIL		935	RFLP	3551	Marino et al. (1996)
F_2		335	RFLP	1079	Dubcovsky et al. (1996)
DH		266	RFLP	1772	Cadalen et al. (1997)
RIL	Chinese Spring	279	SSR	–	Röder et al. (1998)
F_2, DH		53	SSR	–	Stephenson et al. (1998)
DH		620	AFLP, SSR	–	Penner et al. (1998)
RIL		306	RFLP, SSR, AFLP, SSP	3598	Nachit et al. (2001)
RIL	Arina/Forno	401	SSR, AFLP	3086	Paillard et al. (2003)
DH	Courtot/Spring	659	SSR	3685	Sourdille et al. (2003)
RIL	Ning 894,037/Alendra; Wangshuibai/Alendra; Sumai 3/Alendra	216	RAPD, SSR, AFLP	1241.4	Zhou et al. (2003)
ITMI RILs	W 7984/Opata 85	876	eSSRs	–	Peng et al. (2004)
ITMI RILs	W 7984/Opata 85	638		–	Röder et al. (2004)
DH, RIL	Synthetic/Opata;	1235	SSR	2569	Somers et al. (2004)
DH	RL4452/AC Domain				
DH	Wuhan/Maringa				
DH	Superb/BW278				
RIL	W 7984/Opata 85	591	SSR	–	Yu et al. (2004)
RIL	3338/Altgold	231	SSR	4137.8	Xu (2005)

(continued)

Table 3.15 (continued)

Type of population	Parents	No. of loci	Type of marker	Length (cM)	Reference
DH	Chinese Spring /SQ1	567	RFLP, AFLP, SSR	3522	Quarrie et al. (2005)
DH	Fukuho-komugi/Oligoculm	900	RAPD, SSR, AFLP	3948	Suenaga et al. (2005)
RIL	Grandin/BR34	352	SSR, TRAPs	3045	Liu et al. (2005)
RIL	Bainong 64/Jingshuang1	158	SSR,AFLP	3114	Wang et al. (2006)
DH	Kitamoe/Munstertaler	464	SSR	3441	Torada et al. (2006)
DH	Hanxuan10/Lummai14	356	SSR	2536.3	Song et al. (2006)
RIL	Wenmai 6/Shanhongmai	152	SSR	3148	Song et al. (2006)
DH	Arina /NK93604	624	SSR, DArT, AFLP	2595.5	Semagn et al. (2006)
DH	Cranbrook/Halberd	749	SSR, DArT, STM, RFLP, AFLP	2937	Akbari et al. (2006)
DH	O734 (corn)/F1054	38	RGA, AFLP	624	Fayyaz et al. (2007)
RIL	99 G44/Jing771	180	SSR	1532.38	Shi et al. (2007)
RIL	Chuan 35,050/Shannong 483	381	SSR, EST-SSR, ISSR, SRAP, STS, TRAP	3636.7	Li et al. (2007)
RIL	Jing 771/Pm97034	129	SSR	2106.2	Li et al. (2007)
DH	Huapei 3/Yumai 57	305	SSR, EST-SSR	2141.7	Zhang et al. (2008)
RIL	Colosseo/Lloyd	554	SSR, DArT	2022	Mantovani et al.(2008)
RIL	Nanda2419/Wangshuibai	887	SSR, EST-SSR, SCAR, STS	4223.1	Xue et al. (2008)
RIL	Langdon/G18-16	690	SSR, DArT		Peleg et al. (2008)
RIL	Nongda 015/Fuzhuang 3	217	SSR, EST-SSR, SCAR	956.2	Li et al. (2009)
DH	Jinghua 1/Baidongmai	255	SSR	3324	Li et al. (2009)
RIL, DH	P92201D5-2-2/P91193D1; Ajana/WAWHT2074; Cadoux/Reeves; EGA Blanco/Millewa	385/575/ 275/468	SSR, DArT	3013/2825/2198/3058	Francki et al. (2009)

(continued)

Table 3.15 (continued)

Type of population	Parents	No. of loci	Type of marker	Length (cM)	Reference
RIL	Yanzhan 1/Zaosui 30	481	SSR, DArT, ISBP, Function markers	2733.7	Yao et al. (2010)
RIL	Marius/Cajeme71	263	SSR	1113.3	Kerfal et al. (2010)
RIL	Weimai 8/Jimai 20; Weimai 8/Yannong 19	629/681	G-SSR, EST-SSR, STS, ISSR, SRAP, RAPD, DArT	2777.54/3004.62	Cui (2011)
RIL	Weimai 8/Jimai 20; Weimai 8/Yannong 19; Weimai 8/Luohan 2	1113	G-SSR, EST-SSR, STS, ISSR, SRAP, RAPD, DArT	2946.98	Cui (2011)
RIL	Chuan35050/Shannong483	719	DArT, SSR, EST-SSR	4008.4	Wang et al. (2011)
Nature population	94 elite varieties	1171	SSR, DArT	3762.9	Zhang et al. (2013)

References

Akbari M, Wenzl P, Caig V, Carling J, Xia L, Yang S, Uszynski G, Mohler V, Lehmensiek A, Kuchel H, Hayden MJ, Howes N, Sharp P, Vaughan P, Rathmell B, Huttner E, Kilian A. Diversity arrays technology (DArT) for high-throughput profiling of the hexaploid wheat genome. Theor Appl Genet. 2006;113:1409–20.

Akhunov ED, Nicolet C, Dvorak J. Single nucleotide polymorphism genotyping in polyploid wheat with the Illumina GoldenGate assay. Theor Appl Genet. 2009;119:507–17.

Breseghello F, Sorrells ME. Association mapping of kernel size and milling quality in wheat (*Triticum aestivum* L.) cultivars. Genetics. 2006;172:1165–77.

Brookes AJ. The essence of SNP. Gene. 1999;234:177–86.

Cadalen T, Boeuf C, Bernard S, Bernard M. An intervarietal molecular marker map in *Triticum aestivum* L. Em Thell. and comparison with a map from a wide cross. Theor Appl Genet. 1997;94:367–77.

Cavanagh CR, Chao S, Wang S, Huang BE, Stephen S, Kiani S, Forrest K, Saintenac C, Brown-Guedira GL, Akhunova A, See D, Bai G, Pumphrey M, Tomar L, Wong D, Kong S, Reynolds M, da Silva ML, Bockelman H, Talbert L, Anderson JA, Dreisigacker S, Baenziger S, Carter A, Korzun V, Morrell PL, Dubcovsky J, Morell MK, Sorrells ME, Hayden MJ, Akhunov E. Genome-wide comparative diversity uncovers multiple targets of selection for improvement in hexaploid wheat landraces and cultivars. Proc Natl Acad Sci USA. 2013;110:8057–62.

Chen JF, Ren ZL, Gao LF, Jia JZ. Developing new SSR markers from EST in wheat. Acta Agro Sinica. 2005;36:154–8 (in Chinese with English abstract).

Cui F. Construction of high-density wheat molecular genetic map and QTL analysis for yield-related traits. PhD Dissertation of Shandong Agricultural University, 2011 (in Chinese with English abstract).

Doerge RW. Mapping and analysis of quantitative trait loci in experimental populations. Nat Rev. 2002;3:43–52.

Dubcovsky J, Luo MC, Zhong GY, Bransteiter R, Desai A, Kilian A, Kleinhofs A, Dvorak J. Genetic map of diploid wheat, *Triticum monococcum* L., and its comparison with maps of Hordeum vulgare L. Genetics. 1996;143:983–99.

Ellis MH, Rebetzke GJ, Azanza F, Richards RA, Splelmayer W. Molecular mapping of gibberellin-responsive dwarfing genes in bread Wheat. Theor Appl. GeneL. 2005;111:423–30.

Elouafi I, Nachit MM. A genetic linkage map of the Durum × Triticum dicoccoides backcross population based on SSRs and AFLP markers, and QTL analysis for milling traits. Theor Appl Genet. 2004;108:401–13.

Evanno G, Regnaut S, Goudet J. Detecting the number of clusters of individuals using the software STRUCTURE: a simulation study. Mol Ecology. 2005;14:2611–20.

Falush D, Stephens M, Pritchard JK. Inference of population structure using multi locus genotype data: linked loci and correlated allele frequencies. Genetics. 2003;164:1567–87.

Fayyaz E., Shanejat-Bushehri A., Tabatabel B. E. S., Adel J. Constructing a preliminary wheat genetic map using RGA and AFLP Markers. Inter J Agri Bio. 2007;9:863–7.

Francki MG, Walker E, Crawford AC, Broughton S, Ohm HW, Barclay I, Wilson RE, McLean R. Comparison of genetic and cytogenetic maps of hexaploid wheat (*Triticum aestivum* L.) using SSR and DArT markers. Mol Genet Genomics. 2009;281:181–91.

Gale MD, Atkinson MD, Chinoy CN, Harcourt RL, Liu J, Li QY. Genetic maps of hexaploid wheat. In: Li ZS, Xin ZY, editors. Proceedings of the 8th International Wheat Genetic Symposium. Agric Scientech Press, Beijing; 1995. pp. 29–40.

Ganal MW, Durstewitz G, Polley A, Bérard A, Buckler ES, Charcosset A, Clarke JD, Graner EM, Hansen M, Joets J, Le Paslier MC, McMullen MD, Montalent P, Rose M, Schön CC, Sun Q, Walter H, Martin OC, Falque M. A large maize (*Zea mays* L.) SNP genotyping array: development and germplasm genotyping, and genetic mapping to compare with the B73 reference genome. PLoS ONE, 2011; 6: e28334.

Guo CQ, Bai ZA, Liao PA, Jin WK. New high quality and yield wheat variety Yumai 57. China Seed Indus. 2004;4:54 (in Chinese with English abstract).

Gupta M, Chyi YS, Romero SJ, Owen JL. Amplification of DNA markers from evolutionarily diverse genomes using single primers of simple sequence repeats. Theor Appl Genet. 1994;89:998–1006.

Hai Y, Kang MH. Breeding of new wheat variety with high yield and early maturing. Henan Agr Sci. 2007;5:36–7.

Jaccoud D, Peng K, Feinstein D, Kilian A. Diversity arrays: a solid state technology for sequence information independent genotyping. Nucleic Acids Res. 2001;29:25–31.

Karakousis A, Gustafson JP, Chalmers KJ, Barr AR, Langridge P. A consensus map of barley integrating SSR, RFLP, and AFLP markers. Aust J Agric Res. 2003;54:1173–85.

Kerfal S, Giraldo P, Rodriguez-Quijano M, Vazquez JF, Adams K, Lukow OM, Röder MS, Somers DJ, Carrillo JM. Mapping quantitative trait loci (QTLs) associated with dough quality in a soft x hard bread wheat progeny. J Cereal Sci. 2010;52:46–52.

Li SS, Jia JZ, Wei XY, et al. A intervarietal genetic map and QTL analysis for yield traits in wheat. Mol Breeding. 2007;20:167–178.

Li YQ, Su ZF, Wang LX, Ji W, Yao J, Zhao CP. Increasing density of wheat genetic linkage map with molecular markers. Acta Agron Sinica. 2009;35:861–6.

Lillemo M, Joshi AK, Prasad R, Chand R, Singh RP. QTL for spot blotch resistance in bread wheat line Saar co-locate to the biotrophic disease resistance loci Lr34 and Lr46. Theor Appl Genet. 2008;126:711–9.

Lincoln SE, Daly MJ, Lander ES. Mapping genes controlling quantitative traits using MAPMAKER/QTL version 1.1: a tutorial and reference manual. Whitehead Institute for Biometrical Research, Cambridge, Mass; 1993.

Liu ZH, Anderson JA, Hu J, Friesen TL, Rasmussen JB, Faris JD. A wheat intervarietal genetic linkage map based on microsatellite and target region amplified polymorphism markers and its utility for detecting quantitative trait loci. Theor Appl Genet. 2005;111:782–94.

Liu YG, Tsunewaki K. Restriction fragment length polymorphism (RFLP) analysis in wheat. ll. Linkage maps of the RFLP sites in common wheat. Jpn J Genet. 1991;66:617–633.

Lu YL, Yan JB, Guimarâes CT, Taba S, Hao ZF, Gao SB, Chen SJ, Li JS, Zhang SH, Vivek BS, Magorokosho C, Mugo S, Makumbi D, Parentoni SN, Shah T, Rong TZ, Crouch ZH, Xu YB. Molecular characterization of global maize breeding germplasm based on genome-wide single nucleotide polymorphisms. Theor Appl Genet. 2009;120:93–115.

Mantovani P, Maccaferri M, Sanguineti MC, Tuberosa R, Catizone I, Wenzl P, Thomson B, Carling J, Huttner E, Ambrogio ED, Kilian A. An integrated DArT-SSR linkage map of durum wheat. Mol Breeding. 2008;22:629–48.

Marino CL, Nelson JC, Lu YH, Sorrells ME, Leroy P, Lu YH. Molecular genetic maps of the group 6 chromosomes of hexaploid wheat (*Triticum aestivum* L. em. Thell). Genome. 1996;39:359–366.

Messmer MM, Keller M, Zanetti S, Keller B. Genetic linkage map of a wheat-spelt cross. Theor Appl Genet. 1999;98:1163–70.

Nachit MM, et al. Molecular linkage map for an intraspecific recombinant inbred population of durum wheat (*Triticum turgidum* L. var. durum). Theor Appl Genet. 2001;102:177–186.

Penner. An AFLP based genome map of wheat (*Triticum* Agriculture. Plant and Animal Genome VI Conference aestivum) [A]. US Department of [C]. San Diego CA January, 1998, 163.

Paux E, Sourdille P, Salse J, et al. A physical map of the 1-gigabase bread wheat chromosome 3B. science; 2008, 322:101–104.

Peleg Z, Saranga Y, Suprunova T, Ronin YW, Röder MS, Kilian A, Korol AB, Fahima T. High-density genetic map of durum wheat × wild emmer wheat based on SSR and DArT markers. Theor Appl Genet. 2008;117:103–15.

Peng JH, Zadeh H, Lazo GR, Gustafson JP, Chao S, Anderson OD, Qi LL, Echalier B, Gill BS, Dilbirligi M, Sandhu D, Gill KS, Greene RA, Sorrells ME, Akhunov ED, Dvorák J, Linkiewicz AM, Dubcovsky J, Hossain KG, Kalavacharla V, Kianian SF, Mahmoud AA,

Miftahudin, Conley EJ, Anderson JA, Pathan MS, Nguyen HT, McGuire PE, Qualset CO, Lapitan NL. Chromosome bin map of expressed sequence tags in homoeologous group 1 of hexaploid wheat and homoeology with rice and Arabidopsis. Genetics. 2004;168:609–23.

Pestsova E, Ganal MW, Röder MS. Isolation and mapping of microsatellite markers specific for the D genome of bread wheat. Genome. 2000;43:689–97.

Pritchard JK, Wen W, Falush D. Documentation for STRUCTURE software: version 2; 2003.

Quarrie SA, Steed A, Calestani C, Semikhodskii A, Lebreton C, Chinoy C, Steele N, Pljevljakusic D, Waterman E, Weyen J, Schondelmaier J, Habash DZ, Farmer P, Saker L, Clarkson DT, Abugalieva A, Yessimbekova M, Turuspekov Y, Abugalieva S, Tuberosa R, Sanguineti MC, Hollington PA, Aragues R, Royo A, Dodig D. A high-density genetic map of hexaploid wheat (*Triticum aestivum* L.) from the cross Chinese Spring × SQ1 and its use to compare QTLs for grain yield across a range of environments. Theor Appl Genet. 2005;110:865–80.

Rafalski JA. Application of single nucleotide polymorphisms in crop genetics. Curr Opin Plant Biol. 2002;5:94–100.

Röder MS, Huang XQ, Ganal MW. Wheat microsatellites in plant breeding-Potential and Implications. In: Lörz H. & Wenzel G. (Eds.). Molecular Markers in Plant Breeding and Crop Improvement. Springer-Verlag. Heidelberg. 2004.

Röder MS, Korzun V, Wendehake K, Plaschke J, Tixier MH, Leroy P, Ganal MW. A microsatellite map of wheat. Genetics. 1998;149:2007–23.

Semagn K, Bjornstad A. SkimesH, Maroy AG, Tarkegne Y, W illiam M. Distribution of DArT, AFLP, and SSR markers in a genetic linkage map of a doubled-haploid hexaploid wheat population. Genome. 2006;49:545–55.

Shi PC, Wang GL, Zhang Wei, Cao LP. Construction of wheat SSR molecular linkage map and its polymorphism. Xinjiang Agri Sci. 2007;44:71–76.

Somers DJ, Isaac P, Edwards K. A high-density microsatellite consensus map for bread wheat (*Triticum aestivum* L.). Theor Appl Genet; 2004;109:1105–1114.

Song Q, Hyten DL, Jia G, Quigley CV, Fickus EW, Nelson RL, Cregan PB. Development and evaluation of SoySNP50 K, a high-density genotyping array for soybean. PLoS ONE. 2013;8: e54985.

Song YX, Jing RL, Huo NX, Ren ZL, Jia JZ. Detection of QTLs for heading in common wheat (*T. aestivum* L.) using different populations. Sci Agri Sinica. 2006;39:2186–93.

Sourdille P, Cadalen T, Guyomarc'h H, Snape JW, Perretant MR, Charmet G, Boeuf C, Bernard S, Bernard M. An update of the Courtot × Chinese Spring intervarietal molecular marker linkage map for the QTL detection of agronomic traits in wheat. Theor Appl Genet. 2003;106:530–538.

Stephenson P, Bryan G, Kirby J, Collins A, Devos K, Busso C, Gale M. Fifty new microsatellite loci for the wheat genetic map. Theor Appl Genet. 1998;97:946–9

Su ZQ, Hao CY, Wang LF, Dong YC, Zhang XY. Identification and development of a functional marker of TaGW2 associated with grain weight in bread wheat (Triticum aestivum L.). Theor Appl Genet; 2011, 122: 211–223.

Suenaga K, Khairallah M, William HM, Hoisington DA. A new intervarietal linkage map and its application for quantitative trait locus analysis of "gigas" features in bread wheat. Genome. 2005;48:65–75.

Torada A, Koike M, Mochida K, Ogihara Y. SSR-based linkage map with new markers using an intraspecific population of common wheat. Theor Appl Genet. 2006;112:1042–51.

Voorrips RE. MapChart: software for the graphical presentation of linkage maps and QTLs. J Heredity. 2002;93:77–8.

Wang YY, Sun XY, Zhao Y, Kong FM, Guo Y, Zhang GZ, Pu YY, Wu K, Li SS. Enrichment of a common wheat genetic map and QTL mapping for fatty acid content in grain. Plant Sci. 2011;181:65–75.

Wang SC, Wong DB, Forrest K, Allen A, Chao SM, Huang BE, Maccaferri M, Salvi S, Milner SG, Cattivelli L, Mastrangelo AM, Whan A, Stephen S, Barker G, Wieseke R, Plieske J, International Wheat Genome Sequencing Consortium, Lillemo M, Mather D,

Appels R, Dolferus R, Brown-Guedira G, Korol A, Akhunova AR, Feuillet C, Salse J, Morgante M, Pozniak C, Luo MC, Dvorak J, Morell M, Dubcovsky J, Ganal M, Tuberosa R, Lawley C, Mikoulitch I, Cavanagh C, Edwards KJ, Hayden M, Akhunov E. Characterization of polyploid wheat genomic diversity using a high-density 90000 single nucleotide polymorphism array. Plant Biotechnol J; 2014, 12:787–796.

Wenzl P, Li H, Carling J, Zhou M, Raman H, Paul E, Hearnden P, Maier C, Xia L, Caig V, Ovesná J, Cakir M, Poulsen D, Wang J, Raman R, Smith KP, Muehlbauer GJ, Chalmers KJ, Kleinhofs A, Huttner Eric, Kilian A. A high-density consensus map of barley linking DArT markers to SSR, RFLP and STS loci and agricultural traits. BMC Genomics; 2006, 7:206.

Wiedmann RT, Smith TPL, Nonneman DJ. SNP discovery in swine by reduced representation and high throughput pyrosequencing. BMC Genet. 2008;9:81.

Würschum T, Langer SM, Longin CFH, Korzun V, Akhunov E, Ebmeyer E, Schachschneider R, Schacht J, Kazman E, Reif JC. Population structure, genetic diversity and linkage disequilibrium in elite winter wheat assessed with SNP and SSR markers. Theor Appl Genet. 2013;126:1477–86.

Xu SB. Construction of genetic map by SSR marker and location QTL for plant height and heading time in wheat. MD Dissertation of Xinjiang Agricultural University, 2005 (in Chinese with English abstract).

Xue SL, Zhang ZZ, Lin F, Kong ZX, Cao Y, Li CJ, Yi HY, Mei MF, Zhao DM, Zhu HL, Xu HB, Wu JZ, Tian DG, Zhang CQ, Ma ZQ. A high-density intervarietal map of the wheat genome enriched with markers derived from expressed sequence tags. Theor Appl Genet. 2008;117:181–9.

Yao Q, Zhou RH, Pan YM, Fu TH, Jia JZ. Construction of genetic linkage map and QTL analysis of agronomic important traits based on RIL population derived from common wheat variety Yanzhan 1 and Zaosui 30. Sci Agri Sinica. 2010;43:4130–9.

Yu JK, Dake TM, Singh S, Benscher D, Li W, Gill B, Sorrells ME. Development and mapping of EST-derived simple sequence repeat markers for hexaploid wheat. Genome. 2004;47:805–818.

Zhai HM, Tian JC. Development of wheat mutants carrying different null Wx alleles and their starch properties. Acta Agro Sinica. 2007;33(7):1359–1066 (in Chinese with English abstract).

Zhang KP, Wang JJ, Zhang LY, Rong CW, Zhao FW, Peng T, Li HM, Cheng DM, Liu X, Qin HJ, Zhang AM, Tong YP, Wang DW. Association analysis of genomic loci important for grain weight control in elite common wheat varieties cultivated with variable water and fertiliser supply. PLoS ONE. 2013;8:e57853.

Zhang KP, Zhao L, Tian JC, Chen GF, Jiang XL, Liu B. A genetic map constructed using a doubled haploid population derived from two elite Chinese common wheat varieties. J Integrat Plant Bio. 2008;50:941–50.

Zhou MP, Zhang X, Ren LJ, Schol TO, Bai GH, Ma HX, Lu WZ. Construction of wheat genetic linkage map by JoinMap 3.0. J Jiangsu Agri. 2003;19:133–8.

Chapter 4
Genetic Detection of Main Yield Traits in Wheat

Abstract Yield and related traits controlled by multiple genes are the most important goal of wheat breeding. In this chapter, QTL mapping was used to detect yield and related traits, such as thousand-grain weight and spike-related traits (spike length, grain number per spike, spikelets per spike, fertile spikelets per spike, sterile spikelets per spike, compactness, and spike weight). The QTL results may facilitate yield improvement through molecular marker-assisted selection.

Keywords Yield traits · Spike-related traits · Grain-related traits · Tiller character · Biomass yield · QTL mapping · Immortalized F_2 populations · Heterosis loci (HL) · Association mapping

Increasing yield per unit area is the most basic goal of wheat breeding. Wheat yield is determined by three components which are ear number per unit area, grain number per spike, and thousand-grain weight. Yield and related traits are quantitative traits controlled by multiple genes, which are often influenced by environmental factors and exhibit high genotype–environment interactions. QTLs for grain yield and spike-related traits in five different genetic populations were mapped by our research group, which could detect the genetic basis of wheat yield and related traits from the level of a single gene/QTL. The results may facilitate wheat grain yield and related traits improvement through molecular marker-assisted selection in the future.

4.1 Experimental Populations and Methods

4.1.1 Experimental Populations and Field Experimental Design

4.1.1.1 Experimental Populations

The six genetic populations (one DH population, one IF_2 population, two RIL populations, and two natural populations) were created by our group.

J. Tian et al., *Genetic Analyses of Wheat and Molecular Marker-Assisted Breeding, Volume 1*, DOI 10.1007/978-94-017-7390-4_4

Population 1, one DH population of 168 progeny lines was produced by anther culture of wheat F_1 hybrid plants from a cross between two Chinese wheat cultivars Huapei 3 and Yumai 57.

Population 2, one IF_2 population of 168 true F_1 lines, coming from the DH population, was created following the procedures (Hua et al. 2003). The 168 DH lines were randomly divided into two groups through random permutation. Lines in one group were randomly paired with lines in the other group to make crosses under the condition that each of the 168 DH lines was used only once in the crosses. This procedure was repeated two times, resulting in 168 combinations.

Population 3, one population of 256 RILs was generated by single seed descent from the cross of waxy wheat 1 (WN1) and strong gluten Gaocheng 8901.

Population 4, one population of 182 RILs was generated by single seed descent from the cross of Shannong 01-35 and Gaocheng 9411.

Population 5, one population of 109 elite wheat varieties of germplasm "Aimengniu," including three parents, seven sister lines, and bred varieties.

Population 6, one population of 212 varieties mainly coming from Huang-Huai valley regions in China.

The parents of the six populations were differing in many agronomical important traits and quality traits. The six genetic populations were dissected as ear number per unit area, grain number per ear, grain weight, plant height, maturity, and others yield-related traits, and compared and validated important QTL/genes controlling yield-related traits.

4.1.1.2 Field Experimental Design

The lines of population 1 were planted in Tai'an (36° 57′N, 116° 36′E), Shandong Province in 2005 and 2006, and in Suzhou (33° 38′N, 116° 58′E), Anhui Province in 2006 and 2007. The 168 DH lines and parents were planted in three replications at each location with a completely randomized block design. In the autumn of 2005, all DH lines and parents were grown in a plot with three rows in 2 m length and 25 cm between rows. In the autumn of 2006, the lines were grown in a plot with four 2-m rows spaced 25 cm apart.

The lines of population 2 were conducted in Tai'an, Shandong Province (36° 57′N, 116° 36′E) in 2009 and 2010 and in Jiyuan, Henan Province (112° 360′E, 35° 050′N) in 2009 and 2010. The field planting followed a randomized complete block design with the DH population and the IF_2 population in two replications. Each line of DH and IF_2 was grown in a plot with three rows, respectively, and 25 cm between rows. Field management followed standard local practices. Eight mature plants were chosen for trait evaluation. Grain-related traits were measured using 10 main spikes in the different environments. The averages from representative samples were used in data analysis.

The lines of population 3 were planted under three environmental conditions: 2007–2008 and 2010–2011 in Tai'an, Shandong Province, China, and 2010–2011 in Suzhou, Anhui Province, China. The experimental design followed a completely randomized block design with two replications at each location. All lines and parental lines were grown 2 m long by four-row plots, spaced 26 cm apart in every environment. Suzhou was different from Tai'an in climate and soil conditions.

The lines of population 4 were planted in 2008–2011 in Tai'an, Shandong Province, China, and in 2010–2011 in Suzhou, Anhui Province, China. The experimental design followed a completely randomized block design with two replications in each environment. The lines were planted by three-row plots in 2008 and four-row plots in other environments.

The lines of population 5 were planted in Tai'an in 2006–2010, Shandong Province, China, and in 2010–2011 in Suzhou, Anhui Province, China. The experimental design followed a completely randomized block design with two replications. The lines were planted by three-row plots in 2008 and four-row plots in other environments. So environment 1 (E1): Tai'an, 2006; environment 2 (E2): Tai'an, 2007; environment 3 (E3): Tai'an, 2008; environment 4 (E4): Tai'an, 2009; environment 5 (E5): Tai'an, 2010; and environment 6 (E6): Suzhou.

The lines of population 6 were planted in Tai'an and Dezhou Municipal Academy of Agricultural Sciences, Shandong Province, China, in 2011–2014. The trials were two replications with a completely randomized block design, a plot with three rows 2 m long and spaced 25 cm apart.

Managements of populations were conducted in accordance with local practices. There was no lodging, serious pests, and diseases in these growing seasons. The grains were harvested by lines at maturity.

4.1.2 Traits Evaluation and Statistical Analysis

4.1.2.1 Traits Evaluation

Ear number per unit area was evaluated with the number of spikes in uniform 1/500 growth area multiplied by 500. Twenty main stem spikes from the middle of each replication were used for investigating spike-related traits (spike length, grain number per spike, spikelets per spike, fertile spikelets per spike, sterile spikelets per spike, compactness, and spike weight). Thousand-grain weight was measured using three separate samples, each containing 1000 grains. Grain yield was measured as the weight of wheat grains harvested from the entire replication.

4.1.2.2 Data Analyses

The QTL mapping was conducted by software QTL Network 2.0 in population 1. The QTL was named using the abbreviation of trait names plus the corresponding chromosome number. If two and more QTLs were detected on the same chromosome, consecutive Arabic numbers were added to chromosome number and separated by dots.

The software ICIMapping was used in populations 2, 3, and 4. The QTL named method was the same with population 1. The various statistical models in TASSEL 2.1 software (http://www.maizegenetics.net/) were evaluated for genome-wide marker–trait associations in populations 5 and 6, with the mixed linear model (MLM) finally adopted. Both population structure (Q) and kinship (K) were taken into account during the marker–trait association analysis with MLM. The heterosis loci were identified using the software Heterosis (Hua et al. 2003) in population 2.

4.2 QTL Mapping for Main Yield Traits Using Different Populations

4.2.1 QTL Mapping for Spike-Related Traits

4.2.1.1 QTL Analysis of Spike Traits Using a DH Population Derived from "Huapei 3 and Yumai 57"

4.2.1.1.1 Phenotypic Variation

The spike-related traits for the DH population and the parents in three environments were described in Table 4.1 and Fig. 4.1. The parent Yumai 57 had larger values in grain yield, grains per spike, spikelets per spike, compactness, and fertile spikelets per spike. Huapei 3 showed larger values in spike length. Significant transgressive segregation was observed for eight traits among DH lines in the three environments. The spike-related traits of the DH population segregated continuously and followed a normal distribution, indicating its polygenic inheritance and suitability for QTL analysis (Cao et al. 2001).

4.2.1.1.2 Correlation Analysis

Correlations between spike-related traits in three environments were shown in Table 4.2. The highest positive correlation was found between grain per spike

Table 4.1 Phenotypic performance of spike-related traits

Trait	Parent		DH population					
	Huapei 3	Yumai 57	Mean	Maximum	Minimum	SD	Skewness	Kurtosis
Grain number per spike	33.6	44.8	40.4	54.0	27.0	4.9	0.24	0.06
Spike length (cm)	9.8	8.2	8.6	10.9	6.7	0.9	0.08	−0.51
Spikelets per spike	16.4	21.1	19.1	22.0	16.0	1.3	−0.04	0.08
Compactness	17.3	25.7	21.5	28.6	14.3	2.6	0.19	−0.05
Fertile spikelets per spike	16.4	18.1	19.2	23.0	16.0	1.5	0.01	−0.20

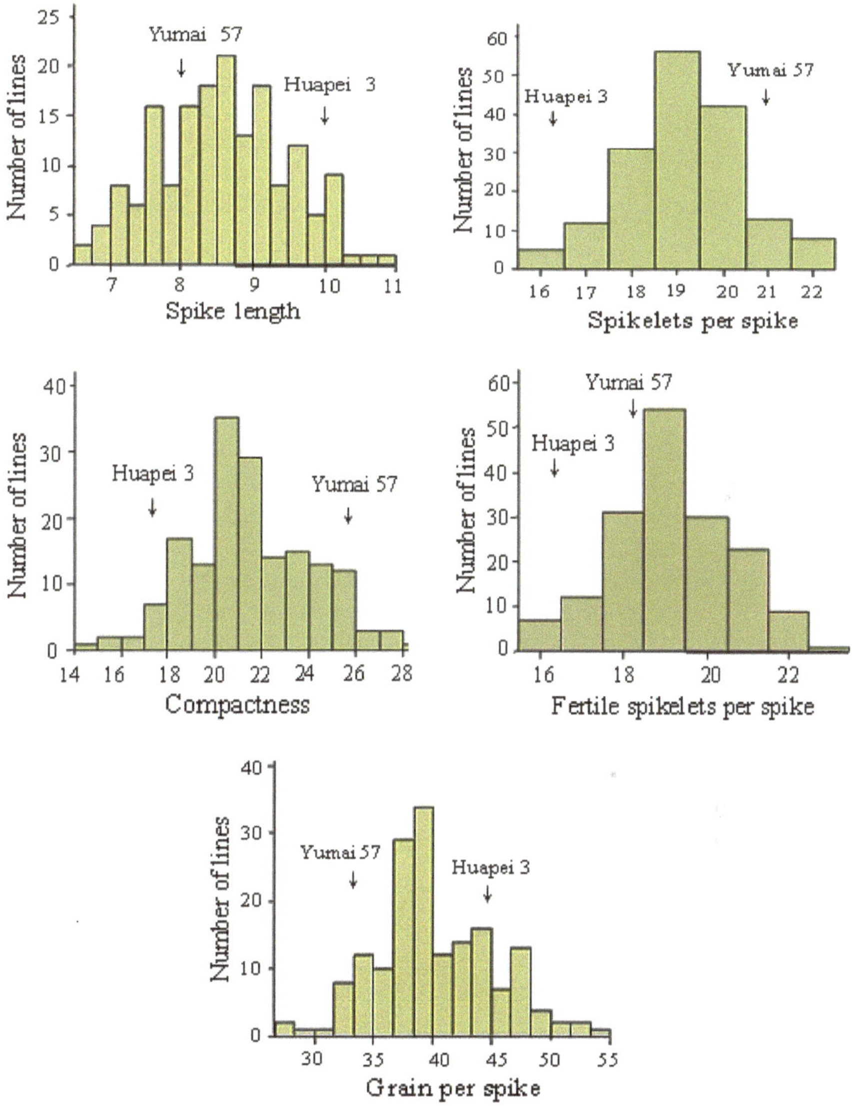

Fig. 4.1 Frequency distribution of spike-related traits in 168 doubled haploid lines derived from the cross of Huapei 3 × Yumai 57 evaluated at 3 environments

and compactness, and the correlation coefficient was 0.895. The spike length was highly negatively correlated with compactness, and the coefficient was −0.80.

Table 4.2 Coefficients of pairwise correlations of the mean values of spike-related traits

Trait	Grains per spike	Spike length	Spikelets per spike	Compactness	Fertile spikelets per spike	Thousand-grain weight
Spike length	−0.795**					
Spikelets per spike	0.481**	0.100				
Compactness	0.895**	−0.803**	0.478**			
Fertile spikelets per spike	0.372**	0.131	0.833**	0.377**		

*P <0.05
**P <0.01

4.2.1.1.3 QTL Mapping for Grain Number Per Spike

Three additive QTLs for grain number per spike were located on chromosomes 2D, 4D, and 5D, accounting for 12.24, 5.06, and 11.67 % of the phenotypic variation, respectively (Table 4.3). The *qSgn2D* had the largest genetic contribution of 12.24 %. The allele of *qSgn4D* was from Huapei 3, and two other alleles were derived from Yumai 57. The *qSgn5D* was also interacted with environments by the contribution rate of 10.25 %.

One pair of epistatic effect for grain number per spike was identified on chromosomes 2A-3A (Table 4.4), accounting for 3.47 % of phenotypic variation. No interaction of epistatic effect with environments was detected.

4.2.1.1.4 QTL Mapping for Spike Length

Five additive QTLs were located on chromosomes 2B, 2D, 4D, 5D, and 6B, ranged from 2.68 to 15.63 % (Table 4.3). The *qSl2D* had the largest genetic contribution of 15.63 %. The allele of *qSl4D* with positive effects came from Yumai 57, and others derived from Huapei 3. The *qSl2B* was interacted with environments, which could account for 6.28 % of variation. There was no epistatic effect for spike length.

4.2.1.1.5 QTL Mapping for Spikelets Per Spike

Four additive QTLs for spikelets per spike were mapped on chromosomes 1B, 4A, 5D, and 7A, accounting for 1.48, 4.37, 13.83, and 1.22 % of phenotypic variation, respectively (Table 4.3). The *qSps5D* contributed the largest variation of 13.83 %. Two loci, *qSps1B* and *qSps5D*, interacted with environments, with the effect 23.60 %.

One pair of epistatic effect for spikelets per spike was mapped on chromosomes 2A-2B (Table 4.4), accounting for 4.99 % of the phenotypic variance. No interaction of epistatic effect with environments was detected.

Table 4.3 Estimated additive (A) and additive × environment (AE) interactions of QTL for spike-related traits

Trait	QTL	Flanking marker	Position (cM)	*F* value	A	H^2 (%)	H^2 AE (%)
Grain number per spike	*qSgn2D*	*Xgwm261–Xgwm296*	0.0	41.13	−2.12	12.24	1.24
	qSgn4D	*Xcfa2173–Xcfe188*	107.8	10.95	1.37	5.06	
	qSgn5D	*Xwmc215–Xgdm63*	74.4	25.38	−2.07	11.67	10.25
Spike length	*qSl2B*	*Xwmc770–Xwmc179*	55.6	7.32	0.18	2.68	6.28
	qSl2D	*Xcfd53–Xwmc18*	1.7	36.89	0.43	15.63	
	qSl4D	*Xcfa2173–Xcfe188*	108.8	15.65	−0.33	8.90	
	qSl5D	*Xbarc1097–Xcfd8*	16.4	7.80	0.21	3.74	
	qSl6B	*Xcfa2187–Xgwm219*	0.0	10.12	0.28	6.51	
Spikelets per spike	*qSps1B*	*Xbarc120.3–Xbarc008*	38.6	5.26	0.23	1.48	0.53
	qSps4A	*Xwmc718–Xwmc262*	3.0	6.69	−0.40	4.37	
	qSps5D	*Xwmc215–Xgdm63*	74.3	34.81	−0.71	13.83	23.07
	qSps7A	*Xwmc607–Xcfa2123*	100.6	5.06	0.21	1.22	
Compactness	*qSc2D*	*Xgwm261–Xgwm296*	0.0	37.86	−1.01	11.41	
	qSc4D	*Xcfa2173–Xcfe188*	108.8	11.30	0.74	5.22	
	qSc5D	*Xwmc215–Xgdm63*	74.4	27.68	−1.13	12.26	13.18
Fertile spikelets per spike	*qFsn4A*	*Xwmc718–Xwmc262*	1.0	9.12	−0.42	5.02	
	qFsn5D	*Xbarc320–Xwmc215*	68.3	32.70	−0.61	10.22	16.22

Table 4.4 Estimated epistasis (AA) of QTL for spike-related traits

Trait	QTL	Flanking marker	Position (cM)	QTL	Flanking marker	Position (cM)	AA	H^2 (%)
Grain number per spike	*qSgn2A*	*Xbarc296–Xwmc582*	69.7	*qSgn3A*	*Xcfd40–Xbarc1097*	2.4	1.13	3.47
Spikelets per spike	*QSps2A*	*Xbarc296–Xwmc582*	69.4	*qSps2B*	*Xbarc373–Xwmc477*	76.6	0.43	4.99
Compactness	*QSc2A*	*Xbarc296–Xwmc582*	69.7	*qSc3A*	*Xcfd40–Xbarc1097*	2.4	0.59	3.33
Fertile spikelets per spike	*qFsn1B*	*Xwmc406–Xbarc156*	24.7	*qFsn1D*	*Xcfd19–Xwmc93*	49.9	−0.45	5.75
	qFsn2B	*Xwmc770–Xwmc179*	58.6	*qFsn6D*	*Xbarc054–Xgwm55*	77.4	−0.37	3.77

4.2.1.1.6 QTL Mapping for Compactness

Three additive QTLs for compactness were mapped on chromosomes 2D, 4D, and 5D, accounting for 11.41, 5.22, and 12.26 % of phenotypic variance, respectively (Table 4.3). The *qSc5D* had the largest genetic contribution of 12.26 %, with the positive allele from Huapei 3. Other alleles came from Yumai 57. The *qSc5D* was interacted with environments, accounting for 13.18 % of phenotypic variance.

One pair of epistatic effect for compactness was mapped on chromosomes 2A-3A, accounting for 3.33 % of phenotypic variance. No interaction of epistatic effect with environment was identified.

4.2.1.1.7 QTL Mapping for Fertile Spikelets Per Spike

Two additive QTLs for fertile spikelets per spike were located on chromosomes 4A and 5D, accounting for 5.02 and 10.22 % of phenotypic variance, respectively. Two positive alleles both came from Yumai 57, and the *qFsn5D* was interacted with environments accounting for 16.22 %.

Two pairs of epistatic effect for fertile spikelets per spike were located on chromosomes 1B-1D and 2B-6D (Table 4.4), accounting for 5.75 and 3.77% of the phenotypic variance, respectively. No interaction of epistatic effect with environments was detected.

4.2.1.2 QTL Analysis of Spike Traits Using a RIL Population Derived from "Gaocheng 8901 and Waxy Wheat 1"

4.2.1.2.1 Phenotypic Variation

Data of phenotypic variations among RIL lines were observed in Table 4.5. It indicated that spike length of Gaocheng 8901 was higher than that of waxy wheat 1, while spikelets per spike, compactness, fertile spikelets per spike, and grain number per spike of Gaocheng 8901 with the higher values under all environmental conditions. Five spike-related traits of the RIL population segregated continuously and fit normal distribution with both absolute values of skewness and kurtosis less than 1.0. Therefore, the data can be used for QTL analysis. Five traits in three environments showed transgressive segregations, indicating that the dominant alleles controlling yield were randomly distributed on parental chromosomes.

4.2.1.2.2 Correlation Analysis

The correlations of the mean of five spike-related traits in three environments were shown in Table 4.6. Except between spike length and grain number per spike, and compactness and fertile spikelets per spike, the correlations of other traits were

Table 4.5 Phenotypic values for spike-related traits of the RIL population in three environments

Trait	Environment	Parent		RIL population			
		Waxy wheat 1	Gaocheng 8901	Range	Mean ± SD	Skewness	Kurtosis
Spike length (cm)	Tai'an, 2008	9.6	8.35	5.80–12.30	9.60 ± 1.14	−0.05	−0.14
	Tai'an, 2011	9.5	8.27	6.23–12.47	8.97 ± 1.16	0.14	−0.10
	Suzhou, 2011	9.33	8.1	6.27–14.00	9.16 ± 1.12	0.51	0.98
Compactness	Tai'an, 2008	1.82	2.93	1.13–3.29	2.17 ± 0.32	0.53	0.80
	Tai'an, 2011	1.83	2.29	1.49–3.30	2.14 ± 0.30	0.77	0.99
	Suzhou, 2011	2.07	2.42	1.38–3.32	2.30 ± 0.17	0.35	0.78
Spikelets per spike	Tai'an, 2008	18	22.67	13.33–25.67	19.46 ± 2.30	−0.25	−0.13
	Tai'an, 2011	17.4	19	15.00–23.00	18.93 ± 1.44	0.09	0.23
	Suzhou, 2011	19.33	20.33	16.67–25.67	20.81 ± 1.64	−0.09	−0.01
Fertile spikelets per spike	Tai'an, 2008	18	20.33	11.50–24.00	18.47 ± 2.26	−0.13	−0.18
	Tai'an, 2011	16.4	17	14.33–22.00	17.86 ± 1.66	−0.2	−0.21
	Suzhou, 2011	17	18.67	14.67–23.00	18.64 ± 1.72	−0.01	−0.15
Grain number per spike	Tai'an, 2008	42	58.3	22.3–68.4	48.29 ± 8.60	−0.1	−0.06
	Tai'an, 2011	42	48	21.0–70.0	53.15 ± 10.39	−0.34	0.10
	Suzhou, 2011	50	53	27.3–69.67	48.47 ± 7.00	0.12	0.37

SD Standard deviation

Table 4.6 Coefficients of pairwise correlations of the mean values of agronomic traits

	Spike length	Spikelets per spike	Compactness	Fertile spikelets per spike
Spikelets per spike	0.333**			
Compactness	−0.686**	0.428**		
Fertile spikelets per spike	0.335**	0.873**	0.343**	
Grain number per spike	0.257**	0.293**	−0.004	0.402**

**Correlation is significant when $p < 0.01$ level

significantly positive, in which the coefficient of spikelets per spike and fertile spikelets per spike was the largest (0.873), followed by the value of spikelets per spike and compactness (0.428). Spike length was negatively correlated with the compactness with the coefficient −0.686.

4.2.1.2.3 QTL Mapping for Spike Length

Eighteen additive QTLs for spike length were mapped on chromosomes 1A, 1B, 2A, 2B, 2D, 3D, 4A, 4B, 6A, and 6B (Table 4.7). Among them, *Qsl-1B* was detected in three environments. The positive alleles came from the female parent waxy wheat 1, with 3.59–5.59 % of phenotypic variance. The *Qsl-6B.5* was detected in E2 and PD, accounting for 20.00, 11.51 % of the phenotypic variance, respectively, with the positive allele derived from Gaocheng 8901. The *Qsl-1A.1* and *Qsl-2B* were detected in E1, with the opposite additive effect. The *Qsl-4A.2* and *Qsl-6A.2* were detected in E3. The *Qsl-4A.2* had 13.14 % of phenotypic variance in E3, while the *Qsl-6B.2* was detected in E2 accounting for 20.55 % of phenotypic variance in E2, with the positive allele coming from waxy wheat 1. Others were detected in one environment.

4.2.1.2.4 QTL Mapping for compactness

Twelve additive QTLs for compactness were mapped on chromosomes 1A, 1B, 2D, 3B, 6A, and 7B (Table 4.7). The *Qsc-1B.1* and *Qsl-6B.1* were stably expressed in three different environments with 5.82–8.25 % of phenotypic variance. The positive alleles came from Gaocheng 8901. The *Qsl-6B.2* was detected in E1 and E2 with 10.07–20.70 % of phenotypic variance, and the positive allele coming from waxy wheat 1. The *Qsl-1A.1* and *Qsl-1B.2* were detected in E2, with the opposite additive effect. Other loci were detected in one environment.

Table 4.7 QTL with significant additive effects for spike-related traits in different environments

Trait	Environment	QTL	Chromesome	Site (cM)	Interval marker	Additive effect	LOD	PVE (%)
Spike length	E1	*Qsl-1A.1*	1A	76.00	*wPt-1171—wPt-664968*	−0.25	3.73	4.82
		Qsl-1B	1B	0.00	*wPt-3563—wPt-8226*	0.27	4.43	5.59
		Qsl-2B	2B	7.00	*wPt-6576—wPt-4199*	0.28	3.70	5.30
		Qsl-6B.4	6B	289.00	*wPt-730273—wPt-6329*	−0.42	10.06	13.59
	E2	*Qsl-1B*	1B	0.00	*wPt-3563—wPt-8226*	0.26	5.43	4.92
		Qsl-2A	2A	213.00	*wPt-8328—wPt-2185*	−0.33	3.07	2.85
		Qsl-2D	2D	8.00	*wPt-664438—wPt-666395*	−0.28	3.97	5.96
		Qsl-4A.1	4A	0.00	*wPt-7524—wPt-2788*	0.19	2.89	2.56
		Qsl-6A.1	6A	86.00	*wPt-730977—wPt-1285*	−0.26	5.22	5.06
		Qsl-6B.1	6B	111.00	*wPt-1852—wPt-3774*	−0.19	2.87	2.57
		Qsl-6B.2	6B	229.00	*wPt-666615—wPt-669607*	0.53	9.90	20.55
		Qsl-6B.5	6B	291.00	*wPt-6329—xgpw-1149*	−0.52	19.16	20.00
	E3	*Qsl-1A.2*	1A	120.00	*wPt-731190—wPt-8347*	−0.22	2.99	3.70
		Qsl-1B	1B	0.00	*wPt-3563—wPt-8226*	0.21	3.22	3.59
		Qsl-4A.2	4A	137.00	*wPt-6440—wPt-8657*	0.41	10.58	13.14
		Qsl-4A.3	4A	146.00	*wPt-4620—wPt-0150*	−0.25	4.52	5.09
	PD	*Qsl-6A.2*	6A	82.00	*wPt-664811—wPt-0228*	−0.26	4.74	5.50
		Qsl-6B.3	6B	248.00	*wPt-669607—wPt-730273*	0.38	5.01	11.51
		Qsl-6B.5	6B	291.00	*wPt-6329—xgpw-1149*	−0.38	9.63	11.51
		Qsl-1A.1	1A	69.00	*wPt-1171—wPt-664968*	−0.22	3.13	5.02
		Qsl-1B	1B	0.00	*wPt-3563—wPt-8226*	0.23	5.85	5.59
		Qsl-2B	2B	0.00	*wPt-6576—wPt-4199*	0.20	4.05	3.80
		Qsl-3D	3D	11.00	*wPt-730886—xgpw-4136*	0.23	2.67	5.58
		Qsl-4A.2	4A	137.00	*wPt-6440—wPt-8657*	0.16	2.78	2.66
		Qsl-4B	4B	71.00	*wPt-3908—wPt-3164*	0.16	2.64	2.47
		Qsl-6A.2	6A	82.00	*wPt-664811—wPt-0228*	−0.21	4.66	4.56
		Qsl-6B.2	6B	235.00	*wPt-666615—wPt-669607*	0.30	6.57	9.31
		Qsl-6B.5	6B	291.00	*wPt-6329—xgpw-1149*	−0.42	17.40	18.77

(continued)

Table 4.7 (continued)

Trait	Environment	QTL	Chromesome	Site (cM)	Interval marker	Additive effect	LOD	PVE (%)
Compactness	E1	*Qsc-1B.1*	1B	0.00	*wPt-3563—wPt-8226*	−0.10	6.40	8.90
		Qsc-6B.1	6B	239.00	*wPt-666615—wPt-669607*	−0.08	4.16	5.82
		Qsc-6B.2	6B	289.00	*wPt-730273—wPt-6329*	0.10	7.09	10.07
	E2	*Qsc-1A.1*	1A	149.00	*wPt-665613—wPt-3904*	0.05	2.57	2.68
		Qsc-1B.1	1B	0.00	*wPt-3563—wPt-8226*	−0.08	5.65	6.00
		Qsc-1B.2	1B	185.00	*wPt-0328—wPt-8971*	−0.06	2.92	4.40
		Qsc-3B	3B	167.00	*wPt-3620—wPt-7906*	0.13	3.19	3.81
		Qsc-6A	6A	88.00	*wPt-1285—xgpw-3087*	0.05	2.54	2.70
		Qsc-6B.1	6B	233.00	*wPt-666615—wPt-669607*	−0.09	4.60	8.25
		Qsc-6B.2	6B	290.00	*wPt-730273—wPt-6329*	0.14	17.20	20.70
	E3	*Qsc-1A.2*	1A	116.00	*wPt-671483—wPt-665250*	0.05	2.71	3.16
		Qsc-1B.1	1B	2.00	*wPt-3563—wPt-8226*	−0.08	6.47	9.11
		Qsc-6B.1	6B	236.00	*wPt-666615—wPt-669607*	−0.07	4.27	6.53
		Qsc-6B.3	6B	291.00	*wPt-6329—xgpw-1149*	0.09	9.64	11.78
		Qsc-7B	7B	35.00	*wPt-4038—wPt-0194*	−0.05	3.01	3.95
	PD	*Qsc-1A.1*	1A	145.00	*wPt-665613—wPt-3904*	0.05	3.28	3.74
		Qsc-1B.1	1B	1.00	*wPt-3563—wPt-8226*	−0.09	11.48	13.66
		Qsc-1B.2	1B	183.00	*wPt-0328—wPt-8971*	−0.05	2.81	3.91
		Qsc-2D	2D	71.00	*wPt-667294—wPt-9963*	0.05	2.94	4.35
		Qsc-6B.4	6B	32.00	*wPt-3581—wPt-3060*	−0.05	3.51	4.14
		Qsc-6B.1	6B	239.00	*wPt-666615—wPt-669607*	−0.07	7.39	7.96
		Qsc-6B.2	6B	289.00	*wPt-730273—wPt-6329*	0.11	16.16	18.12
Spikelets per spike	E1	*Qsps-2D*	2D	168.00	*wPt-974—wPt-4527*	0.44	2.67	3.71
	E3	*Qsps-6A.1*	6A	20.00	*wPt-3091—wPt-0959*	−0.44	2.80	3.87
		Qsps-6A.2	6A	84.00	*wPt-0228—wPt-730977*	−0.33	2.95	4.16

(continued)

Table 4.7 (continued)

Trait	Environment	QTL	Chromesome	Site (cM)	Interval marker	Additive effect	LOD	PVE (%)
Fertile spikelets per spike	E1	*Qsfs-5B*	5B	239.00	*wPt-2041—wPt-9006*	0.56	3.61	6.12
		Qsfs-6A.1	6A	20.00	*wPt-3091—wPt-0959*	−0.51	3.20	5.20
	E3	*Qsfs-3B*	3B	169.00	*wPt-7906—wPt-2559*	0.94	4.37	9.06
		Qsfs-6A.2	6A	85.00	*wPt-730977—wPt-1285*	−0.38	3.36	5.19
		Qsfs-6A.3	6A	130.00	*wPt-729920—wPt-664792*	−0.57	2.85	11.35
Grain number per spike	E1	*Qgps-5B*	5B	239.00	*wPt-2041—wPt-9006*	2.24	4.30	6.86
	E2	*Qgps-2A*	2A	155.00	*wPt-5498—wPt-664128*	−2.07	2.74	4.01
		Qgps-2D.1	2D	155.00	*wPt-8319—wPt-731130*	3.53	6.66	11.65
	E3	*Qgps-1A.1*	1A	2.00	*wPt-5316—wPt-8016*	−1.48	2.84	4.63
		Qgps-6A.1	6A	57.00	*wPt-3468—wPt-9679*	2.26	6.24	10.87
		Qgps-6A.2	6A	84.00	*wPt-0228—wPt-730977*	−3.63	15.03	27.84
	PD	*Qgps-1A.2*	1A	6.00	*wPt-667814—wPt-730885*	−1.47	4.56	5.81
		Qgps-2B	2B	0.00	*wPt-6576—wPt-4199*	1.60	5.09	6.41
		Qgps-2D.2	2D	159.00	*wPt-731130—wPt-730613*	1.63	5.14	7.15
		Qgps-3B.1	3B	61.00	*wPt-664393—wPt-1191*	−2.60	13.12	18.21
		Qgps-3B.2	3B	64.00	*wPt-6216—wPt-9579*	3.49	21.91	32.75
		Qgps-6A.3	6A	20.00	*wPt-3091—wPt-0959*	−1.10	2.60	3.22
		Qgps-6B	6B	2.00	*xgpw-3241—wPt-0315*	−1.29	3.00	4.49

E1: Tai'an, 2008. E2 Tai'an, 2011. E3 Suzhou, 2011. *PD* Pool data
Positive values indicate that waxy wheat 1 alleles increase to corresponding trait value
Negative values indicate that Gaocheng 8901 alleles increase to corresponding trait

4.2.1.2.5 QTL Mapping for Spikelets Per Spike

Three additive-effect QTLs for spikelets per spike were mapped on chromosomes 2D and 6A with 3.71–4.16 % of phenotypic variance, which were detected in E1 and E3 (Table 4.7).

4.2.1.2.6 QTL Mapping for Fertile Spikelets Per Spike

Five additive-effect QTLs for fertile spikelets per spike were mapped on 3B, 5B, and 6A (Table 4.7), which were detected in E1 and E3. Except of the *Qsfs-5B,* other alleles came from Gaocheng 8901, accounting for 5.19–11.35 % of phenotypic variance.

4.2.1.2.7 QTL Mapping for Grain Number Per Spike

Thirteen additive-effect QTLs for grain number per spike were mapped on 1A, 2A, 2B, 2D, 3B, 5B, 6A, and 6B (Table 4.7), which were detected in one environment. The *Qgps-2D.1*, *Qgps-6A.1,* and *Qgps-6A.2* explained 11.65, 10.87, and 27.84 % of the phenotypic variance in E2 and E3, respectively.

4.2.1.3 QTL Analysis of Spike Traits Using a RIL Population Derived from "Shannong01-35 and Gaocheng 9411"

4.2.1.3.1 Phenotypic Variation

The data of phenotypic values for spike-related traits of the RIL population and two parents in four environments were shown in Table 4.8. The spike length and grain weight per spike of Shannong 01-35 were significantly higher than Gaocheng 9411. Spikelets per spike, fertile spikelets per spike, spikelets per spike, and compactness of Gaocheng 9411 were higher. There was no significant difference of sterile spikelets per spike in two parents, different performance in environments, and largely affected by environments. Seven spike-related traits showed continuous variance and with transgressive segregation. Except in single environment, the skewness and kurtosis value of traits were less than 1, which showed a basic normal distribution.

Table 4.8 Phenotypic values for spike-related traits of the RIL population and the parents in different environments

Trait	Environment	Parent		RIL population						
		Shannong 01-35	Gaocheng 9411	Mean	SD	Min	Max	CV (%)	Skewness	Kurtosis
Spike length (cm)	E1	10.43	8.25	9.31	1.32	6.77	13.15	14.14	0.22	−0.33
	E2	10.35	8.22	9.72	1.45	6.63	13.56	14.90	0.49	−0.15
	E3	10.03	8.15	9.14	1.43	6.00	14.13	15.62	0.49	0.70
	E4	10.33	8.67	9.50	1.48	6.00	14.00	15.55	0.58	0.73
Grain number per spike	E1	46.50	59.00	43.19	8.13	25.00	65.67	18.83	0.41	−0.32
	E2	52.67	75.00	54.42	9.16	25.00	78.00	16.84	−0.07	0.15
	E3	39.00	53.60	41.89	6.32	27.60	67.50	15.10	0.61	0.53
	E4	40.67	45.33	43.40	7.52	28.00	67.67	17.33	0.44	0.18
Fertile spikelets per spike	E1	17.67	18.67	16.53	1.52	13.00	20.00	9.20	−0.17	−0.36
	E2	17.00	19.00	16.80	1.45	11.33	20.33	8.63	−0.23	0.34
	E3	15.60	17.20	16.39	1.24	12.00	19.67	7.59	−0.14	0.62
	E4	17.33	18.00	17.26	1.83	10.33	22.33	10.61	−0.17	0.51
Sterile spikelets per spike	E1	1.00	1.67	2.00	0.68	0.00	4.33	34.09	0.21	0.69
	E2	0.00	0.00	0.92	0.68	0.00	4.33	73.76	1.03	2.85
	E3	1.80	1.80	2.08	0.59	0.60	3.75	28.16	0.21	0.30
	E4	2.33	2.00	2.48	0.74	0.33	5.00	30.07	0.36	0.79
Spikelets per spike	E1	18.67	20.33	18.53	1.36	14.50	22.00	7.35	−0.12	−0.18
	E2	17.00	19.00	17.72	1.39	11.33	22.00	7.83	−0.43	2.06
	E3	17.40	19.00	18.47	1.21	14.00	21.33	6.54	−0.27	0.29
	E4	19.67	20.00	19.74	1.72	13.33	24.33	8.73	−0.17	0.36

(continued)

Table 4.8 (continued)

Trait	Environment	Parent		RIL population						
		Shannong 01-35	Gaocheng 9411	Mean	SD	Min	Max	CV (%)	Skewness	Kurtosis
Compactness	E1	178.91	246.46	202.43	29.01	129.28	312.8	14.33	0.66	0.69
	E2	164.20	231.05	185.76	27.29	98	260.06	14.69	0.03	0.06
	E3	173.57	233.13	205.88	32.04	78.47	339.17	15.56	0.22	2.19
	E4	190.32	230.94	211.50	29.59	133.33	333.33	14.00	0.56	1.05
Grain weight per spike (g)	E1	1.48	1.08	1.05	0.31	0.09	1.97	29.36	0.20	0.34
	E2	2.16	1.54	1.61	0.43	0.86	3.01	26.65	0.42	−0.41
	E3	2.04	1.78	1.42	0.32	0.54	2.22	22.42	0.18	−0.49
	E4	2.10	1.50	1.69	0.35	1.05	2.62	20.90	0.24	−0.86

E1: 2008–2009 growing season at Tai'an site. E2: 2009–2010 growing season at Tai'an site. E3: 2010–2011 growing season at Tai'an site. E4: 2010–2011 growing season at Suzhou site

4.2.1.3.2 QTL Mapping for Spike Length

Twenty-three additive QTLs for spike length were mapped on chromosome 1B, 1D, 2A, 2D, 3A, 5A, 6B, 7A, and 7B, accounting for 4.04–40.43 % of phenotypic variance (Table 4.9). Of which, 9 QTLs were main-effect QTLs, with the contribution ratio more than 10 %. The *QSl1B.1-113* and *QSl2A-204* were detected in E2. The *QSl6B.3-2* was identified in E2 and E4. The *QSl1B.1-113* was a main-effect QTL with the genetic contribution ratio more than 20 % in one environment.

4.2.1.3.3 QTL Mapping for Grain Number Per Spike

Twelve QTLs for spikelets per spike were mapped on chromosomes 1B, 2A, 3D, 4B, and 5B, accounting for 5.38–44.10 % of phenotypic variance (Table 4.9). The *QKnps2A-132* was identified in E2, while the *QKnps4B.1-99* was identified in E1 and E3. The *QKnps4B.1-97* (E2), *QKnps4B.1-98* (PD), and *QKnps4B.1-99* were located in the same marker interval.

4.2.1.3.4 QTL Mapping for Fertile Spikelets Per Spike

Thirteen additive QTLs for fertile spikelets per spike were mapped on chromosomes 2D, 4A, 4B, 6A, and 7B, accounting for 5.54–30.02 % of phenotypic variance (Table 4.9). The *QFsn4B.1-99* was identified in E1 and E2, with 5.88 and 6.58 % of genetic contribution. There were 5 main-effect QTLs. The *QFsn6A.1-22* had the largest contribution ratio of 30.02 %. The *QFsn4B.1-97* (PD, 13.20 %) was in the same marker interval with *QFsn4B.1-99* (E1 and E2).

4.2.1.3.5 QTL Mapping for Sterile Spikelets Per Spike

Eight additive QTLs for sterile spikelets per spike were mapped on chromosomes 1B, 3A, 3B, 4B, 4D, 5B, and 6B, accounting for 4.84–12.07 % of phenotypic variance (Table 4.9). Two main-effect QTLs, *QSsn3B.1-215* and *QSsn4D-11*, could explain 12.07 and 10.58 % of the phenotypic variation. The positive alleles were from Gaocheng 9411 and Shannong 01-35, respectively. The *QSsn4B.1-98* and *QSsn6B.1-30* could be detected in E2.

4.2.1.3.6 QTL Mapping for Spikelets Per Spike

Thirteen additive QTLs for spikelets per spike were mapped on chromosomes 1B, 1D, 2B, 2D, 3A, 4B, 5B, 6A, 6D, and 7B, accounting for 5.43–21.87 % of phenotypic variance (Table 4.9). Two main-effect QTLs, *QSnps2B-94* (E1) and

Table 4.9 Additive QTL for spike-related traits in different environments

Trait	Environment	QTL	Left marker	Right marker	A	LOD	PVE (%)
Spike length	E1	*QSl2A-192*	*xgwm294*	*xgwm614*	0.53	2.59	16.10
		QSl5A-51	*wPt-8226*	*Xgpw3217*	−0.37	2.53	7.84
		QSl6B.3-10	*wPt-669607*	*Xgpw1005*	0.54	7.45	15.15
		QSl7B-173	*wPt-1266*	*wPt-0276*	−0.36	2.79	7.12
	E2	*QSl1B.1-66*	*Xgpw2067*	*CFE063*	0.38	3.40	5.65
		QSl1B.1-100	*wPt-3753*	*wPt-1139*	−1.08	24.23	39.52
		QSl1B.1-113	*wPt-2751*	*wPt-3465*	0.86	19.35	29.49
		QSl2A-204	*xgwm294*	*xgwm614*	0.35	3.11	5.93
		QSl2D-21	*wPt-6343*	*wPt-667485*	−0.43	5.98	8.28
		QSl3A-72	*wPt-8779*	*wPt-2938*	0.50	8.43	11.71
		QSl3A-78	*Xgpw7080*	*Xgpw1005*	−0.55	10.54	14.36
		QSl6B.3-2	*wPt-1325*	*wPt-669607*	0.44	6.48	8.86
		QSl7A-183	*wPt-3883*	*wPt-7785*	−0.35	3.98	5.82
	E3	*QSl5A-40*	*wPt-8226*	*Xgpw3217*	−0.44	2.99	6.80
		QSl7B-179	*wPt-1266*	*wPt-0276*	−0.54	2.63	10.10
	E4	*QSl1B.1-9*	*wPt-731490*	*wPt-4555*	0.43	2.70	4.69
		QSl4A-159	*wPt-664749*	*wPt-4064*	−0.35	3.28	5.52
		QSl6B.3-2	*wPt-1325*	*wPt-669607*	0.39	3.63	6.56
		QSl7A-6	*wPt-667038*	*wPt-0961*	−0.37	3.71	6.36
		QSl7A-49	*wPt-6768*	*wPt-6495*	0.53	4.89	12.81
	PD	*QSl1B.1-104*	*wPt-5363*	*wPt-1363*	0.99	22.54	40.43
		QSl1B.1-113	*wPt-2751*	*wPt-3465*	0.73	15.53	25.69
		QSl1D-0	*wPt-730758*	*wPt-4647*	0.27	2.97	4.04
		QSl2A-204	*xgwm294*	*xgwm614*	0.41	4.18	9.70
		QSl2D-20	*wPt-6343*	*wPt-667485*	−0.32	3.75	5.80
		QSl6B.3-6	*wPt-1325*	*wPt-669607*	0.48	7.87	12.29

(continued)

Table 4.9 (continued)

Trait	Environment	QTL	Left marker	Right marker	A	LOD	PVE (%)
Grain numbr per spike	E1	*QKnps4B.1-99*	*wPt-7569*	*wPt-3908*	2.60	4.34	9.88
	E2	*QKnps2A-132*	*xgwm425*	*wPt-3114*	−2.10	2.94	5.38
		QKnps4B.1-97	*wPt-7569*	*wPt-3908*	2.63	4.08	7.94
		QKnps4D-12	*Xgpw3113*	*Xgpw342*	−3.58	3.67	15.48
		QKnps5B.1-50	*wPt-6052*	*wPt-4703*	2.24	3.16	6.02
		QKnps5B.2-12	*wPt-9103*	*wPt-4418*	−2.39	3.83	6.75
	E3	*QKnps3D-154*	*Xgpw7643*	*wPt-731103*	−1.44	2.57	5.15
		QKnps4B.1-99	*wPt-7569*	*wPt-3908*	1.49	2.68	5.40
		QKnps4B.2-36	*wPt-8756*	*CFE149*	1.54	2.62	5.59
	E4	*QKnps1B.1-81*	*wPt-665375*	*wPt-0260*	−5.80	18.49	44.10
		QKnps1B.1-104	*wPt-5363*	*wPt-1363*	5.54	16.38	38.44
		QKnps3D-99	*Xgpw7646*	*Xgpw2046*	−2.22	2.83	8.67
	PD	*QKnps2A-132*	*xgwm425*	*wPt-3114*	−1.79	4.58	8.81
		QKnps4B.1-98	*wPt-7569*	*wPt-3908*	1.86	4.57	9.04
Fertile spikelets per spike	E1	*QFsn4B.1-99*	*wPt-7569*	*wPt-3908*	0.38	2.91	5.88
		QFsn6A.1-31	*wPt-7127*	*wPt-9690*	0.36	2.79	5.66
		QFsn7B-109	*Xgpw8040*	*wPt-5892*	−0.48	4.41	10.00
	E2	*QFsn2D-120*	*wPt-666987*	*wPt-0330*	−0.35	2.78	5.79
		QFsn4B.1-99	*wPt-7569*	*wPt-3908*	0.38	3.19	6.58
	E3	*QFsn2D-119*	*wPt-667294*	*wPt-4144*	−0.29	2.63	5.54
		QFsn4B.2-30	*wPt-8756*	*CFE149*	0.43	3.50	11.33
	E4	*QFsn4A-158*	*wPt-664749*	*wPt-4064*	−0.46	2.85	6.24
		QFsn4B.1-94	*wPt-7569*	*wPt-3908*	0.52	3.01	7.43
	PD	*QFsn4B.1-97*	*wPt-7569*	*wPt-3908*	0.43	5.99	13.20
		QFsn6A.1-14	*wPt-3091*	*wPt-731153*	−0.54	9.99	21.14
		QFsn6A.1-22	*wPt-0959*	*wPt-730631*	0.63	13.08	30.02

(continued)

Table 4.9 (continued)

Trait	Environment	QTL	Left marker	Right marker	A	LOD	PVE (%)
Sterile spikelets per spike	E1	*QSsn1B.1-61*	*Xgpw2067*	*CFE063*	0.19	2.96	6.39
	E2	*QSsn4B.1-98*	*wPt-7569*	*wPt-3908*	−0.20	3.33	8.44
		QSsn6B.1-30	*wPt-0259*	*wPt-2095*	−0.18	3.22	7.00
	E3	*QSsn3A-115*	*wPt-7992*	*wPt-671711*	0.17	2.52	6.59
	E4	*QSsn5B.2-18*	*wPt-9103*	*wPt-4418*	−0.22	2.71	8.49
	PD	*QSsn3B.1-215*	*wPt-1171*	*wPt-5906*	−0.16	6.95	12.07
		QSsn3B.1-222	*wPt-1171*	*wPt-5906*	0.12	3.75	7.10
		QSsn4B.1-98	*wPt-7569*	*wPt-3908*	−0.11	2.66	4.84
		QSsn4D-11	*Xgpw3113*	*Xgpw342*	0.15	2.53	10.58
		QSsn6B.1-30	*wPt-0259*	*wPt-2095*	−0.12	3.99	6.79
Spikelets per spike	E1	*QSnps2B-76*	*wPt-1140*	*wPt-4199*	0.41	3.93	8.47
		QSnps2B-94	*CFE052*	*wPt-5374*	−0.64	9.21	21.87
		QSnps6A.1-20	*wPt-0959*	*wPt-730631*	0.33	3.16	5.94
		QSnps7B-116	*wPt-7887*	*wPt-5343*	−0.34	2.70	5.99
	E2	*QSnps1D-41*	*wPt-3738*	*wPt-7946*	−0.33	2.83	5.80
		QSnps2D-15	*wPt-667765*	*wPt-3757*	−0.36	3.57	6.75
		QSnps5B.2-74	*wPt-666268*	*wPt-3922*	0.42	4.66	8.97
		QSnps5B.2-83	*wPt-7665*	*wPt-3569*	−0.59	8.49	17.79
	E3	*QSnps4B.1-99*	*wPt-7569*	*wPt-3908*	0.31	2.62	6.20
	E4	*QSnps1B.1-20*	*wPt-730156*	*wPt-9925*	−0.56	2.53	8.47
		QSnps3A-1	*Xgpw8072*	*Xgpw2270*	0.22	3.00	6.67
		QSnps4A-156	*wPt-4620*	*wPt-672107*	−0.20	2.62	5.43
	PD	*QSnps6D-132*	*xcfd49*	*wPt-672044*	−0.25	3.07	6.94

(continued)

Table 4.9 (continued)

Trait	Environment	QTL	Left marker	Right marker	A	LOD	PVE (%)
Compactness	E1	*QSc4D-12*	*Xgpw3113*	*Xgpw342*	10.00	2.65	12.50
		QSc6B.3-9	*wPt-669607*	*Xgpw1005*	−10.00	5.19	11.45
	E2	*QSc1B.1-8*	*wPt-731490*	*wPt-4555*	−13.00	5.20	13.05
		QSc6B.3-5	*wPt-1325*	*wPt-669607*	−9.00	4.34	10.01
	E3	*QSc1B.1-9*	*wPt-731490*	*wPt-4555*	−11.00	3.50	7.02
		QSc5A-49	*wPt-8226*	*Xgpw3217*	10.00	3.39	9.15
		QSc6B.3-9	*wPt-669607*	*Xgpw1005*	−8.00	2.94	5.62
	E4	*QSc1B.1-9*	*wPt-731490*	*wPt-4555*	−12.00	4.22	6.96
		QSc2A-214	*xgwm614*	*wPt-666340*	−8.00	3.00	5.84
		QSc2D-18	*wPt-6343*	*wPt-667485*	28.00	31.38	69.50
		QSc3A-116	*wPt-7992*	*wPt-671711*	9.00	2.88	5.35
		QSc6B.3-11	*wPt-669607*	*Xgpw1005*	−9.00	4.07	7.25
		QSc7B-165	*wPt-3723*	*wPt-1266*	10.00	4.10	8.11
	PD	*QSc1B.1-8*	*wPt-731490*	*wPt-4555*	−13.00	7.38	13.53
		QSc2A-203	*xgwm294*	*xgwm614*	−9.00	4.14	11.27
		QSc2D-18	*wPt-6343*	*wPt-667485*	5.00	2.73	3.85
		QSc3A-131	*wPt-671711*	*xcfd35*	7.00	2.95	5.79
		QSc5A-57	*wPt-8226*	*Xgpw3217*	7.00	2.90	5.96
		QSc6B.3-8	*wPt-1325*	*wPt-669607*	−8.00	5.99	9.00
		QSc7B-117	*wPt-7887*	*wPt-5343*	−6.00	3.04	5.06
		QSc7B-165	*wPt-3723*	*wPt-1266*	10.00	7.74	14.28

(continued)

Table 4.9 (continued)

Trait	Environment	QTL	Left marker	Right marker	A	LOD	PVE (%)
Grins weight per spike	E1	*QKwps2B-96*	*CFE052*	*wPt-5374*	−0.09	3.11	8.72
		QKwps4B.1-99	*wPt-7569*	*wPt-3908*	0.07	2.87	5.45
	E2	*QKwps4B.1-99*	*wPt-7569*	*wPt-3908*	0.12	4.64	7.53
		QKwps4D-5	*Xgpw3113*	*Xgpw342Phenotypic variations for grain-related traits of the*	−0.15	5.24	12.49
		QKwps6A.1-135	*TaGw2-CAPS*	*wPt-667618*	0.09	2.75	4.47
	E3	*QKwps2A-130*	*xgwm425*	*wPt-3114*	−0.08	3.29	6.19
		QKwps3D-150	*Xgpw2046*	*Xgpw7643*	−0.09	3.50	7.58
		QKwps4B.1-99	*wPt-7569*	*wPt-3908*	0.13	9.37	16.00
		QKwps6A.1-141	*TaGw2-CAPS*	*wPt-667618*	0.08	3.39	5.72
	E4	*QKwps2D-40*	*wPt-667485*	*wPt-1068*	−0.12	3.91	10.47
		QKwps4B.1-99	*wPt-7569*	*wPt-3908*	0.11	5.80	10.53
		QKwps6A.1-31	*wPt-7127*	*wPt-9690*	0.09	3.86	6.96
		QKwps6B.2-0	*wPt-9881*	*wPt-3060*	−0.10	4.73	8.32
		QKwps7B-170	*wPt-1266*	*wPt-0276*	−0.11	3.99	8.51
	PD	*QKwps2A-132*	*xgwm425*	*wPt-3114*	−0.07	2.91	4.72
		QKwps3D-154	*Xgpw7643*	*wPt-731103*	−0.07	3.09	4.82
		QKwps4B.1-99	*wPt-7569*	*wPt-3908*	0.13	9.72	16.68
		QKwps4D-11	*Xgpw3113*	*Xgpw342*	−0.12	4.72	15.80
		QKwps6A.1-142	*TaGw2-CAPS*	*wPt-667618*	0.07	2.88	4.53
		QKwps6A.2-0	*wPt-671568*	*wPt-665169*	−0.07	2.88	4.44

E1: 2008–2009 growing season at Tai'an site. E2: 2009–2010 growing season at Tai'an site. E3: 2010–2011 growing season at Tai'an site. E4: 2010–2011 growing season at Suzhou site. *PD* Pool data

Positive and negative values of additive effect (EstAdd) indicate that alleles to increase thousand-grain weight are inherited from Shannong 01-35 and Gaocheng 9411, respectively

QSnps5B.2-83(E2), had more than 10 % of phenotypic variance, with the positive alleles coming from Gaocheng 9411.

4.2.1.3.7 QTL Mapping for Compactness

Sixteen additive QTLs for compactness were mapped on chromosomes 1B, 2A, 2D, 3A, 4D, 5A, 6B, and 7B, accounting for 3.85–69.5 % of phenotypic variance (Table 4.9). Of which, 8 were main-effect QTLs. The *QSc2D-18* had the largest contribution ratio of 69.50 %. The *QSc1B.1-8* was identified in E2, with 13.05 and 13.53 % of phenotypic variance. The *QSc6B.3-9* was identified in E1 and E3, with 11.45 and 5.62 % of phenotypic variance, respectively. The *QSc1B.1-9* was detected in E3 and E4, with 7.02 and 6.69 % of phenotypic variance, respectively, which has the same marker interval with the *QSc1B.1-8*. The *QSc6B.3-11* (E4, 7.25 %) was in the same marker interval with the *QSc6B.3-9*. The *QSc7B-165* could be detected in E4.

4.2.1.3.8 QTL Mapping for Grain Weight Per Spike

Eighteen additive QTLs for grain weight per spike were mapped on chromosomes 2A, 2B, 2D, 4B, 4D, 6A, 6B, and 7B, accounting for 4.44–16.68 % of phenotypic variance (Table 4.9). Four main-effect QTLs were detected. The positive alleles of *QKwps4D-5*(E2), *QKwps2D-40*(E4), and *QKwps4D-11*(PD) came from Gaocheng 9411, while *QKwps4B.1-99* derived from Shannong 01-35. The *QKwps4D-5*(E2) was in the same marker interval with the *QKwps4D-11*(PD). The *QKwps4B.1-99* was detected in four environments, with the contribution ratio in E3, E4 more than 10 %. The *QKwps4B.1-99* was a stable main-effect QTL.

4.2.2 QTL Mapping for Grain-Related Traits

4.2.2.1 QTL Analysis of Grain Traits Using a DH Population Derived from "Huapei 3 and Yumai 57"

4.2.2.1.1 Phenotypic Variations

The grain-related traits for the DH population and the parents in three environments were described in Table 4.10. The parent Yumai 57 had larger values in grain yield, but the thousand-grain weight and grain diameter showed lower than that of Huapei 3. Significant transgressive segregation was observed for three traits among DH lines in the three environments. The grain-related traits of the DH population segregated continuously and followed a normal distribution, indicating its polygenic inheritance and suitability for QTL analysis (Cao et al. 2001).

Table 4.10 Phenotypic variations for grain-related traits of the DH population and the parents in three environments

Trait	Parent		DH population					
	Huapei 3	Yumai 57	Mean	Maximum	Minimum	SD	Skewness	Kurtosis
Grain yield (g m^{-2})	821.1	954.0	773.4	1028.9	513.74	93.9	0.18	−0.05
Thousand-grain weight (g)	56.7	40.7	43.9	66.0	30.2	5.2	0.53	1.76
Grain diameter (cm)	2.9	2.4	2.5	3.2	1.9	0.3	−0.22	−0.81

4.2.2.1.2 QTL Mapping for Grain Yield

Three QTLs with additive effects were mapped to chromosome 2D, 4A, and 5D for grain yield, accounting for 14.07, 4.52, and 10.32 % of the phenotypic variance, respectively (Table 4.11). Yumai 57 alleles at three loci increased the grain yield. The *qGY5D* was interacted with environments, accounting for 11.28 % of the phenotypic variance.

Three pairs of epistatic effects were identified for grain yield, which located on chromosomes 2A-4D, 2D-3B, and 6A-6B (Table 4.12), accounting for 2.25, 4.03, and 6.51 % of the phenotypic variation, respectively. No epistatic effects were detected interaction with environments.

4.2.2.1.3 QTL Mapping for Grain Diameter

Four additive QTLs for grain diameter were mapped on chromosomes 3A, 3B, 6A, and 7D, accounting for 1.72, 6.34, 13.80, and 6.35 % of the phenotypic variance, respectively (Table 4.11). The *qGd 6A* had the highest genetic contribution of

Table 4.11 Estimated additive (A) and additive × environment (AE) interactions of QTL for grain yield and spike-related traits

Trait	QTL	Flanking marker	Position (cM)	*F* value	A	H^2 (%)	H^2 AE (%)
Grain yield (g m^{-2})	*qGY2 Da*	*Xcfd53–Xwmc18*	2.8	31.32	−43.96	14.07	
	qGY4A	*Xwmc718–Xwmc262*	0.0	10.99	−24.90	4.52	
	qGY5D	*Xwmc215–Xgdm63*	68.2	32.90	−37.65	10.32	31.49
Thousand-grain weight (g)	*qTgw3B*	*Xgwm533–Xbarc251*	29.0	7.50	1.14	3.36	
	qTgw4B	*Xwmc413–Xcfd39.2*	7.7	10.72	1.30	4.39	
	qTgw6Ab	*Xbarc1055–Xwmc553*	49.0	24.41	2.37	14.64	
Grain diameter (cm)	*qGd3A*	*Xwmc264–Xcfa2193*	143.9	9.28	0.04	1.72	
	qGd3B	*Xgwm389–Xgwm533*	17.6	13.33	0.08	6.34	
	qGd6A	*Xbarc1055–Xwmc553*	43.0	21.67	0.12	13.80	
	qGd7Db	*Xgwm437–Xwmc630.1*	127.6	15.93	0.08	6.35	

Table 4.12 Estimated epistasis (AA) of QTL for grain yield and spike-related traits

Trait	QTL	Flanking marker	Position (cM)	QTL	Flanking marker	Position (cM)	AA	H^2 (%)
Grain yield (g m^{-2})	*Qgy2A*	*Xwmc401–Xbarc353*	70.8	*qGY4D*	*Xgwm34–Xbarc376*	1.0	17.59	2.25
	Qgy2Db	*Xgwm311.2–Xcfd50*	145.1	*qGY3B*	*Xgwm194–Xcfa2173*	67.0	23.52	4.03
	Qgy6A	*Xwmc553–Xgwm732*	58.3	*qGY6B*	*Xcfa2187–Xgwm219*	1.0	29.91	6.51
Thousand-grain weight (g)	*QTgw2D*	*Xbarc129.2–Xcfd50*	127.4	*qTgw5D*	*Xwmc630.2–Xcfd40*	2.0	1.43	5.37
	QTgw6Aa	*Xbarc023–Xbarc1077*	34.5	*qTgw7A*	*Xbarc049–Xwmc530*	75.7	−1.69	7.42
Grain diameter (cm)	*QGd2D*	*Xwmc18–Xwmc170.2*	60.9	*qGd 7Db*	*Xgwm437–Xwmc630.1*	127.6	−0.07	4.10
	QGd6B	*Xcfa2187–Xgwm219*	4.0	*qGd 7A*	*Xbarc259–Xwmc596*	54.7	0.08	5.43
	qGd6D	*Xgwm55–Xgwm133.2*	82.9	*qGd 7 Da*	*Xbarc244–Xbarc352*	56.6	1.44	5.37

13.80 %. Four positive alleles came from Huapei 3, while Huapei 3 with bigger grain diameter than Yumai 57. There was no interaction of additive effect with environment.

Three pairs of epistatic effect for grain diameter were located on chromosomes 2D-7D, 6B-7A, and 6D-7D (Table 4.12), accounting for 4.10, 5.43, and 5.37 % of phenotypic variance, respectively. There was no interaction of epistatic effect with environment, and the total epistatic effect contribution ratio was 15.9 %.

4.2.2.1.4 QTL Mapping for Thousand-Grain Weight

Three additive QTLs for thousand-grain weight were mapped on chromosomes 3B, 4B, and 6A, accounting for 3.36, 4.39, and 14.64 % of phenotypic variance, respectively (Table 4.11). The *qTgw6Ab* had the largest genetic contribution of 14.64 %. Three positive alleles came from Huapei 3, while Huapei 3 had larger thousand-grain weight than Yumai 57. No interaction between additive effects with environments was detected.

Two pairs of epistatic effect for thousand-grain weight were mapped on chromosome 2D-5D and 6A-7A (Table 4.12), accounting for 5.37 and 7.42 % of phenotypic variance, respectively. No interaction of epistatic effect with environment was detected. The total epistatic effect was 12.79 %.

4.2.2.2 QTL Analysis of Grain Traits Using a RIL Population Derived from "Shannong 01-35 and Gaocheng 9411"

4.2.2.2.1 Phenotypic Variation

The grain weight and grain diameter data were shown in Table 4.13, and the traits of Shannong 01-35 were larger than Gaocheng 9411. The difference in grain weight was more than 20 mg, and grain diameter reached to 0.7 mm. The data in the population showed a continuous variation, and transgressive segregation, skewness, and kurtosis were less than 1, showing the normal distribution. These indicated that the two grain traits were quantitative traits controlled by multiple genes.

4.2.2.2.2 QTL Mapping for Grain Weight

Fifteen additive QTLs for grain weight were mapped on chromosomes 1B, 2D, 3B, 4B, 4D, 5B, 6A, and 6D (Table 4.14). The positive alleles came from the female parent Shannong 01-35. Four QTLs, *QGw1B.1-113*, *QGw6A.1-134*, *QGw6D-178,* and *QGw4B.1-9*, were identified in two or more than two environments, with 4.72–15.41 % of the phenotypic variance. The *QGw1B.1-112*(E2) and *QGw1B.1-113* were located in the same marker interval. The *QGw6A.1-132* (E4) was detected in the same marker interval with *QGw6A.1-134* (E1, E2, PD). The *QGw6A.1-136* (E3)

Table 4.13 Phenotypic values for kernel-related traits of the RIL population and the parents in different environments

Trait	Environment	Parent		RIL population						
		Shannong 01-35	Gaocheng 9411	Mean	SD	Min	Max	CV (%)	Skewness	Kurtosis
Grain weight (mg)	E1	54.67	32.43	41.64	5.75	22.86	59.64	13.80	−0.05	0.52
	E2	62.63	33.59	42.74	5.83	29.06	63.52	13.63	0.40	0.51
	E3	62.07	36.85	45.85	5.55	24.93	62.77	12.10	−0.21	1.40
	E4	57.39	34.98	42.30	5.30	26.90	58.13	12.53	0.17	0.00
Grain diameter (mm)	E1	3.35	2.73	3.03	0.19	2.38	3.71	6.35	−0.07	0.97
	E2	3.49	2.78	3.04	0.19	2.62	3.65	6.14	0.29	0.01
	E3	3.51	2.89	3.15	0.18	2.49	3.72	5.76	−0.28	1.10
	E4	3.40	2.78	3.00	0.18	2.47	3.57	5.87	0.13	0.13

E1: 2008–2009 growing season at Tai'an site. E2: 2009–2010 growing season at Tai'an site. E3: 2010–2011 growing season at Tai'an site. E4: 2010–2011 growing season at Suzhou site

Table 4.14 Additive QTLs for kernel-related traits in different environments

Trait	*QTL*	Left marker	Right marker	Environment	A	LOD	PVE (%)
Grain weight (mg)	*QGw1B.1-112*	*wPt-2751*	*wPt-3465*	E2	1.57	4.53	7.16
	QGw1B.1-113	*wPt-2751*	*wPt-3465*	E1/E3/E4/PD	1.59/1.44/1.71/1.50	3.77/3.44/6.01/4.89	7.04/6.48/9.82/8.06
	QGw2D-120	*wPt-666987*	*wPt-0330*	E4	1.02	2.66	4.14
	QGw3B.1-70	*wPt-667746*	*wPt-0021*	E2	−1.09	2.56	3.96
	QGw3B.1-265	*wPt-4209*	*wPt-1191*	E2	1.17	3	4.65
	QGw4B.1-61	*xcfd54–4D*	*Xgpw2172*	E3	1.26	2.98	5.75
	QGw4B.1-99	*wPt-7569*	*wPt-3908*	E2/E4	1.34/1.07	3.94/2.81	5.93/4.37
	QGw4B.2-36	*wPt-8756*	*CFE149*	E3	1.2	2.58	4.94
	QGw4D-12	*Xgpw3113*	*Xgpw342*	E4	−1.51	2.75	9.1
	QGw5B.1-118	*wPt-3503*	*CFE186*	PD	1.17	3.64	5.82
	QGw5B.1-121	*CFE186*	*xgwm639*	E1	1.54	3.96	7.82
	QGw6A.1-132	*CFE043*	*TaGw2-CAPS*	E4	1.18	3.14	5.26
	QGw6A.1-134	*CFE043*	*TaGw2-CAPS*	E1/E2/PD	1.64/2.19/1.52	4.43/9.60/5.55	8.17/15.41/9.01
	QGw6A.1-136	*TaGw2-CAPS*	*wPt-667618*	E3	1.67	4.68	9.61
	QGw6D-178	*wPt-666615*	*wPt-666008*	E1/E3/PD	1.43/1.16/1.09	3.47/2.57/3.04	6.42/4.72/4.86
Grain diameter (mm)	*QGd1B.1-28*	*wPt-9925*	*Xgpw2281*	E1	−0.08	3.81	7.96
	QGd1B.1-27	*wPt-9925*	*Xgpw2281*	E4	−0.08	5.29	9.38
	QGd1B.1-29	*wPt-9925*	*Xgpw2281*	E2/E3/PD	−0.08/-0.08/-0.06	6.93/5.46/5.52	9.32/10.23/7.62
	QGd1D-41	*wPt-3738*	*wPt-7946*	E2	−0.05	6.25	8.45
	QGd3A-191	*wPt-667139*	*wPt-666832*	E1/E4/PD	−0.06/-0.04/-0.06	3.59/3.05/5.39	5.73/4.06/8.25
	QGd3B.1-266	*wPt-4209*	*wPt-1191*	E2/E4	0.04/0.04	3.07/3.34	4.18/5.07
	QGd4B.1-99	*wPt-7569*	*wPt-3908*	E1/E2/E3/E4/PD	0.04/0.06/0.06/0.07/0.06	2.89/9.21/6.13/10.99/9.50	4.57/12.16/10.75/16.24/13.37
	QGd4D-9	*Xgpw3113*	*Xgpw342*	E4	−0.06	4.04	10.75
	QGd5A-142	*wPt-2768*	*CFE019*	E2	−0.04	3.03	4.81
	QGd5B.1-11	*wPt-8604*	*wPt-2041*	PD	−0.05	3.89	6.67
	QGd5B.1-126	*CFE186*	*xgwm639*	E1/E4/PD	0.04/0.04/0.04	3.16/3.82/3.40	5.19/5.35/4.57
	QGd6A.1-132	*CFE043*	*TaGw2-CAPS*	E1	0.04	3.17	5.38
	QGd6A.1-133	*CFE043*	*TaGw2-CAPS*	PD	0.03	2.61	3.49
	QGd6A.1-135	*TaGw2-CAPS*	*wPt-667618*	E2	0.04	3.63	4.76

E1: 2008–2009 growing season at Tai'an site. E2: 2009–2010 growing season at Tai'an site. E3: 2010–2011 growing season at Tai'an site. E4: 2010–2011 growing season at Suzhou site
Positive and negative values of additive effect indicate that alleles to increase thousand-grain weight are inherited from Shannong 01-35 and Gaocheng 9411, respectively

was away from the *QGw6A.1-134* by 2 cM. The *QGw6A.1-134* was the only one main-effect QTL with 15.41 % of genetic contribution.

4.2.2.2.3 QTL Mapping for Grain Diameter

Nineteen additive QTLs for grain diameter were mapped on 13 chromosomes, with the 4.00–29.36 % of phenotypic variance (Table 4.14). Among them, nine additive-effect loci came from Shannong 01-35, and other ten QTLs came from Gaocheng 9411. The *QGd4B.1-99* and *QGd6D-178* were stable QTLs detected in environments. Seven loci, *QGd1B.1-29*, *QGd3A-191*, *QGd7A-226, QGd4B.1-99, QGd5B.1-126, QGd6D-178,* and *QGd3B.1-266*, were identified in two or more than two environments. There were five main-effect QTLs for grain diameter. The *QGd4B.1-99* had the largest contribution ratio in four environments.

In all, 34 QTLs for grain weight and grain diameter in four environments were mapped on wheat fourteen chromosomes except for 1A, 2A, 2B, 3D, 4A, 5D, and 7D in this RIL population.

4.2.3 The Summary and Comparison of QTL Mapping Results for Main Yield Traits Among Three Populations

A total of 175 QTLs s for spike-related traits were detected on 20 chromosomes except for 4D in one DH population and two RIL populations (Table 4.15). Among them, the QTLs for spike length were detected on chromosomes 2D and 6B in the three populations, so as the QTL for compactness on 2D. In population 2, the *QSl6B.2* and *QSc6B.1* for spike length were closely linked to marker *wPt669607*, so as *QSl6B.3-10* and *QSc6B.3-9(-10)* for compactness. The QTL located on 6A near marker *wPt0959* for spikelets per spike was detected in populations 2 and 3.

Fifty-four of 175 QTLs were main-effect QTLs with the more than 10 % of phenotypic contribution (Table 4.16). Four QTLs of *Qsl-1B* (E2), *Qsc-1B.1*(E2), *QSc-6B.1,* and *QKwps4B.1-99* (E3) were stably detected in all environments.

4.3 QTL Mapping for Wheat Tiller Number in Different Period Using a DH population and an Immortalized F_2 Population

Tiller is one of the most important agronomic traits in cereal crops because tiller number per plant determines the number of spikes or panicles per plant, a key component of grain yield and/or biomass. Tiller or the degree of branching

Table 4.15 The distribution of QTLs for spike-related traits detected in the three populations

Chromosome	Number of QTL							
	Grain number per spike	Spike length	Spikelets per spike	Compactness	Fertile spikelets per spike	Sterile spikelets per spike	Grain weight per spike	Grain yield
1A	–/2/–	–/2/–		–/2/–				
1B	–/–/1	–/1/5	1/–/1	–/2/2		–/–/1		
1D		–/–/1	–/–/1					
2A	–/1/1	–/1/2		–/–/2			–/–/2	
2B	–/1/–	1/1/–	–/–/2				–/–/1	
2D	1/2/–	1/1/2	–/1/1	1/1/1	–/–/2		–/–/1	1/–/–
3A		–/–/2	–/–/1	–/–/2		–/–/1		
3B	–/2/–			–/1/–	–/1/–	–/–/2		
3D	–/–/2	–/1–					–/–/2	
4A		–/3/1	1/–/1		–/–/2			1/–/–
4B	–/–/4	–/1/1	–/–/1		1/–/4	–/–/1	–/–/1	
4D	1/–/1	1/–/–		1/–/1		–/–/1	–/–/2	
5A		–/–/2		–/–/2				
5B	–/1/2		–/–/2		–/1/–	–/–/1		
5D	1/–/–	1/–/–	1/–/–	1/–/–	1/–/–			1/–/–
6A	–/3/–	–/2/–	–/2/1	–/4/4	–/3/3		–/–/5	
6B	–/1/–	1/5/3				–/–/1	–/–/1	
6D			–/–/1					
7A		–/–/3	1/–/–	–/1/2				
7B		–/–/2	–/–/1		–/–1		–/–/1	
7D								

For each entry, the three figures refer to population 1, population 2, and population 3, respectively

Table 4.16 The number of QTLs for spike-related traits detected in the three populations

Trait	Number of QTL												
	R^2 %[a]			No. of environments[b]				No. of environments + P[c]					Total
	<5 %	5–10 %	>10 %	1	2	3	4	0 + P	1 + P	2 + P	3 + P	4 + P	
Grain yield	1/–/–	0/–/–	2/–/–	–/–/–	–/–/–	–/–/–	–/–/–	–/–/–	–/–/–	–/–/–	–/–/–	–/–/–	3/–/–
Grain number per spike	0/4/0	1/4/9	2/5/3	–/6/10	–/0/1	–/0/0	–/–/0	–/7/1	–/0/1	–/0/0	–/0/0	–/–/0	3/13/12
Spike length	2/5/2	2/8/11	1/5/10	–/14/17	–/1/1	–/1/0	–/–/0	–/2/4	–/5/2	–/1/0	–/1/0	–/–/0	5/18/23
Spikelets per spike	3/3/0	0/0/11	1/0/2	–/3/12	–/0/0	–/0/0	–/–/0	–/0/1	–/0/0	–/0/0	–/0/0	–/–/0	4/3/13
Compactness	0/8/0	1/1/9	2/3/7	–/7/9	–/1/2	–/2/9	–/–/0	–/2/5	–/2/3	–/1/0	–/2/0	–/–/0	3/12/16
Fertile spikelets per spike	0/0/0	1/4/6	1/1/5	–/5/7	–/0/1	–/0/0	–/–/0	–/0/3	–/0/0	–/0/0	–/0/0	–/–/0	2/5/11
Sterile spikelets per spike	–/–/0	–/–/6	–/–/2	–/–/5	–/–/0	–/–/0	–/–/0	–/–/3	–/–/2	–/–/0	–/–/0	–/–/0	–/–/8
Grain weight per spike	–/–/5	–/–/7	–/–/4	–/–/10	–/–/0	–/–/0	–/–/1	–/–/5	–/–/0	–/–/0	–/–/0	–/–/1	–/–/16

[a]The number of QTL that accounted for 5, 5–10 or >10 % of the phenotypic variance
[b]The number of QTL that showed significance in 1, 2, 3, or 4 different environments
[c]The number of QTL that have been detected in 0, 1, 2, 3, or 4 different environments and in pooled data
For each entry, the three figures refer to population 1, population 2, and population 3, respectively

determines shoot architecture, which affects a plant's light harvesting potential, the synchrony of flowering and seed set, and, ultimately, the reproductive success of a plant. Tillers that grow from the main stem are called primary tillers, and those from primary tillers are secondary tillers. In practice, however, only a few tiller buds grow into a tiller, and only a proportion of these tillers survive to become the ultimate number of tillers, depending on tiller appearance and tiller survival. Tillers of different genotypes show various spatial orientations at different developmental stages, giving rise to morphologically distinct plant types. Before stem elongation, seedling growth habit (SGH) varies from prostrate to semi-prostrate to erect. After anthesis, spikes of the adult plant also differ in their compactness from very spreading to very compact. However, the genetic basis for tillering is not well elucidated. So in this study, we used the two populations, DH and IF_2, to dissect the genetics of tiller number in three different growth stages, which will be useful for developing further selection strategies in wheat breeding for tiller number.

4.3.1 Research Materials and Methods of Wheat Tiller Character

4.3.1.1 Plant Materials

The detailed descriptions of DH and IF_2 were seen in Sect. 4.1.1.1 population 1 and population 2.

4.3.1.2 Field Experiments Design

Field experiments were carried out at (Taian, Shandong Province) and (Jiyuan, Henan Province) in 2008. The field planting followed a randomized complete block design with the DH population and the IF_2 population in each trail. Each plot consisted of 3 rows: one row for a cross in the IF_2 population and two rows for each of its respective parents (DH lines). There were 20 plants in each row, with a distance of 10 cm between plants within each row and 25 cm between rows. The field management followed the local standard practices. Ten plants in the middle of the inner one row were chosen from each plot for trait evaluation, but plants in the plot border were not selected. When there were less than 10 plants in a plot in the IF_2, all plants were selected. Tiller number was measured as the number of effective tillers which arised from axillary at least 1.5 cm on every plant in maximum population of prewinter (MPW), maximum population in spring (MPS), and effective population in harvest (EPH). Trait measurements were averaged ten plants within each plot to statistical analyses.

4.3.1.3 Statistical Analysis

Analysis of variance (ANOVA) and correlation was performed using statistical software SPSS13.0. The heritability (h^2) was calculated as $h^2 = \left(\sigma_b^2 - \sigma_w^2\right)/\left(\sigma_b^2 + (r-1)\sigma_w^2\right)$, where σ_b^2 was between-group variance, σ_w^2 was within-group variance, and r was the number of observation. The estimates of σ_b^2, σ_w^2 were obtained from an ANOVA.

QTL analysis was performed separately for the DH and IF_2 populations. QTLs were mapped by inclusive composite interval mapping (ICIM). The LOD score for declaring a QTL was 2.0 for two populations, and the step of position was 1 cm. The mean QTLs were listed in our study. A QTL names are abbreviations of the trait followed by its respective linkage group number and the chromosome. An alphabetical letter a or b was added if more than one QTL was found in one linkage group.

4.3.2 QTL Mapping for Wheat Tiller Character in Different Period

4.3.2.1 Phenotypic Analysis

The measurements of tiller number in the three periods for both populations as well as two parents were listed in Table 4.17. Yumai 57 had higher phenotypic values than Huapei 3 for tiller number measured. The performance of the IF_2 population was bigger range of variation than those of the DH population in the three stages, but both had similar average which located in the middle of the two parents, which showed transgressive segregation and distribution in the IF_2 population and DH population. The number of tillers was more in Taian than in Jiyuanin at MPW, and MPS had further tillers than EPH in two populations. Both DH population and IF_2 population showed higher heritability from 0.77 to 0.95, which also indicating that environment had small effects on tiller number in the three periods.

4.3.2.2 QTL Mapping of MPW

Seven QTLs were identified for MPW, and both parents carried QTL alleles which increased phenotypic values (Tables 4.18 and 4.19). A major QTL (*QMpw5D*), contributing to this trait, which was detected in two environments in both DH population and IF_2 population, was mapped to the *Xwmc215–Xbarc345* interval on chromosome 5D (Fig. 4.2) and explained 14.18–26.89 % of the phenotypic variation. In the DH population, the QTL on 6A was detected in Taian (2008), but it was not significant in the IF_2 population. At the same time, the QTLs, distributed on

Table 4.17 Phenotypic summary of tiller number in maximum population of prewinter (MPW), maximum population in spring (MPS), and effective population in harvest (EPH) for Huapei 3 (P_1), Yumai 57 (P_2), and the DH population and the IF_2 population at Taian and Jiyuan in 2008

Tiller period	Parents		DH population				IF_2 population			
	Huapei 3	Yumai 57	Mean	Min	Max	h^2	Mean	Min	Max	h^2
MPW[a]	19	30	25	14	42	0.95	25	12	38	0.94
MPW[b]	17	19	18	11	32		18	5	30	
MPS[a]	20	27	27	13	44	0.77	26	11	38	0.81
MPS[b]	22	24	23	10	47		22	8	41	
EPH[a]	13	15	16	9	23	0.90	15	8	28	0.91
EPH[b]	19	22	13	7	22		13	5	24	

[a]Environment: 2008 Taian
[b]Environment: 2008 Jiyuan
Min minimum; *Max* maximum; h^2 heritability

chromosomes 1D, 4D, and 5A.2, were detected in the two trials in the IF_2 population, but it did not show significant effect in the DH population.

Its D/A ratio was near *QMpw5D which* showed overdominance effect in Taian (2008). But showed partial dominance in Jiyuan (2008), with the dominant allele originating from Yumai 57 and conferring lower tillers.

4.3.2.3 QTL Mapping of MPS

A total of ten regions were found to have effects on MPS (Tables 4.18 and 4.19). At these QTLs, Yumai 57 alleles increased tiller number and had additive effects of 0.22 and 3.55 which accounted for 5.53 and 34.96 % of the phenotypic variance.

Table 4.18 Putative QTL for tillering detected in the DH population through ICIM

Trait	QTL	Position	Interval	A	LOD	R^2 (%)[c]
MPW[a]	*QMpw6A*	9	*Xgwm459–Xgwm334*	−1.37	3.12	8.06
MPW[b]	*QMpw5D*	74	*Xwmc215–Xbarc345*	−2.00	7.67	23.19
MPS[a]	*QMps6D*	157	*Xswes679.1–Xcfa2129*	1.74	2.60	8.68
MPS[b]	*QMps4D*	125	*Xcfe188–Xbarc224*	−1.38	2.92	5.53
	QMps5D	73	*Xwmc215–Xbarc345*	−3.51	13.03	34.96
EPH[a]	*QEph5B.2*	28	*Xbarc232–Xwmc235*	0.76	2.60	10.91
EPH[b]	*QEph2B*	6	*Xwmc661–Xbarc200*	−0.58	2.15	4.49
	QEph3A	198	*Xbarc1177–Xbarc276.2*	−0.60	2.20	4.62
	QEph6A	43	*Xgwm82–Xwmc553*	−0.68	2.82	6.03

A Additive effect; positive additive effects indicate that the Huapei 3 allele increases the value of the trait
[a]Environment: 2008 Taian
[b]Environment: 2008 Jiyuan
[c]Proportion of the phenotypic variation explained by the QTL

Table 4.19 Putative QTL for tillering detected in the IF_2 population through ICIM

Trait	QTL	Position	Interval	A	D	LOD	R^2 (%)[c]	D/A[d]
MPW[a]	*QMpw5A.2*	11	*Xcwem32.2–Xwmc59*	2.06	0.67	3.28	8.21	PD
	QMpw5D	79	*Xwmc215–Xbarc345*	−1.92	−2.08	3.79	14.18	OD
MPW[b]	*QMpw1D*	48	*Xcfd19–Xwmc93*	−1.06	2.15	3.93	9.45	OD
	QMpw4D	1	*Xwmc473–Xwmc331*	0.61	2.62	4.53	7.97	OD
	QMpw5D	63	*Xbarc320–Xwmc215*	−2.97	0.37	14.01	26.89	PD
MPS[a]	*QMps2B*	94	*Xwmc445.2–Xgwm111*	−0.23	2.7	2.57	6.43	OD
	QMps5A.2	3	*Xcfe026.1–Xcwem32.2*	2.76	2.75	8.16	15.91	PD
	QMps5D	67	*Xbarc320–Xwmc215*	−2.29	−0.95	6.49	12.71	PD
	QMps6A	67	*Xwmc553–Xgwm732*	0.33	−3.26	2.03	9.51	OD
	QMps6D	64	*Xbarc054–Xgwm55*	−0.22	3.58	3.36	11.28	OD
MPS[b]	*QMps3A*	107	*Xcfa2134–Xwmc527*	0.85	−2.5	3.58	6.73	OD
	QMps5D	68	*Xbarc320–Xwmc215*	−3.55	−0.31	13.08	28.51	PD
EPH[a]	*QEph4D*	94	*Xgwm194–Xcfa2173*	1.75	−0.64	4.45	21.32	PD
	QEph6D	107	*Xgwm133.2–Xswes861.1*	4.67	−2.37	3.9	22.85	PD
	QEph6D	132	*Xswes679.1–Xcfa2129*	2.78	−3.68	3.31	16.28	OD
EPH[b]	*QEph2B*	74	*Xbarc1114–Xwmc175*	1.19	−0.14	3.97	9.12	PD
	QEph2D	0	*Xgwm261–Xwmc112*	0.13	1.14	2.11	4.52	OD
	QEph6A	43	*Xgwm82–Xwmc553*	−1.14	0.22	3.89	8.60	PD

A Additive effect, *D* dominance effect. Positive additive effects indicate that the Huapei 3 allele increases the value of the trait

[a]Environment: 2008 Taian

[b]Environment: 2008 Jiyuan

[c]Proportion of the phenotypic variation explained by the QTL

[d]Ratio of estimated dominant effect to the absolute value of additive effect. When a ratio larger than unity is regarded as overdominance, a ratio falling between 0 and 1 is regarded as partial dominance

QMps5D and *QMps6D* were identified in both DH and IF_2 populations, but *QMps5D* (DH) mapped to the *Xswes679.1–Xcfa2129* interval, while *QMps5D* (IF_2) were identified in the *Xbarc054–Xgwm55* interval; similarly, *QMps6D* (DH) were identified in the *Xwmc215–Xbarc345* interval, while *QMps6D* (IF_2) were identified in the *Xbarc320–Xwmc215* interval (Fig. 4.7). However, the remaining QTLs (*QMps2B, QMps3A, QMps4D et al*) were only detected once in DH or IF_2, likely because of this trait's higher sensitivity to environments.

Both *QMps5A.2* and *QMps5D* showed partial dominance, but dominant allele, respectively, originated from Huapei 3 and Yumai 57. *QMps3A* and *QMps6A* showed overdominance with Huapei 3 contributing the dominant allele, respectively, and both increased tillers.

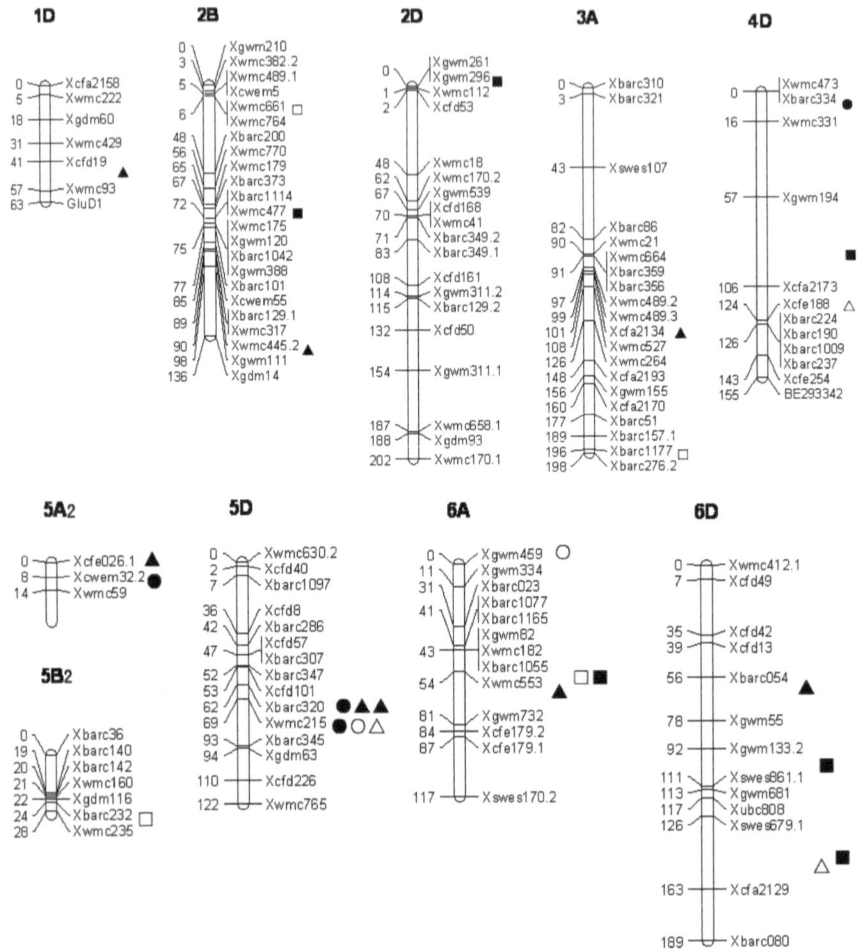

Fig. 4.2 Positions of QTLs associated with tillering at different period in the DH and IF_2 population derived from Huapei 3 × Yumai 57 by ICIM. ○ QTL for MPW in DH; ● QTL for MPW in IF_2; △ QTL for MPS in DH; ▲ QTL for MPS in IF_2 ; □ QTL for EPH in DH; ■ QTL for EPH in IF_2

4.3.2.4 QTL Mapping of EPH

Ten QTLs were detected for EPH. The QTL on chromosome 2B and 6A was both identified in the two populations; besides, *QEph6* was identified in the interval *Xgwm82–Xwmc553* in DH and IF_2, where again the allele had an additive effect of 0.68 and 1.14 on tillers increase which accounted for 6.03–8.60% of the phenotypic variance. Two QTLs (*QEph5B.2*, *QEph3A*) were detected in the DH population (Table 4.18), but neither of them was significant in the IF_2 population. On the contrary, four QTLs on 2D, 4D, and 6D (*QEph2D*, *QEph4D*, and *Qeph6D*) were detected in the IF_2 population (Table 4.19), but they were not detected in the DH population. The Huapei 3 allele of *QEph2D* and *QEph4D* increased tillers, and the former showed overdominance and the latter showed partially dominant.

4.3.3 Research Progress of Wheat Tiller Character QTL Mapping and Comparision of the Results with Previous Studies

4.3.3.1 Research Progress of Wheat Tiller Character QTL Mapping

Although a number of QTLs of tillering in rice, barley, and rye were available, only few studies had been carried out in wheat. Law (1967) discovered that the factor responsible for tiller number on chromosome 7B could be either the marker e_l acting pleiotropically on this character or a factor tightly linked to the marker. Shah et al. (1999) mapped a significant QTL (R^2 = 19.4 %) for tillering on chromosome arm *3AL*. Kato et al. (2000) detected three minor QTLs (*Vrn-A1*, *QTn.ocs-5A.1*, and *QTn.ocs-5A.2*) for tillering associated with the vernalization gene on chromosome 5A of wheat. Li et al. (2002) reported that QTLs with significant effect on tiller number per plant were found to be located on 1D, 2D, and 6A chromosomes in wheat with winter/spring growth characters. Of which, there was one QTL close to *Gli-A2* (Xpsr10) severely affecting the wheat tiller number. Huang et al. (2003) identified eight QTLs for tiller number on 1B, 2A, 2D, 3B, 4D, 5D, 6D, and 7A chromosomes using BC_2F_2 with explaining more than 9 % of the phenotypic variance. Kumar et al. (2007) detected the QTLs for tiller number per plant on 3BL, 4AL, and 6DL using two RILs. Deng et al. (2011) found one QTL, *Qsn.sdau-4B*, on 4BL controlling the maximum tiller number of spring with explaining 67.6 % of the phenotypic variance. Kuraparthy et al (2007) showed a tillering mutant, tiller inhibition (*tin3*) gene, which was placed on 3A and produced one main culm compared to the wild type with many tillers in the F_2 wheat population. Yang et al. (2013) found six QTLs for tiller numbers of prewinter on 2A, 2D, 5D, and 7A chromosomes, and seven QTLs for tiller numbers of spring on 1A, 2D, 4B, 5D, 7A, and 7D chromosomes with explaining from 1.97 to 32.60 % of the phenotypic variance.

4.3.3.2 Comparision of the Results with Previous Studies

Previous studies have ever identified the QTLs of tiller numbers on 1A, 1B, 1D, 2A, 2D, 3A, 3B, 4A, 4B, 4D, 5A, 5D, 6D, and 7A chromosomes. In our study, QTL on 2D and 6A were also detected at similar regions on chromosome arm 2DS and 6AS, and we also identified a QTL on chromosome 1D, but had a different region on 1D comparing with *Xmwg837–Xmwg337* (Li et al. 2002). In view of the above results, there were two QTLs on 3A and 5A.2 controlling tiller number that could be detected under different environments and in different populations. The former QTL was detected on the same chromosome by Shah et al. (1999) and Kuraparthy et al. (2007), and the latter QTL was detected on chromosome 5A.2 by Kato et al. (2000). Kato et al. (2000) also detected a QTL with significant effect on tiller number on 5A.1; Shah et al. (1999) mapped a significant QTL for tillering on chromosome arm 7B, and both Huang et al. (2003) and Narasimhamoorthy et al. (2006) detected a QTL on chromosome 3B; unfortunately, we failed to detect any QTL on chromosome 3B, 5A.1, and 7B in the DH and IF_2 populations by ICIM. Yang et al. (2013) also found the QTLs controlling tiller numbers of prewinter and spring on 5D chromosome with flanking markers between *wms 174 and wms 292.* Fortunately, we found some new QTLs, which affected tiller number in different period in the DH and/or IF_2 populations. For example, a major QTL on 5D chromosome, which was detected at MPW and MPS in two populations and explained 12.71–34.96 % of the phenotypic variation, played an important role to early developing tillers. A QTL on chromosome 6D was detected in MPS and EPH and had a great influence on later developing tillers. At the same time, some QTLs on 2B, 4D, and 5B.2 may affect the entire period of wheat tiller growth phases.

4.4 QTL Mapping for Biomass Yield, Grain Yield, and Straw Yield Using a DH Population

By 2020, wheat production will have to increase by 40 % in order to meet requirements with respect to human food and animal feed, which will mainly be accomplished by increasing yield (Pfeiffer et al. 2000). "Increasing the intensity of production in those ecosystems that lend themselves to sustainable intensification, while decreasing intensity of production in the more fragile ecosystems" may be the only way for agriculture to keep pace with increasing population growth (Borlaug and Dowswell 1997). Hence, future wheat improvement strategies must emphasize biomass yield (BY), grain yield (GY), and straw yield (SY) in concert. BY is an important trait for wheat improvement and for its contribution to the world's economy. GY is a particularly complex trait, being the end product of many processes in the plant, and, in consequence, is very environmentally dependent (Quarrie et al. 2005). GY forms one of the key economic drivers behind successful wheat production and is consequently a major target for wheat breeding programs (Kuchel et al. 2007). SY, an important by-product in the production of agricultural crops, is considered as a potentially considerable source of renewable energy supply with an estimated value of 47×10^{18} J worldwide (Lal et al. 2005). Recently, the straw of crops has received renewed attention, resulting in substituting straw for natural gas or marsh gas instead of simply burning it. The main benefits of this practice lie in the conservation of non-renewable resources and in the reduction of greenhouse gas emissions (Gabrielle and Gagnaire, 2008). Therefore, understanding the genetic architecture for the aerial part of BY, GY, and SY will be of importance to wheat improvement.

So in this study, the purpose of this research was to dissect QTLs for BY, GY, and SY using a wheat DH population comprising 168 lines and to understand the genetic basis for yield potential associated with BY, GY, and SY. The results will be of great significance for helping breeders to enhance the yield of wheat.

4.4.1 Materials and Methods

4.4.1.1 Materials

The detailed descriptions of DH were seen in Sect. 4.1.1.1 population 1.

4.4.1.2 Field Experiments and Linkage Map

Field experiments were conducted in Tai'an, Shandong Province (116° 36′E, 36° 57′N), and in Jiyuan, Henan Province (112° 36′E, 35° 05′N), in 2008. Field planting followed a randomized complete block design with the DH population and the parents in each trial. Each plot consisted of three rows. There were 20 plants in each row, with a distance of 10 cm between plants within each row and 25 cm between rows. The field management followed local standard practices. Ten plants in the middle of the inner two rows of each plot were individually harvested from the soil surface to measure BY per plant in terms of the total dry weight (g) of the entire plant, GY in terms of the total dry weight (g) of grains from the entire plant, and SY in terms of the total dry weight (g) of straws from the entire plant. Trait measurements were averaged over ten plants within each plot for determining differences in statistical analyses.

4.4.1.3 Statistical Analysis

Phenotypic and correlation analyses were conducted using SPSS version 13.0 (SPSS, Chicago, IL).

QTL analysis was performed using the software QTL Network 2.0 (Yang and Zhu 2005) based on the MLM. Composite interval analysis was undertaken using forward–backward stepwise multiple linear regression with a probability in and out of the model ranging from 0.05 to 0.5, and a window size set at 10 cM. Significant thresholds for QTL detection were calculated for each data set using 1000 permutations and a genome-wide error rate of 0.05 (significant). The final genetic model incorporated significant additive and epistatic effects as well as their environmental interactions. A significant QTL was identified if the phenotype was associated with a marker locus at $P < 0.005$.

4.4.2 *QTL Mapping for Biomass Yield, Grain Yield, and Straw Yield*

4.4.2.1 Phenotypic Variation and Correlations

The phenotypic variation among DH lines and the parents of BY, GY, and SY measured under two environments in 2008 are summarized in Table 4.20. HP3 and YM57 differed significantly in the measured traits, with phenotypic values of HP3 for BY, GY, and SY being much higher than those of YM57. Some lines had more extreme values than the parents in both environments, showing substantial transgressive segregation, although the average values of DH lines for those traits were intermediate between the parental values. In addition, the three traits showed considerable phenotypic variation and continuous distributions, indicating their quantitative nature. Both the skew and kurt of BY, GY, and SY were less than 1.0, implying polygenic inheritance and suitability of the data for QTL analysis. Harvest index (grain yield/biomass yield) of DH lines ranged from 0.32 to 0.54 in Tai'an and ranged from 0.34 to 0.53 in Jiyuan. Both parents had a stable harvest index in the two environments.

The correlations among BY, GY, and SY are shown in Table 4.21. The traits were significantly correlated with each other in both environments. BY showed significantly positive correlation with GY ($r_1^2 = 0.87^{**}$ and $r_2^2 = 0.88^{**}$, where subscripts 1 and 2 represent the environments in Tai'an and Jiyuan, respectively), BY showed significantly positive correlation with SY ($P \leq 0.01$), and GY showed significantly positive correlation with SY ($P \leq 0.01$). This showed that both GY and SY played important roles in BY.

4.4.2.2 QTL Mapping for BY

Four additive QTLs controlling BY were located on chromosomes 3A, 4B, 4D, and 5A2 in the two environments, accounting for 2.57–10.87 % of the phenotypic variation (Table 4.22; Fig. 4.3). Both parents carried QTL alleles that increased phenotypic values. A major QTL, *QBy4D*, with its allele originating from HP3, was mapped to the *Xbarc334–Xwmc331* interval on chromosome 4D and explained 10.87 % of the phenotypic variation. The QTL, *QBy3A*, originating from HP3, had positive effects on BY. However, the YM57 alleles of *QBy4B* and *QBy5A2* increased BY.

Table 4.20 Phenotypic summary of BY, GY, and SY for HP3, YM57, and the DH lines at Tai'an and Jiyuan in 2008

Environment	Traits	Parents		DH population						HI
		HP3	YM57	Means	Min	Max	SD	Skew	Kurt	
2008 in Tai'an	BY(g)	48.52	38.68	45.82	23.53	65.34	7.17	−0.14	−0.18	0.32–0.54
	GY(g)	21.55	15.55	19.69	9.48	28.70	3.71	−0.27	−0.48	
	SY(g)	26.97	23.13	26.14	14.05	41.04	4.34	0.34	0.58	
2008 in Henan	BY(g)	58.18	46.87	52.05	26.80	73.00	7.79	−0.28	−0.06	0.34–0.53
	GY(g)	26.43	19.48	22.33	9.55	30.63	3.89	−0.29	−0.12	
	SY(g)	31.75	27.39	29.72	17.25	46.75	4.78	0.45	0.90	

Note: *SD* standard deviation; *Min* minimum; *Max* maximum; *Skew* skewness; *Kurt* kurtosis; *HI* harvest index

Table 4.21 Coefficients of BY, GY, and SY at Tai'an and Jiyuan in 2008

Environment	2008 in Tai'an		2008 in Henan	
	BY	GY	BY	GY
GY	0.87**		0.88**	
SY	0.91**	0.58**	0.92**	0.61**

Note *Significant at 0.05 probability level
**Significant at 0.01 probability level

Table 4.22 Additive QTLs for BY, GY, and SY detected at Tai'an and Jiyuan in 2008

Traits	Chromosome	QTLs	Marker intervals	Position	*F* value	A^a	R^2 (%)[b]
BY	3A	*QBy3A*	*Xwmc264–Xcfa2193*	140.9	7.66	1.45	3.5
	4B	*QBy4B*	*Xwmc48–Xbarc1096*	18.4	14.05	−1.92	6.12
	4D	*QBy4D*	*Xbarc334–Xwmc331*	4.1	16.06	2.56	10.87
	5A2	*QBy5A2*	*Xcfe026.1–Xcwem32.2*	0	7.36	−1.24	2.57
GY	2D	*QGy2D*	*Xgwm539–Xcfd168*	68.4	9.32	0.88	5.26
	4B	*QGy4B*	*Xwmc48–Xbarc1096*	18.4	12.54	−0.77	4.01
	4D	*QGy4D*	*Xbarc334–Xwmc331*	2.1	15.65	1.14	8.74
	7B2	*QGy7B2*	*Xwmc273.1–Xcfd22.1*	10.7	8	−0.8	4.29
SY	2B	*QSy2B*	*Xbarc101–Xcwem55*	77.4	9.2	−0.98	3.88
	3A	*QSy3A*	*Xwmc264–Xcfa2193*	139.9	8.42	0.72	2.1
	4B	*QSy4B*	*Xwmc657–Xwmc48*	17.7	11.27	−1.55	9.76
	4D	*QSy4D*	*Xbarc334–Xwmc331*	3.1	11.68	1.24	6.22
	5A2	*QSy5A2*	*Xcfe026.1–Xcwem32.2*	0	8.99	0.2	2.89

Note [a]Additive effect; a positive value indicates that the allele from HP3 increases the trait value; a negative value indicates that the allele from YM57 increases the trait value
[b]Proportion of the phenotypic variation explained by the QTL

Three pairs of epistatic effects were detected for BY, which were located on chromosomes 2A/2D, 2A/4D, and 3A/4B, respectively (Table 4.23, Fig. 4.3). These QTLs explained 2.22–8.46 % of the phenotypic variation. The QTL pair *QBy2A-2/QBy4D* acted in favor of the parent and had the largest effect, contributing a BY of 2.25 and accounting for 8.46% of the phenotypic variance. The other two QTL pairs in favor of the recombinant type, *QBy2A-1/QBy2D* and *QBy3A-2/QBy4B*, had an effect of 1.15 and 1.81 and together explained 7.69 % of the phenotypic variation.

4.4.2.3 QTL Mapping for GY

Four QTLs with additive effects significantly influencing GY were located on chromosomes 2D, 4B, 4D, and 7B2 (Table 4.22; Fig. 4.3). All four QTLs could

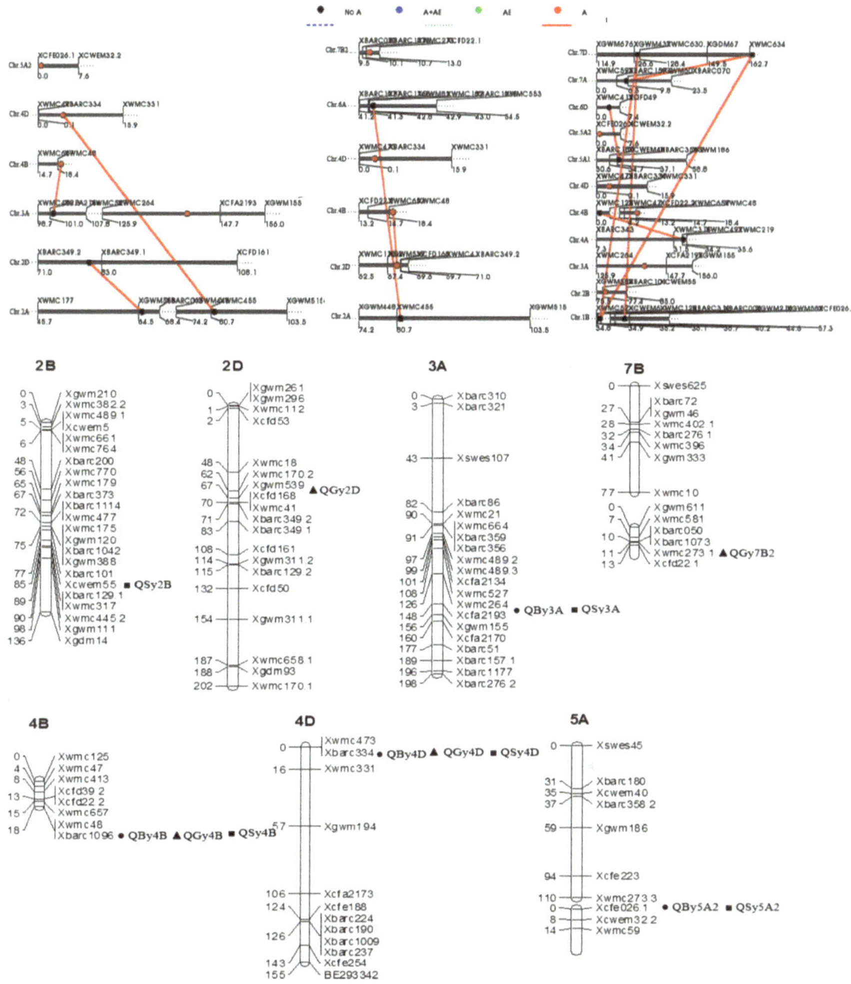

Fig. 4.3 Positions of additive QTLs and epistatic QTLs conferring BY, GY, and SY at Tai'an and Jiyuan in 2008. QTLs listed on the top were epistatic, and those below were additive, respectively. ● QTL for BY▲ QTL for GY ■ QTL for SY

account for 22.30 % of the phenotypic variation. The QTL *QGy4D*, with its allele originating from HP3, made the highest contribution explaining 8.74 % of the phenotypic variation. Two QTLs (*QGy4B* and *QGy7B2*) had negative effects on GY and were contributed by YM57 alleles, while the locus *QGy2D* had positive effects on GY and was transmitted by HP3 alleles. This suggested that alleles increasing GY were dispersed within the parents, resulting in small differences in phenotypic values between the parents and transgressive segregants among the DH lines.

Table 4.23 Additive × additive epistatic QTLs for BY, GY, and SY detected at Tai'an and Jiyuan in 2008

Traits	Chromosome	QTLs	Marker intervals	Position	Chromosome	QTLs	Marker intervals	Position	*AA*[a]	R^2 (%)[b]
BY	2A	*QBy2A-1*	*Xgwm558–Xbarc015*	64.5	2D	*QBy2D*	*Xbarc349.2–Xbarc349.1*	80.0	−1.15	2.22
	2A	*QBy2A-2*	*Xwmc455–Xgwm515*	80.7	4D	*QBy4D*	*Xbarc334–Xwmc331*	4.1	2.25	8.46
	3A	*QBy3A-2*	*Xcfa2134–Xwmc527*	101.0	4B	*QBy4B*	*Xwmc48–Xbarc1096*	18.4	−1.81	5.47
GY	2D	*QGy2D*	*Xgwm539–Xcfd168*	68.4	4B	*QGy4B*	*Xwmc48–Xbarc1096*	18.4	−0.70	3.34
	2A	*QGy3A*	*Xwmc455–Xgwm515*	80.7	6A	*QGy6A*	*Xbarc1055–Xwmc553*	43	0.88	5.28
SY	1B	*QSy1B-1*	*Xcwem6.1–Xwmc128*	34.9	7D	*QSy7D-1*	*Xwm437–Xwmc630.1*	126.6	1.54	9.53
	1B	*QSy1B-1*	Xcwem6.1-Xwmc128	34.9	7D	*QSy7D-2*	*Xgdm67–Xwmc634*	162.5	−0.66	1.78
	1B	*QSy1B-2*	*Xgwm218–xgwm582*	42.2	7D	*QSy7D-1*	*Xwm437–Xwmc630.1*	126.6	−0.75	2.27
	4A	*QSy4A*	*Xwmc313–Xwmc497*	33.5	4B	*QSy4B*	*Xwmc125–Xwmc47*	0.0	−1.05	4.48
	5A1	*QSy5A1*	*Xcwem40–Xbarc358.2*	36.7	6D	*QSy6D*	*Xwmc412.1–Xcfd49*	3.0	−1.22	4.99
	7A	*QSy7A*	*Xbarc157.2–Xgwm60*	8.3	7D	*QSy7D*	*Xgdm67–Xwmc634*	162.5	−0.50	1.00

Note [a]Additive × additive effect; a positive value indicates that the effect of the parents effect is larger than the recombinant effect, and a negative value means that the recombinant effect is larger than the parents' effect

[b]Proportion of the phenotypic variation explained by additive × additive QTL

Two pairs of epistatic interactions were common in both environments for GY (Table 4.23, Fig. 4.3). One interaction (*QGy3A/QGy6A*) acted to increase the values of the parental types, and the other (*QGy2D/QGy4B*) acted in the opposite direction; that is, recombinant effects were larger than parental effects. These QTLs accounted for 5.28 and 3.34 % of the phenotypic variance, respectively.

4.4.2.4 QTL Mapping for SY

Five regions on chromosomes 2B, 3A, 4B, 4D, and 5A2 associated with SY were identified in both Tai'an and Jiyuan (Table 4.22; Fig. 4.3). These loci explained 2.10–9.76 % of the phenotypic variation. Of these loci, two of the favorable alleles (*QSy2B* and *QSy4B*), deriving from YM57, decreased SY and had an additive effect of 0.98 and 1.55, respectively, whereas others (*QSy3A*, *QSy4D*, and *QSy5A2*), originating from HP3, increased SY. Furthermore, *QSy4D*, which was detected in the interval *Xbarc334–Xwmc331* had the most significant effect, accounting for 9.76 % of the phenotypic variance.

Six pairs of epistatic interactions on chromosomes 1B/7D, 4A/4B, 5A1/6D, and 7A/7D, respectively, were identified for SY (Table 4.23; Fig. 4.3). The three additive × additive epistatic interactions of *1B/7D* acted in favor of the parental and recombinant types, respectively. The three other epistatic interactions acted by increasing the values of the recombinant types, which explained 1.00–4.99 % of the phenotypic variation.

4.4.3 Research Progress of QTL Mapping for BY, GY, SY in Wheat and Comparision of the Results with Previous Studies

4.4.3.1 Research Progress of QTL Mapping for BY, GY, SY in Wheat

Genetic dissection of yield components can help to elucidate the physiological route from gene to phenotype for BY, GY, and SY (Kuchel et al. 2007). Ayala et al. (2002) discovered five QTLs for biomass on chromosomes 2B, 2D, 3B, 4D, and 6A under artificial inoculation with a BYDV-PAV-Mex isolate and under disease-free conditions in two wheat populations, Opata × Synthetic (ITMI) and Frontana × INIA66 (F × I). Zhang et al. (2004) detected four QTLs (*qBM-1-1, qBM-1-2, qBM-3,* and *qBM-5*) with R^2 values ranging from 13 to 28 % for biomass in a rice doubled haploid (DH) population. Liu et al. (2006) showed additive and additive by additive epistatic QTLs with significant effects on rice biomass yield and its two component traits (SY and GY), using a population of 125 DH lines from an inter-subspecific cross of IR64 and Azucena. Four QTLs and one pair of epistatic QTLs were also found to be responsible for the positive correlation between BY and SY. Kirigwi et al. (2007) mapped some QTLs for biomass production and biomass production rate on the proximal region of chromosome 4AL in wheat under differing soil moisture regime treatments. Anhalt et al. (2009) mapped QTLs contributing to BY traits, including dry weight and dry matter in an F_2 population consisting of 360 individual genotypes. A major QTL with additive effects was found on LG3 to be largely responsible for BY traits in *Lolium perenne* L. (perennial ryegrass).

4.4.3.2 Comparision of the Results with Previous Studies

In previous studies, many QTLs affecting yield have been reported on all chromosomes, with the exceptions of chromosomes 3D and 5D, and no significant gene by environment interactions was examined (Huang et al. 2006; Cuthbert et al. 2008; McIntyre et al. 2010). The most significant QTLs simultaneously identified for BY, GY, and SY in the current study were located on chromosomes 4B and 4D. In fact, chromosomes 4B and 4D are known to carry a number of major genes affecting plant height, yield productivity, and yield components (Huang et al. 2006). Therefore, the QTLs on chromosomes 4B and 4D should be considered to increase wheat biomass, grain, and straw in wheat molecular breeding.

Marza et al. (2006) and Quarrie et al. (2005) discovered that a number of major genes affecting yield productivity were located on chromosome 5AL in a similar position to chromosome 5A.2. In the present study, QTLs (*QBy5A2* and *QSy5A2*) on 5A2 were detected for BY and SY in the same regions for *Xcfe026.1–Xcwem32.2*. Quarrie et al. (2005) and Cuthbert et al. (2008) showed two grain yield QTL clusters on chromosomes 7A and 7B around the *Xwmc273* locus. We detected a QTL (*QBy7B*) close to *Xwmc273.1* on 7B2 with significant effects on GY, which had a similar interval to that reported by Quarrie et al. (2005). Unfortunately, however, we failed to detect any QTL on chromosome 7A. On chromosome 2D, a QTL that increased GY was detected using 402 DH lines from the spring wheat cross Superb (high yielding)/BW278 (low yielding) by Cuthbert et al. (2008). The QTL on chromosomes 2D affecting GY was also identified in our study. Zhang et al. (2009) also detected a QTL with significant effects on GY on chromosome 2D. At the same time, we found some new QTLs that affected both BY and SY. For example, a QTL detected on chromosome 3A had a great influence on increasing BY and SY in different environments and explained 3.50 and 2.10 % of the phenotypic variation, respectively. These results suggest that QTLs controlling BY, GY, and SY will certainly be helpful to improve biomass potential and future biofuel production through marker-assisted selection.

4.5 QTL Analysis of Heterosis for Number of Grains and Grain Weight Per Spike Using a DH population and an Immortalized F_2 Population

Heterosis, or hybrid vigor, is described as the phenomenon of an F_1, generated by crossing of two genetically different individuals, and is superior to either parent. Heterosis has contributed greatly to the production of high-yielding varieties in some crops during the past century. "Permanent F_2" performed well in identifying genetic locus and composition of heterosis. QTL mapping was made using the software ICIMapping. The heterosis loci were identified using the software Heterosis (Hua et al. 2003). Our group conducted QTL analysis of heterosis for grain-related traits in immortalized F_2 population, which provided reference for molecular marker breeding and methods for selecting hybrid wheat combination with strong heterosis.

4.5.1 QTL-Based Analysis of Heterosis for Grain Number Per Spike

4.5.1.1 Phenotypic Variance

Grain number per spike for the DH population and IF_2 population and the parents in the three environments are presented in Table 4.24. Yumai 57 had larger values than Huapei 3 for grain number per spike, with a difference of about two kernels. The two parents were stable in all three environments, suitable for building genetic groups for analysis of grain number per spike. The heterosis over mid-parent of F_1 had a range of 15.03–16.40 %, indicating that grain number per spike of F_1 hybrid exhibited strong heterosis.

The grain number per spike of the DH and IF_2 populations followed a normal distribution (Fig. 4.4), and absolute values of both skewness and kurtosis were less than 1.0, indicating polygenic inheritance and suitability of the data for QTL analysis (Cao et al. 2001). The F_1 hybrid and IF_2 population exhibited strong heterosis, suitable for HL analysis (Fig. 4.4).

4.5.1.2 QTLs for Grain Number Per Spike

In the DH and IF_2 populations, a total of 17 QTLs were associated with grain number per spike across the three different environments (Tables 4.25 and 4.26; Fig. 4.5). In the DH population, nine additive QTLs were detected on chromosomes 1A, 2B, 2D, 3B, 6A, and 7B, accounting for phenotypic variance ranging from 4.94 to 31.25 %. Eight QTLs in the IF_2 population were detected on chromosomes 1A, 2B, 3B, and 6A, explaining 4.78–46.75 % of phenotypic variance.

Table 4.24 Phenotypic performance of grain number per spike in DH and immortalized F_2 populations at three environments in 2009 and 2010 years

Env.	Parent		F_1		DH population		IF_2 population				Ht (%)	
	P1	P2	Mean	Ht (%)	Mean	Range	Mean	Range	Skewness	Kurtosis	Mean (%)	Range (%)
E1	41.75	44.14	49.40	15.03	42.07	29.50–56.00	45.90	32.33–59.00	−0.058	−0.512	6.87	−24.72 to 37.39
E2	41.00	42.33	48.50	16.40	42.74	30.00–57.20	43.93	28.00–58.00	0.088	0.365	5.20	−18.40 to 30.81
E3	39.05	41.22	46.33	15.44	40.85	29.00–56.60	42.00	28.40–52.80	−0.219	−0.487	5.19	−21.76 to 31.56

P1: Huapei 3, P2: Yumai 57, *Ht* heterosis over mid-parent, E1: Jiyuan site in 2009, E2: Jiyuan site in 2010, E3: Tai'an site in 2010

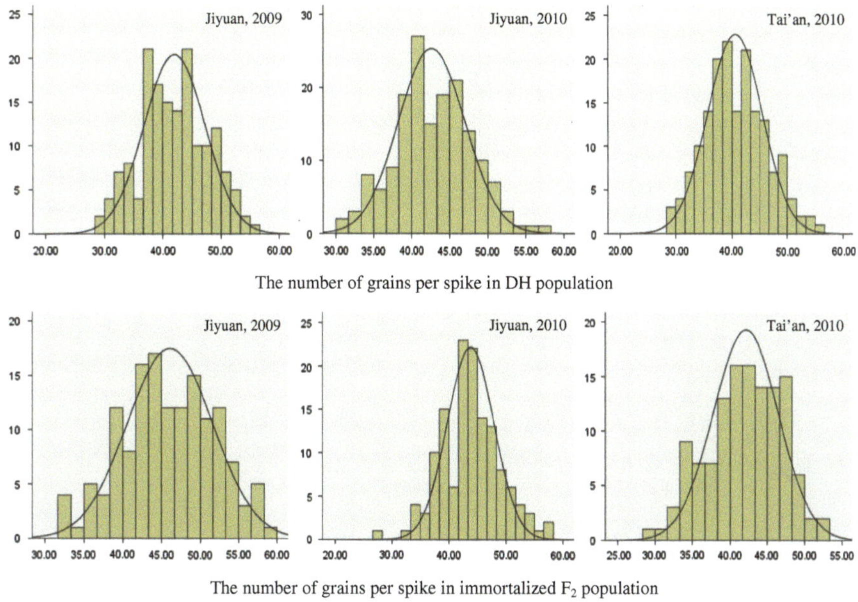

Fig. 4.4 Frequency distribution of grain number per spike in DH and IF_2 populations at three environments in 2009 and 2010 years

4.5.1.2.1 QTL Mapping for Grain Number Per Spike in the DH Population

Three QTLs were detected on chromosomes 1A, 2B, and 7B in E1, accounting for 31.25, 11.67, and 10.77 % of phenotypic variance, respectively (Table 4.25; Fig. 4.5). The *QGns1A-1* had the largest effect in the DH population, explaining

Table 4.25 QTL for number of grains per spike in DH population at three environments in 2009 and 2010 years

Environment	QTL	Position (cM)	Maker interval	*A*	LOD	PVE (%)
E1	*QGns1A-1*	60.0	*Xbarc350–Xwmc120*	3.56	12.74	31.25
	QGns2B-1	75.0	*Xwmc175–Xgwm388*	−2.15	5.01	11.67
	QGns7B₁	34.0	*Xwmc396–Xgwm333*	2.11	4.78	10.77
E2	*QGns1A-2*	64.0	*Xwmc120–Xgwm498*	1.41	3.66	8.58
	QGns2B	68.0	*Xbarc373–Xbarc1114*	−1.28	3.07	7.11
E3	*QGns1A-3*	0.0	*Xgwm259–Xcwem32.1*	−1.29	2.65	4.94
	QGns2D	114.0	*Xcfd161–Xgwm311.2*	1.92	5.82	11.58
	QGns3B	47.0	*Xwmc3–Xwmc1*	1.93	5.44	10.69
	QGns6A-1	41.0	*Xbarc023–Xbarc1077*	−1.29	2.77	5.25

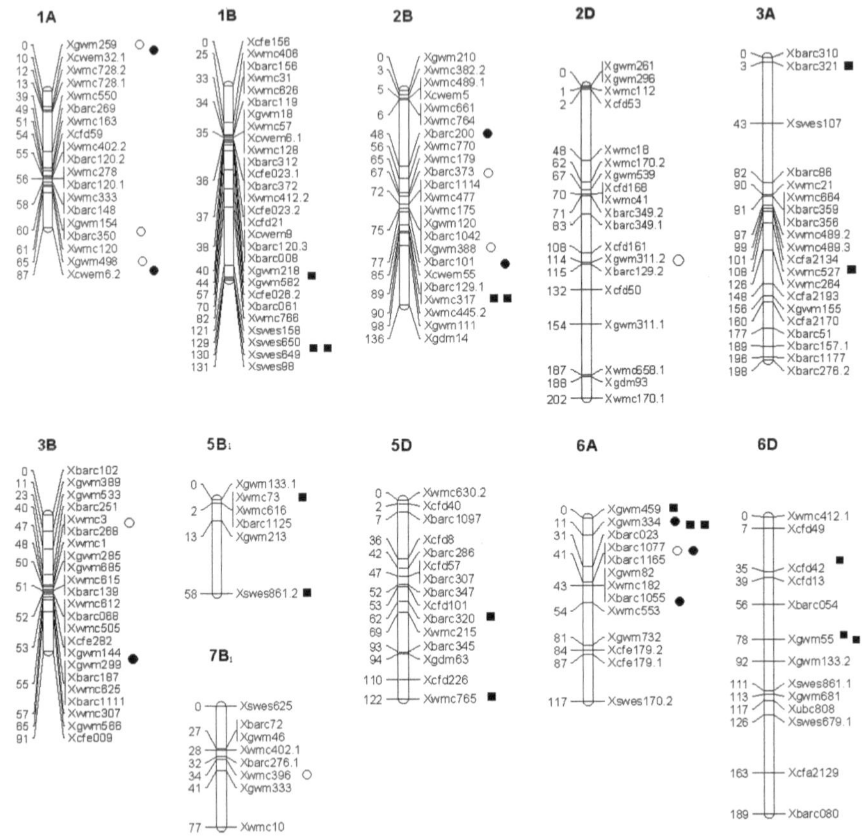

Fig. 4.5 Chromosomal location of heterotic loci (HL) and QTL for grain number per spike. ○ QTL for GNS in DH, ● QTL for GNS in IF_2, ■ HL for GNS in IF_2

31.25 % of variation, and the favorable allele was contributed by Huapei 3. In E2, two QTLs were detected on chromosomes 1A and 2B, accounting for 8.58 and 7.11 % of variance, respectively. The *QGns1A-2* was located on the same chromosome, close to *QGns1A-1*, which was detected in E1. The favorable allele was contributed by Huapei 3, increasing grain per spike by 1.41 kernels due to additive effects. In E3, four additive QTLs were observed on chromosomes 1A, 2D, 3B, and 6A, explaining phenotypic variation ranging from 4.94 to 11.58 %. Of these, two loci (*QGns2D, QGns3B*) were conferred by the allele of Huapei 3 and by the Yumai 57 allele at the other two loci.

4.5.1.2.2 QTL Mapping for Grain Number Per Spike in the IF_2 Population

Four, one, and three additive QTLs were detected in E1, E2, and E3, respectively (Table 4.26; Fig. 4.5). The *QGns2B-2* had the most significant additive effect, explaining 46.75 % of phenotypic variation. The favorable allele came from Yumai 57. In E2, the contribution of the QTL located on chromosome 2B was 6.77 %. In E1 and E3, two and one additive QTLs were observed on chromosome 6A, accounting for 12.29, 17.58, and 4.78 % of phenotypic variance, respectively. Of these eight QTLs, only *QGns1A* and *QGns6A* detected in E2 were contributed by Huapei 3.

4.5.1.2.3 QTL Detected in Both Populations Simultaneously

QGns1A-1, *QGns1A-2,* and *QGns1A-5* observed in the interval *Xbarc350–Xcwem6.2* with totally explaining 55.51 % of the variation. *QGns2B-1* and *QGns2B-2* are located on chromosome 2B, and they were major QTLs. The stable QTL *QGns6A-1* was identified in DH and IF_2 populations.

4.5.1.3 Heterosis Loci (HLs) of Grain Number Per Spike

A total of 17 HLs were identified on chromosomes 1B, 2B, 3A, 5D, 6A, and 6D using the modified composite interval mapping model, explaining from 2.78 to 12.18 % of phenotypic variation of heterosis (Table 4.27; Fig. 4.5). The HL *QHgns1B-2* detected in both E1 and E2 was mapped to the interval *Xswes650–Xswes649* on chromosome 1B, explaining 4.18 and 5.6 %, respectively. The *QHgns3A-1* had the most significant additive effects, explaining 12.18 % of heterosis phenotypic variation and increasing grain number per spike by 4.46 kernels. Three HLs were observed within the interval *Xgwm459–Xbarc023* on chromosome 6A, accounting for 10.83, 3.72, and 4.64 %, respectively. Of these, *QHgns6A-2* was detected in both E1 and E2, reducing grain number per spike by 2.91 and 2.59 kernels. The *QHgns6D-2* and *QHgns6D-3* were detected within the same marker interval *Xbarc054–Xgwm55* but at different positions, explaining 3.1 and 9.82 % of heterosis phenotypic variation, respectively. The *QHgns6D-1 and QHgns6D-2* reduced grain number per spike by 3.18 and 2.86 kernels. The *QHgns6D-3* increased it by 3.21 kernels due to additive effects. The *QHgns2B* was observed in both E2 and E3, accounting for 2.78 and 11.6 %, respectively. These HLs explained 52.62, 30.73, and 29.41 %, respectively, of the total heterosis phenotypic variation in three different environments.

Table 4.26 QTL for number of grains in immortalized F_2 population at three environments in 2009 and 2010 years

Environment	QTL	Position (cM)	Maker interval	A	Dominance	LOD	PVE (%)	Gene action
E1	*QGns1A-4*	7.0	*Xgwm259–Xcwem32.1*	−0.30	−3.08	2.86	6.01	OD
	QGns2B-2	78.0	*Xbarc101–Xcwem55*	−5.69	0.60	21.30	46.75	PD
	QGns6A-1	41.0	*Xbarc023–Xbarc1077*	−0.21	4.17	7.05	12.29	OD
	QGns6A	45.0	*Xbarc1055–Xwmc553*	−0.13	−5.06	7.61	17.58	OD
E2	*QGns2B*	48.0	*Xbarc200–Xwmc770*	−2.15	−0.81	2.75	6.77	PD
E3	*QGns1A-5*	77.0	*Xgwm498–Xcwem6.2*	0.07	−4.09	4.15	15.68	OD
	QGns3B	54.0	*Xgwm144–Xgwm299*	−0.54	3.39	5.16	9.09	OD
	QGns6A	11.0	*Xgwm334–Xbarc023*	1.71	0.04	2.86	4.78	PD

Table 4.27 Heterotic loci for grain number per spike in the immortalized F_2 population

Environment	HL	Position (cM)	Maker interval	LOD	Dominance	PVE (%)
E1	*QHgns1B-1*	42	*Xgwm218–Xgwm582*	7.73	4.49	9.36
	QHgns1B-2	129.0	*Xswes650–Xswes649*	5.73	2.56	4.18
	QHgns3A-1	3.3	*Xbarc310–Xbarc321*	6.28	4.46	12.18
	QHgns5D	61	*Xcfd101–Xbarc320*	5.57	2.35	3.51
	QHgns6A-1	0	*Xgwm459–Xgwm334*	9.55	4.16	10.83
	QHgns6A-2	15	*Xgwm334–Xbarc023*	5.59	−2.91	3.72
	QHgns6D-1	33	*Xcfd49–Xcfd42*	6.6	−3.18	5.74
	Qhgns6D-2	73	*Xbarc054–Xgwm55*	3.98	−2.86	3.1
E2	*QHgns1B-2*	129	*Xswes650–Xswes649*	3.67	−2.49	5.6
	QHgns2B	89	*Xwmc317–Xwmc445.2*	3.54	1.78	2.78
	QHgns3A-2	107	*Xcfa2134–Xwmc527*	5.34	2.34	4.43
	QHgns5B$_1$	1.6	*Xgwm133.1–Xwmc73*	4.25	2.95	8.13
	QHgns5B$_1$	58.2	*Xgwm213–Xswes861.2*	3.98	−3.39	9.79
E3	*QHgns2B*	89	*Xwmc317–Xwmc445.2*	6.44	−3.57	11.6
	QHgns5D	121.9	*Xcfd226–Xwmc765*	4.01	1.90	3.35
	QHgns6A-2	15	*Xgwm334–Xbarc023*	6.47	−2.59	4.64
	QHgns6D-3	77	*Xbarc054–Xgwm55*	7.8	3.21	9.82

4.5.2 QTL-Based Analysis of Heterosis for Grain Weight Per Spike

4.5.2.1 Phenotypic Analysis

The number of grains per spike for the DH population and IF_2 population and the parents in the three environments are presented in Table 4.28. Yumai 57 had larger values than Huapei 3 for grain weight per spike, and stable in all three environments, suitable for building genetic groups for analysis of grain weight per spike. The DH and IF_2 populations had a large range of variation, followed a normal distribution (Fig. 4.6), and the values of both skewness and kurtosis were small, indicating polygenic inheritance and suitability of the data for QTL analysis. The F_1 hybrid and IF_2 population exhibited strong heterosis, and they were suitable for HL analysis.

4.5.2.2 QTL Mapping for Grain Weight Per Spike

Ten additive QTLs for grain weight per spike were detected on chromosomes 1B, 2B, 2D, and 7B, accounting for 3.49–58.58 % of phenotypic variance (Tables 4.29 and 4.23; Fig. 4.7). The *QGws1B-2* had the largest effect in the population,

Table 4.28 Phenotypic performance of grain weight per spike (g) in DH and IF_2 populations in three environments

Env.	Parent		DH population		IF_2 population				Ht (%)	
	P1	P2	Mean	Range	Mean	Range	Skewness	Kurtosis	Mean (%)	Range (%)
E1	2.57	2.95	2.45	1.49–3.65	2.63	1.31–3.35	0.36	0.83	11.90	−37.45 to 67.00
E2	2.51	2.83	2.10	1.03–3.14	2.43	1.37–3.35	−0.32	−0.20	15.64	−21.34 to 58.27
E3	2.57	2.59	2.33	1.38–3.25	2.36	1.42–3.13	−0.32	−0.28	6.40	−27.24 to 45.63

P1: Huapei 3, P2: Yumai 57, *Ht* heterosis over mid-parent, E1: Jiyuan site in 2009, E2: Jiyuan site in 2010, E3: Tai'an site in 2010

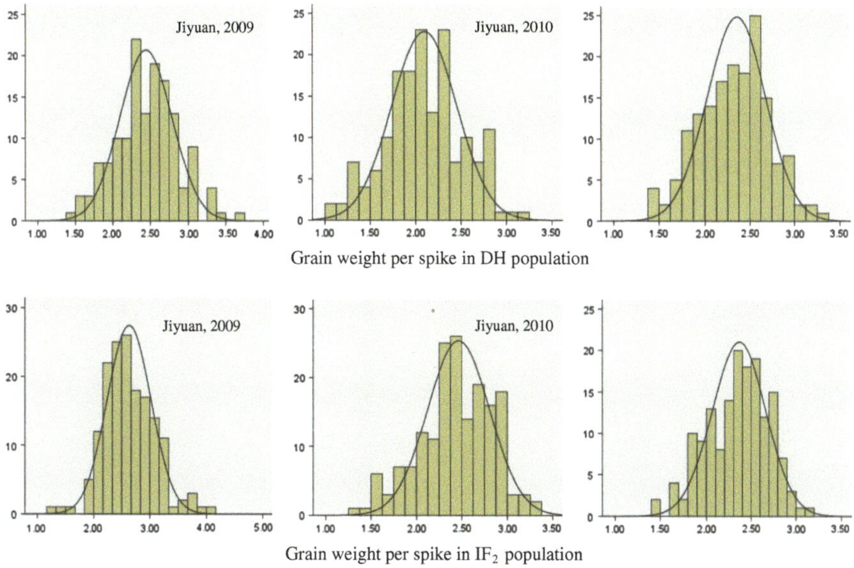

Fig. 4.6 Frequency distribution of grain weight per spike in DH and IF_2 populations at three environments

Table 4.29 Intervals, effects, and contributions of additive QTL for grain weight per spike in the DH population

Env.	QTL	Position (cM)	Maker interval	A	LOD	PVE (%)
E1	*QGws2D*	68	*Xgwm539–Xcfd168*	0.12	3.88	9.09
E2	*QGws2B*	76	*Xgwm388–Xbarc101*	−0.19	9.75	18.23
	QGws2D	69	*Xgwm539–Xcfd168*	0.12	4.72	7.77
	QGws7B$_1$	34	*Xwmc396–Xgwm333*	0.24	14.5	27.86
E3	*QGws2D*	68	*Xgwm539–Xcfd168*	0.09	2.50	5.81

explaining 58.58 % of variance. The favorable allele was contributed by Yumai 57. The *QGws2D* was detected in three environments in DH population, but not identified in the IF_2 population. Two additive QTLs were located on 2B, accounting for 18.23 and 3.49 % of phenotypic variance, respectively.

The *QGws7B$_1$* detected in two populations was located in the marker interval of *Xwmc396–Xgwm333* on chromosome $7B_1$, accounting for 27.86 and 21.01 % of phenotypic variance, respectively (Table 4.30).

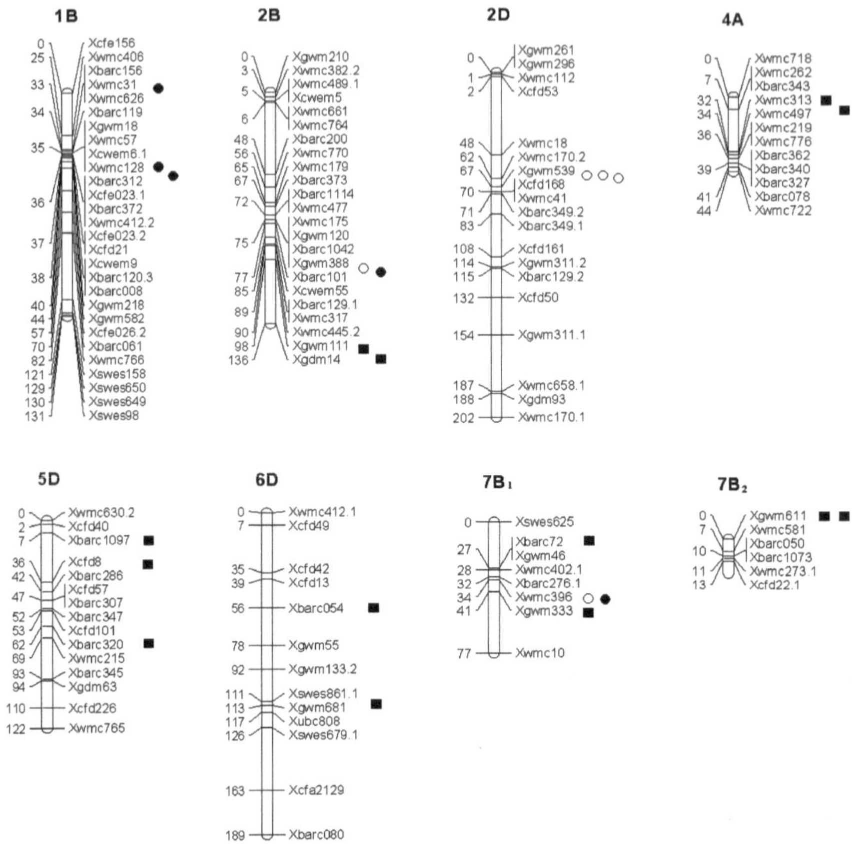

Fig. 4.7 Chromosomal location of heterotic loci (*HL*) and QTL for grain weight per spike (*GWS*) ○ QTL for GWS in DH, ● QTL for GWS in IF_2, ■ HL for GWS in IF_2

4.5.2.3 Heterosis Loci (HL) of Grain Weight Per Spike

A total of 13 HLs was detected in three environments, located on chromosomes 2B, 4A, 5D, 6D, $7B_1$, and $7B_2$, accounting for 2.4–26.0 % of grain weight per spike variance (Table 4.31). Three heterosis loci were located on chromosome 5D. The *QHgws5D-1* had the highest contribution ratio of 26.0 % and could increase 0.42 g of grain weight per spike. The *QHgws2B* and *QHgws7B₁* were detected in E1 and E3. The *QHgws2B* explained 2.9 and 10.8 % of phenotypic variance in two environments and decreased 0.24 and 0.34 g of grain weight per spike, while the *QHgws7B₁* explained 2.49 and 4.49 % of phenotypic variance. The *QHgws7B₂* was detected in E1 and E3, which was located in the marker interval of *Xgwm611–Xwmc581*, accounting for 2.4 and 7.49 % of phenotypic variance, respectively.

Table 4.30 Intervals, effects, and contributions of additive QTL for grain weight per spike in the IF_2 population

Env.	QTL	Position (cM)	Maker interval	Additive	Dominance	LOD	PVE (%)	D/A
E1	*QGws1B-1*	33	*Xwmc31–Xwmc626*	−0.43	−0.007	32.17	35.74	OD
	QGws1B-2	36	*Xwmc128–Xbarc312*	0.56	−0.01	39.96	58.58	OD
	QGws2B	78	*Xbarc101–Xcwem55*	−0.15	−0.02	4.34	3.49	OD
E2	*QGws1B*	35	*Xcwem6.1–Xwmc128*	−0.14	0.05	4.52	6.72	OD
	QGws7B₁	34	*Xwmc396–Xgwm333*	0.01	0.40	12.83	21.01	OD

Table 4.31 Heterotic loci for grain weight per spike in the IF_2 population

Environment	QTL	Maker interval	LOD	Dominance	PVE (%)
E1	*QHgws2B*	*Xgwm111–Xgdm14*	17.78	−0.24	2.9
	QHgws4A	*Xwmc313–Xwmc497*	4.63	0.25	2.9
	QHgws6D	*Xcfd13–Xbarc054*	3.79	0.30	4.41
	QHgws6D	*Xswes861.1–Xgwm681*	19.96	0.51	6.11
	QHgws7B$_1$	*Xbarc72–Xwmc402.1*	9.81	−0.28	2.49
	QHgws7B$_2$	*Xgwm611–Xwmc581*	13.76	−0.22	2.4
E2	*QHgws4A*	*Xwmc313–Xwmc497*	3.92	−0.22	5.79
	QHgws5D-1	*Xcfd40–Xbarc1097*	7.34	0.42	26.0
	QHgws5D	*Xcfd8–Xbarc286*	4.19	−0.34	9.55
	QHgws2B	*Xgwm111–Xgdm14*	4.32	−0.34	10.8
E3	*QHgws5D*	*Xcfd101–Xbarc320*	3.45	0.17	6.03
	QHgws7B$_1$	*Xgwm333–Xwmc10*	3.43	0.15	4.49
	QHgws7B$_2$	*Xgwm611–Xwmc581*	5.67	−0.20	7.49

4.6 QTL Analysis for Thousand-Grain Weight Using the High-Density Genetic Map

The current problem of much QTLs reported was its poor usability, which could be solved by precisely positioning the genes by high-density genetic maps. In this study, a RIL population of 173 lines derived from Shannong 01-35× Gaocheng 9411 was scanned by 90 k SNP chip and DArT microarray. The high-density genetic map with DArT, SSR, and SNP markers was conducted. QTL analysis for grain weight in four environments was conducted. These results could provide the basis for gene mapping, molecular marker development, and fine mapping of functional genes.

4.6.1 QTL Mapping for Thousand-Grain Weight

4.6.1.1 SNP and Genetic Map

A total of 9576 polymorphic markers were screened using the parents, including 9072 SNP, 59 SSR, 442 DArT, 2 gluten loci, and 1 *TaGW2*-CAPS.

According to the molecular information of Kansas University Wheat Genome and Triticarte Pty. Ltd, the high-density genetic map of 61 linkage groups was conducted, including 6241 polymorphic markers (other 3335 markers not within 61 linkage groups) of 6000 SNP (one *TaGW2*-CAPS), 216 DArT, and 25 SSR. The markers were located on 20 chromosomes except 4D, and the total length of chromosomes was 4825.29 cM, with the average distance between markers of 0.77 cM.

4.6.1.2 Phenotypic Variance

The average of thousand-grain weight for Shannong 01-35 in four environments was 61.45 g, significantly greater than 36.1 g of Gaocheng 9411, and the differences in four environments reached to 25.3 g. The variance coefficient was 13.57, 13.91, 10.82, and 11.15 % (Table 4.32). The thousand-grain weight showed a normal distribution and significant transgressive segregants (Fig. 4.8).

4.6.1.3 QTL Mapping

QTL mapping for thousand-grain weight in four environments and the mean were conducted by QTL Network 2.0. Nine additive-effect QTLs for thousand-grain weight were detected on chromosomes 1B, 4B, 5B, and 6A (Table 4.33). Nine QTLs were contributed by Shannong 01-35, and single QTL could increase 1.09–2.97 g grain weight. Among them, five QTLs, *QGW4B-17*(E3, E4, AE, ME), *QGW4B-5*(E1), *QGW4B-2*(E2), *QGW6A-344*(E3), and *QGW6A-137*(E4), were the main-effect QTLs with larger than 10 % contribution. The *QGW4B-17* was detected in E3, E4, AE, and ME. The genetic contribution was more than 16 %. The largest contribution of 33.3 % was detected in Suzhou (2010). The *QGW4B-5*(E1), *QGW4B-2*(E2), *QGW6A-344*(E3), and *QGW6A-137*(E4) could explain 12.2, 12.5, 12.5, and 10.1 % of phenotypic variance, respectively.

Table 4.32 Phenotypic values for thousand-grain weight (g) of the RIL population and the parents in different environments

Environment	Parent		RIL population				
	Shannong 01-35	Gaocheng 9411	Mean ± SD	Range	Variance coefficient (%)	Skewness	Kurtosis
2008 Tai'an	60.3	34.9	44.5 ± 6.0	22.9–66.0	13.57	−0.09	0.71
2009 Tai'an	64.1	38.0	43.1 ± 6.0	25.5–60.8	13.91	0.07	−0.35
2010 Tai'an	62.6	37.0	45.0 ± 4.9	31.1–65.4	10.82	−0.49	1.51
2010 Suzhou	58.8	34.9	43.7 ± 4.9	32.5–59.6	11.15	0.05	1.65
Average	61.5	36.2	44.1 ± 5. 5	26.0–62.9	12.36	−0.12	0.88

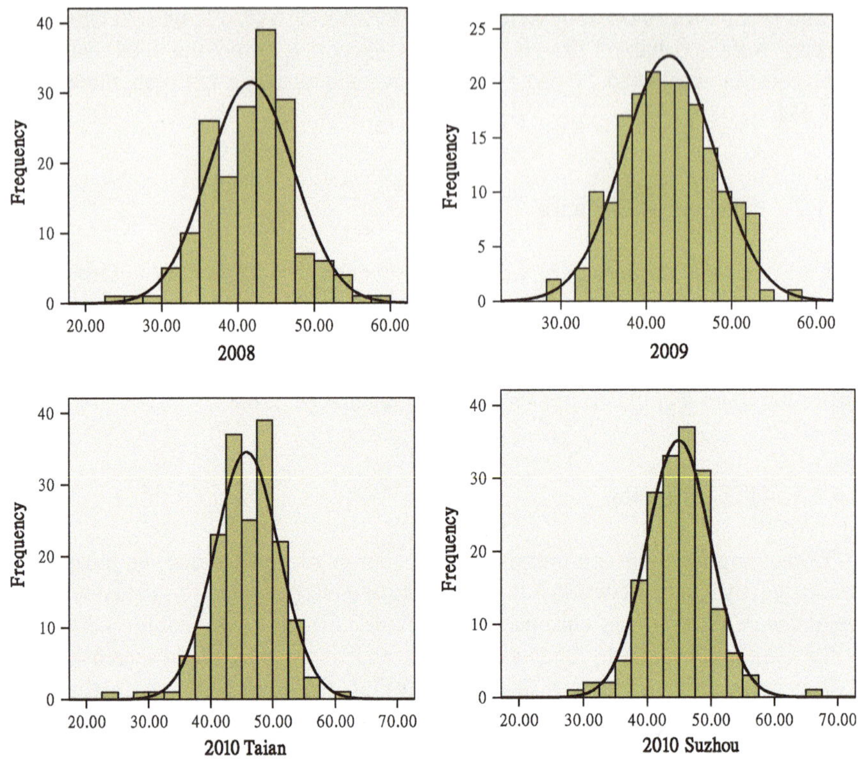

Fig. 4.8 Frequency of thousand-grain weight in RIL population

4.6.2 Research Progress of QTL Mapping for Thousand-Grain Weight and Comparision of the Results with Previous Studies

The QTL mapping for wheat thousand-grain weight has been gained more and more attention, which were detected on all chromosomes. However, few genes for thousand-grain weight in hexaploid wheat were identified, and the molecular-assisted selection markers were lack. Currently, the main bottleneck effect of marker-assisted selection for thousand-grain weight was as follows. Firstly, the most QTLs were not functional markers. Most of QTLs for thousand-grain weight were not validated by breeding. Secondly, the QTL from single mapping population was often with low positioning accuracy and large confidence intervals. The excessive or false QTLs reduced the molecular-assisted selection efficiency. The problems could be solved by fine positioning the gene for thousand-grain weight using high-density genetic maps.

Table 4.33 QTL analysis of thousand-grain weight in the RIL population

Environment	QTL	Chr	Maker interval	Position (cM)	Range	A[a]	*P* value	h^2 (a)[b] (%)
2008 Tai'an (E1)	*QGW4B-5*	4B	*WSNP_KU-TDURUM_CONTIG*	18.6	14.7–35.3	1.6518	0.000027	12.2
	QGW6A-232	6A	*RFL_CONTIG-TAGW2-CAPS*	100.6	97.8–115.6	1.8996	0.000002	9.6
2009 Tai'an (E2)	*QGW4B-2*	4B	*EXCALIBUR_C-CAP*	2.3	0.0–12.0	1.9835	0.000001	12.5
2010 Tai'an (E3)	*QGW4B-17*	4B	*EX_C101685-RAC875_C*	33.3	25.0–36.3	2.5307	0	24.1
	QGW6A-344	6A	*IAAV622-WSNP_EX_C*	165.7	159.0–176.1	1.6754	0.000001	12.5
2010 Suzhou (E4)	*QGW1B-26*	1B	*WPT-8682-TDURUM_CONTIG*	39.2	33.2–42.0	1.0935	0.000083	2.5
	QGW4B-17	4B	*EX_C101685-RAC875*	34.3	31.3–36.3	2.9718	0	33.3
	QGW5B-449	5B	*JD_C-WSNP_EX_C47152*	100.9	96.1–105.6	1.2714	0.000005	5.4
	QGW6A-137	6A	*IAAV1346-WSNP_EX_REP_C*	94.2	90.5–94.9	1.5677	0	10.1
AE	*QGW4B-17*	4B	*EX_C101685-RAC875_C*	34.3	23.0–36.3	2.2046	0	24.0
	QGW6A-136	6A	*TA004558-IAAV1346*	93.7	89.6–94.9	1.4503	0.000001	9.5
ME	*QGW4B -17*	4B	*EX_C101685-RAC875-c*	35.3	32.3–36.3	2.3022	0	16.5
	QGW5B -471	5B	*WSNP_KU-WPT-0103*	130.9	126.1–133.9	1.1883	0	3.6
	QGW6A -168	6A	*BS00012351-BS00089169*	98.5	97.1–101.6	1.6282	0	6.9

[a]Additive effects; positive value indicates that allele from 01-35 and negative value indicates that allele from 9411
[b]Percentage of phenotypic variation explained by QTL with additive effect

Wang et al. (2009a, b) identified QTLs for thousand-grain weight on chromosome 1B using RIL populations. Huang et al. (2004) found grain weight-related QTLs on chromosome 4B in Canadian wheat. Ding et al. (2011) detected QTLs on 4B, 5B, and 6A in two RIL populations. The nine additive-effect QTLs in our study was located on 1B, 4B, 5B, and 6A, which was consistent with the previous reports (Huang et al. 2004; Wang et al. 2008; Ding et al. 2011).

The *QGW6A-232* closely linked with thousand-grain weight was detected in the high-density genetic map, which was in line with the *TaGW2-CAPS*, which was related to rice grain weight and developed by Su et al. (2011). It indicated the genetic map had high-density markers and high-reliability between markers, closed to the fine mapping degree.

4.7 Association Mapping for Spike-Related Traits Using Wheat Backbone Parent "Aimengniu" Population

Association mapping or association analysis was the correlation analysis of markers and traits based on the linkage disequilibrium between alleles in different gene locus. Compared with linkage mapping, association mapping had the following advantages: (1) Association mapping do not need to build a mapping population, which uses the natural population, saving time, and effort. More populations are available for use (Zhang et al. 2006). (2) Association mapping is more accurate mapping, which uses the information in long-term evolution, with a high resolution rate, enabling the fine mapping for quantitative trait loci, evenly directly targeted to the gene. (3) Association mapping could simultaneously detect multiple alleles in a natural population. In recent years, with the improvement of genomics and biostatistical software, exploring quantitative trait loci using association mapping in natural populations have become the hot spots of the international plant genomics research.

4.7.1 Correlation Analysis for Spike-Related Phenotypic Traits

4.7.1.1 Phenotypic Variance

The eight traits of spikelets per spike, fertile spikelets per spike, sterile spikelets per spike, compactness, grain number per spike, grain weight per spike, and thousand-grain weight in six environments were shown in Table 4.34. Phenotypic traits among individuals were significantly different with rich variances, little differences between years and locations. The phenotypic variance explained by population structure was 2.79–11.49 %, and the broad-sense heritability was 55.88–98.05 %, indicating the traits were stability inherited.

Table 4.34 Phenotypic values for spike-related traits of the association population in different environments

Traits	Environment	Min	Max	Mean	SD	R^2 (%)[a]	H^2 (%)[b]
Spike length (cm)	E1: Tai'an, 2006	6.20	16.00	9.70	1.61	11.46	98.05
	E2: Tai'an, 2007	6.73	14.87	9.60	1.75	13.37	
	E3: Tai'an, 2008	6.63	16.00	9.67	1.66	15.36	
	E4: Tai'an, 2009	6.45	16.17	9.73	1.52	9.63	
	E5: Tai'an, 2010	6.60	14.00	9.02	1.39	9.50	
	E6: Suzhou, 2010	6.50	14.93	9.43	1.60	9.62	
Spikelets per spike	E2: Tai'an, 2007	15.3	22.00	18.67	1.86	3.44	96.55
	E3: Tai'an, 2008	15.33	24.00	19.24	1.64	1.95	
	E4: Tai'an, 2009	15.00	25.33	18.44	1.77	3.68	
	E5: Tai'an, 2010	16.00	25.67	19.26	1.95	4.69	
	E6: Suzhou, 2010	16.33	28.33	20.86	2.04	3.92	
Fertile spikelets per spike	E2: Tai'an, 2007	11.67	20.33	15.97	1.54	4.85	93.77
	E3: Tai'an, 2008	9.33	23.00	19.91	1.87	4.90	
	E4: Tai'an, 2009	15.00	24.67	18.19	1.71	4.91	
	E5: Tai'an, 2010	12.33	25.67	18.77	1.86	4.73	
	E6: Suzhou, 2010	14.00	23.67	18.17	1.89	4.68	
Sterile spikelets per spike	E2: Tai'an, 2007	0	4.00	1.84	0.77	2.43	55.88
	E3: Tai'an, 2008	0.33	5.67	2.19	1.11	4.05	
	E4: Tai'an, 2009	0	1.33	0.26	0.37	0.62	
	E5: Tai'an, 2010	0	4.00	1.58	0.82	2.62	
	E6: Suzhou, 2010	0	3.67	1.79	1.02	4.22	
Compactness	E2: Taian, 2007	126.28	263.54	191.33	29.23	6.65	92.02
	E3: Tai'an, 2008	133.31	236.76	212.64	30.89	9.0	
	E4: Tai'an, 2009	138.10	266.10	235.18	30.97	4.63	
	E5: Tai'an, 2010	100.00	258.06	206.25	26.72	6.31	
	E6: Suzhou, 2010	158.68	367.44	227.64	39.63	5.82	
Grain number per spike	E1: Tai'an, 2006	35.00	61.00	47.29	5.88	5.36	96.58
	E2: Tai'an, 2007	18.33	67.67	43.49	9.50	6.75	
	E3: Tai'an, 2008	11.67	68.67	41.29	8.83	10.00	
	E4: Tai'an, 2009	31.67	68.67	48.35	8.61	1.89	
	E5: Tai'an, 2010	27.67	66.33	42.87	7.61	4.19	
	E6: Suzhou, 2010	26.00	66.67	43.86	8.93	3.79	
Grain weight per spike (g)	E4: Tai'an, 2009	1.07	2.68	1.73	0.36	4.29	87.95
	E5: Tai'an, 2010	1.16	2.84	1.80	0.32	0.62	
	E6: Suzhou, 2010	1.01	2.72	1.64	0.33	3.79	
Thousand-grain weight (g)	E1: Tai'an, 2006	21.90	55.90	41.22	6.45	5.59	91.68
	E2: Tai'an, 2007	24.70	56.00	43.27	6.44	4.81	
	E3: Tai'an, 2008	23.10	60.80	45.35	6.74	6.33	
	E4: Tai'an, 2009	24.60	59.40	47.53	6.34	2.96	
	E5: Tai'an, 2010	35.42	61.77	50.58	5.58	7.49	
	E6: Suzhou, 2010	23.67	57.53	44.40	6.22	10.15	

[a]Percentage of phenotypic variation explained by population structure

[b]Broad-sense heritability (the same below)

Table 4.35 Coefficients of pairwise correlations of the mean values of agronomic traits

Trait	TGW	FSN	GNS	SL	CS	SSN	TSN
FSN	−0.287**						
GNS	−0.104	0.194*					
SL	−0.06	0.455**	0.024				
CS	0.066	−0.13	0.142	0.795**			
SSN	0.247**	−0.17	−0.05	0.177	−0.888		
TSN	−0.14	0.895**	0.172	0.524**	−0.156	0.274**	
GWS	0.344**	0.383**	0.197*	0.369**	−0.209*	−0.03	0.377**

*Correlation is significant at p < 0.01
**Correlation is significant at p < 0.01
TGW Thousand-grain weight; *FSN* fertile spikelets per spike; *GNS* grain number per spike; *SL* spike length; *CS* compactness; *SSN* sterile spikelets per spike; *TSN* spikelets per spike; *GWS* grain weight per spike

4.7.1.2 Correlation Analysis for Spike-Related Traits

The mean of spike-related traits in six environments was shown in Table 4.35. The spikelets per spike and fertile spikelets per spike had the highest positive correlation, with the coefficient 0.895**. Followed by spike length and compactness, the correlation coefficient was 0.795**.

4.7.2 Association Mapping for Spike-Related Traits

Association mapping for spike-related traits in six environments by DArT markers was conducted. A total of 132 DArT markers were detected in $P < 0.001$, distributed on 20 chromosomes except 4D. Seventy-two loci had the phenotypic variance of more than 10 % (Table 4.36).

Twenty-eight DArT markers for spike length were detected, which distributed on chromosomes 1D, 2A, 2D, 5B, 5D, 6A, 7A, 7B, and 7D. The range of R^2 was 7.3–17.95 %. No association in 3rd and 4rd environments was detected. The markers *wPt-9814* (5B, 69.7 cM) and *wPt-9814* (7B, 56.7 cM) were detected in two environments and had the highest R^2 value (17.95 %, $p < 0.0001$).

Fifteen associations for spikelets per spike in four environments were detected, with the range of R^2 4.1–14.37 %. The marker *wPt-7576* (6B, 73.6 cM) was identified in three environments, with the largest R^2 (14.37 %, $p < 0.0001$). The markers *wPt-3506* and *tPt-4887* were completely linkage disequilibrium with *wPt-7576* and detected in three environments.

Ten associations for fertile spikelets per spike in four environments were detected. The range of R^2 was 9.57–11.26 %. The marker *wPt-7576* (6B, 73.6 cM) was detected in two environments, with the R^2 of 9.57 and 11.07 %. The markers

Table 4.36 The marker loci associated with spike-related traits and corresponding explained phenotypic variation

Trait	Chromosome	Marker	Position (cM)	R^2 (%)					
				E1	E2	E3	E4	E5	E6
Spike length	1D	*wPt-9380*	49.2						7.36*
	2A	*wPt-7024*	84.6		8.98*				13.71**
		wPt-6711	–		8.84*				
		wPt-0568	–		7.58*				
		wPt-8132	–						
	2D	*wPt-9749*	4.9		11.94**				
		wPt-6847	59.5		8.33*				
		wPt-6704	–		9.1*				
		wPt-6850	–		8.27*				
	5B	*wPt-9666*	41.1		7.3*				8.19*
		wPt-9814	69.7						17.95**
		wPt-5737	69.9						15.67**
	5D	*wPt-2256*	–		8.63*				
	6A	*wPt-666494*	63.7	16.61*					
		wPt-668031	66.3	9.24*					
		wPt-4145	–	17.46					
	7A	*wPt-3425*	–		8.82*				
	7B	*wPt-5737*	56.6		7.3*			11.59*	15.67**
		wPt-9814	56.7						17.95**
		wPt-2878	–						8.3*
	7D	*wPt-663992*	9.4	13.76*	8.44*				
		wPt-663849	9.4	17.32*	8.36*				
		wPt-664384	52.6	9.57*					
		wPt-2592	84.7						
		wPt-4315	–						

(continued)

Table 4.36 (continued)

Trait	Chromosome	Marker	Position (cM)	R^2 (%)					
				E1	E2	E3	E4	E5	E6
Spikelets per spike	1A	*wPt-5160*	–						10.23*
		wPt-7147	–						10.27*
		wPt-6005	83.3						9.95*
	5D	*wPt-2256*	–						9.97*
	6B	*wPt-4858*	66.1				10.13*	11.1*	14.37**
		wPt-7576	73.6				10.01*	13.35**	13.51**
		tPt-3506	–				9.58*	13.41**	12.93**
		tPt-4887	–					12.31*	
	7A	*wPt-4617*	–			4.1**			
Fertile spikelets per spike	5A	*wPt-9094*	–					10.04*	
	5B	*wPt-5346*	62.5						9.62*
		wPt-665267	200.5						10.4*
	6B	*wPt-7576*	73.6				9.57*	11.07*	
		tPt-3506	–				10.24*	11.26*	
		tPt-4887	–				9.81*	9.95*	
	7A	*wPt-6417*	–		9.62*				
Sterile spikelets per spike	2D	*wPt-6574*	8.7		12.3*			11.16*	
		wPt-3728	9.5		9.15*			10.31*	
		wPt-6847	59.5						
		wPt-7466	–						
	3A	*wPt-734141*	87.4				10.45*		
	3B	*wPt-7961*	–				12.4*		
	4B	*wPt-6149*	–						10.78*
	5B	*wPt-5346*	62.5				11.25*		
		wPt-665267	200.5				7.81*		
	6A	*wPt-9474*	136.1		11.72*				
	7D	*wPt-667894*	15.1						13.49*
		wPt-663971	15.1						13.45*

(continued)

Table 4.36 (continued)

Trait	Chromosome	Marker	Position (cM)	R^2 (%)					
				E1	E2	E3	E4	E5	E6
Compactness	1D	*wPt-3738*	36.8		9.79*		13.33*		
		wPt-669498	230.7						
	2A	*wPt-6711*	–		8.89*				
	2D	*wPt-6704*	–		9.13*			9.6*	
		wPt-9749	4.9		12.69**				
	3A	*wPt-0398*	146.4		8.88*			13.1*	
		wPt-666832	191.1					11.89*	
		wPt-9160	–						
	3D	*wPt-4569*	139.5					9.58*	
	5B	*wPt-9814*	69.7						11.12*
		wPt-5737	69.9						10.86*
	6A	*wPt-8266*	67		8.76*		11.29*		
		wPt-0139	–						
	7B	*wPt-5737*	56.6						10.86*
		wPt-9814	56.7						11.12*
	7D	*wPt-3268*	–		13.42**				
Grain number per spike	1B	*wPt-8240*				10.2*			
	2D	*wPt-4527*	203.3	18.51*					
	4B	*wPt-1708*	93.8			11.74*			
		wPt-5265	0			1.1**			
	6B	*wPt-1089*	39			9.5*			
	6D	*wPt-9863*	–				9.83*		
	7A	*wPt-6417*	–			10.49*		10.32*	
	7B	*wPt-0547*	–			9.09*			
		wPt-9515	–			8.04*			
Grain weight per spike	3A	*wPt-0398*	146.4				12.13*		
		wPt-9160	–				13.35*		
	4A	*wPt-6900*	–						10.15*
	6B	*wPt-4893*	–						9.23*
	7D	*wPt-2565*	1.3						12.59*

(continued)

Table 4.36 (continued)

Trait	Chromosome	Marker	Position (cM)	R^2 (%)					
				E1	E2	E3	E4	E5	E6
Thousand-grain weight	1A	*wPt-4134*	–				10.74*		9.26*
		wPt-7241	–				8.09*		9.26*
		wPt-8455	27.1						
		wPt-669484	143.9						
	1B	*wPt-2526*	76.3					9.67*	
	2A	*wPt-9277*	53		13.44*		9.1*		
		wPt-6775	283				9.69*		
		wPt-1142	–						
	2B	*wPt-2425*	206.7				11.22*		
		wPt-4368	–				10.38*		
		wPt-5736	–				11.26*		
	4A	*wPt-6867*	–				11.12*		
	5B	*wPt-1302*	16.3		8.48*		10.72*	15.53**	9.26*
		wPt-9613	94.4						8.74*
		wPt-7237	–						
		wPt-1548	–						
	6A	*wPt-9679*	26.1			7.78*			
		wPt-1695	66.5			9.71*			
		wPt-7840	–			8.61*			
	6B	*wPt-4858*	66.1		10.62*				10.6*
		wPt-7576	73.6		11.87*				12.27*
		tPt-4887	–		11.11*				11.92*
		tPt-3506	–						11.26*
	6D	*wPt-672044*	134.6			8.91*	8.78*		
		wPt-664719	134.9			9.37*			
		wPt-2864	–						
	7B	*wPt-8598*	142.4			12.34*			9.26*
		wPt-8981	–			8.05*			9.26*
		wPt-9665	–						9.26*
		wPt-7471	–						
		wPt-9904	–						

*P <0.001
**P <0.0001

wPt-3506 and *tPt-4887* were completely linkage disequilibrium with *wPt-7576* and detected in two environments, too.

Twelve associations for sterile spikelets per spike in four environments were detected, ranging from R^2 7.81 to 13.49 %. The markers *wPt-667894* and *wPt-663971* (7D, 15.1 cM) had the R^2 of 13.49 and 13.45 %, respectively.

Seventeen loci for compactness in four environments were detected, which were located on 1D, 2A, 2D, 3A, 3D, 5B, 6A, 7B, and 7D, ranging of R^2 from 8.76 to 13.42 %. The marker *wPt-3268* (7D) had the largest R^2 ($p < 0.0001$). The unmapped *wPt-3268* (7D) was detected in two environments with the R^2 of 9.13 and 9.6 % ($p < 0.001$), respectively.

Ten markers for grain number per spike in four environments were located. The R^2 is ranged from 8.04 to 18.51 %. The most loci were detected in the 3rd environment. The marker *wPt-4527* (7D, 203.3 cM) had the largest contribution. The unmapped *wPt-6417* was detected in two environments with the R^2 of 10.49 and 10.32 %, respectively.

Five associations for grain weight per spike in two environments were detected, with R^2 ranging from 9.23 to 13.35 %. The marker *wPt-9160* had the largest genetic contribution, which was strongly linkage disequilibrium ($r^2 = 0.94$) with *wPt-0398* (3A, 146.4 cM).

Thirty-five DArT markers for thousand-grain weight in five environments were detected. The R^2 ranged from 7.78 to 15.53 %. The markers *wPt-1305* (5B, 16.3 cM) and *wPt-7576* (6B, 73.6 cM) were identified in two environments with high R^2.

4.8 Research Progress of QTL Mapping for Wheat Grain Yield Trait and Comparision of the Results with Previous Studies

4.8.1 Overview of QTL Mapping for Wheat Yield-Related Traits

The yield-related traits were quantitative traits with complex genetic mechanism. Detecting the quantitative traits was reliable on the basis of high-density genetic maps. In current, QTLs for wheat yield-related traits were as follows.

4.8.1.1 QTL Mapping for Spike-Related Traits

The spike-related traits included spike length, grain number per spike, spikelets, fertile spikelets, sterile spikelets, grain weight, compactness, and so on. Previous studies have found that the QTLs were distributed on the chromosomes 5A, 2D, and 3D (Table 4.37). QTLs for spikelets per spike and grains per spike were near the centromere on 5B (Mirura et al. 1992, 1994). The QTLs for spikelets and grain weight per spike were detected on 1BS, 4A, 7D, and 5A (Schlegel et al. 1994; Araki et al. 1999; Kato et al. 2000).

Börner et al. (2002) detected four main-effect QTLs for spike length on 1BS, 4AS, 4AL, and 5AL in 11 environments using a RIL population (114 lines). Two main QTLs for grains per spike were located on 2DS and 4AL. Three QTLs for grain weight per spike were mapped on 2DS, 4AL, and 5AL. Huang et al. (2004) identified 8 QTLs for grains per spike on chromosomes 1D, 2A, 3D, 6A, 7A, and 7D using a BC_2F_1 population.

Zhou et al. (2006) constructed a RIL population including 104 lines using Wangshuibai and Alondra and detected 8 QTLs for grain number per main spike in 3 environments (on chromosomes 1B, 1D, 3B, 4A, 5D, and 6B) and 11 QTLs for grain number per spike (on chromosomes 1B, 1D, 2A, 2B, 3B, 4A, 5D, 6B, 7A). Ma et al. (2007) identified 61 QTLs for spike length, spikelets per spike, fertile spikelets per spike, sterile spikelets per spike, and compactness using a RIL and a immortalized F_2 population. Li et al. (2007) detected 7 QTLs for grains per spike on chromosomes 1D, 2A, 3B, 6A, and 6B and 5 QTLs for sterile spikelets per spike on chromosomes 1A, 4A, 6B, 7A, and 7D. Chu et al. (2008) detected 3 QTLs for spike length on chromosomes 3DS, 4AL, and 5AL using a DH population and 2 QTLs for compactness on chromosomes 5AL and 5BL. Wang et al. (2010) detected QTLs for spike length, spikelets per spike, and grains per spike on chromosomes 4AL, 5AL, 7BL, and 6AL.

4.8.1.2 QTL Mapping for Grain-Related Traits

Grain-related traits were also quantitative traits controlled by multigenes, affected by both genotype and environments. The grain-related traits included thousand-grain weight, volume weight, grain length, grain width, and grain thickness. To date, researchers made a few QTL mapping for thousand-grain weight and grain size using different populations (Table 4.37).

Sun et al. (2009) conducted 6 QTLs for grain length, 3 QTLs for grain width, 4 QTLs for grain weight, and 7 QTLs for volume weight using a RIL population derived from "Chuan35050 × Shannong483," and the phenotypic variance ranged from 5.9 to 26.4 %. Ramya et al. (2010) detected 10 QTLs for grain weight on chromosomes 1A, 1D, 2B, 2D, 4B, 5B, and 6B with 4.15–15.53 % of phenotypic variance; single QTL explained 4.15–15.53 % of phenotypic variance; 6 QTLs for grain length on chromosomes 1A, 2B, 2D, 5A, 5B, and 5D with 4.36–10.6 % of phenotypic variance; and 9 QTLs for grain width on chromosomes 1D, 2B, 2D, 4B, 5B, and 5D with 4.42–11.54 % of phenotypic variance using a 185 lines of RIL population derived from Rye Selection 111 × Chinese Spring. Three traits were simultaneously controlled by genes on chromosomes 2B, 2D, and 5B. Wang et al. (2009a, b) identified 17, 16, 18, and 21 and 21 QTLs for grain length, width, thickness, and thousand-grain weight, respectively. Li et al. (2007) identified 9 QTLs for grain weight using a RIL population, a QTL cluster of 8 QTLs on chromosomes

Table 4.37 Summary of QTLs for spike-related traits previously published

Trait	QTL	Chr.	Marker	PVE (%)	Mapping population	References
Spike length	*QSpl.nau-1A*	1A	*Xzmh161–Xbarc28.2*	10.6	Wangshuibai/Nanda2419 RIL	Ma et al. (2007)
	QSpl.nau-1A	1A	*Xzmh161–Xbarc28.2e*	11.3	Wangshuibai/Nanda2419 IF2	Ma et al. (2007)
	QEl	1BS	*Xbcd338*	15	ITMI RIL	Jantasuriyarat et al. (2004)
	QEl	1BS	*Xbcd12*	18	ITMI RIL	Jantasuriyarat et al. (2004)
	QSl.ccsu-2B.5	2BL	*Xwmc272–Xwmc474*	17.01	WL711/PH132 RIL	Kumar et al. (2007)
	QSps.ccsu-2B.2	2BS	*Xgwm1128–Xgwm682*	15.81	WL711/PH132 RIL	Kumar et al. (2007)
	QSpl.nau-2D	2D	*Xgwm261–XRPP5*	20	Wangshuibai/Nanda2419 IF_2	Ma et al. (2007)
	QSl.wa -2DS.e1	2D	*Xwmc112–Xwmc503*	13.61	3228/Jing 4839 $F_{2:3}$	Wang et al. (2010)
	QSl.ccsu-2D.5	2DL	*Xgwm349–Xgwm382*	11.36	WL711/PH132 RIL	Kumar et al. (2007)
	QSpl.nau-4A	4A	*Xcfd2–Xmag1353d*	11.8	Wangshuibai/Nanda2419 RIL	Ma et al. (2007)
	QSl.wa -4AL.e1	4A	*Xbarc170–Xwmc707*	2.13	3228/Jing 4839 $F_{2:3}$	Wang et al. (2010)
	QSl.wa -4AL.e3	4A	*Xbarc1047–Xwmc718*	11.89	3228/Jing 4839 $F_{2:3}$	Wang et al. (2010)
	QEl	4AL	*Xfbb154*	22	ITMI RIL	Jantasuriyarat et al. (2004)
	QEl	4AL	*Xcdo545*	11	ITMI RIL	Jantasuriyarat et al. (2004)
	QSps.ccsu-4A.5	4AL	*Xgwm959–Xgwm832*	12.21	WL711/PH132 RIL	Kumar et al. (2007)
	QEl.fcu-4A	4AL	*Xbarc26–Xbarc236*	11.14	TA4152-60/ND495 DH	Chu et al. (2008)
	QEl	4DL	*Xbcd1117*	11	ITMI RIL	Jantasuriyarat et al. (2004)
	QSpl.nau-5A	5A	*Xwmc96–Xgwm304*	13.4	Wangshuibai/Nanda2419 IF2	Ma et al. (2007)
	QEl.fcu-5A	5AL	*Xwmc96–Xgwm410.1*	19	TA4152-60/ND495 DH	Chu et al. (2008)
	QSpl.nau-5B	5B	*Xmag959–Xwmc421.3*	10.2	Wangshuibai/Nanda2419 RIL	Ma et al. (2007)
	QSps.ccsu-6A.3	6AL	*Xgwm570–Xgwm1040*	12.46	WL711/PH132 RIL	Kumar et al. (2007)
	QSl.wa -6BL.e2	6B	*Xbarc24–Xbarc361*	11.09	3228/Jing 4839 $F_{2:3}$	Wang et al. (2010)
	QEl	7AS	*Xabc158*	18	ITMI RIL	Jantasuriyarat et al. (2004)
	QSpl.nau-7D	7D	*Xcfd46–Xwmc702*	29.7	Wangshuibai/Nanda2419 RIL	Ma et al. (2007)
	QSpl.nau-7D	7D	*Xcfd46–Xwmc702*	31.4	Wangshuibai/Nanda2419 IF2	Ma et al. (2007)

(continued)

Table 4.37 (continued)

Trait	QTL	Chr.	Marker	PVE (%)	Mapping population	References
Compactness	*QCpt.nau-2D*	2D	*Xgwm261–XRPP5*	11.7	Wangshuibai/Nanda2419 RIL	Ma et al. (2007)
	QCpt.nau-2D	2D	*Xgwm261–XRPP5c*	23.2	Wangshuibai/Nanda2419 IF_2	Ma et al. (2007)
	QCpt.nau-4A	4A	*Xcfd2–Xmag1353*	15.9	Wangshuibai/Nanda2419 RIL	Ma et al. (2007)
	QCpt.nau-4A	4A	*Xcfd2–Xmag1353*	18.1	Wangshuibai/Nanda2419 RIL	Ma et al. (2007)
	QCpt	4AL	*Xfbb114*	17	ITMI RIL	Jantasuriyarat et al. (2004)
	QCpt.nau-5A2	5A	*Xmag1281–Xwmc96*	14.9	Wangshuibai/Nanda2419 IF_2	Ma et al. (2007)
	QCpt	5AL	*Xrz395*	14	ITMI RIL	Jantasuriyarat et al. (2004)
	QCpt.nau-5B1	5B	*Xmag959–Xwmc421.3*	10.5	Wangshuibai/Nanda2419 RIL	Ma et al. (2007)
	QCpt.nau-5B2	5B	*Xwmc28–Xbarc59*	10.8	Wangshuibai/Nanda2419 RIL	Ma et al. (2007)
	QCpt.nau-5B2	5B	*Xwmc28–Xbarc59*	15.5	Wangshuibai/Nanda2419 IF_2	Ma et al. (2007)
	QEc.fcu-5B	5BL	*Xfcp593–Xbarc142*	13	TA4152-60/ND495 DH	Chu et al. (2008)
	QCpt	6AS	*Xcdo1428*	25	ITMI RIL	Jantasuriyarat et al. (2004)
	QCpt	6AS	*Xgwm494*	34	ITMI RIL	Jantasuriyarat et al. (2004)
	QCpt.nau-7D	7D	*Xcfd46–Xwmc702c*	11.3	Wangshuibai/Nanda2419 RIL	Ma et al. (2007)
	QCpt.nau-7D	7D	*Xcfd46–Xwmc702c*	15.5	Wangshuibai/Nanda2419 RIL	Ma et al. (2007)
	QCpt.nau-7D	7D	*Xcfd46–Xwmc702e*	13.5	Wangshuibai/Nanda2419 IF_2	Ma et al. (2007)

(continued)

Table 4.37 (continued)

Trait	QTL	Chr.	Marker	PVE (%)	Mapping population	References
Spikelets per Spike	*QSpn.nau-2D*	2D	*Xaf12–Xcfd239*	20.8	Wangshuibai/Nanda2419 RIL	Ma et al. (2007)
	QSpn.nau-2D	2D	*Xwmc181.1–Xaf12d*	15.1	Wangshuibai/Nanda2419 IF_2	Ma et al. (2007)
	QSps.ccsu-2D.2	2DS	*Xgwm261–Xcdo1379*	13.46	Opata/W7984 RIL	Kumar et al. (2007)
	QSpn	3AS	*XATPase*	11	ITMI RIL	Jantasuriyarat et al. (2004)
	QSpn.nau-3B	3B	*Xmag980–Xwmc510*	13.8	Wangshuibai/Nanda2419 RIL	Ma et al. (2007)
	QSpn	3DL	*Xfbb269*	16	ITMI RIL	Jantasuriyarat et al. (2004)
	QSpn	4AL	*Xbcd1670*	12	ITMI RIL	Jantasuriyarat et al. (2004)
	QSps.ccsu-4A.1	4AL	*Xgwm637–Xmwg549*	11.22	Opata/W7984 RIL	Kumar et al. (2007)
	QSps.ccsu-4D.1	4DS	*Xmwg634–Xbcd265*	16.56	Opata/W7984 RIL	Kumar et al. (2007)
	QSpn.nau-5A	5A	*Xzmh1368–Xcfa2163*	14.6	Wangshuibai/Nanda2419 RIL	Ma et al. (2007)
	QSpn.nau-5A	5A	*Xcfa2163–Xcfa2185.2d*	12	Wangshuibai/Nanda2419 RIL	Ma et al. (2007)
	QSpn.nau-5A	5A	*Xcfa2185.2–Xmag1494e*	12.7	Wangshuibai/Nanda2419 IF_2	Ma et al. (2007)
	QSpn.nau-5B	5B	*Xwmc723.1–Xmag2152*	15.4	Wangshuibai/Nanda2419 RIL	Ma et al. (2007)
	QTss.sdau-5D.e3	5D	*Xswes197–Xgwm272*	51.79	Chuan35050/Shannong483 RIL	Li et al. (2007)
	QSpn.nau-7A	7A	*Xbarc154–Xwmc83e*	10	Wangshuibai/Nanda2419 RIL	Ma et al. (2007)
	QSpn.nau-7A	7A	*Xwmc83–Xwmc17*	12.2	Wangshuibai/Nanda2419 IF_2	Ma et al. (2007)
	QSpn	7AL	*Xfba69*	18	ITMI RIL	Jantasuriyarat et al. (2004)
	QSpn	7AL	*Xfbb18*	18	ITMI RIL	Jantasuriyarat et al. (2004)
	QSpn.nau-7D	7D	*Xcfd46–Xwmc702c*	12.5	Wangshuibai/Nanda2419 RIL	Ma et al. (2007)
	QSpn.nau-7D	7D	*Xwmc702–Xcfd14c*	26.4	Wangshuibai/Nanda2419 RIL	Ma et al. (2007)
	QSpn.nau-7D	7D	*Xwmc702–Xcfd14c*	21.1	Wangshuibai/Nanda2419 IF_2	Ma et al. (2007)

(continued)

Table 4.37 (continued)

Trait	QTL	Chr.	Marker	PVE (%)	Mapping population	References
Fertile Spikelets	*QFspn.nau-2D*	2D	*Xaf12–Xcfd239d*	13	Wangshuibai/Nanda2419 RIL	Ma et al. (2007)
	QFspn.nau-2D	2D	*Xwmc181.1–Xaf12e*	11.3	Wangshuibai/Nanda2419 IF_2	Ma et al. (2007)
	QFspn.nau-3B1	3B	*Xmag980–Xwmc510*	13	Wangshuibai/Nanda2419 RIL	Ma et al. (2007)
	QFspn.nau-3B2	3B	*Xgwm533.1–Xgwm493e*	10.9	Wangshuibai/Nanda2419 IF_2	Ma et al. (2007)
	QFss.sdau-5D.e3	5D	*Xswes197–Xgwm272*	70.25	Chuan35050/Shannong483 RIL	Li et al. (2007)
	QFss.sdau-6B.e4	6B	*Xwmc487–Xwmc737*	10.99	Chuan35050/Shannong483 RIL	Li et al. (2007)
	QFspn.nau-7A	7A	*Xwmc525–Xmag1759.2*	11.3	Wangshuibai/Nanda2419 RIL	Ma et al. (2007)
	QFspn.nau-7D	7D	*Xwmc702–Xcfd14c*	23.5	Wangshuibai/Nanda2419 RIL	Ma et al. (2007)
	QFspn.nau-7D	7D	*Xwmc702–Xcfd14c*	17.4	Wangshuibai/Nanda2419 IF_2	Ma et al. (2007)
	QFss.sdau-7D.e1	7D	*Xgdm67–Xwmc31*	44.32	Chuan35050/Shannong483 RIL	Li et al. (2007)
	QFss.sdau-7D.e2	7D	*Xgwm428–Xgdm67*	67.63	Chuan35050/Shannong483 RIL	Li et al. (2007)
	QSspn.nau-1B	1B	*Xcfa2292–Xbarc80*	10.4	Wangshuibai/Nanda2419 RIL	Ma et al. (2007)
Sterile spikelets per spike	*QSspn.nau-4A*	4A	*Xwmc161–Xzmh991*	21.8	Wangshuibai/Nanda2419 RIL	Ma et al. (2007)
	QSspn.nau-4A	4A	*Xwmc161–Xzmh991*	12.5	Wangshuibai/Nanda2419 RIL	Ma et al. (2007)
	QSspn.nau-5A	5A	*Xzmh1368–Xcfa2163*	11.2	Wangshuibai/Nanda2419 RIL	Ma et al. (2007)
	QSspn.nau-5A	5A	*Xzmh1368–Xcfa2163*	12	Wangshuibai/Nanda2419 RIL	Ma et al. (2007)
	QSss.sdau-6B.e2	6B	*Xgwm132b–Xwmc487*	11.72	Chuan35050/Shannong483 RIL	Li et al. (2007)
	QSss.sdau-7A.e4	7A	*Xswes430b–Xubc840a*	16.17	Chuan35050/Shannong483 RIL	Li et al. (2007)
	QSss.sdau-7D.e2	7D	*Xgdm67–Xwmc31*	43.66	Chuan35050/Shannong483 RIL	Li et al. (2007)
	QSss.sdau-7D.e3	7D	*Xgdm67–Xwmc31*	27.53	Chuan35050/Shannong483 RIL	Li et al. (2007)
	QKns.sdau-2A.e1	2A	*Xubc873a–Xwmc63*	20.68	Chuan35050/Shannong483 RIL	Li et al. (2007)

(continued)

Table 4.37 (continued)

Trait	QTL	Chr.	Marker	PVE (%)	Mapping population	References
Grain number per spike	*QKns.sdau-2A.e2*	2A	*Xgwm71b–Xgwm71c*	11.26	Chuan35050/Shannong483 RIL	Li et al. (2007)
	QKns.sdau-2A.e3	2A	*Xwmc296–Xgwm339*	15.61	Chuan35050/Shannong483 RIL	Li et al. (2007)
	QGps.ccsu-2A.3	2AS	*Xgwm1115–Xgwm1052*	10.87	WL711/PH132 RIL	Kumar et al. (2007)
	QGps.ccsu-2B.1	2BS	*Xrz444–Xgwm410*	11.59	Opata/W7984 RIL	Kumar et al. (2007)
	QGps.ccsu-2D.3	2DL	*Xcdo1008–Xcdo36*	14.01	Opata/W7984 RIL	Kumar et al. (2007)
	QGps.ccsu-2D.2	2DS	*Xgwm296–Xgwm261*	12.55	Opata/W7984 RIL	Kumar et al. (2007)
	QKns.sdau-3B.e1	3B	*Xswes240–Xubc815a*	17.14	Chuan35050/Shannong483 RIL	Li et al. (2007)
	QGps.ccsu-3B.2	3BL	*Xgwm-376–Xgwm285*	19.82	Opata/W7984 RIL	Kumar et al. (2007)
	Qker.ksu-3D	3D	*Xgwm161*	11	Karl92 / TA4152-4 $B_C2F_{2:4}$	Narasimhamoorthy et al. (2006)
	QGps.ccsu-3D.1	3DL	*Xabc176–XgbxG305*	14.42	Opata/W7984 RIL	Kumar et al. (2007)
	QGps.ccsu-3D.2	3DL	*Xksud19–Xksuh15*	11.7	Opata/W7984 RIL	Kumar et al. (2007)
	QGns.wa-4BL.e1	4B	*Xbarc20–Xwmc238*	18.21	3228/Jing 4839 $F_{2:3}$	Wang et al. (2010)
	QGps.ccsu-4B.4	4BS	*Xwmc42–Xgwm898*	14.28	WL711/PH132 RIL	Kumar et al. (2007)
	QSpn.fcu-4D	4D	*Xbarc48–Xgwm19*	29	TA4152-60/ND495 DH	Chu et al. (2008)
	QKns.sdau-6A.e4	6A	*Xswes96–Xubc810a*	18.48	Chuan35050/Shannong483 RIL	Li et al. (2007)
	QKns.sdau-6B.e4	6B	*Xwmc398-Xtrap5*	11.89	Chuan35050/Shannong483 RIL	Li et al. (2007)
	QGps.ccsu-7A.1	7AL	*Xgwm276-XksuH9*	23.80	Opata/W7984 RIL	Kumar et al. (2007)
	QGps.ccsu-7A.2	7AS	*Xgwm1171-Xwmc9*	17.61/8.92	WL711/PH132 RIL	Kumar et al. (2007)

7D for thousand-grain weight, grain number per spike, sterile spikelets per spike, and fertile spikelets per spike. Ma et al. (2012) detected a QTL for thousand-grain weight on 4B with 5.15 % of phenotypic variance using a RIL population, positive alleles coming from the parent CSCR6. Wang et al. (2009a, b) detected 21 QTLs for thousand-grain weight on chromosomes 1B, 2A, 2D, 3B, 4A, 4D, 5A, 6D, and 7D using a RIL population in four environments. Kumar et al. (2006) conducted a RIL population having 100 lines by patents of Rye Selection (high grain weight) and Chinese Spring (low grain weight) and detected 12 QTLs for grain weight with 6.57–10.76 % of phenotypic variance. Dholakia et al. (2003) found 9, 2, and 7 markers for grain length, grain width, and grain weight using a RIL population of 106 pedigrees. Huang et al. (2003) identified 8 QTLs for grain weight on chromosomes 2A, 2D, 4D, 5B, 7A, 7B, and 7D using a BC_2F_2 population by AB-QTL backcrossing.

In all, QTL mapping for wheat grain-related traits was conducted using different populations. The loci for grain-related traits were almost located on 21 chromosomes. Most of the QTLs were located on chromosomes 1A, 2D, 3A, 5A, 5B, 5D, 6A, and 7D. Some QTLs controlled multiple traits on chromosomes 2A, 2B, 2D, 4B, 5A, 5B, 5D, 6A, 6B, and 7A.

4.8.1.3 QTL Mapping for Spike Number Per Plant

Spike number per plant was the most important yield trait. Kato et al. (2000) detected 4 QTLs for spike number per plant on 5A chromosome using recombination inbred lines. Huang et al. (2003) detected 8 QTLs for spike number per plant on chromosomes 1B, 2A, 2D, 3B, 4D, 5D, 6D, and 7A using a BC_2F_2 population. Huang et al. (2004) detected 2 QTLs on chromosomes 1B and 7A. Zhou et al. (2006) conducted a RIL population derived from "Wangbaishui × Alondra" (104 lines), and found 1 QTL for spike number per plant on chromosomes 5A, accounting for 10.3–18.8 % of phenotypic variance in 3 environments. Marza et al. (2006) detected 1 QTL for spike number per plant on 3BS, accounting for 12 % of phenotypic variance. Kumar et al. (2007) made QTL mapping for yield-related traits using two RIL populations and detected QTLs for spike number per plant on chromosomes 3AL, 7AL, 7BL, 1AL, 1BS, 3BL, 3DL, 4AL, and 6DL. Liao et al. (2008) used a $F_{2:3}$ population of 85 lines deriving from 5 backcrosses between Am3 (donor parent) and Laizhou 953 (recurrent parent) and one self-cross to QTL mapping for spike number per plant. Two QTLs for spike number per plant were located on 4B and 7B. And the QTL for spike number per plant on chromosomes 4B was detected in 3 environments. Wang et al. (2010) detected 7 QTLs for spike number per plant on chromosomes 1AL, 1BL, 3BL, 3DL, 4DL, 5BL, and 6AL using 273 $F_{2:3}$ lines in one environment (Table 4.38).

Table 4.38 The summary of QTL for kernel-related traits previously published

Trait	QTL	Chr.	Marker	PVE (%)	Mapping population	References
Grain length	*QKl.sdau-1A*	1A	*Xsrap21-Xwmc44*	10.8	Chuan35050/Shannong483 RIL	Sun et al. (2009)
	QKl.sdau-1B.1	1B	*Xswes15–Xubc811b*	10.6	Chuan35050/Shannong483 RIL	Sun et al. (2009)
	QGl.nfcri-2A.2	2A	*Xbarc1165-Xbarc124*	16.53	Yu8679/Heshangmai RIL	Wang et al. (2009a, b)
	QKl.sdau-2B	2B	*Xsrap1a-Xgwm120*	26.4	Chuan35050/Shannong483 RIL	Sun et al. (2009)
	QKl.ncl-2D.1	2D	*Xgwm349-Xcfd161*	10.6	Rye Selection 111/Chinese spring RIL	Ramya et al. (2010)
	QGl.nfcri-2D	2D	*Xbarc228-Xcfd168*	19.02	Yu8679/Heshangmai RIL	Wang et al. (2009a, b)
	QKl.sdau-4A	4A	*Xswes24a–Xswes24c*	14.6	Chuan35050/Shannong483 RIL	Sun et al. (2009)
	QKl.sdau-4B	4B	*Xgwm495-Xwmc238*	18.4	Chuan35050/Shannong483 RIL	Sun et al. (2009)
	QGl.nfcri-5D	5D	*Xcfd3-Xbarc93*	10.58	Yu8679/Heshangmai RIL	Wang et al. 2009a, b
	QGw.nfcri-1B.2	1B	*Xwmc269-Xgwm33*	10.91	Yu8679/Heshangmai RIL	Wang et al. (2009a, b)
Grain width	*QKw.sdau-2A*	2A	*Xwmc179a-Xw181a*	19.4	Chuan35050/Shannong483 RIL	Sun et al. (2009)
	QGw.nfcri-2A.3	2A	*Xbarc1165-Xbarc124*	29.87	Yu8679/Heshangmai RIL	Wang et al. (2009a, b)
	QKw.ncl-2B.1	2B	*Xbarc183-Xbarc13*	11.54	Rye Selection 111/Chinese spring RIL	Ramya et al. (2010)
	QGw.nfcri-5A	5A	*Xbarc165-Xbarc186*	10.82	Yu8679/Heshangmai RIL	Wang et al. (2009a, b)
	QKw.sdau-5D	5D	*Xswes342a–Xswes342b*	12.1	Chuan35050/Shannong483 RIL	Sun et al. (2009)
	QGw.nfcri-5D	5D	*Xbarc93-Xgdm63*	11.03	Yu8679/Heshangmai RIL	Wang et al. (2009a, b)
	QKw.sdau-6A	6A	*Xswes123b–Xswes332a*	20	Chuan35050/Shannong483 RIL	Sun et al. (2009)
	QTgw.nfcri-1B.1	1B	*Xwmc156-Xwmc269*	13.96	Yu8679/Heshangmai RIL	Wang et al. (2009a, b)

(continued)

Table 4.38 (continued)

Trait	QTL	Chr.	Marker	PVE (%)	Mapping population	References
Thousand-grain weight	*QTgw.nfcri-1B.2*	1B	*Xwmc419-Xbarc181*	13.96	Yu8679/Heshangmai RIL	Wang et al. (2009a, b)
	QTkw.sdau-1D	1D	*Xwmc222-Xwmc432a*	13.1	Chuan35050/Shannong483 RIL	Sun et al. (2009)
	QTkw.sdau-1D.e3	1D	*Xwmc222-Xwmc432a*	16.88	Chuan35050/Shannong483 RIL	Li et al. (2007)
	QTkw.sdau-1D.e4	1D	*Xwmc432a-Xwmc336c*	12.76	Chuan35050/Shannong483 RIL	Li et al. (2007)
	QTkw.sdau-2A	2A	*Xwmc181a-Xubc840c*	20.1	Chuan35050/Shannong483 RIL	Sun et al. (2009)
	QTgw.nfcri-2A.1	2A	*Xbarc1165-Xbarc124*	16.8	Yu8679/Heshangmai RIL	Wang et al. (2009a, b)
	QTkw.ncl-2B.1	2B	*Xbarc183-Xbarc7*	15.53	Rye Selection 111/Chinese spring RIL	Ramya et al. (2010)
	QTkw.sdau-3B.e1	3B	*Xubc880c-Xubc880b*	11.02	Chuan35050/Shannong483 RIL	Li et al. (2007)
	QTkw.sdau-5D.e3	5D	*Xswes340a–Xswes342a*	15.98	Chuan35050/Shannong483 RIL	Li et al. 2007
	QTkw.sdau-6A	6A	*Xswes332a–Xgwm617*	13.2	Chuan35050/Shannong483 RIL	Sun et al. (2009)
	QTkw.sdau-6A.e2	6A	*Xswes332a–Xgwm617*	16.82	Chuan35050/Shannong483 RIL	Li et al. (2007)
	QTkw.sdau-6A.e2	6A	*Xswes96–Xubc810a*	18.48	Chuan35050/Shannong483 RIL	Li et al. (2007)
	QTgw.nfcri-6D	6D	*Xbarc196*	10.9	Yu8679/Heshangmai RIL	Wang et al. (2009a, b)
	QTkw.sdau-7D.e1	7D	*Xgdm67-Xwmc31*	44.15	Chuan35050/Shannong483 RIL	Li et al. (2007)

4.8.2 Comparision of the Results with Previous Studies

Our group conducted some QTLs for grain yield and related traits using 5 populations (partial results not shown in this chapter). Compared with other studies, most of the QTLs detected in our study seemed to have similar or the same chromosomal locations with different mapping populations as compared with previous reports. Some new QTL loci were also detected.

The *qGY2Da* for grain yield was detected on chromosome 2D in population 1 with the largest genetic contribution. The *qGY2Da* stably expressed in 3 environments, which was mapped in the similar location of *QHD.ksu-2D* (Narasimhamoorthy et al. 2006) and *QYld.ipk-2D.1* (Huang et al. 2006) with a little larger additive effect. They may be the same QTL for grain yield on chromosome 2D. The *qGY2Da* was also located on the similar position of *Rht8* on chromosome 2DS. Rebetzke et al. (1999) demonstrated that *Rht8* could increase grain number per pike and grain yield, which was in accordance with results in our study. Some QTLs or association mapping loci for spike length, fertile spikelets per spike, and grain yield were identified on chromosome 4A, which was in line with the previous reports (Ma et al. 2012; Kumar et al. 2007).

The QTLs for grain weight per spike were identified on chromosomes 2A, 2B, 2D, 3A, and 4D, which were in accordance with the report by Cui et al. (2011) using RIL populations. In addition, no report about QTL for grain weight per spike on chromosome 4B was shown, and our group identified *QKwps4B.1-99* for grain weight per spike in E1, E2, E3, and E4. The genetic effects were more than 10 %. The *QKwps4B.1-99* may be new locus controlling grain weight per spike on chromosome 4B, which can be used for grain weight per spike by marker-assisted selection.

References

Anhalt UCM, Heslop-Harrison (Pat) JS, Piepho HP, Byrne et al. Quantitative trait loci mapping for biomass yield traits in a Lolium inbred line derived F_2 population. Euphytica. 2009;170(1–2): 99–107.

Araki E, Miura H, Sawada S. Identification of genetic loci affecting amylose content and agronomic traits on chromosome 4A of wheat. Theor Appl Genet. 1999;98:977–84.

Ayala L, Henry M, Ginkel MV, Singh R, et al. Identification of QTLs for BYDV tolerance in bread wheat. Euphytica. 2002;128(2):249–59.

Borlaug NE, Dowswell CR. The acid lands: one of agriculture's last frontiers. In: Moniz AC et al, editors. Plant-soil interactions at low pH. Brazil: Brazilian Soil Science Society; 1997. pp. 5–15.

Börner A, Schumann E, Fürste A, Cluster H, Leithold B, Roder MS, Weber WE. Mapping of quantitative trait loci determining agronomic important characters in hexaploid wheat (*Triticum aestivum* L.). Theor Appl Genet. 2002;105:921–36.

Cao G, Zhu J, He C, Gao Y, Yan J, Wu P. Impact of epistasis and QTL × environment interaction on the developmental behavior of plant height in rice (*Oryza sativa* L.). Theor Appl Genet. 2001;103:153–60.

Chu CG, Xu SS, Friesen TL, Faris JD. Whole genome mapping in a wheat doubled haploid population using SSRs and TRAPs and the identification of QTL for agronomic traits. Mol Breeding. 2008;22:251–66.

Cuthbert JL, Somers DJ, Brule-Babel AL, Brown PD, et al. Molecular mapping of quantitative trait loci for yield and yield components in spring wheat (*Triticum aestivum* L.). Theor Appl Genet. 2008;117(4):595–608.

Cui F. Construction of high-density wheat molecular genetic map and QTL analysis for yield-related traits. PhD dissertation of Shandong Agricultural University; 2011 (in Chinese with English abstract).

Deng S, Wu X, Wu Y, et al. Characterization and precise mapping of a QTL increasing spike number with pleiotropic effects in wheat. Theoret Appl Genet. 2011;122:281–289.

Ding AM, Li J, Cui F, Zhao CH, Ma HY, Wang HG. QTL mapping for yield related traits using two associated RIL populations of Wheat. Acta Agron Sin. 2011;37:1511–1524 (in Chinese with English abstract).

Dholakia BB, Ammiraju JSS, Singh H, Lagu MD, Röder MS, Rao VS, Dhaliwal HS, Ranjekar PK, Gupta VS, Weber WE. Molecular marker analysis of kernel size and shape in bread wheat. Plant Breeding. 2003;122:392–5.

Gabrielle B, Gagnaire N. Life-cycle assessment of straw use in bio-ethanol production: a case study based on biophysical modeling. Biomass Bioenergy. 2008;32(5):431–41.

Hua J, Xing Y, Wu W, Xu C, Sun X, Yu S, Zhang Q. Single-locus heterotic effects and dominance by dominance interactions can adequately explain the genetic basis of heterosis in an elite rice hybrid. Proc Natl Acad Sci USA. 2003;100:2574–9.

Huang XQ, Coster H, Ganal MW, Röder MS. Advanced backcross QTL analysis for the identification of quantitative trait loci alleles from wild relatives of wheat (*Triticum aestivum* L.). Theor Appl Genet. 2003;106:1379–89.

Huang XQ, Kempf H, Ganal MW, Röder MS. Advanced backcross QTL analysis in progenies derived from a cross between a German elite winter wheat variety and a synthetic wheat (*Triticum aestivum* L.). Theor Appl Genet. 2004;109:933–43.

Huang XQ, Cloutier S, Lycar L, Radovanovic N, et al. Molecular detection of QTLs for agronomic and quality traits in a doubled haploid population derived from two Canadian wheats (*Triticum aestivum* L.). Theor Appl Genet. 2006;113(4):753–66.

Jantasuriyarat C, Vales MI, Watson CJW, Riera-Lizarazu O. Identification and mapping of genetic loci affecting the free-threshing habit and spike compactness in wheat (*Triticum aestivum* L.). Theor Appl Genet. 2004;108:261–73.

Kato K, Miura H, Sawada S. Mapping QTL controlling grain yield and its components on chromosome 5A of wheat. Theor Appl Genet. 2000;101:1114–21.

Kirigwi FM, Ginkel MV, Guedira GB, Gill BS, et al. Markers associated with a QTL for grain yield in wheat under drought. Mol Breeding. 2007;20:401–13.

Kuchel H, Williams KJ, Langridge P, Eagles HA, et al. Genetic dissection of grain yield in bread wheat. I. QTL analysis. Theor Appl Genet. 2007;115(8):1029–41.

Kumar N, Kulwal PL, Gaur A, Tyagi AK, Khurana JP, Khurana P, Balyan HS, Gupta PK. QTL analysis for grain weight in common wheat. Euphytica. 2006;151:135–144.

Kumar N, Kulwal PL, Balyan HS, Gupta PK. QTL mapping for yield and yield contributing traits in two mapping populations of bread wheat. Mol Breeding. 2007;19:167–77.

Kuraparthy V, Sood S, Dhaliwal HS, Chhuneja P, Gill BS. Identification and mapping of a tiller inhibition gene (*tin3*) in wheat. Theor Appl Genet. 2007;114:285–94.

Lal R. World crop residues production and implications of its use as a biofuel. Environ Int. 2005;31(4):575–84.

Law CN. The location of genetic factors controlling a number of quantitative characters in wheat. Genetics. 1967;56:445–461.

Li HH, Ye G, Wang J. A modified algorithm for the improvement of composite interval mapping. Genetics. 2007;175:361–374.

Li SS, Jia JZ, Wei XY, Zhang XC, Li LZ, Chen HM, Fan YD, Sun HY, Zhao XH, Lei TD, Xu YF, Jiang FS, Wang HG, Li LH. A intervarietal genetic map and QTL analysis for yield traits in wheat. Mol Breeding. 2007;20:167–78.

Li WL, Nelson JC, Chu CY, Shi LH, Huang SH, Liu DJ. Chromosomal locations and genetic relationships of tiller and spike characters in wheat. Euphytica. 2002;125:357–66.

Liao XZ, Wang J, Zhou RH, Li RZ, Jia JZ. Mining favorable alleles of QTLs conferring 1000-grain weight from synthetic wheat. Acta Agronomica Sin. 2008;34(11):1877–84 (in Chinese with English abstract).

Liu GF, Yang J, Zhu J. Mapping QTL for biomass yield and its components in rice (*Oryza sativa* L.). J Genet Genomics. 2006;33(7):607–16.

Ma J. G J Yan, C J Liu. Development of near-isogenic lines for a major QTL on 3BL conferring *Fusarium*crown rot resistance in hexaploid wheat. Euphytica. 2012;183:147–52.

Ma ZQ, Zhao D, Zhang C, Zhang Z, Xue S, Lin F, Kong Z, Tian D, Luo Q. Molecular genetic analysis of five spike-related traits in wheat using RIL and immortalized F2 populations. Mol Gen Genomics. 2007;277:31–42.

Marza F, Bai GH, Carver BF, Zhou WC. Quantitative trait locus for yield and related traits in the wheat population Ning7840 × Clark. Theor Appl Genet. 2006;112:688–98.

McIntyre CL, Mathews KL, Rattey A, Chapman SC, et al. Molecular detection of genomic regions associated with grain yield and yield-related components in an elite bread wheat cross evaluated under irrigated and rainfed conditions. Theor Appl Genet. 2010;120(3):527–41.

Miura H, Parker BB, Snape JW. The location of major genes and associated quantitative trait loci on chromosome arm 5BL of wheat. Theor Appl Genet. 1992;85:197–204.

Miura H, Worland AJ. Genetic control of vernalization, day length response and earlinessper se by homoeologous group-3 chromosomes in wheat. Plant Breeding. 1994;113:160–9.

Narasimhamoorthy B, Gill BS, Fritz AK, Nelson JC. Brown G L Guedira Advanced backcross QTL analysis of a hard winter wheat synthetic wheat population. Theor Appl Genet. 2006;112:787–96.

Pfeiffer WH, Sayre KD, Reynolds MP. Enhancing genetic gain yield potential and yield stability in durum wheat. In: Royo C, Nachit MM, Di Fonzo N, Araus JL, editors. Durum wheat improvement in the Mediterranean Region: new challenges. Zaragoza: IAMZ. Options mediterraneennes; 2000. Vol. 40, pp. 83–93.

Quarrie SA, Steed A, Calestani C, Semikhodskii A, et al. A high-density genetic map of hexaploid wheat (*Triticum aestivum* L.) from the cross Chinese Spring × SQ1 and its use to compare QTLs for grain yield across a range of environments. Theor Appl Genet. 2005;110(5):865–80.

Ramya P, Chaubal A, Kulkarni K, Gupta L, Kadoo N, Dhaliwal HS, Chhuneja P, Lagu M, Gupta V. QTL mapping of 1000-kernel weight, kernel length, and kernel width inbread wheat (*Triticum aestivum* L.). J Appl Genet. 2010;51:421–429.

Rebetzke GJ, Richards RA, Fischer VM, Mickelson BJ. Breeding long coleoptile, reduced height wheats. Euphytica. 1999;106:159–68.

Schlegel R, Meinel A. A quantitative trait locus (QTL) on chromosome arm 1RS of rye and its effect on yield performance of hexaploid wheat. Cereal Res Commun. 1994;22:7–13.

Shah MM, Gill KS, Baenziger PS, Yen Y, Kaeppler SM, Ariyarathne HM. Molecular mapping of loci for agronomic traits on chromosome 3A of bread wheat. Crop Sci. 1999;39:1728–32.

Su Y, Rao YC, Hu SK, Yang YL, Gao ZY, Zhang GH, Liu J, Hu J, Yan MX, Dong GJ, Zhu L, Gao LB, Qian Q, Zeng DL. Map-based cloning proves qGC-6, a major QTL for gel consistency of japonica/indica cross, respods by Wary in rice (Oryza sativa L.). Theoret Appl Genet. 2011;123(5):859–67.

Sun XY, Wu K, Zhao Y, Kong FM, Han GZ, Jiang HM, Huang XJ, Li RJ, Wang HG, Li SS. QTL analysis of kernel shape and weight using recombinant inbred lines in wheat. Euphytica. 2009;16:615–24.

Wang JS, Liu WH, Wang H, Li LH. Wu J, Yang XM, Li XQ, Gao AN. QTL mapping of yield-related traits in the wheat germplasm 3228. Euphytica. 2010;177:277–292.

Wang RX, Zhang XY, Wu L, Wang R, Hai L, You GX, Yan CS, Xiao SH. QTL analysis of grain size and related traits in winter wheat under different ecological environments. Sci Agri Sin. 2009a;42:398–407 (in Chinese with English abstract).
Wang RX, Hai L, Zhang XY, You GX, Yan CS, Xiao SH. QTL mapping for grain filling rate and yield-related traits in RILs of the Chinese winter wheat population Heshangmai × Yu8679. Theor Appl Genet. 2009b;118:313–25.
Yang J, Zhu J. Methods for predicting superior genotypes under multiple environments based on QTL effects. Theor Appl Genet. 2005;110:1268–74.
Yang L, Shao H, Wu QX, Yu J, Rang CF, Li LQ, Li XJ. QTLs mapping and epistasis analysis for the number of tillers and spike number per plant in wheat. J Triticeae Crops. 2013;33:875–882.
Zhang KP, Xu XB, Tian JC. QTL mapping for grain yield and spike related traits in common wheat. Acta Agron Sin. 2009;35:270–278 (in Chinese with English abstract).
Zhang K, Zhang Y, Chen G, Tian J. Genetic analysis of grain yield and leaf chlorophyll content in common wheat. Cereal Res Commun. 2009;37(4):499–511.
Zhang XY, Tong YP, You GX, Hao CY, Ge HM, Wang LF, Li B, Dong YC, Li ZS. Hitchhiking effect mapping: a new approach for discovering agronomic important genes. Sci Agric Sin. 2006;39:1526–35 (in Chinese with English abstract).
Zhang ZH, Li P, Wan LX, Hu ZL, et al. Genetic dissection of the relationships of biomass production and partitioning with yield and yield related traits in rice. Plant Sci. 2004;167(1):1–8.
Zhou MP, Ren LJ, Zhang X, Yu GH, Ma HX, Lu WZ. Analysis of QTLs for yield traits of wheat. J Triticeae Crops. 2006;26:35–40 (in Chinese with English abstract).

Chapter 5
Genetic Detection of Main Quality Traits in Wheat

Abstract Wheat quality is one of the important breeding objectives for wheat breeders, but most of the major genes controlling these traits are unclear. So in this chapter, grain quality traits, nutritional quality traits, flour quality traits, dough quality traits, and processing quality traits were genetically dissected by QTL mapping. Of which, grain quality included grain weight, grain length, diameter, and hardness; nutritional quality presented protein content, beneficial mineral elements, amino acid content and components, and carotenoid pigments; flour quality contained gluten content and index, flour whiteness and color, PPO activity, sedimentation volume, paste viscosity parameters, falling number, starch content and components; for dough quality, farinograph, mixograph and alveograph parameters were involved; and processing quality mainly discussed Chinese noodle and steamed bread quality. Some major QTLs identified for wheat quality traits provided important genetic and molecular information for marker-assisted selection breeding.

Keywords Quality traits · Grain quality · Nutritional quality · Flour quality · Dough quality · Processing quality · Noodle cooking quality · Steamed bread texture properties · QTL mapping

Most of the important quality traits of the wheat are quantitative traits. They represent continuous variation and are much sensitive to environmental factors. The traditional quantitative genetics only detected genetic effect and environmental effect by regarding multigenes controlling quantitative traits as a whole, but this method cannot determine the number of QTLs and the individual genetic effect of single QTL. Although Sax (1923) proposed the concept of genetic markers for the detection of QTL for quantitative trait, there was almost no progress in the research because of the limit number of genetic markers that can be used before 1980s. Since the 1990s, with the rapid development of molecular biology and computer technology, the new identification methods and statistical analysis tools have been explored for the quantitative traits' genetic research. There was a great progress on the study of the inheritance essence of quantitative traits, determining their positions on the chromosomes and interactions between the genes.

J. Tian et al., *Genetic Analyses of Wheat and Molecular Marker-Assisted Breeding, Volume 1*, DOI 10.1007/978-94-017-7390-4_5

Till now, the great progress has been achieved on the QTL mapping of wheat quality. Because the wheat quality was influenced by lots of factors and existed multiallelic variations, the QTL results showed to be much different under different materials. Moreover, there was no precise contribution effect of each locus and some gene loci influencing quality and yield have not been detected. In addition, most of the quality traits were influenced by genotype, environment, and the interactions between them, and it was not clear about the degree of the influence of environment. Therefore, QTL mapping for main quality traits was studied using three genetic populations for dissecting the special excellent quality gene in wheat resource, which will lay a good foundation for molecular marker-assisted selection and molecular pyramiding breeding for high-quality wheat, and provide new strategies and methods for efficient utilization of germplasm resources to breed high-yield and good-quality wheat varieties.

5.1 QTL Mapping of Wheat Kernel Quality Traits

Grain characters, such as grain weight, grain shape, and grain hardness, greatly affect wheat yield and quality, which are controlled by multigenes. Grain weight is one of the "three components" of yield. Grain shape and its uniformity affect not only the wheat yield, but also the quality of milling and processing. In general, big kernel, thin skinned, and nearly circular grain shape have high flour extraction. The kernel hardness also influences wheat milling and processing quality, which is one of the most important indicators determining trade price and end use. Therefore, it is of great significance to increase wheat yield and quality by studying the grain weight, grain type, and grain hardness at molecular level, which will reveal the impact of molecular genetic mechanism of wheat quality and yield.

5.1.1 QTL Mapping of Wheat Grain Quality Traits

Grain quality of wheat includes grain weight, grain length, diameter, and hardness.

5.1.1.1 Materials and Methods

5.1.1.1.1 Plant Materials

Doubled haploid lines derived from a cross between two Chinese wheat cultivars Huapei 3 (Hp3) and Yumai 57 (Ym57) were used for the construction of a linkage map. The DH population and parents were kindly provided by Professor Yan Hai, Henan Academy of Agricultural Sciences, Zhengzhou, China. Hp3 and Ym57 were released by Henan Province in 2006 (Hai and Kang 2007) and the State (China) in

2003 (Guo et al. 2004), respectively. The two parents, which are cultivated over a large area in the Huang-Huai wheat region in China, differ for several agronomically important traits as well as for baking quality traits (Guo et al. 2004; Hai and Kang 2007).

The IF_2 population was created following the design of Hua et al. (2003). In this design, crossing was done among DH chosen by random permutations of 168 lines. In each round of permutation, the 168 DHs were randomly divided into two groups, and the lines in two groups were paired up at random without replacement to provide parents for 84 crosses. Each of 168 lines was used only once in each round of pairing and crossing. This procedure was repeated twice, resulting in a population consisting of 168 crosses, which constituted the IF_2 population used.

The DH population map built in 2008 had 323 molecular markers, including 284 SSR loci, 37 EST-SSR loci, 1 ISSR loci, and 1 HMW loci, which were mapped on 24 linkage groups, covering 2485.7 cM according to the Zhang et al. (2009a, b, c, d). The genotypes of the IF_2 population were deduced on the basis of DH line genotypes.

5.1.1.1.2 Field Trials and Kernel Measurement

The field trials were conducted in five environments: at Tai'an (36°18′N, 117°13′ E), Shandong Province, in 2008 and 2009; at Jiyuan (35°05′N, 112°36′E), Henan Province, in 2009. The experimental design followed a randomized complete block design with two replications at each location. All lines and parental lines were grown in three-row plots (length 2 m, between-row spacing 25 cm).

In 2008–2009, three environment treatments were conducted in Tai'an: E1 (2008): normal irrigation and fertilizer, irrigated once at the stage of overwintering period, jointing, anthesis, and grain filling, and fertilized 225 and 112.5 kg/ha urea at the stage of jointing and anthesis, respectively; E2 (2008): only fertilizer without irrigation, that is, fertilized 225 and 112.5 kg/ha urea at the stage of jointing and anthesis, respectively, without irrigation; and E3 (2008): only irrigation without fertilizer, that is, irrigated once at the stage of overwintering period, jointing, anthesis, and grain filling without fertilizer.

In 2009–2010, the field trials were conducted in two environments: at Tai'an (36°18'N, 117°13'E), Shandong Province, and at Jiyuan (35°05′N, 112°36′E), Henan Province. The experimental design followed a randomized complete block design with three replications at each location. All lines and parental lines were grown in three-row plots (length 2 m, between-row spacing 25 cm). The field management followed the local standard practices. E3 stands at Tai'an in 2009; E4 represents at Jiyuan in 2009.

The repeated lines were mixed as one unit after harvest. The grain weight, grain diameter, and hardness were measured by SKCS 4100. The average value of the two replications was calculated for data analysis.

5.1.1.2 Results and Analysis

5.1.1.2.1 Phenotypic Analysis

The parents Yumai 57 and Huapei 3 have obvious difference in thousand-grain weight, grain length, grain size, and hardness (Table 5.1). DH population and IF_2 population showed obvious phenotypic variation, and the phenotypic values segregated continuously. Kernel traits were typically quantitative traits and showed bitransgressive segregation. Besides, in E4 and E5, the kernel traits of IF_2 were significantly higher than those of DH population and showed high heterosis phenomenon.

Table 5.1 Phenotypic value of 1000-kernel weight, grain size and hardness for parents, DH population and IF_2 population in different environment

Population	E	1000-kernel weight (g)	Grain length (mm)	Grain diameter (mm)	Hardness (HI)
HP 3	1[a]	57.06	4.82	3.38	62.50
	2[b]	54.34	4.79	3.35	61.45
	3[c]	50.42	4.6	3.22	61.05
	4[d]	52.47	4.70	3.29	48.14
	5[e]	59.20	4.95	3.46	41.70
YM 57	1	45.00	4.60	3.22	27.06
	2	43.32	4.52	3.16	23.61
	3	47.14	4.66	3.26	25.25
	4	47.71	4.66	3.26	29.18
	5	51.90	4.90	3.43	23.94
DH	1	45.60 ± 6.70[f]	4.44 ± 0.33	3.10 ± 0.23	48.51 ± 20.96
		30.47-59.28[g]	3.73 − 5.19	2.62 − 3.64	7.99 − 83.08
	2	43.81 ± 6.02	4.34 ± 0.30	3.04 ± 0.21	45.69 ± 20.12
		29.4 − 58.26	3.68-4.98	2.59-3.48	3.24 − 76.66
	3	46.29 ± 6.01	4.47 ± 0.30	3.13 ± 0.21	48.18 ± 20.60
		31.27 − 59.88	3.67 − 5.15	2.58 − 3.61	8.44 − 80.86
	4	45.07 ± 6.35	4.42 ± 0.30	3.09 ± 0.21	47.61 ± 21.15
		28.94 − 57.27	3.72 − 5.10	2.61 − 3.57	7.11 − 81.49
	5	51.56 ± 7.08	4.72 ± 0.34	3.31 ± 0.24	39.4 ± 18.41
		28.70 − 63.38	3.79 − 5.26	2.66 − 3.68	3.02 − 68.86
IF_2	4	46.89 ± 5.59	4.49 ± 0.26	3.14 ± 0.18	46.18 ± 15.21
		31.76 − 57.60	3.86 − 5.14	2.71 − 3.60	12.06 − 74.93
	5	53.48 ± 5.53	4.79 ± 0.26	3.35 ± 0.18	39.34 ± 13.66
		34.09 − 62.92	3.97 − 5.30	2.78 − 3.72	5.24 − 68.63

[a]Natural irrigation and N fertilization in 2008, Tai'an; [b]No natural irrigation but *N* fertilization in 2008, Tai'an; [c]Natural irrigation but no *N* fertilization in 2008, Tai'an; [d]Natural irrigation and *N* fertilization in 2009, Jiyuan; [e]Natural irrigation and *N* fertilization in 2009, Tai'an; [f]Mean value ± standard deviation; [g]Range

5.1.1.2.2 QTL Analysis of Grain Weight, Grain Length, and Hardness

Comprehensive analysis was conducted based on the phenotypic data of DH population under five environments and IF_2 population under two environments. Totally 32 additive effects QTLs and 18 pairs of epistatic effects QTLs for grain weight, grain length, grain diameter and grain hardness were detected (Tables 5.2, 5.3, 5.4 and 5.5; Fig. 5.1). Among them, the QTL located on chromosome 6A controlling grain weight with additive effect and epistatic effect could be detected at the same time in two populations. Moreover the QTL of grain length was closely linked with that of grain diameter.

Table 5.2 QTL for 1000-kernel weight, grain size, and hardness in DH population

Trait	QTL	Flanking marker	Position (cM)	A[a]	H^2 (%)[b]
1000-kernel weight	*Qtkw2Da*	*XCFD53–XWMC18*	3.6	0.9536	2.82
	Qtkw2Db	*XGWM539–XCFD168*	67.2	0.7995	5.40
	Qtkw3A	*XWMC264–XCFA2193*	175.6	1.8215	3.73
	Qtkw3B	*XGPW7774–XGWM533*	34.0	1.3332	3.61
	Qtkw6A	*XGWM82–XWMC182*	74.2	1.7513	9.39
	Qtkw7B.2	*XGPW2224–XGPW3256*	28.5	−1.6969	4.22
	Qtkw7D	*XGWM676–XGWM437*	121.9	1.6504	5.55
Grain length	*Qgl1B*	*XGWM582–XGPW7388*	57.7	−6.7065	5.23
	Qgl2D	*XWMC112–XCFD53*	0.9	2.6348	1.01
	Qgl3Aa	*XWMC264–XCFA2193*	169.6	7.4054	2.85
	Qgl3Ab	*XGPW1108–XGPW1107*	249.1	−3.1339	3.30
	Qgl3B	*XGWM389–XGPW2344*	14.0	8.2874	5.10
	Qgl6A	*XBARC1055–XWMC553*	75.2	8.6047	14.96
	Qgl7B.2	*XGPW3226–XGPW2224*	21.5	−6.4065	5.45
	Qgl7D	*XGWM676–XGWM437*	121.9	6.9417	6.98
Grain diameter	*Qgd1B*	*XGWM582–XGPW7388*	57.7	−0.0448	5.21
	Qgd2D	*XWMC112–XCFD53*	0.9	0.0174	1.00
	Qgd3Aa	*XWMC264–XCFA2193*	170.6	0.0480	2.88
	Qgd3Ab	*XGPW1108–XGPW1107*	249.1	−0.0190	3.29
	Qgd3B	*XGWM389–XGPW2344*	14.0	0.0561	5.05
	Qgd6A	*XBARC1055–XWMC553*	75.2	0.0679	15.02
	Qgd7B.2	*XGPW3226–XGPW2224*	21.5	−0.0435	5.40
	Qgd7D	*XGWM676–XGWM437*	122.9	0.0457	6.91
	Qhd1Ba	*XGWM582–XGPW7388*	50.7	−7.5933	7.51
Hardness	*Qhd1Bb*	*XWMC766–XSWES98*	129.3	4.4118	0.33
	Qhd4B	*XWMC48–XBARC1096*	18.3	−4.4475	6.43
	Qhd5A	*XBARC358.2–XGWM186*	47.3	4.0207	4.34
	Qhd6A	*XGWM459–XGWM334*	38.8	3.4650	2.36

[a]Additive effects; positive additive effects indicate that the Huapei 3 allele increases the value of the trait, and negative additive effects indicate that the Yumai 57 allele increases the value of the trait; [b]percentage of phenotypic variation explained by QTL with additive effect

Table 5.3 Estimated epistasis for 1000-kernel weight, grain size, and hardness in DH population

Trait	QTL	Flanking marker	Position (cM)	QTL	Flanking marker	Position (cM)	AA[a]	H^2 (%)
1000-kernel weight	*Qtkw2Da*	*XCFD53–XWMC18*	3.6	*Qtkw7D*	*XGWM676–XGWM437*	121.9	−0.86	1.05
	Qtkw3A	*XWMC264–XCFA2193*	175.6	*Qtkw7B.2*	*XGPW2224–XGPW3256*	28.5	−0.77	1.13
	Qtkw6A	*XGWM82–XWMC182*	74.2	*Qtkw7B.2*	*XGPW2224–XGPW3256*	28.5	0.8	1.14
Grain length	*Qgl1B*	*XGWM582–XGPW7388*	57.7	*Qgl2D*	*XWMC112–XCFD53*	0.9	4.77	1.51
	Qgl2D	*XWMC112–XCFD53*	0.9	*Qgl7D*	*XGWM676–XGWM437*	121.9	−5.23	1.16
	Qgl3Aa	*XWMC264–XCFA2193*	169.6	*Qgl6A*	*XBARC1055–XWMC553*	75.2	3.98	0.22
	Qgl3B	*XGWM389–XGPW2344*	14	*Qgl7B.2*	*XGPW3226–XGPW2224*	21.5	4.06	0.56
	Qgl6A	*XBARC1055–XWMC553*	75.2	*Qgl7B.2*	*XGPW3226–XGPW2224*	21.5	3.37	0.86
Grain diameter	*Qgd1B*	*XGWM582–XGPW7388*	57.7	*Qgd2D*	*XWMC112–XCFD53*	0.9	0.0352	1.55
	Qgd2D	*XWMC112–XCFD53*	0.9	*Qgd7D*	*XGWM676–XGWM437*	122.9	−0.0381	1.26
	Qgd3Aa	*XWMC264–XCFA2193*	170.6	*Qgd6A*	*XBARC1055–XWMC553*	75.2	0.0301	0.19
	Qgd3B	*XGWM389–XGPW2344*	14	*Qgd7B.2*	*XGPW3226–XGPW2224*	21.5	0.0303	0.54
	Qgd6A	*XBARC1055–XWMC553*	75.2	*Qgd7B.2*	*XGPW3226–XGPW2224*	21.5	0.0255	0.86
	Qgd6A	*XBARC1055–XWMC553*	75.2	*Qgd7D*	*XGWM676–XGWM437*	122.9	0.0248	0.2
Hardness	*Qhd5A*	*XBARC358.2–XGWM186*	47.3	*Qhd6A*	*XGWM459–XGWM334*	38.8	−3.01	1.77

[a]Positive value represents that parent-type effect is bigger than recombinant-type effect, and negative value represents that recombinant-type effect is bigger than parent-type effect

Table 5.4 Putative QTL for 1000-kernel weight, grain size, and hardness in IF_2 population

Trait	QTL	Flanking marker	Position (cM)	A	H^2 (A) (%)	D[a]	H^2 (D)[b] (%)
1000-kernel weight	*qtkw6A*	*XGWM82–XWMC182*	74.2	2.9409	11.75	1.5154	1.37
Grain length	*qgl6A*	*XBARC1055–XWMC553*	75.2	16.84	15.10		
	qgl7D	*XGWM437–XWMC630.1*	126.6	9.6997	8.24		
Grain diameter	*qgd6A*	*XBARC1055–XWMC553*	75.2	0.1123	15.03		
	qgd7D	*XGWM437–XWMC630.1*	126.6	0.0673	8.26		
Hardness	*qhd2A*	*XGPW2321–XGWM558*	78.8	−5.3829	3.77		
	qhd7D	*XGWM295–XGWM676*	97.3	8.4948	7.90		

[a]Dominant effect, positive dominant effect indicate that the phenotype value of heterozygote was higher than that of the homozygote, negative dominant effect indicate that the phenotype value of homozygote was higher than that of the heterozygote. [b]Percentage of phenotypic variation explained by QTL with additive effect

5.1.1.2.2.1 QTL Analysis for Grain Weight, Grain Length, and Hardness Using DH Populations

Totally seven additive effect QTLs for grain weight were located on chromosomes 2D, 3A, 3B, 6A, 7B, and 7D (Table 5.2; Fig. 5.1), explaining total of 34.72 % phenotypic variance. Of which, *Qtkw6A* has the largest genetic contribution, which explained 9.39 % of the phenotypic variation. *Qtkw7B.2* explained 4.22 % of the phenotypic variation, and its efficiency allele came from the paternal Yumai 57, but other additive positive gene loci were from Huapei 3. Three pairs of epistatic QTLs for grain weight were detected, which located on chromosomes 2D-7D, 3A-7B.2, and 6A-7B.2, which explained 1.05, 1.13, and 1.14 % phenotypic variation, respectively. Eight additive effect QTLs controlling grain length were detected, which located on chromosomes 1B, 2D, 3A, 3B, 6A, 7B.2, and 7D. Single QTL explained 1.01–14.96 % of phenotypic variation. Of which, *Qgl6A* on chromosome 6A had the largest genetic contribution, which explained 14.96 % of the phenotypic variation and its positive gene was from the female parent Huapei 3. Five pairs of epistatic effects on grain length were.

Five epistatic QTLs of grain length on chromosomes 1B-2D, 2D-7D, 3A-6A, 3B-7B.2, and 6A-7B.2, explained 1.51, 1.16, 0.22, 0.56 and 0.86 % of the phenotypic variation, respectively (Table 5.3). Eight additive QTLs for grain diameter were detected on chromosomes 1B, 2D, 3A, 3B, 6A, 7B.2, and 7D, and the single QTL explained the phenotypic variance of 1.00–15.02 % (Table 5.3; Fig. 5.1). *Qgd6A* had the largest genetic contribution, explaining 15.02 % of the phenotypic variation with its positive gene from the female parent Huapei 3. Six pairs of epistatic QTLs for grain diameter loci were located on chromosomes 1B-2D, 2D-7D, 3A-6A, 3B-7B.2, 6A-7B.2, and 6A-7D, explaining 0.19–1.55 % of PVE.

Totally five additive effect loci for grain hardness were detected, which located on the chromosomes 1B, 4B, 5A, and 6A, with single QTL explaining 0.33–7.51 % of PVE (Table 5.3). Of which, *Qhd1Ba* located on chromosome 1B had the largest genetic contribution, explaining 7.51 % of the phenotypic variation with the positive gene contributed by Yumai 57. One pair of epistatic effect QTL for grain

Table 5.5 Estimated epistasis for 1000-kernel weight and grain length in IF_2 population

Trait	QTL	Flanking marker	Position (cM)	QTL	Flanking marker	Position (cM)	AA[a]		AD		DA		DD	
							AA H^2 (%)		AD H^2 (%)		DA H^2 (%)		DD H^2 (%)	
1000-kernel	*qtkw2A*	*XWMC455–XGWM515*	112	*qtkw4B*	*XWMC47–XWMC413*	4.2	−3.02	6.42			1.98	1.73		
Weight	*qtkw2B*	*XGPW2279–XGPW2228*	82.3	*qtkw7B.2*	*XGPW2224–XGPW3256*	30.5			5.05	1.74				
Grain length	*qgl1A*	*XGWM498–XGPW7412*	102.5	*qgl6B*	*XCFA2257–XCFD48*	50.4	19.7	0.37	46.86	0.34	−35.15	0.14	−66.94	1.58

[a]Epistasis includes interactions of additive-by-additive (AA), additive-by-dominance (AD), dominance-by-additive (DA), and dominance-by-dominance (DD)

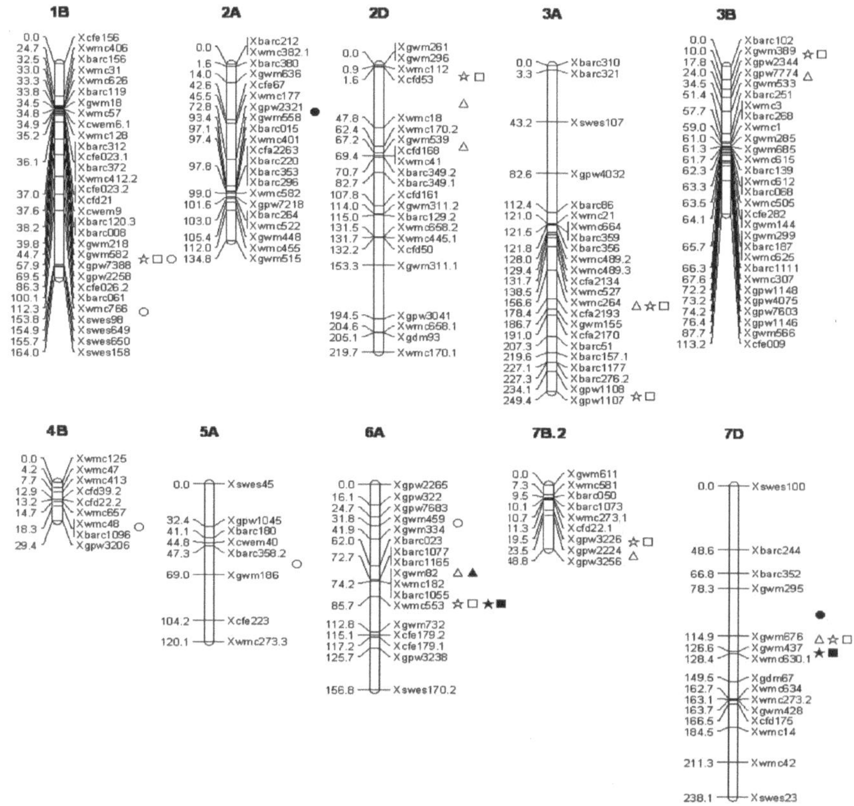

Fig. 5.1 QTL for TKW, GZ, and HD in DH and IF_2 populations. △☆□○ QTL for TKW, GL, GD, and HD in DH population; ▲★■● QTL for TKW, GL, GD, and HD in IF_2 population

hardness is located on chromosome 5A-6A, explaining 1.77 % of the phenotypic variation (Table 5.3).

5.1.1.2.2.2 QTL Analysis for Grain Weight, Grain Length, and Grain Hardness Using IF_2 Population

The loci *Qtkw6A* is located on 6A with both additive and epistatic effects (Table 5.4), explaining 1.37 and 11.75 % of the phenotypic variance, respectively, and its positive alleles were from Huapei 3. Besides, *Qtkw6A* was also detected in DH population. Two pairs of QTLs located on chromosomes 2A-4B and 2B-7B.2 were detected with dominance effects, including additive × additive (A × A), additive × dominance (A × D), dominance × additive (D × A), and dominance × dominance (D × D). Two additive effect sites for grain length were detected, which located on chromosomes 6A and 7D, explaining 23.34 % of PVE and its positive allele was from Huapei 3. *Qgl6A* had the greatest genetic contribution, explaining 15.10 % of PVE, and *Qgl6A* was the same site in DH population. One pair of epistatic loci on 1A-6B also includes additive × additive (A × A), additive × dominance (A × D), dominance × additive (D × A), and dominance × dominance (D × D) effects, explaining 0.37, 0.34, 0.14, and 1.58 % of the phenotypic variation, respectively (Table 5.5). Two additive effect loci of the grain diameter were located on chromosomes 6A and 7D, explaining 23.29 % of the total phenotypic variance, and its positive alleles were from Huapei 3. *Qgd6A* with additive effect had the greatest genetic contribution, explaining 15.10 % of the phenotypic variation. Besides, *Qgd6A* was the same loci in DH population. Two additive effect loci, *Qhd2A* and *Qhd7D*, for hardness were located on chromosomes 2A and 7D, explaining 3.773–7.90 % of the phenotypic variation, respectively. Their positive alleles were from Yumai 57 and Huapei 3, respectively.

5.1.2 Research Progress of Wheat Grain Quality QTL Mapping and Comparison with Previous Studies

5.1.2.1 Research Progress of Wheat Grain Quality QTL Mapping

Grain characteristics generally include grain weight, grain shape, grain color, grain hardness, and other traits. Wheat grain hardness is one of the most important quality traits and an important basis for market grading and pricing. With the development of test methods for hardness, study for molecular genetic basis has been gradually accelerating. Grain hardness was mainly controlled by a major gene (Ha) on the short arm of chromosome 5D and a number of minor effect genes. The major gene was dominant inheritance (Campbell et al. 1999; Chen et al. 2005). Pina and Pinb are the basis of wheat grain hardness. Sourdille et al. (1996) pointed out that grain hardness was controlled by multigenes. Of which, a major gene (Ha) was located on 5DS, and 4 minor genes were located on 2A, 2D, 5B, and 6D chromosomes, and another three minor genes were located on 5A, 6D, and 7A. Li et al. (2012) conducted QTL analysis for grain weight, grain shape, and grain hardness in DH183 and IF_2 populations, and he totally detected 35 additive effects QTLs and 18 pairs of epistatic effects QTLs. Of which, there were eight additive QTLs and five epistatic QTLs for grain weight, and 10 additive QTLs and six epistatic QTLs for grain shape, and seven additive QTLs and one epistatic QTL for grain hardness. Among them, *Qtkw6A* could be detected with additive effects and dominance effects in DH and IF_2 populations, and the phenotypic variance of the additive effect in the two populations was 9.39 and 11.75 %, respectively, and the contribution of dominance effects was 1.37 %. *Qgd6A* also detected in DH and IF_2 populations with 15.02 and 15.03 % of the phenotypic variance, respectively. It was the same gene locus for controlling grain length in DH and IF_2 populations with 14.96 and 15.10 % of the phenotypic variance, respectively.

5.1.2.2 Comparison of This Result with Previous Studies

Previous studies have identified the QTLs of wheat grain weight on chromosomes 1A, 1B, 1D, 2A, 2B, 2D, 3A, 3B, 3D, 4B, 5A, 5B, 5D, 6A, 6D, 7A, and 7D (Huang et al. 2004; Campbell et al. 2003; Wang et al. 2009; Yan et al. 2011). In this study, the QTLs for grain weight and grain shape including additive and epistatic loci were detected in both groups, indicating that wheat grain weight and grain type are complex quantitative traits. The genetic contribution rate of *Qtkw7b.2* on 7B.2 was 5.55 %, which was detected in DH population. In addition, *Qtkw7b.2* has the epistatic effects with *Qtkw3A* and *Qtkw6A*, respectively, and the total contribution was 2.27 %. It is a new QTL for grain weight in the present study. *Qtkw6A* with both additive and dominance effects explained 11.75 and 1.37 % of the phenotypic variation, respectively, which was the same loci detected in DH population. In the neighborhood of the *Qtkw6A*, Huang et al. (2004) detected *QTgw.ipk-6A.2*, and Groos et al. (2003) was also detected the QTL for grain weight in the same position. Su et al. (2011) cloned *TaGW2* based on homology GW2 in rice, which related to wheat grain weight and grain width, and the gene was located between the two SSR marker intervals barc1165-barc1055 on chromosome of 6A using the Chinese spring nulli tetrasomic lines, which was in the same marker interval of *Qtkw6A* in the present study. Therefore, *Qtkw6A* is an important QTL for marker-assisted selection breeding and cross-breeding.

The QTLs in this study for grain hardness were minor QTLs located on chromosomes 1B, 2A, 4B, 5A, 6A, and 7D in these two populations. The additive QTL, *Qhd7D,* in IF_2 population explained the highest of genetic contribution. So, the grain hardness is not only controlled by the major gene (Ha) on 5DS, but also affected by minor genes from other chromosomes (Sourdille et al. 1996).

5.2 QTL Mapping of Wheat Nutritional Quality Traits

Wheat is the world's second largest crop and contains large amounts of essential nutrients with 60–80 % carbohydrates, 8–15 % proteins, 1.5 % fat, 2–2.5 % of vitamins and minerals as well as other trace substances. The nutritional quality includes the content and quality of these above-mentioned nutrients, such as the content and quality of grain proteins, amino acids (especially lysine, threonine, and other limiting amino acid), carbohydrates, fats, minerals, and their components. The nutritional quality determined the nutritional value of wheat, which is directly related to the improvement of people's living standards and health. Therefore, people had analyzed the nutritional content of wheat grain many years ago and conducted the genetic analysis on grain protein content, amino acids, and starch. However, QTL analysis of the nutritional quality of wheat just began with the development of science and technology after the 1980s. Our research group has carried out the QTL analysis of nutritional quality traits in wheat based on a good foundation and technical advantages of wheat quality breeding.

5.2.1 QTL Mapping for the Protein Content of Wheat Grain and Flour

The protein content and quality is one of the important factors affecting the wheat processing quality. There were high correlations between wheat grain protein content and wet gluten content, dough rheological characteristics, and food processing quality. Different foods require different flours with different protein contents. For example, the flour for steamed bread generally needs the (12.5 ± 1) % of protein content, while the Chinese yellow alkaline noodle requires about 12–13 % of flour protein content. Genetic analysis indicated that the protein content was an additive–dominance model, and its genetic effect was overdominant. The dominance effect showed to be higher than additive effect. Meanwhile, the expression of protein content was strongly affected by environments. Therefore, the grain protein content was a quantitative trait controlled by multigenes. Previous researches indicated that about 13 chromosomes affected the protein content in different degrees, such as 1A, 2A, 3A, 5A, 7A, 2B, 3B, 4B, 6B, 7B, 4D, 5D, and 7D chromosomes. So the QTL of grain and flour protein content was studied using DH population in different environments.

5.2.1.1 Materials and Methods

5.2.1.1.1 GPC Measurement

Grain protein content and flour protein content were determined by near-infrared reflectance (NIR) using DA7200 type instrument (approved method 39-11 and 39-25, AACC 2004).

5.2.1.1.2 Plant Materials

Doubled haploid lines derived from a cross between two Chinese wheat cultivars Huapei 3 (Hp3) and Yumai 57 (Ym57) were used for the construction of a linkage map. The DH population and parents were kindly provided by Professor Yan Hai, Henan Academy of Agricultural Sciences, Zhengzhou, China. Hp3 and Ym57 were released by Henan Province in 2006 (Hai and Kang 2007) and the State (China) in 2003 (Guo et al. 2004), respectively. The two parents, which are cultivated over a large area in the Huang-Huai wheat region in China, differ for several agronomically important traits as well as for baking quality traits (Guo et al. 2004; Hai and Kang 2007).

5.2.1.1.3 Field Trials

The field trials were conducted in the three environments: at Tai'an (36°18′N, 117° 13′E), Shandong Province, in 2005 and 2006, and at Suzhou (31°3′2 N, 120°62′E),

Anhui Province, in 2006. The experimental design followed a randomized complete block design with two replications at each location. In autumn 2005, all lines and parental lines were grown in three-row plots (length 2 m, between-row spacing 25 cm); in autumn 2006, the lines were grown in four-row plots (length 2 m, between-row spacing 25 cm). The soil was brown earth that contained 40.2, 51.3, and 70.8 mg/kg of N, P, and K, respectively, in the top 20 cm. Before planting, 27,500 kg/ha farmyard manure or barnyard manure (N content 0.05–0.1 %), 225 kg/ha urea, 300 kg/ha P diamine fertilizer, 225 kg/ha KCl, and 15 kg/ha zinc sulfate were added as fertilizers. In Tai'an, the rainfall during the growth cycles was 165 and 172.5 mm in 2005–2006 and 2006–2007, respectively; in Suzhou, it was 207.5 mm in 2006–2007. Crop management followed local practices. Plots were irrigated during the winter (December 1, 2005) and at jointing (April 3, 2006), anthesis (May 7, 2006), and grain filling (May 20, 2006). They were top-dressed with 225 and 75 kg/ha urea at the jointing stage (April 3, 2006) and anthesis stage (May 7, 2006) with irrigation, respectively. The lines were harvested individually at maturity. Harvested grain samples were cleaned prior to conditioning, and flour milling was performed in a flour processing mill (Quadrumat Senior, Brabender, Germany) at flour extraction rates of around 70 %. Prior to milling, the hard, medium hard (mixtures of hard and soft wheats), and soft wheats were tempered to a moisture content of approximately 16, 15, and 14 %, respectively.

5.2.1.1.4 Statistical Analysis

The normal distribution test, correlation analysis, and the paired-samples t-test were carried out using SPSS version 13.0 (SPSS, Chicago, IL). QTLs with additive effects and epistatic effects as well as QE interaction in the DH population were mapped by the software QTL Network version 2.0 (Yang and Zhu 2005) based on the mixed linear model (Wang et al. 1999). Composite interval mapping was performed using the forward–backward stepwise multiple linear regression model (Eberly 2007) with a probability into and out of the model of 0.05 and a window size set at 10 cM. Significant thresholds for QTL detection were calculated for each data set using 1000 permutations and a genome-wide error rate of 0.10 (suggestive) and 0.05 (significant). The final genetic model incorporated significant additive effects and epistatic effects as well as their environmental interaction.

QTLs for grain protein content and flour protein content were denoted as Gpc and Fpc, respectively, followed by the relevant chromosome number. If there were more than one QTL on a chromosome, the serial number was added after the chromosome number, separated by a hyphen. For example, *QFpc5A-2* refers to a QTL on chromosome 5A. The positions of these QTLs were indicated by the marker interval flanking the concerned QTL with the estimated distance (cM) from the left marker.

5.2.1.2 Genetic Analysis of Protein Content in DH Population

5.2.1.2.1 Phenotypic Variation

The frequency distributions of grain protein content and flour protein content in the 168 DH lines derived from the cross of Hp3 and Ym57 in the three environments are shown in Fig. 5.2. The mean grain protein content was higher in the DH population than in the parents, Hp3 and Ym57, in all environments, and the mean flour protein content was higher in the DH population than in Hp3 and Ym57 except in 2005 in Tai'an. As such, the DH population expressed large-scale transgressive segregation. Both grain protein content and flour protein content segregated continuously and approximately fit normal distributions with absolute values of both skewness and kurtosis of <1.0, indicating that both traits were suitable for QTL mapping. Grain protein content was significantly correlated with flour protein content in the DH population ($r = 0.896$, $P < 0.01$).

5.2.1.2.2 QTL with Additive Effects and Additive X Environment (AE) Interaction

For grain protein content, four QTLs with additive effects were mapped to chromosomes 3A, 3B, 5D, and 6D, respectively (Table 5.6; Fig. 5.3). All four QTLs showed significant additive effects. Each of these QTLs explained 3.09 to 8.40 % of the phenotypic variance, and *QGpc3A* having the most significant effect accounted for 8.40 % of the phenotypic variance. Three additive loci, *QGpc3A*, *QGpc3B*, and *QGpc5D*, increased grain protein content by 0.29, 0.18, and 0.18 %, respectively, whose alleles come from Ym57. *QGpc6D* locus increased the grain protein content by 0.19 %, accounting for 3.45 % of the phenotypic variance, whose allele was from Hp3. These results suggested that the alleles that increased grain protein content were dispersed within the two parents, resulting in small differences in phenotypic values between the parents and transgressive segregants among the DH population. The total additive QTLs detected for grain protein content accounted for 18.25 % of the phenotypic variance.

Two additive effects were involved in AE interaction (Table 5.6; Fig. 5.3), explaining 3.64 and 5.63 % of the phenotypic variance, respectively. The *QGpc5D* locus reduced grain protein content by 0.19 % owing to AE effects but increased grain protein content by 0.24 %, correspondingly contributing 5.63 % of the phenotypic variance. The general contribution of all three AE effects on grain protein content was 9.27 %.

Four QTLs were detected for flour protein content on chromosomes 3A, 5D, 6D, and 7D, respectively (Table 5.6; Fig. 5.3). All four QTLs were identified with significant additive effects. The part of PVE explained by these QTLs ranged from 1.55 to 15.11 %. The strongest QTL on chromosome 3A explained up to 15.11 % of the variation of the trait, and the positive allele for this QTL was from Ym57. For *QFpc5D* and *QFpc7D*, the favorable alleles came from Ym57. The Hp3 allele

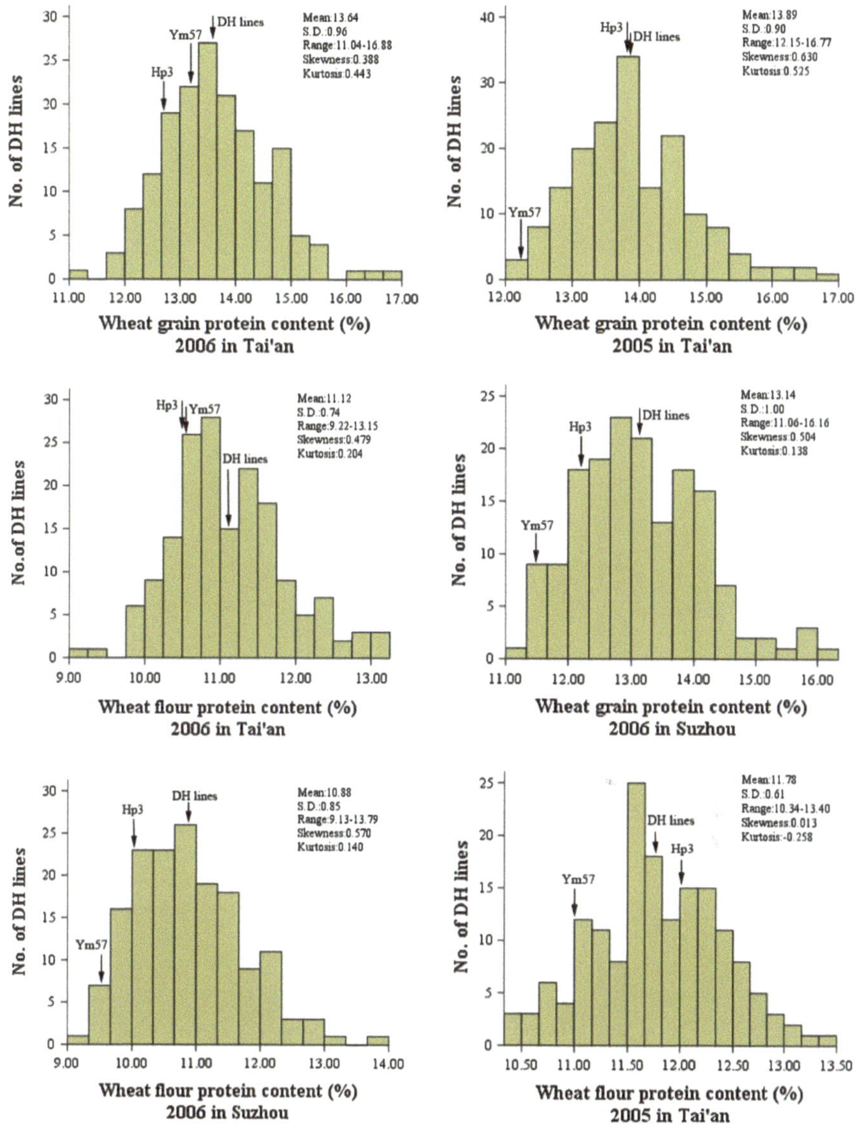

Fig. 5.2 Frequency distributions of grain protein content and flour protein content in 168 DH lines derived from a cross of Hp3 and Ym57 evaluated at three environments in the 2005 and 2006 cropping seasons. The means of trait values for the DH lines and both parents are indicated by *arrows*. Several statistics for the traits in the DH lines are shown in the *top right corner* of each graph

increased flour protein content at the *QFpc6D* by 0.20 %, accounting for 6.81 % of the phenotypic variance. The total additive QTLs detected for flour protein content accounted for 30.98 % of the phenotypic variance.

Table 5.6 Estimated additive (A) and additive × environment (AE) interactions of QTL for grain protein content and flour protein content

Traits	QTL	Flanking marker	Site/cM	A	H^2 (A)/%	(AE1)	H^2 (AE1)/%	(AE2)	H^2 (AE2)/%
Grain protein content	*QGpc3A*	*Xbarc86–Xwmc21*	8.2	−0.29**	8.40				
	QGpc3B	*Xgwm685–Xwmc615*	1.3	−0.18**	3.31				
	QGpc5D	*Xbarc320–Xwmc215*	9.7	−0.18**	3.09	0.24**	5.63	−0.19*	3.64
	QGpc6D	*Xcfd42–Xcfd13*	9.6	0.19**	3.45				
Flour protein content	*QFpc3A*	*Xbarc86–Xwmc21*	7.2	−0.29**	15.11				
	QFpc5D	*Xbarc320X–wmc215*	11.7	−0.21**	7.51	0.16	4.70		
	QFpc6D	*Xcfd42X–cfd13*	7.6	0.20**	6.81				
	QFpc7D	*Xwmc634–Xwmc273.2*	4.6	−0.09**	1.55				

E1: Tai'an, 2005; E2: Tai'an, 2006

* $P < 0.05$; ** $P < 0.005$; *** $P < 0.001$

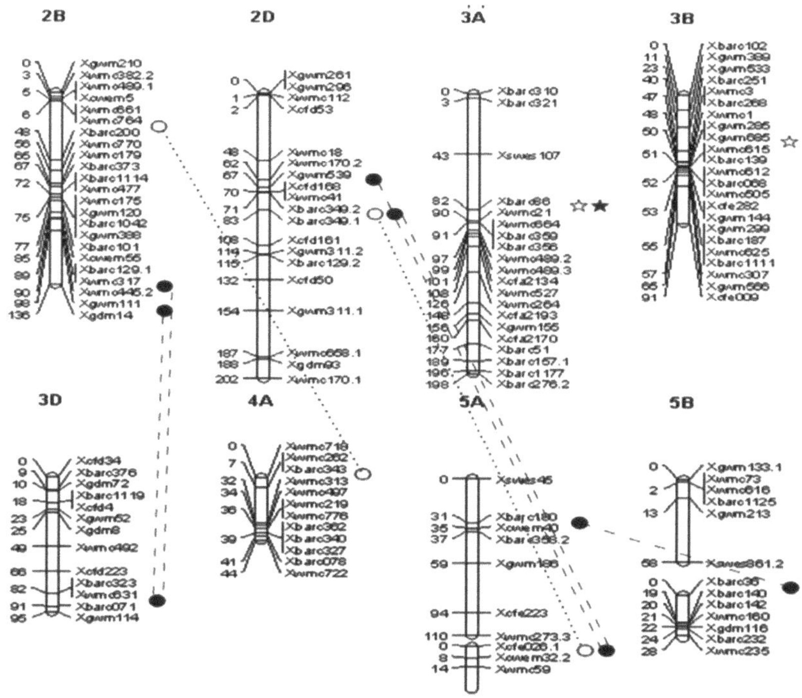

Fig. 5.3 Genetic linkage map of wheat showing mapping QTL with additive effects, epistatic effects, and QE interaction for grain protein content and flour protein content. For grain protein content: ☆ locus involved in additive effects, △ locus involved in AE, ○···○ locus involved in epistasis; for flour protein content: ★ locus involved in additive effects, ▲ locus involved in AE, ●- - -● locus involved in epistasis

One additive effect was involved in AE interaction (Table 5.6; Fig. 5.3). The Ym57 alleles at one locus, *QFpc5D*, increased the flour protein content by 0.16 %, and their corresponding contribution to the phenotypic variance was 4.70 %.

5.2.1.2.3 QTLs with Epistasis Effects and Epistasis × Environment (AAE) Interaction

For grain protein content, two pairs of epistatic effects were identified (Table 5.7; Fig. 5.3). One pair of epistasis QTLs, *QGpc2B/QGpc4A*, had the largest effect, contributing 0.38 % to grain protein content and accounting for 14.12 % of the phenotypic variance. Two pairs of epistatic effects both increased grain protein content and explained 24.00 % of the phenotypic variance. Two pairs of epistatic effects were non-main-effect QTL.

Five pairs of epistatic effects were identified for flour protein content (Table 5.7; Fig. 5.3). These QTLs had corresponding contributions ranging from 1.51 to 5.07 %.

Table 5.7 Estimated epistasis effects and epistasis × environment (AAE) interaction for grain protein content QTL and flour protein content QTL at three environments in 2005 and 2006 cropping seasons

Traits	QTL	Flanking marker	Position (cM)	QTL	Flanking marker	Position (cM)	AA[a]	H^2 (AA, %)[b]
Grain protein content	*QGpc2B*	*Xwmc764–Xbarc200*	12.0	*QGpc4A*	*Xbarc343–Xwmc313*	6.8	0.38***	14.12
	QGpc2D	*Xbarc349.2–Xbarc349.1*	9.0	*QGpc5A*	*Xcfe026.1–Xcwem32.2*	5.0	0.31***	9.88
Flour protein content	*QFpc2B-1*	*Xwmc317–Xwmc445.2*	4.0	*QFpc3D*	*Xwmc631–Xbarc071*	6.9	0.14***	3.30
	QFpc2B-2	*Xgwm111–Xgwm14*	22.0	*QFpc3D*	*Xwmc631–Xbarc071*	6.9	0.10*	1.61
	QFpc2D-1	*Xgwm539–Xcfd168*	4.9	*QFpc5A-2*	*Xcfe026.1–Xcwem32.2*	4.0	0.09***	1.51
	QFpc2D-2	*Xbarc349.2–Xbarc349.1*	9.0	*QFpc5A-2*	*Xcfe026.1–Xcwem32.2*	4.0	0.10***	1.93
	QFpc5A-1	*Xbarc180–Xcwem40*	8.6	*QFpc5B*	*Xbarc36–Xbarc140*	6.0	0.17***	5.07

[a]The epistatic effect. A positive value means that the parent-type effect is greater than the recombinant-type effect, and a negative value means that the parent-type effect is less than the recombinant-type effect

[b]H^2 (AA, %) indicates the contribution explained by putative epistatic QTL $*P < 0.05$, $**P < 0.005$, $***P < 0.001$, respectively

Epistasis occurring between loci *QFpc5A1* and *QFpc5B* had the largest effect, contributing 0.17 % to flour protein content and accounting for 5.07 % of the phenotypic variance. All five pairs of epistatic effects increased flour protein content and collectively explained 10.12 % of the phenotypic variance. All the epistatic effects were non-main-effect QTL.

No QTL was detected in AAE interaction for both grain protein content and flour protein content.

5.2.1.3 Genetic Analysis of Grain Protein Content in DH and IF_2 Populations

5.2.1.3.1 Phenotypic Variation for the Grain Protein

Crude protein of parents of Huapei 3 and Yumai 57 remained stable in different years and different locations (Table 5.8). Crude protein in Huapei 3 is slightly higher than that of Yumai 57. Descendants of DH and IF_2 populations showed continuous variation with the larger variation range, and there was an obvious two-way transgressive segregation phenomenon, so it can be used for QTL mapping.

5.2.1.3.2 QTL for Grain Protein Content in IF_2 Population

Genetic analysis for the grain protein content of IF_2 population was conducted by ICIM software. Totally nine QTLs on different chromosomes were detected in the three environments (Table 5.9), and among them, *Qgpc3B* was located on 3B and stably expressed in the three environments with the 5.32–6.21 % of the phenotypic variance, and its positive allele came from male parent Yumai 57. *Qgpc5A* detected in the environment 5 had the greatest genetic contribution (8.49 %), and its positive allele was from female parent Huapei 3.

Table 5.8 Phenotypic value of grain protein content for parents and DH and IF_2 populations in different environments

E	HP3	YM57	DH	IF_2
5[a]	14.80	13.44	14.76 ± 1.02[d]	14.23 ± 0.84
			12.53 − 18.09[e]	12.03 − 16.20
6[b]	14.94	13.94	13.36 ± 1.30	12.59 ± 1.06
			9.67 − 16.66	10.32 − 14.93
7[c]	14.02	13.63	12.38 ± 1.43	12.56 ± 1.40
			9.04 − 16.88	9.40 − 17.42

[a]Jiyuan, 2009; [b]Tai'an, 2010; [c]Jiyuan, 2010; [d]mean value ± standard deviation; [e]range

Table 5.9 Putative QTL for grain protein content in IF_2 population

E	Chromosome	Position (cM)	Flanking marker	A[a]	D[b]	H^2 (%)[c]
5[d]	1B	37	*Xcfd21–Xcwem9*	−0.2496	0.0539	5.5989
	1D	16	*Xwmc222–Xgdm60*	0.2846	−0.0791	7.6019
	3B	73	*Xgpw1148–Xgpw4075*	−0.3354	0.0455	5.324
	5A.2	9	*Xcwem32.2–Xwmc59*	0.1024	−0.4262	8.4905
	5D	0	*Xwmc630.2–Xcfd40*	−0.0312	−0.411	6.089
	6A	38	*Xgwm459–Xgwm334*	−0.0161	0.4673	8.3388
6[e]	1B	108	*Xbarc061–Xwmc766*	−0.002	−0.5104	6.2866
	1D	50	*Xcfd19–Xwmc93*	0.1494	0.4472	5.5284
	2A	106	*Xgwm448–Xwmc455*	0.346	0.0005	5.9051
	3B	75	*Xgpw7603–Xgpw1146*	−0.0354	0.5267	6.2097
7[f]	2A	130	*Xwmc455–Xgwm515*	−0.4533	−0.1359	6.1607
	2B	1	*Xgwm210–Xwmc382.2*	0.4807	−0.2352	6.0404
	2D	24	*Xcfd53–Xwmc18*	−0.4133	−0.5612	8.4124
	3B	73	*Xgpw1148–Xgpw4075*	−0.636	0.1942	6.1737

[a]Additive effects. Positive additive effects indicate that the Huapei 3 allele increases the value of the trait, and negative additive effects indicate that the Yumai 57 allele increases the value of the trait; [b]a positive value of dominant effect D denotes the heterozygote higher than the homozygote; [c]percentage of phenotypic variation explained by QTL with additive effect. [d]2009, Jiyuan; [e]2010, Tai'an; [f]2010, Jiyuan

5.2.1.4 Genetic Analysis for Protein Content Based on the "Gaocheng 8901 × Nuomai 1" RIL Population

5.2.1.4.1 Phenotypic Variation Analysis of Flour Protein Content

Phenotypic data of wheat flour protein content in the parent and the RIL population are shown in Table 5.10; Fig. 5.4. Results showed that the flour protein content of the female parent Nuomai is slightly higher than that of the male parent Gaocheng 8901. The RIL population showed transgressive segregation in protein content, and the absolute value of kurtosis and skewness were lower than 1.0, which indicated the normal distribution, so that the flour protein content was quantitative trait controlled by multigenes.

Table 5.10 Phenotypic values for wheat flour protein content of two parents and the RIL population in the three environments

Environment	Parents		RIL population			
	Nuomai 1 (%)	Gaocheng 8901 (%)	Range (%)	Mean ± SD	Skewness	Kurtosis
Tai'an, 2008	14.65	14.28	11.79–18.17	14.78 ± 1.12	0.082	0.253
Tai'an, 2009	13.71	13.64	11.59–15.96	13.73 ± 0.69	0.132	0.107
Anhui, 2011	13.62	13.3	11.38–15.33	13.49 ± 0.65	0.092	0.792

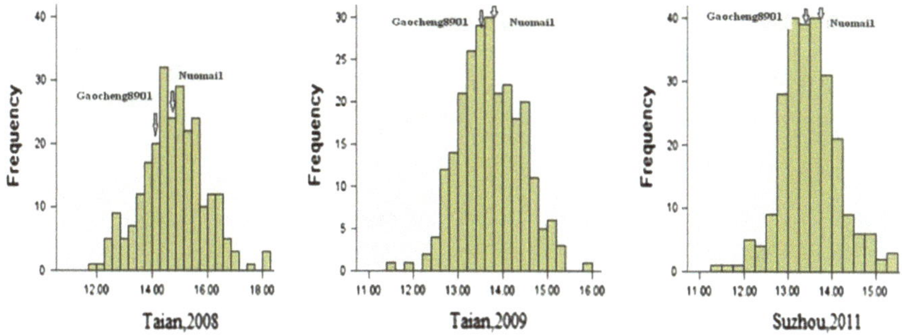

Fig. 5.4 Frequency distribution of wheat flour protein content

Table 5.11 QTL with significant additive effects for flour protein content detected in different environments

Environment	QTL	Site (cM)	Interval markers	Additive effect	Phenotypic variance explained %
E1	*QFPr-1A.1*	133.00	*wPt-6654–wPt-2872*	−0.28	6.40
	QFPr-1A.2	2.00	*wPt-5316–wPt-8016*	0.20	3.06
	QFPr-1B	234.00	*wPt-7273–wPt-3566*	0.21	3.40
	QFPr-4A	101.00	*wPt-664948–Wx-B1*	0.26	5.23
	QFPr-6B.1	21.00	*wPt-1730–wPt-9881*	−0.29	6.70
	QFPr-6B.2	84.00	*wPt-666793–wPt-663764*	0.22	3.80
	QFPr-7A.1	141.00	*wPt-5590–wPt-2780*	0.31	7.73
E2′	*QFPr-4A*	98.00	*wPt-664948–Wx-B1*	0.17	6.57
E3	*QFPr-4A*	101.00	*wPt-664948–Wx-B1*	0.14	4.90
	QFPr-7A.2	157.00	*wPt-731311–Wx-A1*	0.15	5.81
	QFPr-7B	193.00	*wPt-669158–wPt-7009*	−0.14	5.12
PD	*QFPr-1A.1*	135.00	*wPt-6654–wPt-2872*	−0.16	5.20
	QFPr-3B.1	164.00	*wPt-5870–wPt-3620*	0.58	7.06
	QFPr-3B.2	166.00	*wPt-3620–wPt-7906*	0.59	7.82
	QFPr-4A	101.00	*wPt-664948–Wx-B1*	0.23	10.38
	QFPr-7A.3	142.00	*wPt-2780–wPt-9207*	0.18	6.29

E1: Tai'an, 2008; E2′: Tai'an, 2009; E3: Suzhou, 2011; PD: pool data; positive values indicate that Nuomai 1 alleles increase corresponding trait value; negative values indicate that Gaocheng 8901 alleles increase corresponding trait value

5.2.1.4.2 QTL Analysis for Flour Protein Content

Twelve additive QTLs for wheat flour protein content were detected, which located on chromosomes 1A, 1B, 3B, 4A, 6A, 7A, and 7B (Table 5.11). Of which, *QFPr-4A* could be detected in E1, E2, E3, and average *P*, with the increasing effect variation coming from the female parent Nuomai 1, which explained from 4.90 to

10.38 % of the phenotypic variation. The locus was closely linked with *Wx-B1*. *QFPr-1A.1* was detected in E1 and P, and its increasing effect gene came from Gaocheng 8901, which explained 6.40 and 5.20 % of the phenotypic variation, respectively. The remaining QTL only detected in single environment.

5.2.1.5 Genetic Analysis for Protein Content Based on the "01-35 × Gaocheng 9411" Population

5.2.1.5.1 Phenotypic Variation Analysis for Grain and Flour Protein Content

Phenotypic data of grain protein content in the three environments and flour protein content in two environments are shown in Table 5.12. The large variation of the grain protein content and flour protein content in population showed continuous distribution, and there is an obvious transgressive separation. The coefficient of variation was more than 5.00 %, and the absolute value of skewness and kurtosis was less than 1, which indicated that protein contents of grain and flour were typically quantitative traits controlled by multigenes. In addition, grain protein content is positively correlated with the flour protein content with the correlation coefficient of 0.905.

5.2.1.5.2 QTL Analysis of the Protein Content of Grain and Flour

QTL mapping for grain protein content in the three environments E1, E2, and E4 based on the phenotypic value and average value (Table 5.13). Twelve additive QTLs controlling the grain protein content were detected, distributing on chromosomes 2B, 2D, 4B, 5B, 5D, 6D, and 7 B (Table 5.13). Of these, three major additive QTLs, including *QGpc2D-108*(E1), *QGpc6D-171*(E2), and *QGpc4B.1-62* (PD), derived from the maternal Shannon 01-35, accounting for 11.81, 14.58, and 10.35 % of the phenotypic variation, respectively, and they were expressed in specific environment. *QGpc4B.1-62* (PD, 10.355) and *QGpc4B.1-67* (E2, 7.91 %) shared the same marker intervals. Three QTLs, including *QGpc5B.2-12* (E1, PD),

Table 5.12 Phenotypic values for grain protein content and flour protein content of the RIL population and the parents in different environments

Trait	Environment	Parents		RIL population						
		01-35	9411	Mean	SD	Min.	Max.	CV (%)	Skewness	Kurtosis
GPC	E1	13.48	13.32	14.17	1.08	11.82	18.69	7.65	0.58	1.23
	E2	14.7	13.44	13.72	0.70	12.27	15.76	5.10	0.47	−0.27
	E4	14.77	14.84	14.39	1.01	11.77	17.29	7.00	0.07	−0.16
FPC	E1	13.33	14.04	14.09	0.99	11.56	16.43	7.03	0.14	−0.18
	E2	14.36	13.36	13.64	0.70	12.15	15.67	5.13	0.39	−0.24

E1: 2008–2009 growing season at Tai'an site, E2: 2009–2010 growing season at Tai'an site, E4: 2010–2011 growing season at Suzhou site

Table 5.13 Additive QTLs for grain protein content and flour protein content in different environments

Trait	Environment	QTL	Marker intervals	EstAdd	PVE (%)
Grain protein content	E1	*QGpc2D-108*	*wPt-4413-wPt-667294*	0.37	11.81
		QGpc5B.2-12	*wPt-9103-wPt-4418*	0.31	7.81
		QGpc5B.2-51	*wPt-4418-wPt-5168*	−0.24	4.91
		QGpc7B-0	*wPt-5463-wPt-4230*	−0.33	9.42
	E2	*QGpc2B-70*	*wPt-3132-wPt-1140*	−0.22	8.79
		QGpc2D-183	*wPt-2360-wPt-2644*	0.13	3.63
		QGpc4B.1-67	*Xgpw2172-wPt-1505*	0.20	7.91
		QGpc5B.1-117	*wPt-4628-wPt-3503*	0.19	7.27
		QGpc5D-0	*CFE291-wPt-0419*	0.19	7.23
		QGpc6D-171	*wPt-666615-wPt-666008*	0.27	14.58
	E4	*QGpc2D-117*	*wPt-665545-wPt-665228*	−0.28	7.79
		QGpc5D-0	*CFE291-wPt-0419*	0.23	5.15
	PD	*QGpc2D-183*	*wPt-2360-wPt-2644*	0.16	4.54
		QGpc4B.1-62	*Xgpw2172-wPt-1505*	0.24	10.35
		QGpc5B.2-12	*wPt-9103-wPt-4418*	0.18	5.87
		QGpc5D-0	*CFE291-wPt-0419*	0.21	8.19
Flour protein content	E1	*QFpc2A-283*	*wPt-5887-wPt-669199*	−0.30	9.18
		QFpc2B-75	*wPt-1140-wPt-4199*	−0.21	4.48
		QFpc2B-203	*wPt-1454-wPt-2425*	0.55	32.66
		QFpc2D-117	*wPt-4413-wPt-667294*	0.27	7.83
		QFpc2D-180	*wPt-2360-wPt-2644*	0.32	10.43
		QFpc3A-14	*Xgpw2270-wPt-1681*	−0.22	5.29
		QFpc3B.1-215	*wPt-1171-wPt-5906*	0.37	14.31
		QFpc4A-38	*Xgpw4040-wPt-7001*	−0.26	7.23

(continued)

Table 5.13 (continued)

Trait	Environment	QTL	Marker intervals	EstAdd	PVE (%)
		QFpc5B.2-12	*wPt-9103-wPt-4418*	0.66	45.61
		QFpc5B.2-51	*wPt-4418-wPt-5168*	−0.24	5.92
		QFpc7A-223	*Xgpw2139-wPt-7151*	−0.31	10.02
		QFpc7B-202	*wPt-0276-wPt-8920*	−0.26	7.00
	E2	*QFpc1B.1-80*	*wPt-2230-wPt-665375*	0.50	38.42
		QFpc1B.1-104	*wPt-5363-wPt-1363*	−0.38	21.04
		QFpc2D-98	*wPt-4413-wPt-667294*	0.28	15.62
		QFpc5B.1-118	*wPt-3503-CFE186*	0.18	6.45
		QFpc5B.2-12	*wPt-9103-wPt-4418*	0.16	5.04
		QFpc5B.2-52	*wPt-4418-wPt-5168*	−0.18	6.49
		QFpc5D-0	*CFE291-wPt-0419*	0.16	5.31
	PD	*QFpc2A-277*	*wPt-1871-wPt-3281*	−0.25	9.85
		QFpc2B-74	*wPt-1140-wPt-4199*	−0.22	6.61
		QFpc2D-127	*wPt-666987-wPt-0330*	0.19	5.48
		QFpc2D-183	*wPt-2360-wPt-2644*	0.21	6.59
		QFpc3A-25	*Xgpw2270-wPt-1681*	−0.19	5.53
		QFpc5B.2-12	*wPt-9103-wPt-4418*	0.18	5.00
		QFpc5B.2-52	*wPt-4418-wPt-5168*	−0.17	4.79
		QFpc7A-222	*wPt-3572-Xgpw2139*	−0.18	4.92
		QFpc7B-202	*wPt-0276-wPt-8920*	−0.26	10.20

E1: 2008–2009 growing season at Tai'an site, E2: 2009–2010 growing season at Tai'an site, E4: 2010–2011 growing season at Suzhou site. Positive and negative values of additive effect (EstAdd) indicate that alleles to increase thousand-grain weight are inherited from Shannong 01-35 and Gaocheng 9411, respectively

QGpc2D-183 (E2, PD), and *QGpc5D-0* (E2, E3 and PD), detected in single and average environments.

QTL mapping for flour protein content based on the phenotypic value and its average data in two environments was conducted. Totally 24 additive QTLs were detected to be associated with flour protein content, distributing on 1B, 2A, 2B, 2D, 3A, 3B, 4A, 5B, 5D, 7A, and 7B chromosomes, explaining 4.48 to 45.61 % of the phenotypic variation. Nine main QTLs on chromosomes 1B, 2B, 2D and 3B, 5B, 7A, and 7B, explaining 10.20–45.61 % of the phenotypic variation. *QFpc5B. 2-12* was detected in E1 with accounting for 45.61 % of phenotypic variation (PVE), which also be detected in the E2 and average with accounting for 5.04 and 5.00 % of the phenotypic variation, respectively. *QFpc7B - 202* was detected in E1 and average with 7.00 and 10.20 % of PVE, respectively. *QFpc5B. 2-52* detected in E2 and average perhaps was the same locus with *QFpc5B. 2-51* in E1 because they had the same marker intervals.

5.2.1.6 Research Progress of Protein Content QTL Mapping and Comparison with Previous Studies

5.2.1.6.1 Research Progress of Protein Content QTL Mapping

Grain protein is one of the main quality traits influencing the final application of durum wheat and bread wheat. It has been reported in 1948 (Finney and Barrimore 1948) that there exists a linear relationship between protein content and bread volume in bread wheat. At the same time, it was confirmed that wheat with high protein content affects the quality of the pasta (Marchylo et al. 2001). Due to the influence of environmental factors and other factors, the target of improved protein content by conventional breeding methods is very difficult, but it is available for the use of aneuploidy (Snape et al. 1995) and the molecular marker tool to improve the properties. The gene explored in wild tetraploid *Triticum turgidum* ssp *dicoccoides* can improve protein content from 13 to 89 % in common wheat (Avivi 1978; Grama et al. 1984). Particularly, Avivi (1978) imported the gene of FA15-3 derived from *T. turgidum* ssp *dicoccoides* t. to the durum wheat and increased the protein content of bread wheat.

In recent years, the detection of genetic basis of wheat protein traits has been made great progress. Nelson et al. (2006) conducted QTL mapping for a series of wheat quality traits, including the protein composition and quality, the strength of the flour, and kneading characteristics. It was pointed out that many quality traits shared similar QTL in all environments. Blanco et al. (2006) conducted QTL analysis for hard wheat grain protein content and discovered three QTLs that have a major effect on grain protein content, located on the 2AS, 6AS, and 7B, respectively. Charmet et al. (2005) conducted QTL mapping for the accumulation of dry matter, nitrogen, and protein components. A total of 7 QTLs were detected, and among them, five are related to the accumulation of dry matter and nitrogen, in which two of five QTLs affected protein component and other two QTLs were only related to protein content.

So far, domestic and overseas researchers carried out QTL mapping on the grain protein content and flour protein content traits with the RIL and DH genetic populations under different environmental conditions, and 64 QTLs were detected. The effect of 14 QTLs of these was more than 10 % with the greatest effect value up to 32.40 % (Table 5.14). The QTL loci involve 19 chromosomes. Of which, 6B and 7B chromosomes captured the most of QTL loci, each up to 6, and the following was 2A (5 QTL) and 6A (5 QTL). Therefore, 6B and 7B chromosomes are very important for wheat protein.

5.2.1.6.2 Comparison of This Result with Previous Studies

The results of QTL analysis for the grain protein content in DH population, IF_2 population, and RIL population showed that these QTLs involved in 14 chromosomes (1B, 1D, 2A, 2B, 2D, 3A, 3B, 4B, 5A, 5B, 5D, 6A, 6D, and 7B), and among them, QTLs were detected on 3B, 2D, 2B, and 6D chromosomes in the two populations. QTLs for protein content were detected on 5D in the three populations, indicating that some important QTLs/genes of controlling the grain protein content were located on 5D chromosome.

In DH population and two of RIL populations, QTLs for flour protein content were detected on 15 chromosomes (1A, 1B, 2A, 2B, 2D, 3A, 3B, 4A, 5B, 5D, 6B, 6D, 7A, 7B, and 7D), including the QTLs related to wheat flour protein content on 3A and 5D that were detected both in DH population and in 01-35/Gaocheng 9411 RIL population, and QTLs for flour protein content were detected on 1B, 4A, 7A, 7B, and 3B chromosomes in only two RIL populations.

Blanco et al. (2006) detected QTL for protein content in durum wheat and found that 3 QTLs had the major effect on grain protein content, which located on the 2AS, 6AS, and 7BL. Charmet et al. (2005) conducted QTL analysis on the wheat seed dry matter, nitrogen accumulation, and protein components, and a total of 7 QTLs were identified. Kuchel et al. (2006) located the QTL for flour protein content and flour yield on chromosome 6A. Huang et al. (2006) detected 5 QTLs for wheat grain and flour protein content on 2D, 4D, and 7B. Zhao et al. (2009) detected the grain protein and flour protein content on chromosomes 3A, 3B, 5D, 6D, and 7D. Previous researches showed that QTLs are distributed on many chromosomes (1A, 1B, 2A, 2B, 2D, 3A, 3B, 4A, 5B, 5D, 6B, 6D, 7A, 7B, 7D). In the present study, 4 stable QTLs were detected in the DH population, which affected both grain protein content and flour protein content. Among them, *QGpc3A* (*QFpc3A*) is one of the most important of 4 QTLs in grain protein content, explaining 8.4 % of the phenotypic variation (flour protein content of 15.11 %). *QGpc5D* (*QFpc3A*) was detected with additive effect and additive-by-environment interaction effects. The epistatic effect of 4 QTLs explained 9.88 % variation for grain protein content. The *QGpc6D*/*QFpc6D* explained 3.45 % of variation for grain protein content and 6.81 % of variation for flour protein content. Although most of previous researchers

Table 5.14 Summary of QTL results of wheat protein content (PVE > 10 %)

QTL	Flanking markers	PVE (%)	Population	References
QGpc.ccsu-2B.1	*Xgwm1249-2B*	13.40	RIL	Prasad et al. (2003)
QGpc.ccsu-2D.1	*Xgwm1264-2D*	14	RIL	Prasad et al. (2003)
QGpc.ccsu-3D.1	*Xgwm456-3D*	16.30	RIL	Prasad et al. (2003)
QGpc.ccsu-3D.2	*Xgwm892-3D*	14	RIL	Prasad et al. (2003)
QGpc.ccsu-7A.1	*Xgwm1171-7A*	32.40	RIL	Prasad et al. (2003)
QGpc.ndsu-5B.1	*Xgwm604-5B*	–	RIL	Gonzalez-Hernandez et al. (2004)
QGpc.ndsu-5B.2	*Xbarc310-5B*	–	RIL	Gonzalez-Hernandez et al. (2004)
QGpc.ndsu-5B.3	*Xwg909-5B*	–	RIL	Gonzalez-Hernandez et al. (2004)
Gpc-B1b	*Xucw79-6B-Xucw71-6B*	–	–	Distelfeld et al. (2004)
Gpc-6B1	*Xucw79-6B-Xucw71-8B*	–	–	Distelfeld et al. (2004)
Gpc-B1	*Xuhw89-6B*	–	–	Uauy et al. (2006)
Gpc-B1a,QGpc. ndsu.6Ba		–	–	Uauy et al. (2006)
Gpc-B1b	*A NAC transcription factor*	–	–	Uauy et al. (2006)
QGpc.ink.7B	*Ppd-B2*	–	–	Khlestkina et al. (2009)
3AS	*Xwmc749-3AS–Xgwm 369-3AS*	11	RIL	Sun et al. (2010)
4B	*Xgwm368-4B–Xwmc617*	11	RIL	Sun et al. (2010)
QFpc.agt-6A	*Xgdm141–Xgwm325*	13	DH	Kuchel et al. (2006)
	Xcfd31–Xgwm44	13	DH	Kuchel et al. (2006)
QGpc.sdau-3B	*Xwmc418–Xubc834a*	13.34	RIL	Sun et al. (2008)
QGpc.sdau-5A	*Xsrap27–Xwmc524*	21.23	RIL	Sun et al. (2008)
QGpc.caas-1A	*wms136*		IL	Li et al. (2012)
QGpc.caas-6A	*wms334*	11.5	IL	Li et al. (2012)
6A	*XE38M60200*	17.1	DH	Perretant et al. (2000)
QGpc.caas-3B	*wmc3-wmc418*	14.5	RIL	Li et al. (2009)
QGpc.caas-4D.1	*cfd71a-wmc457*	14.1	RIL	Li et al. (2009)
QGpc.caas-2B.2	*cwm13-wms71b*	12	RIL	Li et al. (2009)
QGpc.crc-4D	Xwmc617–Xwmc48	29.8	DH	McCartney et al. (2006)
QFpc.crc-4D	Xwmc617–Xwmc48	28.7	DH	McCartney et al. (2006)
QFpc.crc-2B	Xgwm210–Xwmc25	16.7	DH	McCartney et al. (2006)

analyzed QTL for grain protein content and flour protein content separately (Snape et al. 1995; Prasad et al. 1999; Groos et al. 2003; Kulwal et al. 2005), it was first reported stable QTL that affected grain protein content and flour protein content in the present study.

In the population of waxy wheat 1 and Gaocheng 8901, *QFPr-4A* related to flour content was stable expressed in the three environments and located on the range of *wPt-664948–Wx-B1*, indicating that the absence of Wx, especially Wx-B1, had some influence on the content of protein.

5.2.2 QTL Mapping for Beneficial Mineral Elements

Breeding the crops that are rich in mineral elements is an effective way to increase human food nutrition, and like other major food crops, the content of Fe, Zn, and other beneficial elements in wheat cannot meet the needs of the human body. So, it is of great significance for studying the genetic mechanism of mineral elements in wheat. At present, the studies for the mineral elements in wheat were only limited on content, but for the genetic mechanism of wheat, they are relatively less. Recently, the development of molecular marker technology and high-density molecular linkage map provides effective methods for the identification of the QTLs that control mineral elements. Therefore, in this study, DH population derived from the cross of Huapei 3 and Yumai 57 was used for QTL mapping for the content of Fe, Zn, Cu, and Mn. The results in present study laid the foundation for molecular marker-assisted selection in mineral element content of wheat.

5.2.2.1 Materials and Methods

5.2.2.1.1 Water Content Determination

The method was according to the GB/T 5009.3-2003.

5.2.2.1.2 Content of Fe, Zn, Cu, and Mn in Wheat Determination

Grains were milled into flour by Perten 3100 and sieved by 80-mesh polyurethane screen mesh, and flour is placed in sealed bags to save. Accurately weighed 1 g sample in a 250 mL beaker (covered with a Petri dish), then poured into 25 mL digestive solution (nitric acid perchloric acid: = 4:1), heated on an electric heating plate until the solution became clear, then transferred to a 100-mL volumetric flask, constant volume and mixed at last. At the same time, the above operations are conducted for the blank sample.

The contents of Fe, Zn, Cu and Mn were measured by flame Spectrophotometry method using the 1 Z-2000 type polarization of Zeeman atomic absorption spectro-photometer.

5.2.2.2 Results and Analysis

5.2.2.2.1 Contents of Fe, Zn, Cu, and Mn in Two Parents and DH Population

As we can be seen from the contents of Fe, Zn, Cu, and Mn in two parents and DH population (Table 5.15), the contents of Fe, Zn, Cu, and Mn have large range variability in the DH population. Transgressive segregation was obviously found in these four traits, which indicated the genes controlling these traits were dispersed in the two parents.

The distribution frequency of Fe, Zn, Cu, and Mn content is shown in Fig. 5.5. As can be seen from the graph, the contents of Fe, Zn, Cu, and Mn were quantitative traits and showed the basic continuous distribution in H population and indicated that these traits controlled by multiple genes and QTL mapping can be conducted on them.

5.2.2.2.2 Heritability of Fe, Zn, Cu, and Mn Contents

The heritabilities of Fe, Zn, Cu, and Mn contents of DH population are shown in Table 5.16. Of these, the heritability of Fe content in grains was the highest as 38.68 %, followed by Zn with the heritability of 30.26 %; the heritability of Cu and Mn contents was the lowest as 12.95 and 11.01 %, respectively. The results indicated that Fe, Zn, Cu, and Mn contents in wheat grain are slightly influenced by genetics and mostly influenced by the environment and the interaction between genotype and environment.

Table 5.15 Performance of characters of flour color from parents and DH population (μg g^{-1})

Trait	Parents		DH population				
	HP3	YM57	Average	Range	SD	Skewness	Kurtosis
Fe	48.57	36.45	47.68	28.64–66.39	6.17	0.2538	0.7849
Zn	37.82	28.33	32.94	20.51–47.95	5.44	0.2847	0.0986
Cu	4.71	5.96	5.87	4.35–7.77	0.60	0.3024	0.1726
Mn	36.75	33.54	34.59	25.01–52.47	4.75	0.6301*	0.7416

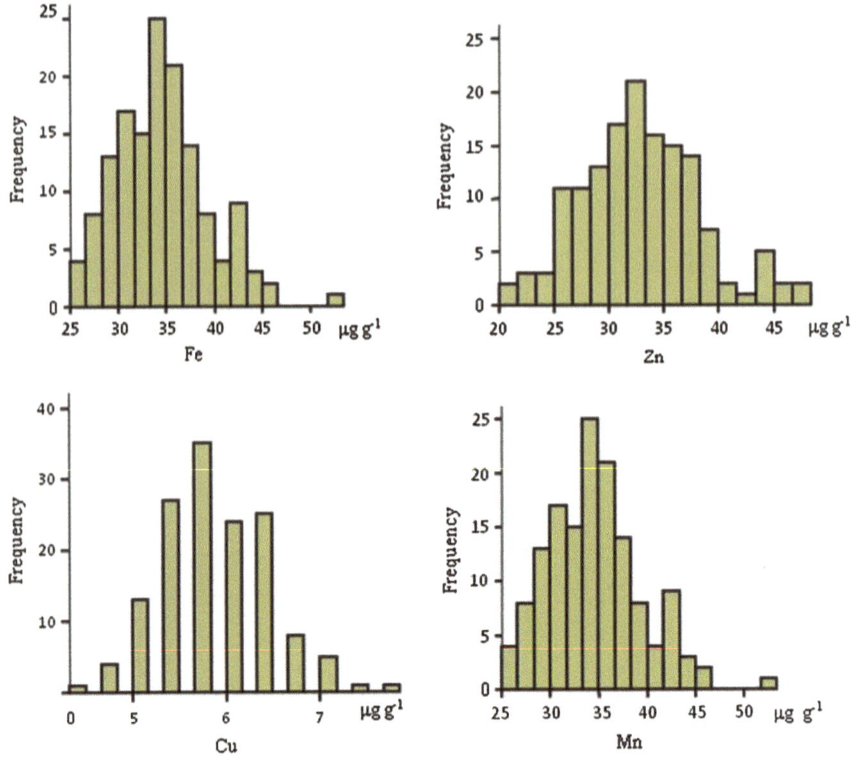

Fig. 5.5 Distribution of Fe, Zn, Cu, and Mn contents of wheat grain in the DH population

Table 5.16 Heritability (h^2 N) % of genetic factors involved in the traits

Characters	Fe	Zn	Cu	Mn
h^2 N (%)	38.68	30.26	12.95	11.01

5.2.2.2.3 Additive QTL Analysis for Fe, Zn, Cu, and Mn Contents

QTL analysis for Fe, Zn, Cu, and Mn contents was studied in three different environments in DH population (Table 5.17; Fig. 5.6), and three QTLs that controlled the content of Fe were detected on 2A, 5D, and 1B chromosomes, explaining 11.12 % of the phenotypic variation. Single QTL explained about 1.90–5.26 % of the phenotypic variation, and the additive effect was between 0.69 and 1.87. The QTL *QFE2A* with the highest value explained 5.26 % of the phenotypic variation, the additive effect was 1.19, and the positive grain allele came from the male parent Yumai 57. The additive effect value of *QFE1B* was 1.87, indicating that the positive gene that increased Fe content was from Yumai 57; *QFE5D* with

Table 5.17 QTL detection on Fe, Zn, Cu, and Mn contents in wheat grain

Element	QTL	Marker flanking	Position (cM)	Confidence interval	Additive	Variation (%)	Resource of allele
Fe	*QFE1B*	*Xbarc061–Xwmc766*	74.20	67.80–81.20	1.87	3.96	YM57
	QFE2A	*Xwmc582–Xbarc264*	69.30	69.30–71.30	1.19	5.26	YM57
	QFE5D	*Xwmc630.2–Xcfd40*	0.00	0.00–3.40	−0.69	1.90	HP3
Zn	*QZN3D*	*Xgdm8–Xwmc492*	27.7	23.00–27.70	−0.64	2.81	HP3
	QZN5A	*Xcewm40–Xbarc358.2*	34.7	30.60–38.10	−0.74	2.49	HP3
	QZN5B	*Xbarc232–Xwmc235*	36.4	29.40–44.40	−1.55	5.45	HP3
Cu	*QCU1D*	*Xwmc93-Glud*	72.0	58.90–61.90	−0.14	2.77	YM57
	QCU2B	*Xwmc477–Xwmc175*	61.9	64.50–74.00	−0.05	0.40	YM57
	QCU4D	*Xcfe254-Be293342*	194.2	169.40–194.20	−0.08	2.06	YM57
Mn	*QMN3A*	*Xwmc527–Xwmc264*	116.70	96.00–129.80	0.57	1.95	YM57

1.90 % of PVE had the additive effect value of −0.69, which suggested the positive allele came from Huapei 3. This would explain the transgressive segregation phenomenon in this population.

Three QTLs for Zn content of kernel were detected on 3D, 5A, and 5B chromosomes, respectively, and totally explained 10.75 % of the phenotypic variation. The single QTL explained the 2.49–5.45 % of the phenotypic variation. *QZN5B* explained 5.45 % of phenotypic variation, and its additive effect value was −1.55. The additive effect of the remaining two loci was smaller with negative value and showed that the increase of Zn content in grain of alleles was derived from the female parent Huapei 3.

Three QTLs for kernel Cu content were detected on 2B, 4D, and 1D chromosomes, respectively, and totally explained 5.23 % of the phenotypic variation. Single QTL explained 0.40–2.77 % variation with additive effect value from 0.05 to 0.14. The additive effect value of these three loci was negative, which indicated that the alleles related to the increase of Cu content were derived from the male parent Yumai 57.

One QTL for Mn content was detected to be located on chromosome 3A, and the additive effect was positive with effect value 0.57, which indicated that the positive allele came from the male parent Yumai 57.

5.2.2.2.4 Epistatic Effect Analysis of Fe, Zn, Cu, and Mn Contents

Epistatic effect analysis for the content of Fe, Zn, Cu, and Mn was conducted. The number of epistatic QTLs was different with different elements, and there were differences in the effects of epistatic QTL (Tables 5.18 and 5.19). Seven pairs of epistatic QTLs that controlled Fe content were detected, and total contributions were 27.57 %. The single QTL explained about from 0.42 to 6.85 % of the phenotypic variation. Among them, epistatic interaction between *QFE1B-3* and *QFE2A* with the highest contribution explained up to 6.85 % variation. *QFE1B-3/QFE2A*, *QFE2D/QFE4A*, and *QFE4A/QFE6B* with positive interaction values indicated that

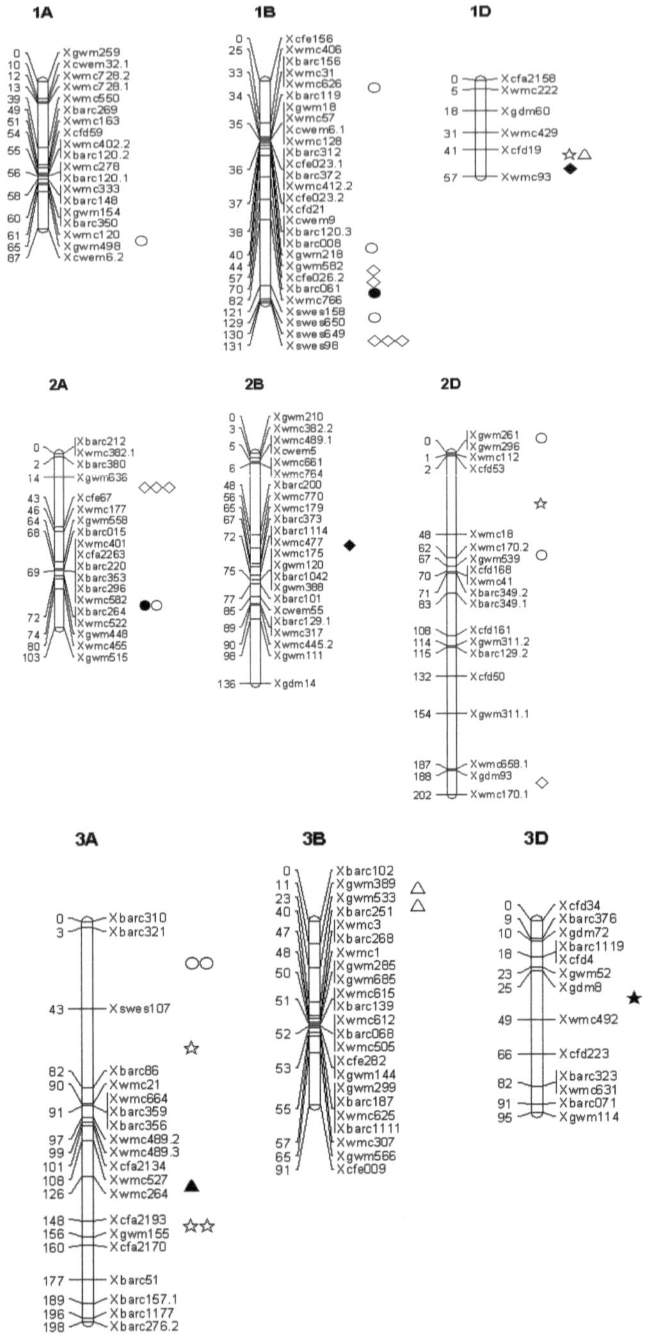

Fig. 5.6 A genetic linkage map of wheat showing mapping QTLs of Fe, Zn, Cu, and Mn with additive and epistatic effects. ●additive effect of Fe; ○epistatic effect of Fe; ★additive effect of Zn; ☆epistatic effect of Zn; ◆additive effect of Cu; ◇epistatic effect of Cu; ▲additive effect of Mn; △epistatic effect of Mn

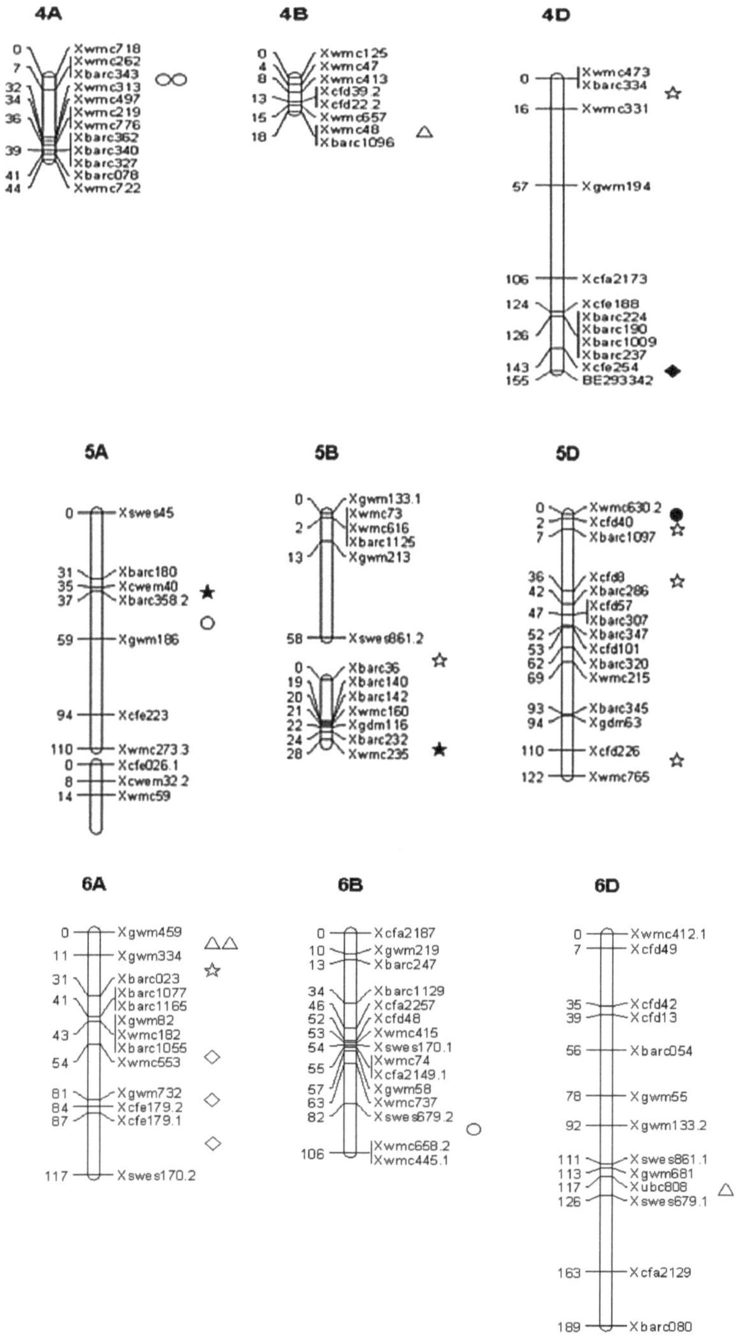

Fig. 5.6 (continued)

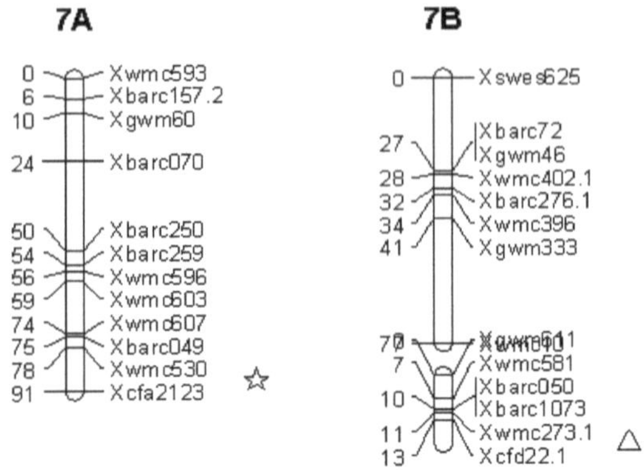

Fig. 5.6 (continued)

parent-type epistatic effect was higher than recombinant-type epistatic effect. The other four pairs of epistatic QTLs with negative effect values showed that the effect of recombinant type was larger than the parent type. Among seven loci with epistatic effect, only *QFE2A* also had additive effect. Each locus would not affect other when two loci alone existed, but when the two sites combined together, then the different additive and additive interaction effects will be produced.

Six pairs of epistatic QTLs that controlled Zn content were detected, and total contribution for Zn content was 19.50 %. The contribution of single loci was between 0.81 and 7.78 %. Among them, the contribution of *QZN4D/QZN6A* was the highest, explaining 7.78 % of the phenotypic variation. The interaction effect values of *QZN2D/QZN5B* and *QZN3A-2/QZN5D-2* showed negative, and the values were −1.38 and −0.52, respectively, which showed that the recombinant-type effect was larger than the parent-type effect. The rest of four pairs of interaction loci with positive effect value indicated that the parent-type effect was larger than recombinant-type effect. Six pairs of QTLs for Cu content were detected and total explained 7.80 % of the phenotypic variation. The contribution of single loci was 0.10–3.48 %. Four pairs of epistatic loci for Mn content were detected, and the contribution was increased to 9.05 % with the single site explaining about 1.52–2.65 % of the phenotypic variation.

5.2.2.3 Research Progress of Kernel Mineral Elements' QTL Mapping and Comparison with Previous Studies

5.2.2.3.1 Research Progress in Kernel Mineral Elements' QTL Mapping

There were fewer researches on the QTL analysis for the mineral element content of wheat grain. Zhao (2005) found three additive loci for Mn content in wheat grain, which located on chromosomes 1B, 5D, and 6A, explaining 11.75, 9.18, and 6.98 % of the phenotypic variation, respectively. Two pairs of epistatic QTL for Mn

Table 5.18 Estimated epistasis (AA) of QTLs on Fe and Zn contents in wheat grain

Element	QTL	Marker flanking	Position (cM)	QTL	Marker flanking	Position (cM)	AA	H^2 (AA, %)
Fe	*QFE1A*	*Xwmc120–Xgwm498*	63.20	*QFE2D*	*Xwmc170.2–Xgwm539*	62.40	−1.67	4.72
	QFE1B-1	*Xwmc626–Xbarc119*	33.30	*QFE3A*	*Xbarc321–Xswes107*	3.30	−0.07	1.16
	QFE1B-2	*Xbarc008–Xgwm218*	38.20	*QFE3A*	*Xbarc321–Xswes107*	3.30	−0.92	0.42
	QFE1B-3	*Xswes158–Xswes650*	124.10	*QFE2A*	*Xwmc382.1–Xbarc380*	1.00	1.96	6.85
	QFE2A	*Xwmc582–Xbarc264*	69.30	*QFE5A*	*Xbarc358.2–Xgwm186*	42.10	−1.77	5.48
	QFE2D	*Xgwm261–Xgwm296*	0.00	*QFE4A*	*Xbarc343–Xwmc313*	7.30	1.58	6.49
	QFE4A	*Xbarc343–Xwmc313*	7.30	*QFE6B*	*Xswes679.2–Xwmc658.2*	94.50	1.30	2.45
Zn	*QZN1D*	*Xcfd19–Xwmc93*	40.90	*QZN5D*	*Xcfd226–Xwmc765*	121.30	0.66	1.73
	QZN2D	*Xcfd53–Xwmc18*	1.60	*QZN5B*	*Xbarc36–Xbarc140*	10.00	−1.38	5.02
	QZN3A-1	*Xswes107–Xbarc86*	43.10	*QZN7A*	*Xwmc530–Xcfa2123*	90.50	0.95	2.72
	QZN3A-2	*Xcfa2193–Xgwm155*	154.60	*QZN5D-1*	*Xcfd8–Xbarc286*	40.60	0.61	1.44
	QZN3A-2	*Xcfa2193–Xgwm155*	154.60	*QZN5D-2*	*Xcfd40–Xbarc1097*	3.40	−0.52	0.81
	QZN4D	*Xbarc334–Xwmc331*	10.00	*QZN6A*	*Xgwm334–Xbarc023*	20.50	1.97	7.78

Table 5.19 Estimated epistasis (AA) of QTLs on Cu and Mn contents in wheat grain

Element	QTL	Marker flanking	Position (cM)	QTL	Marker flanking	Position (cM)	AA	H^2 (AA, %)
Cu	*QCU1B-1*	*Xgwm582–Xcfe026.2*	44.10	*QCU1B*	*Xswes649–Xswes98*	130.60	0.63	0.10
	QCU1B-2	*Xcfe026.2–Xbarc061*	56.80	*QCU1B*	*Xswes649–Xswes98*	130.60	−0.11	1.54
	QCU1B-3	*Xswes649–Xswes98*	130.60	*QCU2D*	*Xgmd93–Xwmc170.1*	187.60	−0.98	1.78
	QCU2A	*Xgwm636–Xcfe67*	16.00	*QCU6A-1*	*Xbarc1055–Xwmc553*	52.70	0.85	0.37
	QCU2A	*Xgwm636–Xcfe67*	16.00	*QCU6A-2*	*Xcfe179.1–Xswes170.2*	116.70	−0.13	3.48
	QCU2A	*Xgwm636–Xcfe67*	42.00	*QCU6A-3*	*Xgwm732–Xcfe179.2*	82.40	−0.07	0.53
Mn	*QMN1D*	*Xcfd19–Xwmc93*	40.90	*QMN7B*	*Xwmc273.1–Xcfd22.1*	12.70	−0.66	1.52
	QMN3B-1	*Xgwm389–Xgwm533*	10.60	*QMN6A*	*Xgwm459–Xgwm334*	0.00	−1.27	2.65
	QMN3B-2	*Xgwm533–Xbarc251*	23.00	*QMN6A*	*Xgwm459–Xgwm334*	0.00	0.90	2.37
	QMN4B	*Xwmc48–Xbarc1096*	18.40	*QMN6D*	*Xubc808–Xswes679.1*	117.30	0.83	2.51

content were detected to be located on 2BL/3D and 7AS/7BL, explaining 14.31 and 9.91 % of the phenotypic variance, respectively. One additive QTL for Zn was detected on 3BS and explained 10.94 % of the phenotypic variation. One pair of epistatic QTL for Cu content was detected on chromosome 3BL/5BL and explained 12.95 % of the phenotypic variation. One pair of epistatic QTL for Fe content was detected on 6AS/7DS, explaining 3.36 % of the phenotypic variation.

5.2.2.3.2 Comparison of This Result with Previous Studies

The additive effect contributions of the QTL for Fe, Zn, Cu, and Mn contents were 11.12, 10.75, 5.23, and 1.95 % of the variation, respectively, and the total contributions of epistasis effect showed 27.57, 19.50, 7.80, and 9.05 % of variation, respectively. These indicated that the contributions of epistatic effects on the four mineral elements were higher than those of their additive effect, which indicated that the additive effect and epistatic effect both played important roles in the variation of phenotype. Of which, only *QFE2A* was detected with both additive effect and epistatic effect. Other epistasis effect QTL was not detected with additive effect, but was detected with large genetic effects on wheat grain Fe, Zn, Cu, and Mn contents with the influence of other loci. Therefore, both the additive effect and the epistatic effect should be paid attention to the genetic breeding of wheat grain mineral nutrition.

5.2.3 QTL Mapping of Amino Acid Content and Components in Wheat Grain

Wheat, the most important cereal crop in the world (Peňa et al. 2006), is the principal source of energy, protein, and dietary fiber for most of the world's population (Abdel-Aal and Huclw 2002). Wheat protein quality is mainly influenced by protein content and the balance of amino acid composition in the wheat (Liu et al. 2002; Li et al. 2000). Amino acid composition in wheat protein is unbalanced, especially for the lysine and threonine contents with the average of 3.0 and 1.4 %, respectively, which is showed to be lower than the ideal content released by FAO/WHO. Therefore, it is of great significance to improve the wheat nutritional quality by increasing the amino acid content, especially lysine content.

The wheat grain amino acid content is controlled by multigenes and has the complex genetic basis. Meanwhile, there was no obvious corresponding relationship between phenotype and genotype, and their testing program is complicated with the high cost, so it is difficult to improve the amino acid content according to the phenotypic selection using traditional breeding. With the development of quantitative genetics and molecular marker technology, the amino acid content and components can be improved by using molecular marker-assisted selection. So in this study, the QTL mapping of amino acid content was studied to detect the major QTL for improving the amino acid content by MAS.

5.2.3.1 Materials and Methods

5.2.3.1.1 Plant Materials

A population of 168 DH lines generated from a cross between two Chinese bread wheat cultivars "Huapei 3"/"Yumai 57" was used for the construction of a genetic linkage map. The DH population and parents were provided by Henan Academy of Agricultural Sciences, Zhengzhou, China. "Huapei 3" and "Yumai 57," winter wheat, were registered by Henan Province in 2006 (Hai and Kang 2007) and by China in 2003 (Guo et al. 2004).

5.2.3.1.2 Field Experiment

The field experiment was conducted under different environmental conditions on the experimental farm at Shandong Agricultural University, Tai'an, Shandong Province, China, in 2008 and 2009. A completely randomized block design was used with three replications in two years, and all lines and parental lines were grown in 2-m-long four-row plots (25 cm apart). Crop managements were carried out following the local practices. The lines were harvested individually at maturity and cleaned prior to milling.

5.2.3.1.3 Milling and Determination of the Amino Acid Content

The whole wheat meals were prepared by milling the wheat with a 3100 Mill (Perten, Sweden). Wheat (200 g) was added into the feeder, and then, the mill was started and continued to run for 1 min after all the wheat pasted the roller.

Amino acid composition was obtained using an amino acid analyzer (Biochram 30; Amersham, Britain) using Chinese standard method GB7649-87 1987. In a test tube, 10 mL of 6 N HCl was added to a 50 mg sample. The test tube was evacuated and flushed with nitrogen, sealed and placed in an oven at 110 °C for 24 h and then cooled to room temperature. The hydrolysate was filtered to remove the visible sediments and evaporated under vacuum at 60 °C. The hydrolysate was dissolved in 1 mL of buffer (pH 2.2). A known volume (20 μL) was injected into the amino acid analyzer to estimate the amino acid profile for each sample. Each sample was replicated.

5.2.3.1.4 Data Analysis

The result of each material was presented on a dry matter basis. The normal distribution test and paired-samples t-test for 17 individual amino acids and total amino acid content of DH lines and parents were analyzed using the SPSS 16.0.

5.2.3.1.5 QTL Analysis

A genetic linkage map of the DH population with 323 markers, including 284 simple sequence repeat (SSR) loci, 37 expressed sequence tag (EST) loci, one inter-simple sequence repeat (ISSR) locus, and one high molecular weight glutenin subunit locus, was used in this study. This linkage map covered a total length of 2485.7 cM, with an average distance of 7.67 cM between adjacent markers.

The inclusive composite interval mapping (ICIM) was applied by means of the QTL IciMapping v2.2 (Li et al. 2007) to identify additive QTL for amino acid contents. A logarithm of odds (LOD) of 2.0 was set to declare QTL as significant. QTL effects were estimated as the proportion of phenotypic variance (R^2) explained by the QTL.

5.2.3.2 Results and Analysis

5.2.3.2.1 Phenotypic Trait Analysis

A total of 18 traits including 17 individual amino acids, TAA of the DH population, and the parents under two environments are described in Table 5.20. The most of AAC was significantly ($P < 0.05$) different between the two parents. "Huapei 3" had higher concentration for the majority of amino acids. Differences were also found for the 18 components of AAC among DH lines under two environments. The DH population showed transgressive segregations in both directions for all traits in this study.

5.2.3.2.2 QTL Analysis

5.2.3.2.2.1 Additive QTL for AAC of Wheat in 2008

Thirty-two individual additive QTLs (LOD > 2.0) were detected for the 18 components of AAC in the DH population in 2008, ranging from one to four QTLs for each trait (Table 5.21; Fig. 5.7). The total phenotypic variation explained by the individual QTL for the 18 components of AAC varied from 4.86 % for Asp content to 30.95 % for Ser content. Most of the 32 QTLs are mainly distributed on chromosomes 3A, 6D, and 7D.

Table 5.20 Amino acid content of wheat for DH population and parents in 2008 and 2009 (mg/g)

Trait	2008					2009				
	Parents		DH population			Parents		DH population		
	"Huapei 3"	"Yumai 57"	Mean	Min.	Max.	"Huapei 3"	"Yumai 57"	Mean	Min.	Max.
Lys	2.66a	2.67a	2.80	1.86	3.66	2.59a	2.38a	2.53	1.58	3.12
Thr	3.04a	3.00a	3.20	2.15	3.90	3.06a	2.82a	2.92	1.77	3.78
Ile	6.72a	6.28b	6.68	3.69	7.99	6.68a	6.14b	6.20	3.04	9.46
Phe	4.47a	4.12b	4.93	3.36	6.27	4.80a	4.27b	4.61	2.52	8.69
Val	8.02a	7.55b	7.86	4.66	9.27	7.61a	6.79b	6.86	2.42	9.95
Met	7.42a	6.70b	6.59	2.16	8.11	6.67a	5.87b	6.45	2.21	11.81
Leu	10.65a	10.80a	10.28	6.28	12.26	10.38a	9.01b	9.76	6.53	12.99
Glu	29.28b	30.28a	32.23	22.72	40.03	28.72a	26.27b	28.40	17.85	36.39
Pro	10.88a	9.68b	10.58	6.98	15.10	10.48b	11.17a	8.40	4.23	12.55
Cys	7.15a	6.74b	6.85	3.20	8.90	5.44a	5.24a	5.99	2.87	10.62
Asp	5.85a	5.98a	6.28	4.24	7.40	5.53a	5.23b	5.73	3.31	6.99
Ser	4.42a	4.44a	4.67	3.05	6.13	4.47a	4.08b	4.56	2.63	5.97
Gly	4.16a	4.36a	4.34	2.81	5.08	4.28a	4.17a	4.19	2.43	5.15
Ala	4.46a	4.28b	4.43	2.82	5.53	4.68a	4.36b	4.60	2.62	5.83
Tyr	3.66a	3.53a	3.60	2.34	4.43	2.70a	2.47a	3.41	1.71	5.66
His	2.54a	2.43a	2.73	1.80	4.96	2.46a	2.22b	2.21	1.36	3.15
Arg	5.25a	4.81b	5.48	3.38	6.64	4.71a	4.91a	4.61	2.68	6.23
TAA	120.63a	117.66a	123.52	83.39	146.29	115.25a	107.40b	111.40	70.34	147.37

The different alphabets in the right side of the same line show significant difference ($P < 0.05$)

Table 5.21 Additive QTL for amino acid content in the DH population in 2008

Trait	QTL	Chr.	Site (cM)	Flanking marker	A	LOD	R^2 (%)
Lys	*QLys3A*	3A	96	Xbarc356-Xwmc489.2	−0.05	2.12	5.96
Thr	*QThr3A*	3A	91	Xbarc356-Xwmc489.2	−0.07	2.86	7.31
Ile	*QIle6D*	6D	112	Xswes861.1-Xgwm681	−0.99	3.93	18.12
Phe	*QPhe6D-1*	6D	38	Xcfd42-Xcfd13	0.13	2.16	5.48
	QPhe6D-2	6D	113	Xswes861.1-Xgwm681	−0.51	2.19	8.68
Val	*QVal2A*	2A	74	Xgwm448-Xwmc455	−0.22	2.55	6.23
	QVal3A	3A	97	Xbarc356-Xwmc489.2	−0.22	2.54	6.12
Mat	*QMet3D*	3D	10	Xgdm72-Xbarc1119	0.29	2.10	5.59
Leu	*QLeu3A*	3A	86	Xbarc86-Xwmc21	−0.30	2.74	8.06
	QLeu6D	6D	112	Xswes861.1-Xgwm681	−1.31	2.61	14.18
Glu	*QGlu3A*	3A	72	Xswes107-Xbarc86	−0.82	3.04	10.20
	QGlu7D	7D	164	Xgwm428-Xcfd175	−0.57	2.21	5.04
Pro	*QPro3A*	3A	91	Xbarc356-Xwmc489.2	−0.37	3.18	7.66
	QPro6A	6A	107	Xcfe179.1-Xswes170.2	0.37	2.35	8.03
Cys	*QCys3D*	3D	10	Xgdm72-Xbarc1119	0.32	2.82	7.44
Asp	*QAsp3A*	3A	91	Xbarc356-Xwmc489.2	−0.12	2.58	6.05
	QAsp6B	6B	0	Xcfa2187-Xgwm219	0.11	2.13	4.86
	QAsp6D	6D	113	Xswes861.1-Xgwm681	−0.43	2.04	7.61
Ser	*QSer3A*	3A	76	Xswes107-Xbarc86	−0.15	4.60	13.12
	QSer7D	7D	164	Xgwm428-Xcfd175	−0.23	12.03	30.95
Gly	*QGly3A*	3A	92	Xbarc356-Xwmc489.2	−0.10	3.95	10.55
	QGly6B	6B	56	Xwmc74-Xgwm58	0.65	3.06	18.20
	QGly7D	7D	152	Xgdm67-Xwmc634	−0.08	2.47	6.63
Ala	*QAla3A*	3A	92	Xbarc356-Xwmc489.2	−0.12	3.36	8.89
Tyr	*QTyr3A*	3A	85	Xbarc86-Xwmc21	−0.09	2.28	6.55
His	*QHis1B*	1B	50	Xgwm582-Xcfe026.2	−0.10	2.35	7.94
	QHis6B	6B	54	Xwmc415-Xswes170.1	−0.87	5.08	30.18
Arg	*QArg2B-1*	2B	42	Xwmc661-Xbarc200	−0.20	5.51	16.03
	QArg2B-2	2B	76	Xgwm388-Xbarc101	0.14	3.32	8.14
	QArg3A	3A	95	Xbarc356-Xwmc489.2	−0.18	4.76	12.24
	QArg7D	7D	151	Xgdm67-Xwmc634	−0.11	2.17	4.94
Total	*QTotal3A*	3A	91	Xbarc356-Xwmc489.2	−2.50	2.48	6.58

Most of the QTLs detected tended to be colocalized within the genome, and thus, 16 chromosome regions were identified in the study in 2008, including seven QTL clusters and nine single QTL. The allelic effects of each QTL in the same QTL cluster were all in the same direction. The QTL cluster, flanked by Xbarc356-Xwmc489.2 on chromosome 3A, was the largest in number, which consisted of nine individual QTLs (associated with Lys, Thr, Val, Pro, Asp, Gly, Ala, Arg, and TAA), followed by the QTL cluster near Xswes861.1-Xgwm681 on chromosome 6D, consisting of four amino acids (Ile, Phe, Leu, and Asp). However, according to the average

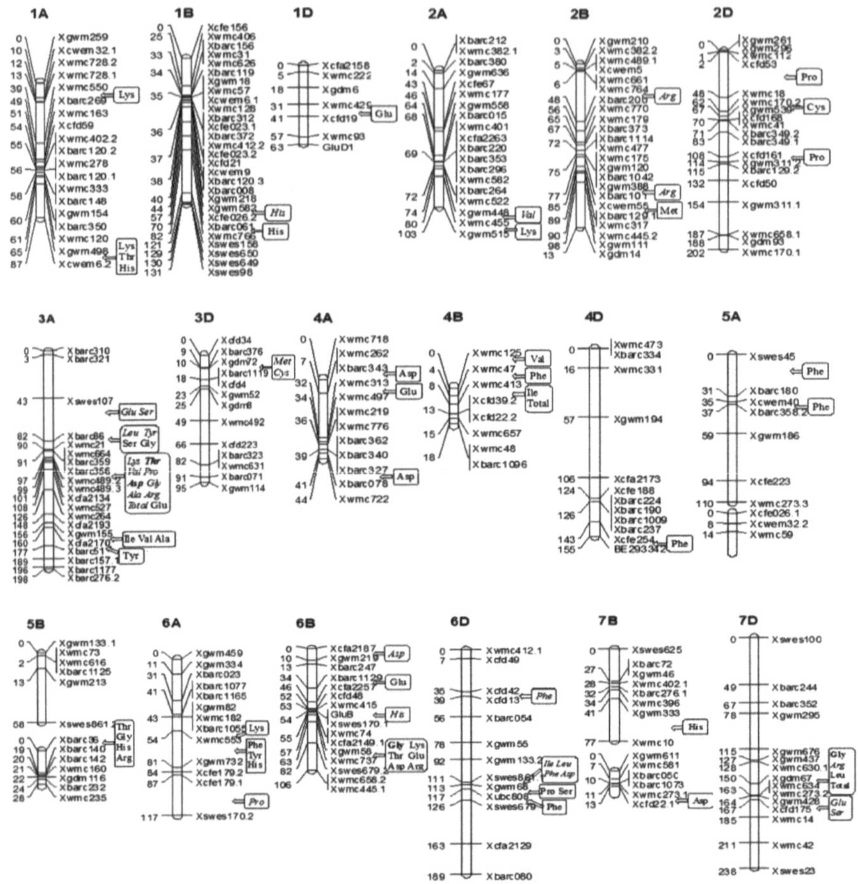

Fig. 5.7 QTL for amino acid content in the DH population in 2008 and 2009 wheat. The *italic* and *normal* format means the QTL for amino acid contents detected in 2008 and 2009 wheat, respectively; the **bold** format means the QTL for amino acid contents detected in two years

phenotypic variation explained by individual QTL (R^2), the QTL cluster flanked by Xgwm428-Xcfd175 on chromosome 7D was the largest, accounting for 17.99 % of total variation, followed by the QTL clusters near Xswes861.1-Xgwm681 on chromosome 6D and near Xswes107-Xbarc86 on chromosome 3A (R^2 = 12.15, 11.66 %, respectively).

Ten major QTLs (R^2 > 10 %) associated with Gly, Ile, Leu, Ser, His, and Arg contents were identified in the population in 2008. Particularly, the QTL for Ser flanked by Xgwm428-Xcfd175 on 7D chromosome could explain 30.95 % of total phenotypic variation and the other QTL for His linked to Xwmc415-Xswes170.1 on chromosome 6B, whose R^2 was 30.18 %.

For Lys and Thr contents, only one QTL distributed on chromosome 3A linked with Xbarc356-Xwmc489.2 was detected, explaining 5.96 and 7.31 % of total

Table 5.22 Additive QTL for amino acid content in the DH population in 2009 wheat

Trait	QTL	Chr.	Site (cM)	Flanking marker	A	LOD	R^2 (%)
Lys	*QLys1A-1*	1A	49	Xwmc550–Xbarc269	0.07	4.41	9.79
	QLys1A-2	1A	65	Xgwm498–Xcwem6.2	−0.10	9.02	21.08
	QLys2A	2A	101	Xwmc455–Xgwm515	0.06	2.64	6.24
	QLys6A	6A	43	Xgwm82–Xwmc553	−0.05	2.53	5.43
	QLys6B	6B	55	Xwmc74–Xgwm58	0.29	3.00	8.47
Thr	*QThr1A*	1A	65	Xgwm498–Xcwem6.2	−0.07	2.66	5.83
	QThr3A	3A	91	Xbarc356–Xwmc489.2	−0.07	3.14	7.03
	QThr5B	5B	2	Xbarc36–Xbarc140	0.08	3.22	8.10
	QThr6B	6B	55	Xwmc74–Xgwm58	0.28	3.07	7.07
Ile	*QIle3A*	3A	159	Xgwm155–Xcfa2170	−0.37	2.04	5.56
	QIle4B	4B	8	Xwmc413–Xcfd39.2	0.42	2.79	7.18
Phe	*QPhe4B*	4B	7	Xwmc47–Xwmc413	0.38	4.06	8.46
	QPhe4D	4D	155	Xcfe254–BE293342	−0.30	2.68	5.39
	QPhe5A-1	5A	6	Xswes45–Xbarc180	−0.46	4.48	12.54
	QPhe5A-2	5A	35	Xcwem40–Xbarc358.2	0.29	2.39	4.82
	QPhe6A	6A	73	Xwmc553–Xgwm732	−0.34	2.42	6.92
	QPhe6D	6D	124	Xubc808–Xswes679.1	−0.50	2.73	7.06
Val	*QVal3A*	3A	159	Xgwm155–Xcfa2170	−0.34	2.44	6.62
	QVal4B	4B	0	Xwmc125–Xwmc47	0.33	2.56	6.42
Mat	*QMet2B*	2B	85	Xcwem55–Xbarc129.1	0.43	2.42	6.46
Leu	*QLeu7D*	7D	162	Xgdm67–Xwmc634	−0.39	3.50	9.48
Glu	*QGlu1D*	1D	32	Xwmc429–Xcfd19	0.54	2.37	4.84
	QGlu3A	3A	91	Xbarc356–Xwmc489.2	−0.80	5.25	10.80
	QGlu4A	4A	32	Xwmc313–Xwmc497	0.52	2.33	4.63
	QGlu6B-1	6B	42	Xbarc1129–Xcfa2257	0.90	2.22	6.75
	QGlu6B-2	6B	55	Xwmc74–Xgwm58	2.25	2.85	5.80
Pro	*QPro2D-1*	2D	45	Xcfd53–Xwmc18	−0.40	3.19	8.53
	QPro2D-2	2D	114	Xcfd161–Xgwm311.2	0.41	3.84	9.10
	QPro6D	6D	114	Xgwm681–Xubc808	−0.89	2.34	8.10
Cys	*QCys2D*	2D	67	Xwmc170.2–Xgwm539	−0.42	2.41	6.08
Asp	*QAsp3A*	3A	92	Xbarc356–Xwmc489.2	−0.13	4.34	9.15
	QAsp4A-1	4A	10	Xbarc343–Xwmc313	−0.12	3.02	7.44
	QAsp4A-2	4A	39	Xbarc362–Xbarc078	0.14	5.09	10.36
	QAsp6B	6B	56	Xwmc74–Xgwm58	0.96	7.35	23.87
	QAsp7B	7B	12	Xwmc273.1–Xcfd22.1	0.09	2.22	4.39
Ser	*QSer3A*	3A	84	Xbarc86–Xwmc21	−0.14	5.10	13.08
	QSer6D	6D	114	Xgwm681–Xubc808	−0.31	2.74	9.40
Gly	*QGly3A*	3A	90	Xbarc86–Xwmc21	−0.12	4.90	10.77
	QGly5B	5B	5	Xbarc36–Xbarc140	0.10	2.71	7.71
	QGly6B	6B	56	Xwmc74–Xgwm58	0.74	3.02	14.55
	QGly7D	7D	162	Xgdm67–Xwmc634	−0.09	2.90	6.27

(continued)

Table 5.22 (continued)

Trait	QTL	Chr.	Site (cM)	Flanking marker	A	LOD	R^2 (%)
Ala	*QAla3A*	3A	160	Xgwm155–Xcfa2170	−0.15	2.61	6.79
Tyr	*QTyr3A*	3A	162	Xcfa2170–Xbarc51	−0.23	3.04	7.74
	QTyr6A	6A	64	Xwmc553–Xgwm732	−0.25	2.87	9.25
His	*QHis1A*	1A	65	Xgwm498–Xcwem6.2	−0.08	3.07	6.23
	QHis1B	1B	82	Xbarc061–Xwmc766	−0.11	5.55	11.94
	QHis5B	5B	3	Xbarc36–Xbarc140	0.09	3.47	8.64
	QHis6A	6A	72	Xwmc553–Xgwm732	−0.09	3.08	8.63
	QHis7B	7B	69	Xgwm333–Xwmc10	0.09	2.88	8.30
Arg	*QArg5B*	5B	0	Xbarc36–Xbarc140	0.13	2.12	5.15
	QArg6B	6B	55	Xwmc74–Xgwm58	0.54	2.00	6.21
Total	*QTotal4B*	4B	9	Xwmc413–Xcfd39.2	2.86	2.42	5.62
	QTotal7D	7D	162	Xgdm67–Xwmc634	−3.48	3.56	8.29

variation, respectively, and the positive alleles originated from "Yumai 57." For TAA, only one QTL flanked by Xbarc356-Xwmc489.2 on chromosome 3A was identified, explaining 6.58 % of total variation.

5.2.3.2.2.2 Additive QTL for AAC in Wheat Grain in 2009

Fifty-three individual additive QTLs (LOD > 2.0) were detected for the 18 components of AAC in the DH population in 2009, ranging from one to six QTLs for each trait (Table 5.22; Fig. 5.7). The total phenotypic variation explained by the individual QTL for the 18 components of AAC varied from 4.39 % (Asp) to 23.87 % (Ser). Most of the QTLs are mainly distributed on chromosomes 1A, 3A, 4A, 4B, 5B, 6A, 6B, 6D, and 7D.

These QTLs detected also tended to be colocalized within the genome, and thus, ten QTL clusters were identified in the population. The QTL cluster, flanked by Xwmc74-Xgwm58 on chromosome 6B, was the largest in number, which consisted of six individual QTLs, followed by the QTL cluster, near Xbarc36-Xbarc140 on chromosome 5B, consisting four individual QTLs.

Nine main-effect QTLs (R^2 > 10 %) associated with Asp, Ser, Glu, Gly, Phe, His, and Lys contents were identified in the DH population in 2009 wheat. Particularly, the QTL for Asp flanked by Xwmc74-Xgwm58 on chromosome 6B could explain 23.87 % of total phenotypic variation and the other QTLs for Lys linked to Xwmc415-Xswes170.1 on chromosome 6B, whose R^2 was 21.08 %, and the positive alleles originated from Yumai 57.

For Lys content, five QTLs distributed on chromosomes 1A, 6B, 6A, and 2A were detected, jointly explaining 51.00 % of the total variation, and particularly, *QLys1A-2* had the highest phenotypic contribution (R^2 = 9.02 %). Two QTLs were identified for TAA, located on chromosomes 4B and 7D, explaining 13.91 % of total variation.

5.2.3.2.2.3 Comparison of QTL Detected in 2008 and 2009

In total, 13 QTL clusters for the 18 components of AAC were identified in the population, with three specific to 2008 wheat and six specific to 2009 wheat. And, four QTL clusters flanked by Xbarc356-Xwmc489.2, Xbarc86-Xwmc21, Xwmc74-Xgwm58, and Xgdm67-Xwmc634, which distributed on chromosomes 3A, 6B, and 7D, respectively, were detected consistently in two years. The QTL clusters in the interval Xbarc356-Xwmc489.2 and Xbarc86-Xwmc21 on chromosome 3A were 7 cM apart from each other, this region is associated with 13 amino acid contents, and the positive alleles came from "Yumai 57."

5.2.3.3 Research Progress of Wheat Grain Amino Acid Content QTL Mapping and Comparison with Previous Studies

5.2.3.3.1 Research Progress of Wheat Grain Amino Acid Content QTL Mapping

Improving nutritional quality of wheat is one of the major objectives of wheat breeding. Most researchers focused on content of protein and starch for nutritional quality of wheat (Cavanagh et al. 2010; Hristov et al. 2010; Sun et al. 2010; Zhang et al. 2011). Amino acids are the materials for protein synthesis and decomposed products and are the main form of using protein in human and animal bodies. Wheat protein quality is manifested in the type and ratio of amino acids. Amino acid composition of the most cereal crop protein is unbalanced for people's needs. For example, lysine is mostly lacking in wheat protein, which seriously affects absorption and utilization of wheat protein. Therefore, it has an important significance in genetic controlling and improving of amino acid composition in wheat protein. Since determination of AAC using an amino acid analyzer is operationally complex, time-consuming, and expensive, few researches focused on QTL mapping for AAC by using genetic population. To our knowledge, there were several researches on QTL mapping for AAC in rice (Zheng et al. 2008; Wang et al. 2008; Tang 2007) and soybean (Panthee et al. 2006a, b); however, similar studies in wheat have not been reported up to the present. The loci and markers associated with AAC detected in the study can be easily used in screening and identification of wheat with high AAC and can be used in marker-assisted selection of hybrids, which has significance for breeding new wheat varieties with high nutritional value.

5.2.3.3.2 Comparison of This Result with Previous Studies

In our research, the QTL mapping of amino acid was studied in 2008 and 2009 using the software QTL Network 2.0 and found that only Thr, Phe, Asp, Ser, and Gly had been identified in the QTLs on the same or similar location in the two years, but there was no same loci for the residue amino acid content in the two years. This was perhaps caused by the environments (Tang 2007).

The additive QTLs were detected for Thr, Ile, Phe, Leu, Glu, Pro, Asp, Ser, and Gly contents by analyzing the combination data using QTL Network 2.0. Of which, the major QTLs on 3A chromosome with the flanking marker *Xbarc86–Xwmc21* were identified controlling the Glu and Ser contents accounting for 10.1 % and 12.4 % of the phenotypic variance, respectively. Their positive alleles were from YM57. So the chromosome fragment between the markers *Xbarc86* and *Xwmc21* on 3A chromosome should be paid more attention.

In addition, there were no interaction effects identified between QTL and environments, which perhaps was caused by the data at one location in two years. Therefore, the QTL analysis of amino acid should be further analyzed in different locations and different years to discuss the additive, epistatic, and interaction effects with environments.

5.2.4 QTL Mapping of Carotenoid Pigments and Other Pigments

5.2.4.1 QTL Mapping of Carotenoid Pigments

5.2.4.1.1 Materials and Methods

5.2.4.1.1.1 Plant Materials

The population of 182 lines was derived from a cross between Shannong 01-35 (39-1/Hesheng 2) (SN 01-35) and Gaocheng 9411 (77546/Linzhang) (GC 9411). This population was also developed by single seed descent, to the $F_{8\text{-}9}$ generation. The grains of SN 01-35 appeared larger than those of GC 9411, but the quality of GC 9411 is better than that of SN 01-35. Thus, the population showed large variations in yield and quality.

5.2.4.1.1.2 Field Trials

The RIL population, along with two corresponding parents, was grown in two distinct locations in 2011 and 2012. E1 and E3 represent Tai'an, Shandong Province, China, respectively; E2 and E4 refer to Jinan, Shandong Province, China, respectively. These lines were sown in a randomized block design with two replicates at each location. Each replication was designed based on a three-row plot with 2 m long and 25 cm row-to-row distance.

All recommended local crop management practices were followed, and damages attributed to lodging, disease, or pests were not observed during the growing seasons.

5.2.4.1.1.3 Testing Method of Flour Carotenoid Pigments

Grains were milled into flour by Perten 3100 and sieved by 60-mesh polyurethane screen mesh, and flour is placed in sealed bags to save. Accurately weighed 2 g sample in stoppered tubes, and then poured into 10 mL water-saturated n-butanol, shaking 1 min, and then stand 30 min; during this time, every 5 min the sample will be shaken once. After shaking, the samples will be centrifuged at 4500 rpm for 15 min, and then, the supernatant will be filtered by Whatman No.1 filter paper. The absorbance value is tested at 440 nm by using the filtered liquid, and the water-saturated n-butanol is as control. According to the absorbance value of samples, the content of carotenoid pigments should be found by the standard curve, and then, the total carotenoid pigment content of samples will be calculated by the formula:

$$\text{Carotenoid pigments content } (\%) = \frac{C \times 10}{W \times 10^6} \times 100$$

C—the content of carotenoid pigment from standard curve (ug/ml);
W—sample weight (g);
10—the volume of extraction liquid;
10^6—the conversion multiples between g and ug.

5.2.4.1.1.4 Statistical Analyses

Statistical analyses (e.g., normal distribution and correlation) were performed using the software SPSS 17.0 (SPSS, Chicago, USA) and Excel 2010. The ICIM was applied by means of the QTL IciMapping v3.2 using the constructed map. For designations of the examined QTLs, refer to the method by MeCoueh.

5.2.4.1.2 Results and Analysis

5.2.4.1.2.1 Phenotypic Data Analysis

A large difference between two parents was found (Table 5.23), and the carotenoid pigment content of Gaocheng 941 is higher than that of 01-35. The evaluated trait exhibited approximately continuous variation in each of the environments, and transgressive segregation was observed, indicating that alleles with positive effects were contributed from both parents. The absolute values of skewness and kurtosis were less than 1.0. These indicated that the flour carotenoid pigment content was controlled by multigenes.

Table 5.23 Phenotypic performance of the flour carotenoid content

Environment		Parent		RIL population					
		Shannong 01-35	Gaocheng 9411	Mean	Max.	Min.	SD	Skewness	Kurtosis
2011	E1	1.06	1.80	1.32	2.23	0.83	0.29	0.69	0.27
	E2	1.11	1.78	1.30	2.25	0.81	0.27	0.66	0.30
2012	E3	1.14	1.40	1.34	2.57	0.61	0.34	0.44	0.40
	E4	1.06	1.44	1.36	2.58	0.58	0.39	0.36	0.46

E1: 2011, Tai'an; E2: 2011, Jinan; E3: 2012, Tai'an; E4: 2012, Jinan

Table 5.24 Positions, effects, and contribution rates of additive QTLs for the flour carotenoid content under the four environments

Environment		QTL	Position (cM)	Flanking marker	A	LOD	H^2 (%)
2011	E1	*QCC1B*	104	wPt-5363–wPt-1363	−0.10	2.80	6.57
		QCC3D	48	Xgpw2046–Xgpw7643	−0.13	2.08	15.08
		QCC7A	20	wPt-6768–wPt-6495	0.07	2.07	4.64
	E2	*QCC3A*	224.0	wPt-666853–wPt-4933	0.24	1.61	8.06
		QCC7A	20	wPt-6768–wPt-6495	0.07	1.73	4.16
		QCC7B	70	wPt-669693–wPt-0600	−0.10	3.17	8.65
2012	E3	*QCC1B*	105	wPt-1781–wPt-0974	−0.13	7.01	13.81
		QCC5A	139	wPt-2768–CFE019	−0.05	1.56	2.74
		QCC5B.1	19	wPt-2586–wPt-9814	−0.05	1.73	3.13
		QCC5B.2	117	wPt-4628–wPt-3503	0.09	5.32	10.04
		QCC7B	69	wPt-669693–wPt-0600	−0.08	4.35	8.14
	E4	*QCC1B*	105	wPt-1781–wPt-0974	−0.12	6.00	13.28
		QCC3B	136	wPt-7984–wPt-4220	0.06	2.54	4.72
		QCC4A	22.0	wPt-8479–Xgpw4040	−0.08	1.89	7.01
		QCC4D	20.0	Xgpw3113–Xgpw342	−0.11	1.54	17.03
		QCC5B.2	117	wPt-4628–wPt-3503	0.08	4.74	8.96
		QCC6B	62.0	wPt-3733–wPt-8721	−0.06	2.16	3.85
		QCC7B	69	wPt-669693–wPt-0600	−0.08	3.94	7.42

E1: 2011, Tai'an; E2: 2011, Jinan; E3: 2012, Tai'an; E4: 2012, Jinan

5.2.4.1.2.2 QTL Mapping of the Flour Carotenoid Pigment Content

There were 14 additive QTLs and 13 epistatic QTLs detected in four environments in two years. These QTLs were mainly distributed on 1B, 3A, 3D, 4A, 5A, and 7B chromosomes (Tables 5.24 and 5.25). Single QTL could explain from 2.74 to 17.03 % of the phenotypic variation.

5.2.4.1.2.2.1 Additive QTL Analysis

Of 14 additive QTLs, each of the three QTLs was detected in E1 and E2. In E1, the QTLs were distributed on 1B, 3D, and 7A chromosomes, explaining 6.57, 15.08, and 4.64 % of the phenotypic variation, respectively; while in E2, they were detected on 3A, 7A, and 7B chromosomes, explaining 4.16 to 8.65 % of the phenotypic variation. Of which, *QCC7A* found in these two environments was a stable QTL, whose positive allele was from Shannong 01-35. However, in E3, five QTLs detected were mapped on 1B, 5A, 5B, and 7B chromosomes, explaining 2.74 to 13.81 % of the phenotypic variation; in E4, eight QTLs were found to be distributed on 1B, 3B, 4A, and 7B chromosomes, explaining 3.85 to 17.03 % of the phenotypic variation. Three QTLs, *QCC1B*, *QCC5B.2*, and *QQCC7B*, were found in these two environments and seemed to be stable. Of which, *QCC1B* and *QCC5B.2* were major QTLs, and the positive allele of *QCC1B* is derived from Gaocheng 9411, while *QCC5B.2* was from Shannong 01-35.

Table 5.25 Estimated epistasis (AA) of QTLs for the flour carotenoid content under the four environments

Environment		QTL	Position (cM)	Flanking marker	QTL	Position (cM)	Flanking marker	AA	H^2 (%)
2011	E1	*QCC1A*	60	wPt-6654–wPt-665784	*qCC 6D*	45	Xgpw3186–wPt-668181	−0.01	21.80
		QCC1A	75	wPt-6654–wPt-665784	*qCC 1A*	90	wPt-2872–wPt-730408	0.06	14.35
		QCC4A	165	wPt-669526–wPt-668307	*qCC 6B*	10	wPt-669607–Xgpw1005	0.03	9.53
	E2	*QCC1B*	90	wPt-8971–wPt-2052	*qCC 2B*	110	CFE052–wPt-5374	−0.07	20.93
		QCC3D	0	wPt-7705–wPt-6909	*qCC 6A*	95	wPt-0832–M181CFE043	0.06	17.06
		QCC2A	130	wPt-0832–M181CFE043	*qCC 5B*	85	wPt-7665–wPt-3569	−0.05	10.19
		QCC2B	150	wPt-666395–wPt-8492	*qCC 6D*	75	wPt-668152–wPt-3350	0.02	9.44
2012	E3	*QCC1A*	50	wPt-6005-wPt-730172	*qCC 1B*	95	wPt-2052–wPt-3753	0.09	19.13
		QCC2A	285	wPt-5887–wPt-669199	*qCC 3A*	45	wPt-7608–wPt-0836	−0.03	17.43
	E4	*QCC1D*	10	wPt-665480–wPt-671415	*qCC 6B*	25	wPt-3060–wPt-3733	0.04	10.92
		QCC1B	80	wPt-2230–wPt-665375	*qCC 6D*	45	Xgpw3186–wPt-668181	0.01	15.72
		QCC1A	105	wPt-667558–wPt-667155	*qCC 5B*	60	wPt-4418–wPt-5168	0.01	11.03
		QCC1B	95	wPt-2052–wPt-3753	*qCC 4A*	25	wPt-8479–Xgpw4040	−0.01	17.34

E1: 2011, Tai'an; E2: 2011, Jinan; E3: 2012, Tai'an; E4: 2012, Jinan

5.2.4.1.2.2.2 Epistatic QTL Analysis

There were 13 epistatic QTLs found in four environments (Table 5.25). Of which, three epistatic QTLs were located on 1A-6D, 1A-1A, and 4A-6B, explaining 21.80, 14.35, and 9.53 % of the phenotypic variation, respectively. The first two QTLs had positive effect. In E2, four epistatic QTLs detected were distributed on 1B-2B, 3D-6A, 2A-5B, and 2B-6D, explaining 20.93, 17.86, 10.19, and 9.44 % of the phenotypic variation, respectively. In E3, two QTLs located on 1A-1B and 2A-3A explained 19.13 and 17.43 % of the phenotypic variation, respectively. In E4, the PVE of four QTLs on 1D-6B, 1B-6D, 1A-5B, and 1B-4A showed higher than 10 %. Of which, three QTLs had positive effects.

These results indicated that the genetic effect of the carotenoid pigment content was not only affected by additive effect but also affected by epistatic effect.

5.2.4.2 QTL Analysis for Flour Yellow Pigment Content

Many studies reported that wheat yellow pigment content was highly related to the surface color of the product. The main component of yellow pigment was lutein, esters, carotene and anthocyanin pigments and minimal flavonoids. In recent years, there was a great progress in the gene locating of the factors of wheat yellow pigment content. Many studies showed that seventh homology groups, especially genes on 7A and 7B chromosomes, played great role in grain yellow pigment content. Mares and Campbell (2001) located QTL for yellow pigment content in chromosomes 3B and 7A using 3 hybrid combinations of diploid wheat. Zhang (2006) located major QTL for grain yellow pigment content in chromosomes 2DL, 3DL, 4A, 5A, and 7AL using DH population of 9507/CA9632. The contribution rate was 12.1–33.9 %. The QTL for flour yellowness value b* was located on chromosomes 1DS, 2DL, 3A, 4D, 5D, 6AL, 6D, and 7AL, and the contribution rate was 12.9–37.6 %. Among them, the QTL effect on 7AL was the largest. Francki et al. (2004) reported that gene controlling flour color was on 3BS. Patil et al. (2008) showed that QTL for the grain yellow pigment content located on 1A, 3B, 5B, 7A, and 7B, and the largest contribution rates were on 7AL, explaining 55.22 % of the phenotypic variation. Zhang et al. (2009a, b, c, d) believed that the yellow pigment content gene was on chromosomes 1A and 4A. Zhang et al. (2009a, b, c, d) conducted QTL analysis for flour color using 168 doubled haploid population and detected 18 additive QTLs and 24 pairs of epistatic QTLs distributed on nineteenth chromosome of wheat. Poznia et al. (2007) located QTL for endosperm color on chromosomes 2A, 4B, 6B, and 7B using doubled haploid population.

Parker et al. (1998) detected major QTL for yellow pigment content on 7A and 3A chromosomes in 150 SSD system of Schomburgk/Yarralinka, which explained 60 and 13 % of the phenotypic variation, respectively. Mares and Campbell (2001) conducted QTL for yellow pigment content using Sunco/Tasman DH group, detected the main-effect QTL on 7A and 3B chromosomes, and explained 27 and 20 % of the phenotypic variance, respectively.

Marais (1992) detected QTL for flour yellow pigment gene on 7DL. Knott et al. (2007) found that QTL for high yellow pigment content was linked to Lr19, which contributed to resistant leaf rust and high yield. But the high yellow pigment content was adverse to flour color, which should be paid attention in breeding. Parker et al. (1998) located QTL for yellowness b* on 7AL and transformed AFLP makers to SCAR markers.

Kuchel et al. (2007) analyzed the gene loci controlling flour b* using DH population derived from the progeny of two Australian wheat cultivars Trident and Molineux and detected QTL for b* on the 7B chromosome in different years, which explained 48–61 % of the phenotypic variation. Elouafi (2001) located 3 yellow QTLs in durum wheat, among them, the loci on 7BL explained 53 % of the phenotypic variation, and the contribution rates of the other two QTLs on the 7AL were 13 and 6 %, respectively. Pozniak et al. (2007) detected 4 QTLs located on 2A, 4B, 6B, and 7B in durum wheat. In addition, other QTLs were also detected in the homology group of 4A, 5A, and 2D, 4D chromosomes. In conclusion, yellow pigment content is regulated by multiple genetic loci and QTLs in seventh homologous play major role. In addition, QTL mapping for carotene content used DH population in his study. QCx5D-10 was detected in 5D chromosome, explaining 27.25 % of the phenotypic variation. At present, functional markers for phytoene synthase and a speed limit of carotenoid pigment biosynthesis have been developed, which laid the foundation for molecular marker-assisted selection for low yellow pigment varieties.

5.3 QTL Mapping for Flour Quality Traits

Flour quality, including flour yield, flour ash, flour whiteness, gluten, sedimentation value, falling number, and other indicators, was related not only to wheat grain properties, hardness, and other physical qualities, but also to the milling process and precision, which furthermore influences the baking or cooking quality. Therefore, the genetic improvement of wheat flour quality traits has an important significance. Similarly, there are many detection analyses and breeding researches for flour quality, but there were few QTL analyses for flour quality because of the large test samples and high cost. However, in the present study, the QTL analysis for the main flour quality trait was conducted using the multiple genetic populations, and some meaningful results can be used in marker-assisted selection.

5.3.1 QTL Mapping of Gluten Content and Gluten Index

Gluten is a special protein which only exists in wheat flour. Because gluten is a protein complex containing the gliadin and glutenin with viscous and elastic characteristics, it is the material basis for making many kinds of foods.

Gluten is composed of gliadin and glutenin, and the content of gliadin has positive correlation with the extension of dough, while the glutenin content determines the dough elasticity. Gluten is formed by the interaction between gliadin and glutenin. Only kneading the gliadin with water, a kind of material without plastic and elastic can be formed, but when there is only the glutenin, the gluten is also not formed, and it is just dough without sticky and toughness. The gliadin contributes to gluten extensibility, and glutenin contributes to gluten elasticity, so the properties of gluten depend on the combination of these two kinds of protein. Gluten is closely related to the wheat quality, especially the second processing quality. In addition, the wheat gluten content is closely related to dough water absorption, dough development time, stability time, bread volume, and texture. Gluten elasticity and extensibility allow making steamed buns, bread, cake, and other foods, maintaining a large volume with full shape and elastic. The properties of these foods are positively correlated with the quality of gluten, so the content of wet gluten, especially the gluten index, is one of the important indexes to appraise the quality of wheat flour.

Wet gluten content and gluten index are quantitative traits controlled by multigenes, not only affected by additive and dominant effects, but also influenced by non-allelic interactions. So QTL mapping for gluten characteristics such as gluten content and the gluten index will be very important for finding the linked markers used in MAS.

5.3.1.1 Determination Methods

The type of 2200 gluten instrument and type of 2015 gluten centrifuge were used to determine wet gluten content and gluten index according to AACC 38-11 standard.

5.3.1.2 QTL Analysis Results

5.3.1.2.1 Phenotypic Analysis

The gluten characters in two parents and DH population are shown in Table 5.26; Fig. 5.8. Two parents show large differences in two traits, and Huapei 3 had high wet gluten, while Yumai 57 had high gluten index. The DH population had the large variation in these two quality traits, showing a continuous distribution and transgressive segregation phenomenon, which indicated that gluten content and gluten index are quantitative traits controlled by multigenes.

Table 5.26 Performance of characters of wet gluten and gluten index from parents and DH population

Trait	Parents		DH population					
	Huapei 3	Yumai 57	Mean	Max.	Min.	SD	Skewness	Kurtosis
Wet gluten	37.9	30.4	36.71	44.00	30.10	2.35	0.32	0.19
Gluten index	47.5	83.6	65.16	87.00	43.00	8.00	0.23	−0.16

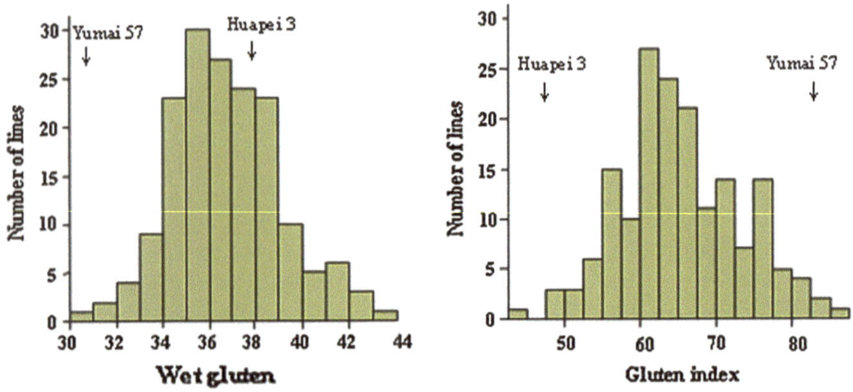

Fig. 5.8 Distribution of wet gluten and gluten index in the DH population

Table 5.27 Coefficients of pairwise correlations of wheat quality traits

Trait	Wet gluten	Gluten index
Wet gluten		0.167*
Gluten index	0.167*	

* The wet gluten was significantly correlated with gluten index

Gluten index and wet gluten content are two different concepts that evaluate the quality and quantity of wheat gluten, and the correlation coefficient was 0.167 at $P = 0.05$ levels (Table 5.27).

5.3.1.2.2 QTL Analysis for Wet Content

Six additive QTLs for wet gluten content were detected to be distributed on 1D, 2B, 3A, 3B, 6D, and 7D chromosomes with a single QTL, explaining 1.36–10.25 % of the phenotypic variation (Table 5.28). The genetic contribution of *QGlu3A* was the highest, explaining 10.25 % of the phenotypic variation. The positive alleles of *QGlu1D*, *QGlu2Ba*, and *QGlu6D* came from Huapei 3, and those of the other 3 additive QTLs came from Yumai 57. Of which, the interaction effect between *QGlu1D* and environment was 6.28 %.

Table 5.28 Estimated additive (A) and additive × environment (AE) interactions of QTLs for wheat quality traits

Trait	QTL	Flanking marker	Position (cM)	A	H^2 (%)	A × E1		A × E2		A × E3	
						AE1	H^2(%)	AE2	H^2 (%)	AE3	H^2 (%)
Wet gluten	*QGlu1D*	*Xcfd19–Xwmc93*	60.9	0.52	3.75						
	QGlu2Ba	*Xwmc477–Xwmc175*	72.2	0.32	1.36						
	QGlu3A	*Xbarc86–Xwmc21*	87.5	−0.87	10.25						
	QGlu3B	*Xgwm566–Xcfe009*	74.8	−0.49	3.33						
	QGlu6D	*Xcfd42–Xcfd13*	35.0	0.64	5.58						
	QGlu7D	*Xwmc634–Xwmc273.2*	162.7	−0.46	2.87						
Gluten index	*QGluin2D*	*Xgwm261–Xgwm296*	0.0	−3.17	11.07						
	QGluin4A	*Xwmc718–Xwmc262*	0.0	−1.41	2.19						
	QGluin5D	*Xwmc215–Xgdm63*	73.4	−3.03	10.11			−1.94	4.14	2.88	9.13

E1: Suzhou, 2006; E2: Tai'an, 2006; E3: Tai'an, 2005

Table 5.29 Estimated epistasis (AA) and epistasis × environment (AAE) interactions of quality traits

Trait	QTL	Flanking marker	Position (cM)	QTL	Flanking marker	Position (cM)	AA	H^2 (%) AA	H^2 (%) AAE
Wet gluten	*qGlu1B*	*Xcfe02.2–Xcfd021*	37.3	*qGlu2D*	*Xwmc170.2–Xgwm539*	62.5	0.60	4.90	
	qGlu2Bb	*Xcwem55–Xbarc129.1*	88.0	*qGlu3D*	*Xbarc071–Xgwm114*	92.0	0.51	3.52	
	qGlu2Bc	*Xgwm111–Xgdm14*	114.4	*qGlu3D*	*Xbarc071–Xgwm114*	92.0	0.30	1.24	
Gluten index	*qGluin6A*	*Xwmc553–Xgwm732*	54.5	*qGluin7D*	*Xgdm–Xwmc634*	155.5	−2.60	7.43	

Three pairs of epistatic QTLs were detected, which located on chromosomes 1B-2D and 2B-3D, accounting for 4.90, 3.52, and 1.24 % of the phenotypic variation, respectively. No epistatic-by-environment effects were detected (Table 5.29).

5.3.1.2.3 QTL Mapping for Gluten Index

Three additive QTLs for gluten index were detected and located on 2D, 4A, and 5D chromosomes, explaining 11.07, 2.19, and 10.11 % of the phenotypic variation (Table 5.28), respectively. The positive alleles of these three loci were from Yumai 57, which was consistent with Yumai 57 having high gluten index content. There was the interaction effect between *QGluin5D* and the environment with explaining 13.27 % of the phenotypic variation.

One pair of epistatic QTL was detected for gluten index, located on chromosome 6A-7D (Table 5.29), explaining 7.43 % of the phenotypic variation, and no additive-by-environment effects were detected.

5.3.1.3 Research Progress of Gluten Quality QTL Mapping and Comparison with Previous Studies

5.3.1.3.1 Research Progress of Gluten Content QTL Mapping

Few researches on gluten content and index of wheat are reported. Tang et al. (2011) studied the genetics of wet gluten content and gluten index using 21 hybrid combinations from different wheat varieties (lines) with different gluten qualities created according to Griffing's diallel crossing method II. Wet gluten content and gluten index are quantitative traits controlled by multiple genes, not only affected by additive and dominant effects, but also influenced by non-allelic interaction. High values of wet gluten content are controlled by dominant genes and expressed in early generation that can be easily selected; while the high value of gluten index was controlled by recessive genes, it should not be chosen in early generation. The narrow heritability of wet gluten content is lower than that of the gluten index, and there is no strong relevance between them. At present, the previous QTL studies for gluten-related traits had been located on 14 chromosomes using DH, RIL, or other populations, and the highest PVE was up to 16.43 % (Table 5.30).

5.3.1.3.2 Comparison of This Result with Previous Studies

The content and quality of gluten are important factors affecting processing quality and are influenced by genotype and environment conditions. When different researchers use different populations to conduct QTL analysis for gluten content,

Table 5.30 Summary of QTL results of wheat gluten characteristics

QTL	Marker intervals	PVE (%)	Mapping population	References
QGlc.ipk-5B	*Xtam72*	–	ITMI population	Pshenichnikova et al. (2008)
QGlc.ipk-7A	*Xcdo475b*	–	ITMI population	Pshenichnikova et al. (2008)
QWgc.sdau-6D	*Xswes426b–Xubc807d*	16.43	RIL	Sun et al. (2008)
QDgc.sdau-6D	*Xswes426b–Xubc807d*	15.55	RIL	Sun et al. (2008)
QDgc.sdau-2D	*Xissr23a–Xwmc181b*	7.99	RIL	Sun et al. (2008)
QWgc.caas-2B.1	*barc55-wmc175*	5	RILs	Li et al. (2009)
QWgc.caas-2B.2	*wms120-barc1042*	8.8	RIL	Li et al. (2009)
QWgc.caas-3A	*wmc388b-wmc343*	8.5/3.8	RIL	Li et al. (2009)
QWgc.caas-4D.1	*cfd71a-wmc457*	4.1/7.6	RIL	Li et al. (2009)
QWgc.caas-7B	*wmc276-barc50/wmc311-wmc276*	5.6/4.8	RIL	Li et al. (2009)
QWgc.caas-7D	*wmc506-cfd66*	10.3	RIL	Li et al. (2009)
QWgc.caas-4D.2	*gwp93025-wmc331*	5.1	RIL	Li et al. (2009)
QWgc.caas-5B.1	*wms67-wms371*	7.7	RIL	Li et al. (2009)
QWgc.caas-5B.2	*wms408-wmc235*	5.3	RIL	Li et al. (2009)
QWgc.caas-5D	*wmc215-vrnD*	7.2	RIL	Li et al. (2009)
QWgc.caas-2A	*pm4-wms356*	6	RIL	Li et al. (2009)
QWgc.caas-2B.3	*wms257-cwm13*	11.3	RIL	Li et al. (2009)

the results are not consistent. Previous research found (Table 5.30) that many chromosomes have been involved affecting gluten content, of which chromosomes 5B and 6D in different studies were detected with QTL for gluten content. Present research also detected gluten content QTL on chromosome 6D indicated that there are some important loci controlling wet gluten content on 6D. Former studies detected QTL for dry gluten content on chromosomes 2D and 6D, the present study also detected QTL for gluten index on 2D chromosome, and the PVE was more than 10 %. Therefore, 2D and 6D are very important for gluten content and quality.

5.3.2 *QTL Mapping of Flour Whiteness, Color, and PPO Activity*

Flour color is an important trait in the assessment of flour quality for the production of many end products (Parker and Langridge 2000). Flour with high levels of yellow pigmentation is preferred for Chinese and Japanese alkaline noodle production. In many Asian countries, noodle are prepared from flour specifically selected to enhance the color of the final product (Kruger et al. 1994). Miskelly (1984) detected a significant positive correlation between flour color and noodle sheet yellowness in both Chinese and Japanese noodle. However, the other end products such as bread, steamed bread, dry white noodle, and dumplings require white flour with extremely low levels of yellow pigmentation. Color is usually expressed as L* (lightness), a* (red-green chromaticity), and b* (yellow-blue chromaticity) of the flour sample. Theoretically, a pure white flour should have zero values for a* and b* and hundred for L (Sun et al. 2002). Low values of L* with high positive values of a* result in a gray and dull color. Significant variations in flour color existed among different genotypes of wheat. Moreover, environment and management practices may have impacts on flour color. Grain protein content, hardness, virtuousness, seed coat color, kernel size, and shape may all contribute to variations in flour color (Zhang and Tian 2008). The flour color of low-protein-content wheat is whiter than that of high-protein-content wheat at the same milling extraction ratio. Flour from wheat with a red seed coat has higher a* values than flour from wheat with a white seed coat (Zhang and Tian 2008). Some recently released varieties showed good agronomic characteristics and are high yielding, but contained high levels of flour carotenoids, which limited the end product uses (Parker and Langridge 2000). It is of great value to understand the wheat cultivar's molecular genetic regulation and to select them effectively for genetic and plant breeding purposes. Flour color behaves as a typical quantitative trait. The selection efficiency of conventional breeding methods for flour color is very low.

There are several reports for flour color associated with markers. Parker et al. (1998) identified ten RFLP marker loci that showed significant associations with QTLs for flour color, located on chromosomes 3A and 7A, by using 150 single seed descent (SSD) lines. The estimated heritability of flour color was calculated at 0.68, indicating that a large part of the expression of this trait is genetically controlled, making it easier to manipulate at the genetic level in a breeding program. Mares and Campbell (2001) reported that xanthophyll content was very strongly associated with QTLs located on chromosomes 3B and 7A in 163 DH lines derived from Sunco × Tasman. Zhang et al. (2006) identified eight QTLs for b* located on chromosomes 1DS, 2DL, 3A, 4D, 5D, 6AL, 6D, and 7AL. Knot (1984) reported that a gene for b* was linked to Lr19. One SSR and sequence-related amplified polymorphism (SRAP) marker closely linked to a Lr19 resistance trait were obtained and were named wms44 and M73 with genetic distances 0 and 2.6 cM, respectively (Li et al. 2005; Liu et al. 2007). A codominant amplified fragment length polymorphism (AFLP) marker linked to a major locus controlling flour color in wheat has been converted to sequence-tagged site (STS) marker for wider applicability in MAS (Parker and Langridge 2000). However, most previous studies have focused on b*, and few studies have reported on a* and L* values, which are also very important components of flour color in wheat.

5.3.2.1 Materials and Methods

5.3.2.1.1 Plant Materials

Materials were same as ones of Sect. 5.2.3.1.1 in this chapter.

5.3.2.1.2 Field Trials and Flour Color Evaluation

Field trials were conducted under three environments during 2005 and 2006 in Tai'an, Shandong Province, and in 2006 in Suzhou, Anhui Province. The experimental design followed a completely randomized block design with three replications at each location. In autumn 2005, all lines and parental lines were grown in 2-m-long three-row plots (25 cm apart), whereas in autumn 2006, they were grown in 2-m-long four-row plots (25 cm apart). Suzhou and Tai'an showed large differences in climate, soil conditions, and day length. At Tai'an, there were remarkable differences in temperature and soil conditions between the years 2005 and 2006. Management was in accordance with local practice. The lines were harvested individually at maturity to prevent yield loss from over-ripening. Harvested grain samples per line from the three replicates at each environment were mixed and cleaned prior to conditioning and milling in order to maintain a manageable number of samples for quality testing.

5.3.2.1.3 Statistical Analysis

Analysis of variance (ANOVA) was carried out using the SPSS version 13.0 (SPSS, Chicago, USA) program. QTL analyses were performed using QTL Network software version 2.0 (Yang and Zhu 2005) based on a mixed linear model (Wang et al. 1999). Composite interval analysis was undertaken using forward–backward stepwise multiple linear regression with a probability into and out of the model $P = 0.05$ and window size set at 10 cM. QTL was declared if the phenotype was associated with marker locus at $P = 0.005$. The final genetic model incorporated significant additive effects and epistatic effects as well as their environment interactions.

5.3.2.2 Results and Analysis

5.3.2.2.1 Statistical Analysis of the Phenotypic Assessments in DH Population

Mean values of flour color for the parents Huapei and Yumai 57, as well as the 168 DH lines under three environments in 2005 and 2006 cropping seasons, are shown in Table 5.31. Huapei had a higher value for b*, while Yumai 57 had higher values

Table 5.31 Phenotypic performance of quality traits

Trait	Parents		DH population					
	Huapei 3	Yumai 57	Mean	Maximum	Minimum	SD	Skewness	Kurtosis
Flour whiteness	74.60	80.10	77.70	81.70	74.40	1.89	0.03	−0.99
a* value	−1.24	−1.06	−1.17	−0.63	−1.49	0.16	−0.05	−0.32
b* value	9.38	7.28	8.33	10.34	6.23	0.89	0.04	−0.58
L* value	90.42	91.42	90.75	93.03	90.33	0.57	−0.48	−0.05
Polyphenol oxidase	0.07	0.09	0.08	0.11	0.07	0.00	0.08	−0.51

Table 5.32 Coefficients of pairwise correlations of wheat quality traits

Trait	Flour whiteness	a* value	b* value	L* value	Polyphenol oxidase
a* value	0.201**				
b* value	−0.904**	−0.526**			
L* value	0.832**	−0.182*	−0.570**		
Polyphenol oxidase	0.041	0.019	0.002	0.106	

*Significant at 0.05 level of probability and **significant at 0.01 probability level

for both a* and L*. Transgressive segregants were observed for the three traits among the DH line in the three environments. Three traits of the DH population segregated continuously and followed normal distribution, and both absolute values of skewness and kurtosis were less than 1.0 (Table 5.31), indicating its polygenic inheritance and suitability of the data for QTL analysis (Cao et al. 2001).

The correlations among a*, b*, and L* values are shown in Table 5.32. Significantly negative correlations were detected between L* and b* ($r = -0.559$) and between a* and b* ($r = -0.494$).

5.3.2.2.2 QTLs for Flour Color Traits

Eighteen QTLs with additive effects and/or additive–environment (AE) effects were detected for flour color in the three environments (Table 5.33) ranging from four to eight QTLs for each trait and were distributed on 12 of the 21 chromosomes.

Twenty-four pairs of QTLs with epistatic effect and/or epistasis 9 environment (AAE) effects were detected for flour color in the three environments (Table 5.34), ranging from four to ten QTLs for each trait. The highest numbers of QTLs with epistatic effects were found in b* and L* with ten pairs of QTLs. In contrast, only four pairs of QTLs were detected for a*.

Table 5.33 Estimated additive (A) and additive × environment (AE) interactions of QTLs

Trait	QTL	Flanking marker	Site/cM	A	H^2/%	(A × E1)		(A × E2)		(A × E3)	
						AE1	H^2/%	AE2	H^2/%	AE3	H^2/%
Flour whiteness	*QWH1B*	*Xbarc061–Xwmc766*	7.0	−0.60	1.00	−1.18	3.85	0.78	1.66	0.43	0.50
	QWH 2D	*Xwmc18–Xwmc170.2*	48.6	1.21	4.04						
	QWH 4 Da	*Xwmc331–Xgwm194*	19.0	0.97	2.56	0.43	0.51				
	QWH 4Db	*Xcfa2173–Xcfe188*	148.8	−1.40	5.37						
a* value	*Qa1B*	*Xbarc372–Xwmc412.2*	36.3	−0.10	25.64						
	Qa3B	*Xbarc102–Xgwm389*	3.0	0.04	3.94						
	Qa5D	*Xwmc215–Xgdm63*	80.3	−0.03	2.24	0.02	1.68	−0.02	1.22		
	Qa6A	*Xgwm459–Xgwm334*	4.0	0.03	2.49						
	Qa6D	*Xcfa2129–X barc080*	164.0	−0.04	3.97						
	Qa7D	*Xgwm295–Xgwm676*	80.3	0.04	3.01						
b* value	*Qb2Ba*	*Xwmc764–Xwmc382.2*	1.7	0.07	0.51						
	Qb3Db	*Xwmc631–Xbarc323*	81.5	0.17	2.95						
	Qb4 Da	*Xcfa2173–Xcfe188*	127.8	0.17	3.45						
	Qb5A	*Xbarc358.2–Xgwm186*	37.1	0.19	4.30						
L* value	*Ql1Ba*	*Xbarc119–Xgwm18*	33.8	0.12	3.48	0.07	1.19				
	Ql2Bb	*Xwmc179–Xbarc373*	73.5	−0.11	2.67						
	Ql3D	*Xwmc631–Xbarc323*	81.5	−0.12	3.49	−0.09	1.82				
	Ql4B	*Xwmc48–Xbarc1096*	18.4	0.06	0.82						
	Ql4Db	*Xcfa2173–Xcfe188*	148.8	−0.16	6.14						
	Ql5Db	*Xcfd226–Xwmc765*	118.5	−0.07	0.98						
	Ql7B	*Xgwm333–Xwmc10*	49.6	0.11	2.83						
	Ql7D	*Xgwm295–Xgwm676*	84.3	−0.09	1.87						

(continued)

Table 5.33 (continued)

Trait	QTL	Flanking marker	Site/cM	A	H^2/%	(A × E1)		(A × E2)		(A × E3)	
						AE1	H^2/%	AE2	H^2/%	AE3	H^2/%
Polyphenol oxidase	*QPpo2B*	*Xwmc770–Xwmc179*	55.6	0.002	2.68						
	QPpo2D	*Xcfd53–Xwmc18*	1.7	0.004	15.64						
	QPpo4D	*Xcfa2173–Xcfe188*	158.8	−0.003	8.89						
	QPpo5D	*Xbarc1097–Xcfd8*	16.4	0.002	3.74						
	QPpo6B	*Xcfa2187–Xgwm219*	0.0	0.003	6.51						

E1: Suzhou, 2006; E2: Tai'an, 2006; E3: Tai'an, 2005

Table 5.34 Estimated epistasis (AA) and epistasis × environment (AAE) interactions of quality traits

Trait	QTL	Flanking marker	Site/cM	QTL	Flanking marker	Site/cM	(AA)	AA H^2/%	AAE H^2/%
Flour whiteness	*QWH1B*	*Xbarc061–Xwmc766*	77.0	*qWH2D*	*Xwmc18–Xwmc170.2*	61	−0.20	0.11	4.89
	QWH1D	*Xgdm60–Xwmc429*	25.0	*qWH 6A*	*Xgwm182–Xbarc1055*	43.1	1.09	3.23	0.72
a* value	*Qa2A*	*Xgwm448–Xwmc455*	75.5	*qa4B*	*Xwmc125–Xwmc 47*	0.0	−0.03	2.35	
	Qa4D	*Xbarc190–Xbarc224*	126.3	*qa6A*	*Xcfe179–Xcfe179.1*	86.0	0.03	2.03	
	Qa5B	*Xbarc140–Xbarc142*	1.4	*qa7B*	*Xwmc273.1–Xcfd22.1*	12.0	−0.02	0.60	
b* value	*Qb3Db*	*Xbarc323–Xwmc631*	82.3	*qb5B*	*Xbarc36–Xbarc140*	0.0	−0.07	0.60	
	Qb1Aa	*Xgwm259–Xcwem32.1*	4.0	*qb1Ba*	*Xcwem6.1–Xwmc128*	35.4	0.11	1.54	
	Qb1Ab	*Xwmc728.1–Xwmc550*	20.0	*qb1Bb*	*Xcfe023.2–Xbcfd21*	37.9	0.11	1.37	
	Qb1Ab	*Xwmc728.1–Xwmc550*	23.0	*qb1Bc*	*Xgwm582–Xcfe 026.2*	49.0	0.12	1.85	
	Qb2A	*Xbarc264–Xgwm448*	70.2	*qb3Db*	*Xbarc323–Xwmc631*	82.2	0.17	3.48	
	Qb2A	*Xbarc264–Xgwm448*	70.2	*qb3 Da*	*Xbarc376–Xgwm72*	9.4	0.18	3.74	
	Qb2Bb	*Xwmc445.2–Xwmc317*	91.5	*qb7A*	*Xgwm60–Xbarc070*	15.0	0.18	3.96	
	Qb3A	*Xcfa2193–Xgwm155*	105.0	*qb5B*	*Xbarc36–Xbarc140*	0.0	−0.23	6.15	
	Qb3Db	*Xbarc323–Xwmc631*	5.2	*qb4Db*	*Xcfe188–Xbarc224*	125.6	0.18	3.94	
	Qb6B	*Xbarc1129–Xwmc658.2*	43.0	*qb7D*	*Xbarc244–Xbarc352*	49.0	−0.21	5.25	
L* value	*Ql2Bb*	*Xwmc179–Xbarc373*	66.9	*ql7B*	*Xgwm333–Xwmc10*	52.0	0.09	1.82	
	Ql3D	*Xbarc323–Xwmc631*	82.2	*ql7D*	*Xgwm295–Xgwm676*	91.0	−0.11	2.82	
	Ql1Bb	*Xcfe023.1–Xbarc372*	36.1	*ql2Ba*	*Xwmc382.2–Xgwm210*	2.9	−0.07	1.01	
	Ql1Bb	*Xcfe023.1–Xbarc372*	36.1	*ql6A*	*Xwmc553–Xgwm732*	60.2	−0.07	1.25	
	Ql1Bc	*Xcfe026.2–Xbarc061*	37.0	*ql2Ba*	*Xwmc382.2–Xgwm210*	4.9	−0.09	1.89	
	Ql2Aa	*Xgwm558–Xbarc015*	65.9	*ql3D*	*Xbarc323–Xwmc631*	82.2	−0.09	1.78	
	Ql2Aa	*Xgwm558–Xbarc015*	65.9	*ql6D*	*Xwmc412.1–Xcfd49*	4.0	0.06	0.80	
	Ql2Ab	*Xbarc264–Xwmc552*	73.2	*ql3D*	*Xbarc323–Xwmc631*	82.2	−0.08	1.45	0.92
	Ql2D	*Xgwm311.2–Xcfd50*	11.6	*ql5Da*	*Xbarc347–Xcfd101*	52.6	0.10	2.20	
	Ql4Da	*Xwmc331–Xgwm194*	28.8	*ql5A*	*Xbarc358.2–Xgwm186*	46.0	−0.17	6.34	2.96

5.3.2.2.2.1 QTLs for Flour Whiteness

Four QTLs distributed on 1B, 2D, and 4D were detected with single QTL, explaining 1.00 to 5.37 % of the phenotypic variation. Their additive effects were between 0.60 and 1.40 (Table 5.33). Of which, the positive alleles of the two QTLs, *QWH1B* and *QWH4Db*, came from Yumai 57, but those of the other two QTLs, *QWH2D* and *QWH4D,* came from Huapei 3. These indicated that the increasing alleles were from the two parents. There were AE interactions for *QWH1B* and *QWH4 Da*, explaining 6.52 % of the phenotypic variation.

Two epistatic QTLs were found (Table 5.34), which distributed on 1D-6A and 1B-2D. They had the AAE interactions, explaining 7.14 % of the phenotypic variation.

*5.3.2.2.2.2 QTLs for a**

Six additive QTLs were detected for a*, located on chromosomes 1B, 3B, 5D, 6A, 6D, and 7D. They increased a* from 0.03 to 0.10, explaining phenotypic variance from 2.24 to 25.64 %. *Qa1B* had the most significant effect accounting for 25.64 % of the phenotypic variance. Three QTLs (*Qa1B*, *Qa5D,* and *Qa6D*) had negative effects on a* and were contributed by Yumai 57 alleles, while the other loci had positive effects on a* and were transmitted by Huapei 3 alleles. This suggested that alleles that increased a* were dispersed within the two parents, resulting in small differences of phenotypic values between the parents and the transgressive segregants among the DH population. The total contribution of all the six QTLs to a* was 41.29 %. *Qa5D* was also involved in AE interactions in two environments. The AE interactions explained 2.90 % of the phenotypic variance for a*.

Four pairs of epistatic effects were identified for a* and located on chromosomes 1A–7B, 2A–4B, 4D–6A, and 5B–7B. They explained the phenotypic variance ranging from 0.60 to 2.35 %. *Qa4D/Qa6A* had positive effect of 0.03. However, the other three pairs showed negative effects on a*. The general contribution of four pairs of epistatic QTLs was 7.10 %, while no-main-effect QTL was detected in epistatic effects. No AAE interactions were identified for a*.

For a*, the total QE interactions could explain 2.90 % of the phenotypic variance.

*5.3.2.2.2.3 QTLs for b**

Four QTLs with additive effects significantly influenced b* were located on chromosomes 2B, 3D, 4D, and 5A. The *Qb5A* made the highest contribution, explaining 4.30 % of the phenotypic variance. All four QTLs could account for 11.21 % of the phenotypic variance and were derived from Huapei 3 alleles which were in accordance with Huapei 3 having much larger b*. No AE interactions were identified for b*.

Ten pairs of epistatic effects were discovered for b* and were located on chromosomes 3D–5B, 1A–1B, 2A–3D, 2B–7A, 3A–5B, 3D–4D, and 6B–7D.

Among them, *Qb3A/Qb5B* had the highest contribution, accounting for 6.15 % of the phenotypic variance. Three pairs of epistatic QTLs (*Qb3D-2/Qb5B*, *Qb3A/Qb5B*, and *Qb6B/Qb7D*) had negative effects on b*. However, the other seven pairs showed positive effects on b*. The total contribution of ten pairs of epistatic QTLs was 31.88 %, so epistasis played strong effect on b*. One main-effect QTL *Qb3D-2* was involved in three pairs of epistatic effects. No AE interactions were identified for b*. No QE interactions for b* were detected in this study.

*5.3.2.2.2.4 QTLs for L**

Eight main-effect QTLs were identified for L*. These QTLs increased the L* from 0.06 to 0.16 and accounted for the phenotypic variance ranging from 0.82 to 6.14 %. The total contribution of the main-effect QTLs could explain 22.28 % of the phenotypic variance of L*. Among them, three alleles (*Ql1B-1ql4B* and *Ql7B*) came from the parent Huapei 3 and the rest derived from the parent Yumai 57 alleles. This suggested that alleles for increased L* were dispersed in the two parents. This result was also in accordance with the presence of a wide range of variation and transgressive segregations of L* in the DH population. Two QTLs were involved in AE interactions, which could explain 3.01 % of the phenotypic variance of L*.

Ten pairs of epistatic QTLs for L* were resolved, explaining the phenotypic variance ranging from 0.80 to 6.24 %. *Ql4D-1/Ql5A* had the largest effect with a negative effect of -0.17, accounting for 6.34 of the phenotypic variance. The total contribution of epistatic QTLs was 21.36 %. Three main-effect QTLs (*Ql3D*, *Ql7B*, and *Ql7D*) were detected in epistatic effects. Two pairs of epistatic QTLs were also involved in AAE interactions, accounting for 3.88 % of the phenotypic variance.

For L*, the total QE interactions could explain 6.89 % of the phenotypic variance.

5.3.2.2.2.5 QTLs for PPO Activity

Five additive QTLs were detected to be distributed on 2B, 2D, 4D, 5D, and 6B chromosomes, explaining 2.68 to 15.64 % of the phenotypic variation for each QTL (Table 5.33). *QPpo2D* had the highest PVE with 15.64 %. The positive allele of *QPpo4D* was from Yumai 57, and the others were from Huapei 3. No epistatic effect and interactions with environments were found.

5.3.2.2.3 QTL Mapping for Flour Whiteness Using RIL Population Derived from Nuomai 1 and Gaocheng 8901

5.3.2.2.3.1 Phenotypic Variation

The whiteness of Nuomai 1 is significantly higher than that of Gaocheng 8901, and the continuous variation was found in this RIL population (Table 5.35). Except the

Table 5.35 Phenotypic values for wheat flour whiteness of two parents and the RIL population in the three environments

Environment	Parents		RIL population			
	Nuomai 1	Gaocheng 8901	Range	Mean ± SD	Skewness	Kurtosis
Tai'an, 2008	82.6	78.4	71.0–84.0	78.95 ± 2.97	−0.296	0.819
Tai'an, 2009	82.2	77.4	71.1–84.9	78.63 ± 3.38	0.017	−1.286
Anhui, 2011	80.4	75.3	71.0–83.5	77.01 ± 3.42	0.042	−1.38

kurtosis in Tai'an 2009 and Anhui 2011, the skewness and kurtosis in other environments are less than 1.0.

5.3.2.2.3.2 QTL Analysis

Three additive QTLs were found on 1D, 4A, and 7B chromosomes, and the positive alleles came from Nuomai 1. Of which, *Qfwh-1D* was detected in the three environments and explained the 43.08–51.97 % of the phenotypic variation on average, which indicated that it was a stable major QTL. However, *Qfwh-4A* and *Qfwh-7B* were identified in single environment (Table 5.36).

5.3.2.3 Research Progress of Flour Whiteness, Color, and PPO Activity QTL Mapping and Comparison with Previous Studies

5.3.2.3.1 The Research Progress of Flour Whiteness, Color, and PPO Activity

Parke et al. (1998) detected one QTL (*Xbcd828*) on 3A, explaining 13 % of the phenotypic variation of flour color and 9 QTLs on 7A, totally accounting for 60 % with the estimated genetic heritability of 0.67 using RFLP molecular marker.

Table 5.36 QTL with significant additive effects for flour whiteness detected in different environments

Environment	QTL	Chromosome	Site/cM	Interval markers	A	Phenotypic variance explained/%
E1	*Qfwh-1D*	1D	8.00	cfd-183—wPt-729773	2.04	43.08
	Qfwh-7B	7B	245.00	wPt-6936—wPt-8920	0.55	3.56
E2	*Qfwh-1D*	1D	6.00	cfd-183—wPt-729773	2.53	51.97
E3	*Qfwh-1D*	1D	9.00	cfd-183—wPt-729773	2.71	45.07
	Qfwh-4A	4A	119.00	wPt-5749—wPt-7280	0.60	3.16
PD	*Qfwh-1D*	1D	8.00	cfd-183—wPt-729773	2.56	44.22

E1: Tai'an, 2008; E2′: Tai'an, 2009; E3: Suzhou, 2011; PD: pool data; positive values indicate that Nuomai 1 alleles increase corresponding trait value; negative values indicate that Gaocheng 8901 alleles increase corresponding trait value

Mares et al. (2001) found the QTLs controlling leaf yellow carotenoid pigment on 3B and 7A chromosomes and also detected QTLs on 2D and 2A chromosomes for noodle color using 163 DH lines. Demeke et al. (2001) studied the QTLs of PPO activity using three RILs and found the major QTLs on 2D for PPO activity and some minor genes on 2A, 2B, 3B, 3D, and 6B chromosomes. Zhang (2003) detected major QTLs on 4D, 4A, and 5D chromosomes for controlling flour color lightness, and for b* value, the major QTLs were identified on 7A and 4D chromosomes and other QTLs on 1BS, 1DS, 2DL, and 3A chromosomes. There were three QTLs for grain yellow pigment content found on 7AL, 3DL, and 2DL, of which the major QTL was on 7AL, explaining 25.7 to 33.9 % of the phenotypic variation; the major QTLs for PPO activity were detected on 2AL and 2DL, explaining 37.9–50 % and 25.0–29 % of the phenotypic variation, respectively. He et al. (2007) identified one QTL for PPO on 2DL. Li et al. (2007a, b) also found the QTLs controlling a* and L* at the similar position and detected the epistatic effect for b^* value. The QTLs for L* were detected on 4D and 5D chromosomes.

At present, most of the QTL researches focused on 1A, 1B, 2A, 2B, 2D, 3A, 3B, 3D, 4A, 4D, 5A, 5D, 6B, 7A, and 7B chromosomes. Of which, 7A chromosome had the most QTLs (Table 5.37), and the second is 1B. The highest PVE reached to 35.9 %.

5.3.2.3.2 Comparison of This Result with Previous Studies

In this study, *QPpo2D* controlling the PPO activity was detected, explaining phenotypic variation of 15.64 %, which was located closely to the result of the researches of Demeke et al. (2001) (Xfba 314) and Zhang (2003) (Xwmc41); *QPpo2B (Xwmc770–Xwmc179)* was close to the *Xgwm410* (Zhang 2003), and *QPpo6B* has the same position as in the Demeke' (2001) research result. Knott (1984) reported a gene for b* linked to Lr19 in "Agatha" wheat. One SSR and SRAP marker closely linked to Lr19 resistance trait were obtained and were named wms44 and M73 on chromosome 7D with genetic distances 0.9 and 2.6 cM, respectively. At a similar location, additive QTLs associated with a* and L* were also detected in this study. Near to this location, one epistatic QTL was detected for b*.

We could not found the same chromosome for mapping flour whiteness using two populations, but the QTLs were involved 1B, 1D, 2D, 4D, 7B, and 4A chromosomes. In the related research, the QTLs for a* were all detected on 1B and 3B chromosomes, which indicated that these two chromosomes were important for a* and other chromosomes such as 1A, 4A, 7A, 5D, 6A, 6D, and 7D also were involved. For b* value, 1B, 3D, 4A, 4D, 5A, 5B, 6A, 7A, and 7B were involved, of which the QTLs on 5A were identified in common wheat and durum wheat. In this study, we detected the QTLs for L value involving 1B, 2B, 3D, 4B, 4D, 5D, 7B, and 7D chromosomes. From the above research, 1B, 3B, 4A, 4D, 5D, 7A, 7B, and 7D chromosomes perhaps contained some important gene loci controlling flour whiteness and color.

Table 5.37 Summary of QTL results of wheat flour color characteristics (PVE > 10 %)

QTL	Flanking marker	PVE/%	Mapping population	References
QFa-1B	*Sec1-HVM23*	26.1	RILs	Zhang et al.(2009)
QFa-7A	*Xwmc809-YP7A*	35.9	Same as above	Same as above
QFb-7A	*Xwmc809-YP7A*	12.6	Same as above	Same as above
QYpc-1B	*Sec1-HVM23*	31.9	Same as above	Same as above
QYpc-7A	*Xwmc809-YP7A*	33.9	Same as above	Same as above
QKj-1B	*Sec1-HVM23*	17.2	Same as above	Same as above
QKj-7A	*Xwmc809-YP7A*	19.3	Same as above	Same as above
QNb-1B	*Sec1-HVM23*	13.7	Same as above	Same as above
QNb-7A	*Xwmc809-YP7A*	22	Same as above	Same as above
QFb.cerz-4AL.2	*wmc219–psr573.2*	6.7–10.6	RILs	Roncallo et al. (2012)
QFb.cerz-5AS	*wmc350–gwm47*	12.9–16.2	Same as above	Same as above
QFb.cerz-5BL.1	*barc74–gwm371*	12.2	Same as above	Same as above
QFb.cerz-6AL.1	*barc146–gwm132*	16.1–21.4	Same as above	Same as above
QFb.cerz-6AL.2	*barc113–wmc553*	17.1–28.3	Same as above	Same as above
QFb.cerz-6AL.3	*barc353–gwm169*	12.4	Same as above	Same as above
QFb.cerz-6AL.3	*gwm169–BE483091_472*	10.4	Same as above	Same as above
QFb.cerz-7AS	*wmc168–barc219*	12.6	Same as above	Same as above
QFb.cerz-7AS	*BQ170462_176–barc174*	11.7	Same as above	Same as above
QFb.cerz-7AL	*wmc116–cfd6*	9.8–22.5	Same as above	Same as above
QFb.cerz-7BS.2	*barc72–gwm297*	12.8	Same as above	Same as above
QFb.cerz-7BL.1	*Psy-B1–cfa2257*	12.1	Same as above	Same as above
QFb.cerz-7BL.2	*cfa2040–barc1073*	9.5–14	Same as above	Same as above

5.3.3 QTL Mapping of Sedimentation Volume

The Zeleny sedimentation value (ZSV) has been proven to be useful in wheat breeding programs for the estimation of wheat eating and cooking quality (Mesdag 1964; Kneževic et al. 1993; Liu et al. 2003; He et al. 2004; Zhang et al. 2005). There is a positive correlation between sedimentation volume and gluten strength or loaf volume. The ZSV is often used as a screening test in wheat breeding. Mesdag (1964) showed that the value of ZSV is a measure for the quantity and quality of the gluten. Because the baking value of wheat flour is largely determined by these components, the ZSV is also considered as a useful predictor for the baking value. Liu et al. (2003) detected that the associations between ZSV and DWCN's (dry white Chinese noodle) appearance and taste also significantly fit quadratic regression model. The gluten quality-related parameter of sedimentation value was significantly associated with pan bread quality score (He et al. 2004). Özberk et al. (2006)

found that the only quality analyses showing significant correlations with market price were Zeleny sedimentation value and hectoliter weights (kg hL1). Ozturk et al. (2008) reported that the cookie diameter gave highly significant correlations with ZSV.

The advent and utilization of molecular markers have provided powerful tools for elucidating the genetic basis of quantitatively inherited traits. However, only a few studies have reported genetic loci that influence ZSV in wheat (Rousset et al. 2001; Kunert et al. 2007; Sun et al. 2008). Rousset et al. (2001) reported that one strong QTL for ZSV was mapped on the long arm of chromosome 1A around *Glu-Al*. A distally located QTL for ZSV was mapped on 1BS. And a major QTL for ZSV, clearly corresponding to the *Glu-Dl* locus, was detected on the chromosome arm 1DL. Kunert net al. (2007) found four putative QTLs for ZSV. Sun et al. (2008) identified three QTLs for ZSV using a RIL derived from the cross between Chuan 35050 and Shannong 483.

QTLs for ZSV were investigated using three populations involving additive and epistatic analysis in this study. The objective of this study was to comprehensively characterize the genetic basis for ZSV of wheat in order to facilitate the future breeding of high-quality wheat varieties using the molecular markers in MAS.

5.3.3.1 Measurements of ZSV

Zeleny sedimentation volume was determined using the testing instrument made by China Agricultural University according to the AACC method 56-61 A.

5.3.3.2 QTL Mapping for ZSV Using DH Population

5.3.3.2.1 Phenotypic Variation for DH Lines and Parents

ZSV of Ym57 showed higher values than that of Hp3, and the continuous variations and transgressive segregation were found in the DH population (Fig. 5.9). ZSV segregated continuously and approximately fit normal distributions with absolute values of both skewness and kurtosis less than 1.0, indicating that this trait was suitable for QTL mapping.

5.3.3.2.2 QTLs with Additive Effects and Additive × Environment (AE) Interactions

Four QTLs with significant additive effects were identified on chromosomes 1B, 1D, 5A, and 5D (Table 5.38; Fig. 5.10). These QTLs explained 2.66 to 14.39 % of the phenotypic variance. The Qzsv-1B had the most significant effect, accounting

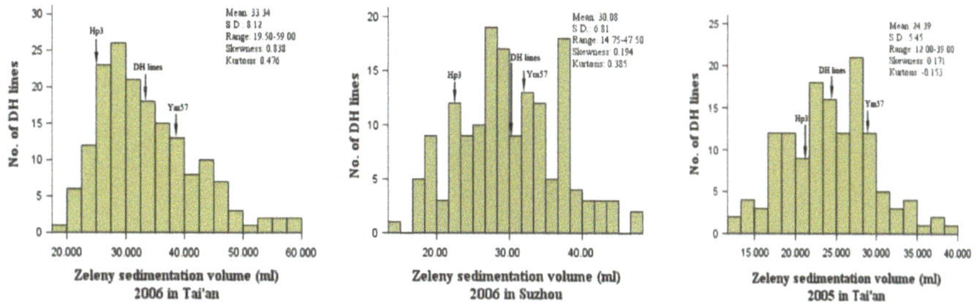

Fig. 5.9 Frequency distributions of ZSV in 168 DH lines derived from a cross of Hp3 × Ym57 evaluated at three environments in 2005 and 2006 cropping seasons. The means of trait values for the DH lines and both parents are indicated by *arrows*. Several statistics for the traits in the DH lines are shown on the *right* of each *plot*

for 14.39 % of the phenotypic variance. The Ym57 alleles at three loci, *Qzsv-1B*, *Qzsv-1D*, and *Qzsv-5D*, increased ZSV by 2.52, 1.98, and 1.20 mL, respectively, owing to additive effects. The Hp3 allele increased ZSV at the *Qzsv-5A* by 1.08 mL, accounting for 2.66 % of the phenotypic variance. This suggested that alleles, which increased ZSV, were dispersed within the two parents, resulting in small differences of phenotypic values between the parents and transgressive segregants among the DH population. The total additive QTLs detected for ZSV accounted for 29.23 % of the phenotypic variance.

One additive effect was involved in AE interactions (Table 5.38; Fig. 5.10). The Ym57 alleles at one locus, *Qzsv-5D*, increased the ZSV by 1.04 mL, correspondingly contributing 2.44 % of the phenotypic variance.

5.3.3.2.3 QTLs with Epistasis Effects and Epistasis × Environment (AAE) Interactions

Four pairs of epistatic QTLs were identified for ZSV and were located on chromosomes 1A, 2A, 3A, 7A, and 7D chromosomes (Table 5.39; Fig. 5.10). These QTLs had corresponding contributions ranging from 0.64 to 6.79 %. One pair of epistasis, occurring between the loci *Qzsv-2A/Qzsv-7A*, had the largest effect, which contributed ZSV of 1.73 mL and accounted for 6.79 % of the phenotypic variance. The four pairs of epistatic QTLs explained 12.11 % of the phenotypic variance. All the epistatic effects were non-main-effect QTLs.

One pair of epistatic QTL was detected in AAE interactions for ZSV (Table 5.39; Fig. 5.10). The AAE effects explained 2.33 % of the phenotypic variance, and this QTL, *Qzsv3A.2/Qzsv7D.1*, increased ZSV by 1.01 mL owing to AAE effects, and simultaneously, the positive value means that the parent-type effect is greater than the recombinant-type effect.

Table 5.38 Estimated additive effects and environment interaction (AE) effects of QTLs for ZSV at three environments in 2005 and 2006 cropping seasons

QTL	Flanking marker[a]	Position[b] (cM)	*F*	*P*	A[c]	H^2 (A, %)[d]	AE1	H^2 (AE1, %)[e]	AE2	H^2 (AE2, %)	AE3	H^2 (AE3, %)
Qzsv-1B	*Xwmc412.2–Xcfe023.2*	36.4	25.22	0.000	−2.52	14.39						
Qzsv-1D	*Xwmc93-GLUD*	61.9	15.91	0.000	−1.98	8.93						
Qzsv-5A	*Xbarc358.2–Xgwm186*	38.1	8.10	0.000	1.08	2.66						
Qzsv-5D	*Xcfd101–Xbarc320*	60.6	12.69	0.000	−1.20	3.25			−1.04	2.44		

[a]"Flanking marker" means the interval of *F* peak value for QTL
[b]"position" means the location of *F* peak value for QTL in "flanking marker"
[c]additive effects: A positive value indicates that the allele from Hp3 increased ZSV, and a negative value indicates that the allele from Ym57 increased ZSV
[d]H^2(A, %) indicates the contribution explained by putative additive QTL
[e]H^2(AE1, %) indicates the contribution explained by additive QTL × environment 1 interaction
E1: Tai'an, 2005; E2: Tai'an, 2006; E3: Suzhou, 2006

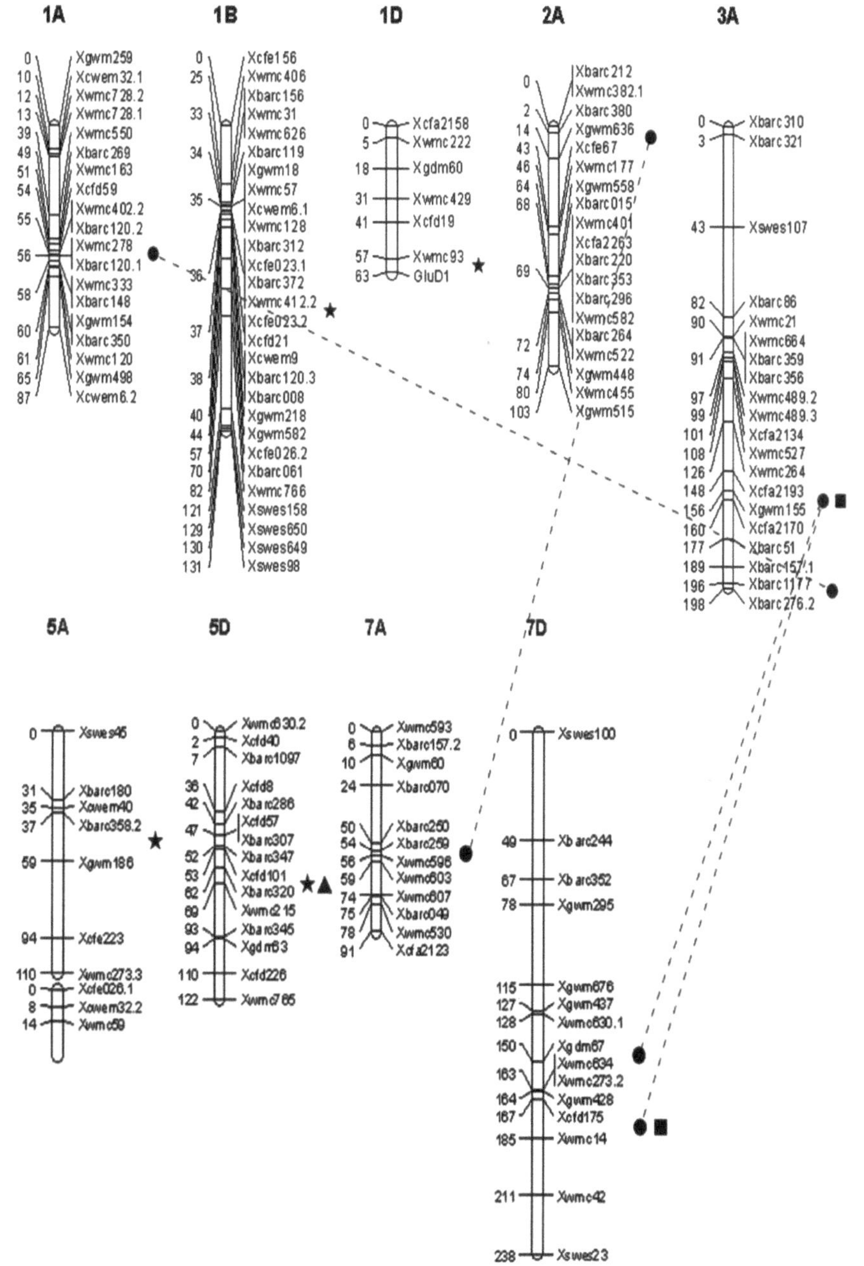

Fig. 5.10 Genetic linkage map of wheat showing mapping QTLs with additive effects, epistatic effects, AE, and AAE for ZSV. ★ locus involved in additive effects, ▲ locus involved in AE, ●– – – ● locus involved in epistasis, ■ locus involved in AAE

Table 5.39 Estimated epistatic effects and epistasis × environment interaction (AAE) effects of QTLs for ZSV at three environments in the 2005 and 2006 cropping seasons

QTL	Flanking marker[a]	Position[b] (cM)	QTL	Flanking marker	Position (cM)	AA[c]	H^2 (AA, %)[d]	AAE1	H^2 (AAE1, %)[e]	AAE2	H^2 (AAE2, %)	AAE3	H^2 (AAE3, %)
Qzsv-1A	*Xwmc278–Xbarc120.1*	56.3	*Qzsv-3A.1*	*Xbarc1177–Xbarc276.2*	196.3	−0.94	1.99						
Qzsv-2A	*Xgwm636–Xcfe67*	29.1	*Qzsv-7A*	*Xbarc259–Xwmc596*	53.7	−1.73	6.79						
Qzsv-3A.2	*Xcfa2193–Xgwm155*	152.7	*Qzsv-7D.1*	*Xcfd175–Xwmc14*	181.5	−1.09	2.69	1.01	2.33				
Qzsv-3A.2	*Xcfa2193–Xgwm155*	152.7	*Qzsv-7D.2*	*Xgdm67–Xwmc634*	161.5	−0.53	0.64						

[a, b]Explanation as in Table 5.1
[c]epistatic effect: A positive value means that the parent-type effect is greater than the recombinant-type effect, and a negative value means that the parent-type effect is less than the recombinant-type effect
[d]H^2(AA, %) indicates the contribution explained by putative epistatic QTL
[e]H^2(AE1, %) indicates the contribution explained by epistatic QTL × environment 1 interaction
E1: Tai'an, 2005; E2: Tai'an, 2006; E3: Suzhou, 2006

Table 5.40 Phenotypic values for Zeleny sedimentation value of the RIL population and the parents in different environments

Environment	Parents		RIL population						
	Shannong 01-35	Gaocheng 9411	Mean	SD	Minimum	Maximum	*CV*/%	Skewness	Kurtosis
E1	30.17	40.65	37.66	5.36	21.51	54.28	14.22	0.10	0.78
E2	27.32	36.77	33.73	3.56	22.43	43.98	10.54	−0.12	0.56

E1: 2009 Tai'an; E2: 2010 Tai'an

5.3.3.3 QTL Mapping of ZSV Using RIL Population Derived from Shannong 01-35 and Gaocheng 9411

5.3.3.3.1 Phenotypic Variation

The ZSV of Gaocheng 9411 is higher than that of Shannong 01-35, and the mean value of the population was between the ZSV values of two parents (Table 5.40). Continuous variation was found in this population, and the variation ranged from 21.51 to 54.28 and 22.43 to 43.98 in two environments, respectively. The coefficient of variation was higher than that of 10.0 %. The skewness and kurtosis of this population is less than 1, and normal distribution was found, which indicated that ZSV belonged to the quantitative trait.

5.3.3.3.2 QTL Analysis of ZSV

Sixteen additive QTLs were found, explaining 3.51 to 58.64 % of the phenotypic variation, which involved eleven chromosomes (1B, 1D, 2A, 3A, 3B, 3D, 4A, 4D, 6A, 7A, and 7B) (Table 5.41). Of which, there were five major QTLs, explaining 11.67 to 58.64 % of the phenotypic variation. The additive effect of *QZsv1B.1-109* was from Shannong 01-35, which increased 2.00 ml, and the other four major QTLs were from Gaocheng 9411, which increased the ZSV from 1.21 to 3.70 ml. In PD, the PVE of *QZsv1B.1-104* had the highest value of 58.64 %, which was from Gaocheng 9411. This locus can be also found in E2, explaining 19.71 %, which increased the ZSV to 1.85 ml. *QZsv1D-108* was detected in E1, E2, and PD environments, and in E2, it has the highest PVE with 11.67 %, which was from Gaocheng 9411. This locus increased ZSV from 1.11 to 1.44 ml.

Table 5.41 Additive QTLs for Zeleny sedimentation value in different environments

Environment	QTL	Left marker	Right marker	EstAdd	PVE/%
E1	*QZsv1B.1-80*	*wPt-2230*	*wPt-665375*	−2.11	12.52
	QZsv1D-108	*wPt-6963*	*wPt-667287*	−1.44	7.58
	QZsv3A-22	*Xgpw2270*	*wPt-1681*	−1.38	7.01
	QZsv4A-43	*Xgpw4040*	*wPt-7001*	−1.03	3.91
	QZsv7A-30	*wPt-6768*	*wPt-6495*	−1.35	6.70
E2	*QZsv1B.1-104*	*wPt-5363*	*wPt-1363*	−1.85	19.71
	QZsv1D-108	*wPt-6963*	*wPt-667287*	−1.21	11.67
	QZsv3B.1-62	*wPt-5295*	*wPt-667746*	−1.52	18.17
	QZsv3B.1-71	*wPt-0021*	*wPt-5704*	1.02	7.93
	QZsv4A-138	*wPt-664948*	*wPt-0538*	−0.86	5.91
	QZsv4D-80	*Xgpw342*	*wPt-2379*	−0.66	3.51
	QZsv6A.1-140	*TaGw2-CAPS*	*wPt-667618*	−0.88	5.97
	QZsv6D-116	*wPt-3127*	*xcfd49*	−1.01	8.15
	QZsv7B-2	*wPt-5463*	*wPt-4230*	−0.78	4.82
PD	*QZsv1B.1-104*	*wPt-5363*	*wPt-1363*	−3.70	58.64
	QZsv1B.1-109	*wPt-1911*	*wPt-2751*	2.00	19.26
	QZsv1D-108	*wPt-6963*	*wPt-667287*	−1.11	7.43
	QZsv2A-327	*wPt-5887*	*wPt-669199*	0.93	4.85
	QZsv3D-269	*xgwm456*	*wPt-668308*	0.88	4.63

E1: 2009 at Tai'an site; E2: 2010 at Tai'an site. Positive and negative values of additive effect (EstAdd) indicate that alleles to increase thousand-grain weight are inherited from Shannong 01-35 and Gaocheng 9411, respectively

5.3.3.4 Research Progress of Sedimentation Volume QTL Mapping and Comparison with Previous Studies

5.3.3.4.1 Research Progress of Sedimentation Volume QTL Mapping

Sedimentation volume is an important indirect parameter for appraising the gluten quality and quantity, which was closely related to high molecular weight glutenin subunits (Zhang et al. 2004; Zhang et al. 2005; Witkowski et al. 2008).

Previous researchers found the QTLs on 1B and 1D chromosomes (Rousset et al. 2001; Kunert et al. 2007; Wu et al. 2008; Zhao et al. 2009). In our study, one QTL (*Qzsv-1D)* was located on 1D chromosome with the flanking marker *Xwmc93* and *GluD1*, and the distance between them was 5.9 cM. It explained the 8.93 % of the phenotypic variation with increasing ZSV 1.98 ml. Rousset et al. (2001) also detected a major QTL on 1D close to the *Glu-D1* locus. Kunert et al. (2007) also identified a major QTL on 1DL linked marker *Xgwm642* close to the *Glu-D1* locus. Knežević et al. (1993) found a significant correlation between the score of *Glu-1* and sedimentation volume ($r = 0.553$), and sedimentation volume was significantly correlated with *Glu-lAa*, *Glu-1Ac*, *Glu-Ba*, and *Glu-1Bc* loci. Witkowski et al.

Table 5.42 Summary of QTL results of wheat flour sedimentation volume characteristics (PVE > 5 %)

QTL	Flanking marker	PVE/%	Mapping population	References
QZsv.sdau-1D.1	*Xwmc432a–Xwmc336c*	16.48	RIL	Sun et al.(2008)
QZsv.sdau-1D.2	*Glu-D1-Xsrap19*	15.48	Same as above	Same as above
QZsv.sdau-3B	*Xwmc418–Xubc834a*	10.3	Same as above	Same as above
QSsd.caas-1A	*bar148*	8.9–9.1	IL	Li et al. (2012)
QSsd.caas-1B	*wmc128*	2.0–15.2	Same as above	Same as above
QSsd.caas-1D	*wmc222*	6.8	Same as above	Same as above
QSsd.caas-2B	*wms374*	9.4–10.9	Same as above	Same as above
QSsd.caas-2D	*barc159*	11.2–11.5	Same as above	Same as above
QSsd.caas-4B	*wms495/wms513*	11.5–21.9	Same as above	Same as above
QSsd.caas-5A	*wms186*	9.3–14.8	Same as above	Same as above
QSsd.caas-5B	*Barc4*	1.8–9.2	Same as above	Same as above
QSsd.caas-6A	*wms334*	2.9–11.1	Same as above	Same as above
QSsd.caas-xa	*barc1031*	1.9–8.3	Same as above	Same as above
QSsd.caas-1B.1	*barc1057-barc187*	13.2–31.5	RILs	Li et al. (2009)
QSsd.caas-1A	*glu-A1-wmc93*	12.1–15.3	Same as above	Same as above
QSsd.caas-1D	*glu-D1-cfd48b*	9.0–19.3	Same as above	Same as above
QSsd.caas-6A.1	*cfd1-wms518a*	7.3–10.1	Same as above	Same as above
QSsd.caas-6A.2	*wms169-wms617*	3.7	Same as above	Same as above
QSsd.caas-3D	*wms191c-wmc533*	6.2	Same as above	Same as above
QSsd.caas-2B	*psp3035-wms148*	4.4–6.1	Same as above	Same as above
QSsd.caas-2D	*wms157-cfd62a1*	2.7–7.5	Same as above	Same as above
QSsd.caas-2A	*wmc177-wms71a*	2.2	Same as above	Same as above
QSsd.caas-1B.2	*wmc134-wms191b*	5.8	Same as above	Same as above
QSsd.caas-4A	*wms160-wmc497*	8.1	Same as above	Same as above
QSsd.crc-1B	*Xgwm131–Glu-B1*	20.6	DH	Mccartney et al.(2006)
QSsd.crc-2A	*Xgwm296*	3.3	Same as above	Same as above
QSsd.crc-6A	*Xbarc23–Xwmc398*	5.6	Same as above	Same as above

(2008) also found this phenomenon using statistics method. At present, the QTLs had been located on 13 chromosomes including 1A, 1B, 1D, 2A, 2B, 2D, 3B, 3D, 4A, 4B, 5A, 5B, and 6A. Of which, most of the QTLs were on 1B and 1D chromosomes. The highest PVE arrived at 31.5 % (Table 5.42).

5.3.3.4.2 Comparison of This Result with Previous Studies

In our study, there were four additive QTLs and four epistatic QTLs detected using DH population. Of which, *Qzsv-1B* explaining 14.39 % of the phenotypic variation was close to *Glu-1D* locus with the distance of 0.9 cM, so it can be used in MAS.

While in RIL population, the four major QTLs, *QZsv1B.1-80*, *QZsv1B.1-104*, *QZsv1B.1-109*, and *QZsv1D-108*, were detected on 1B and 1D chromosomes, which were related to Glu-1D locus. Of which, *QZsv1B.1-104* and *QZsv1D-108* were stable QTLs. From Table 5.42, most of the researchers detected QTLs on 1B and 1D which were close to *Glu-1D* and *Glu-1B*. Therefore, these results further testified that the high molecular weight glutenin subunit is an important factor affecting sedimentation volume.

5.3.4 QTL Mapping of Paste Viscosity Characteristics

Wheat grain (*Triticum aestivum* L.) contains 60–70 % starch and is the major source for commercial starch in China. A commercially important property of starch is its viscosity upon heating in water. The starch viscosity profile, also termed the RVA profile (because it is tested on the rapid visco analyzer, RVA), can reflect the process of either swelling or reintegrating of starch (Bao et al. 2002). The RVA profile has proven useful in wheat breeding programs for the estimation of wheat eating and cooking quality (He et al. 2003; Liu et al. 2003; Zhang et al. 2004; He et al. 2004; Zhang et al. 2005). Peak viscosity, pasting temperature, and setback during cooling may be valuable predictors of bread firming during baked bread storage. A high peak viscosity was found desirable for Chinese steamed bread quality, regardless of the processing method used (He et al. 2003). The correlation coefficients (r) between starch paste breakdown (RVA) and noodle viscoelasticity and smoothness were 0.63 and 0.59, respectively (Zhang et al. 2005).

QTLs for the paste viscosity characteristics have been identified in rice. Bao et al. (2002) showed that RVA profiles were mainly controlled by waxy locus gene (*wx*) located on rice chromosome 6, which encodes the granule-bound starch synthase. Bao et al. (2002) identified a QTL flanked by *Amy2A* and *RG433* on the end of the long arm of chromosome 6. It was identified for setback and consistency (viscosity) and might cover the gene encoding for starch branching enzyme I. The numbers of QTLs with additive effects on rice grain for peak viscosity, breakdown, final viscosity, setback, and pasting temperature were 5, 2, 3, 2, and 2, respectively (Wang et al. 2007). Only, a few studies have reported genetic loci that influence RVA profile parameters in wheat (Kuchel et al. 2006; Sun et al. 2008).

In this study, QTLs for RVA profile parameters were investigated based on the mixed linear model in different populations. The objective of this study was to comprehensively characterize the genetic basis for paste viscosity characteristics of wheat in order to facilitate our future breeding of high-quality wheat varieties.

5.3.4.1 Measurements of RVA Profile Parameters

RVA profile parameters were determined with a rapid visco analyzer (RVA-Super 3, Newport Scientific, Australia) using the LS/T 6101-2002 standard. Wheat paste viscosity characteristics were described by six parameters: peak viscosity (PV), trough viscosity (TV), breakdown (BD), final viscosity (FV), setback (SB), and pasting temperature (PT) according to Sun et al. (2008).

5.3.4.2 QTL for RVA Based on DH Population

5.3.4.2.1 Phenotypic Variation for DH Lines and Parents

As is shown in Table 5.43, RVA profile parameters of Yumai 57 showed higher values than those of Huapei 3 except for BD in 2006 in Suzhou; the means of the RVA profile parameters fell between the two parent's values except for BD. The continuous distribution of each trait was observed with transgressive segregation in the DH lines. The skewness value and the kurtosis value of PK, TV, and FV in 2006 in Suzhou were a slightly larger than 1.000, and the skewness values of FV in 2005 in Tai'an and the kurtosis values of SB were a slightly larger than 1.000, indicating that their distributions were skewed to some extent, while the distributions for the other traits were normal (Table 5.43).

5.3.4.2.2 Correlation Coefficients Between RVA Profile Parameters

Pairwise correlation coefficients between RVA profile parameters are listed in Table 5.44. They are significantly positively correlated with each other. The results also indicated that RVA profile parameters were strongly correlated with each other. The relationships between RVA profile parameters were different from the results of other populations (Sun et al. 2008).

5.3.4.2.3 Results and Analysis

Mapping analysis produced a total of 35 QTLs for 6 RVA profile parameters, with a single QTL explaining 0.91–21.34 % of the phenotypic variation (Tables 5.45 and 5.46). The 35 QTLs were distributed on 15 chromosomes (Fig. 5.11). Among the six traits, all PVE over 30 % and the part of PVE explained by these QTLs of a single trait ranged from 31.30 to 52.00 %.

5.3.4.2.3.1 QTLs with Additive Effects and Additive × Environment (AE) Interactions

The numbers of QTLs with additive effects on PV, TV, BD, FV, SB, and PT were 3, 3, 2, 4, 4, and 1, respectively. Thirteen QTLs (76.5 %) were located on the A genome. The A genome seemed to play a predominant role in RVA profile parameters. The additive effects mainly came from the parent Yumai 57 (13 QTLs, 76.5 %). The *QBd-4A* had the most significant additive effect, accounting for

Table 5.43 Phenotypic performance for RVA profile parameters in 168 DH lines derived from a cross of Huapei 3 × Yumai 57 at three environments

Trait	Environment	Parents		DH population					
		Huapei 3	Yumai 57	Mean	Min.	Max.	SD[a]	Skewness	Kurtosis
Peak viscosity (RVU)	Tai'an, 2005	123.58	231.75	189.35	64.25	276.42	48.57	−0.818	0.056
	Tai'an, 2006	138.00	217.08	168.32	70.75	272.17	39.52	−0.209	−0.150
	Suzhou, 2006	223.75	241.58	236.24	95.08	302.33	37.74	−1.113	1.676
Trough viscosity (RVU)	Tai'an, 2005	39.17	136.42	102.45	0.00	170.58	39.16	−0.962	0.086
	Tai'an, 2006	50.58	122.50	82.93	5.67	201.25	32.72	−0.108	0.096
	Suzhou, 2006	116.92	148.00	140.09	17.58	199.92	31.32	−1.378	2.256
Breakdown (RVU)	Tai'an, 2005	84.42	95.33	86.90	54.00	132.08	18.65	0.382	−0.740
	Tai'an, 2006	87.42	94.58	85.39	55.33	123.33	14.82	0.316	−0.448
	Suzhou, 2006	106.83	93.58	96.15	58.75	138.58	18.33	0.249	−0.576
Final viscosity (RVU)	Tai'an, 2005	96.67	258.67	200.12	14.58	317.25	63.47	−1.186	0.862
	Tai'an, 2006	113.00	234.75	166.65	23.67	389.83	54.03	−0.124	0.989
	Suzhou, 2006	199.33	265.00	242.26	50.17	335.25	45.44	−1.227	2.284
Setback (RVU)	Tai'an, 2005	57.50	122.25	97.67	15.00	146.67	25.49	−1.409	1.989
	Tai'an, 2006	62.42	112.25	83.73	18.00	188.58	22.09	−0.014	2.656
	Suzhou, 2006	82.42	117.00	102.18	32.58	138.92	16.36	−0.555	1.109
Pasting temperature (°C)	Tai'an, 2005	61.75	67.15	64.22	55.10	73.45	3.23	0.549	0.256
	Tai'an, 2006	62.80	74.35	67.61	59.15	75.90	3.51	0.577	−0.697
	Suzhou, 2006	63.55	68.65	67.71	62.65	77.55	3.73	0.739	−0.517

[a]"SD" means standard deviation

Table 5.44 Pairwise correlation coefficients for RVA profile parameters

	PV	TV	BD	FV	SB	PT
PV	1.000					
TV	0.943**	1.000				
BD	0.646**	0.355**	1.000			
FV	0.905**	0.983**	0.288**	1.000		
SB	0.759**	0.868**	0.141**	0.944**	1.000	
PT	0.288**	0.212**	0.323**	0.189**	0.130**	1.000

**Represent significance at 1 % confidence level

21.34 % of the phenotypic variance. For a single trait, the part of PVE explained by these QTLs with additive effects ranged from 3.86 % of PT to 33.27 % of SB.

One additive effect was involved in AE interactions (Table 5.45; Fig. 5.11). The Yumai 57 alleles at one locus, *QPv-2A*, increased the PV by 5.26 RVU with correspondingly contributing 1.48 % of the phenotypic variance.

5.3.4.2.3.2 QTLs with Epistasis Effects and Epistasis × Environment (AAE) Interactions

Eighteen pairs of epistatic effects were identified for RVA profile parameters and were located on 13 chromosomes (Table 5.46; Fig. 5.11). These QTLs had corresponding contributions ranging from 0.91 % to 14.95 %. One pair of epistasis for BD occurring between the loci *Xgdm63–Xcfd226* had the largest effect, which contributed BD of 7.34 RVU and accounted for 14.95 % of the phenotypic variance. The two loci of epistatic effects mainly came from the parent Huapei 3 and Yumai 57, respectively (15 QTLs, 83.3 %). For a single trait, the part of PVE explained by these QTLs with epistatic effects ranged from 9.02 % of FV to 27.86 % of PT. Sixteen QTLs with epistatic effects (88.9 %) were located on A and D genomes. No QTL was detected in AAE interactions for RVA profile parameters.

5.3.4.2.4 Research Progress of Wheat Grain Quality QTL Mapping and Comparison with Previous Studies

5.3.4.2.4.1 Relationship Between Additive Effects and Epistatic Effects on RVA Profile Parameters

It would be interesting to study the relationships between the additive QTLs and epistatic QTLs identified. The majority (94.4 %) of loci involved in the epistatic interactions did not appear to have significant additive effects on RVA profile parameters in wheat. Four pairs of epistatic effects were identified for BD; one single loci (*QBd-2A.2/QBd-4A.2*, PVE 0.91 %) involved in epistatic effects had additive effects (*QBd-4A*, PVE 21.34 %). Similarly, Ma et al. (2007) observed that 37 % of the main-effect QTLs were involved in the epistatic interactions on maize

Table 5.45 Estimated additive and environment interaction (AE) effects of QTLs for RVA profile parameters at three environments

Trait	QTL	Flanking marker[a]	Position[b] (cM)	*F*	*P*	A[c]	H^2 (A, %)[d]	AE1	H^2 (AE1, %)[e]
PV	*QPv-1B*	*Xswes158–Xswes650*	121.6	6.70	0.000	8.15	3.54		
	QPv-2A	*Xgwm448–Xwmc455*	76.2	7.76	0.001	−5.94	1.88	−5.26	1.48
	QPv-6A	*Xgwm82–Xwmc182*	42.8	13.82	0.000	−11.01	6.46		
TV	*QTv-2A*	*Xwmc582–Xbarc264*	71.8	8.79	0.000	−5.70	2.64		
	QTv-4A	*Xwmc262–Xbarc343*	6.5	8.19	0.000	−6.51	3.45		
	QTv-6A	*Xbarc1055–Xwmc553*	46.0	18.36	0.000	−10.17	8.41		
BD	*QBd-4A*	*Xwmc718–Xwmc262*	5.0	50.81	0.000	8.77	21.34		
	QBd-7A	*Xwmc596–Xwmc603*	57.1	11.28	0.000	−5.16	7.39		
FV	*QFv-2A*	*Xwmc582–Xbarc264*	71.8	8.95	0.000	−7.54	1.88		
	QFv-4A	*Xwmc262–Xbarc343*	6.5	15.85	0.000	−16.95	9.52		
	QFv-5D	*Xbarc345–Xgdm63*	92.7	7.77	0.003	6.10	1.23		
	QFv-6A	*Xbarc1055–Xwmc553*	46.0	19.60	0.000	−18.68	11.56		
SB	*QSb-2A*	*Xwmc582–Xbarc264*	70.8	9.25	0.000	−3.21	2.23		
	QSb-4A	*Xwmc262–Xbarc343*	6.5	30.40	0.000	−8.49	15.52		
	QSb-6A	*Xbarc023–Xbarc1077*	39.5	13.48	0.000	−6.16	8.17		
	QSb-7D	*Xgwm676–Xgwm437*	120.9	9.26	0.000	−5.84	7.35		
PT	*QPt-1B*	*Xbarc312–Xcfe023.1*	36.1	10.98	0.000	0.69	3.86		

[a]"Flanking marker" means the interval of F peak value for QTL
[b]"Position" means the location of F peak value for QTL in "flanking marker"
[c]Additive effects: A positive value indicates that the allele from Hp3 increased RVA profile parameters, and a negative value indicates that the allele from Ym57 increased RVA profile parameters
[d]H^2 (A, %) indicates the contribution explained by putative additive QTL
[e]H^2 (AE1, %) indicates the contribution explained by additive QTL × environment 1 interaction
E1: Tai'an, 2005; E2: Tai'an, 2006; E3: Suzhou, 2006

Table 5.46 Estimated epistatic and epistasis × environment interaction (AAE) effects of QTLs for RVA profile parameters at three environments

Trait	QTL	Flanking marker[a]	Position[b] (cM)	QTL	Flanking marker	Position (cM)	AA[c]	H^2 (AA, %)[d]
PV	*QPv-1A*	*Xcfd59–Xwmc402.2*	55.0	*QPv-6D.1*	*Xcfd49–Xcfd42*	25.4	−9.35	4.66
	QPv-1D	*Xgdm60–Xwmc429*	19.5	*QPv-6D.1*	*Xcfd49–Xcfd42*	7.4	5.34	1.52
	QPv-1D	*Xgdm60–Xwmc429*	19.5	*QPv-6D.2*	*Xbarc054–Xgwm55*	57.4	−14.85	11.76
TV	*QTv-1D*	*Xgdm60–Xwmc429*	18.5	*QTv-6D*	*Xbarc054–Xgwm55*	61.4	−13.37	14.55
	QTv-2B	*Xbarc101–Xcwem55*	77.4	*QTv-6B*	*Xcfa2187–Xgwm219*	0.0	7.46	4.52
BD	*QBd-2A.1*	*Xgwm558–Xbarc015*	64.5	*QBd-4A.1*	*Xbarc343–Xwmc313*	21.3	−2.32	1.49
	QBd-2A.2	*Xgwm448–Xwmc455*	74.2	*QBd-4A.2*	*Xwmc718–Xwmc262*	5.0	−1.81	0.91
	QBd-5D.1	*Xbarc286–Xcfd57*	41.6	*QBd-6D.1*	*Xcfd49–Xcfd42*	34.4	−4.62	5.92
	QBd-5D.2	*Xgdm63–Xcfd226*	93.6	*QBd-6D.2*	*Xcfa2129–Xbarc080*	173.1	−7.34	14.95
FV	*QFv-1A*	*Xbarc148–Xgwm154*	58.1	*QFv-6D*	*Xcfd49–Xcfd42*	34.4	−12.86	5.47
	QFv-3A	*Xgwm155–Xcfa2170*	156.0	*QFv-7A*	*Xwmc596–Xwmc603*	56.1	−10.35	3.55
SB	*QSb-1A.1*	*Xbarc148–Xgwm154*	58.1	*QSb-6D*	*Xcfd42–Xcfd13*	35.0	−3.66	2.88
	QSb-1A.2	*Xgwm498–Xcwem6.2*	71.1	*QSb-6D*	*Xcfd42–Xcfd13*	35.0	−3.34	2.41
	QSb-3A	*Xgwm155–Xcfa2170*	158.0	*QSb-7A*	*Xwmc596–Xwmc603*	58.1	−4.31	3.99
PT	*QPt-2A*	*Xbarc264–Xwmc522*	72.2	*QPt-3D*	*Xwmc631–Xbarc071*	82.1	−0.88	6.32
	QPt-3A	*Xwmc527–Xwmc264*	119.8	*QPt-7D*	*Xgwm676–Xgwm437*	114.9	−1.16	10.99
	QPt-5B	*Xgwm213–Xswem861.2*	13.1	*QPt-7D*	*Xgdm67–Xwmc634*	149.5	−0.70	3.95
	QPt-7A	*Xbarc157.2–Xgwm60*	7.3	*QPt-7D*	*Xcfd175–Xwmc14*	166.5	0.90	6.60

[a, b]Explanation as in Table 5.1

[c]The epistatic effect: A positive value means that the parent-type effect is greater than the recombinant-type effect, and a negative value means that the parent-type effect is less than the recombinant-type effect

[d]H^2 (AA, %) indicates the contribution explained by putative epistatic QTL

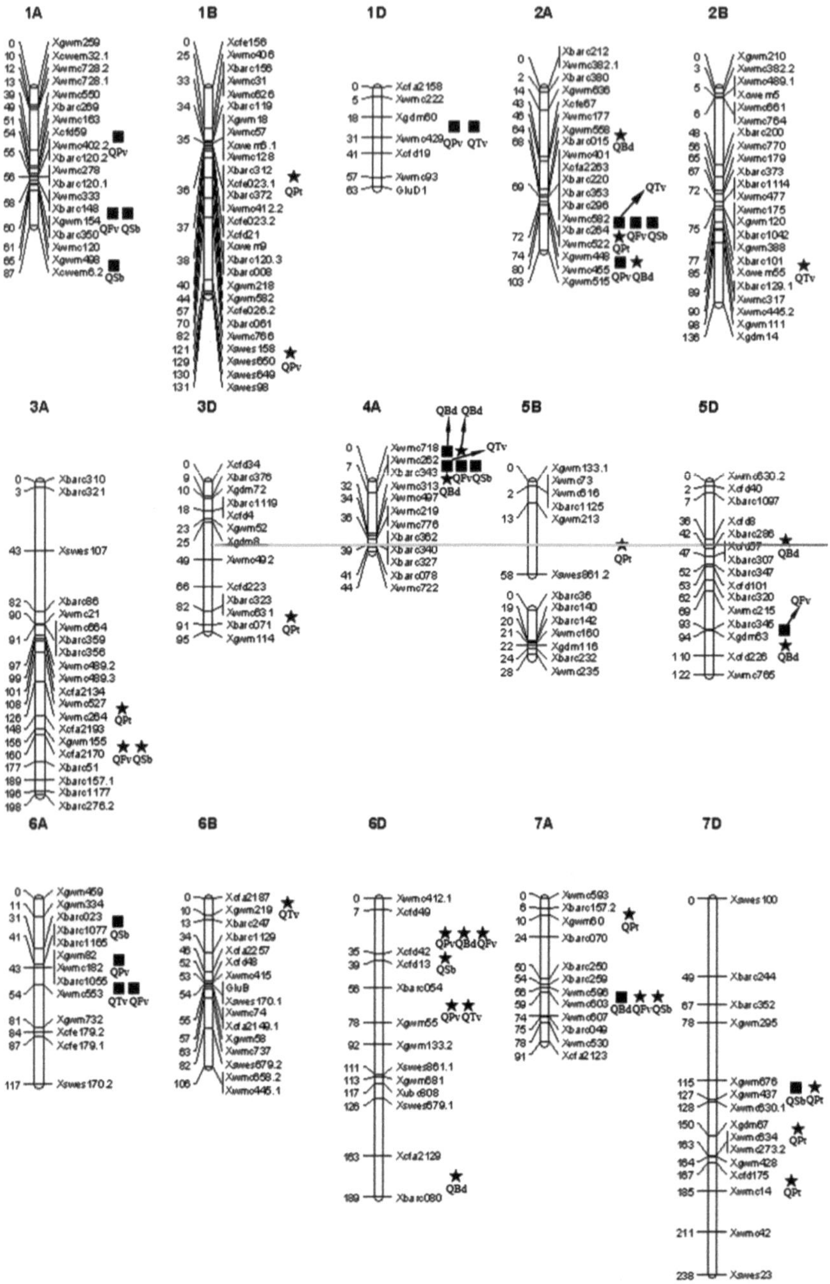

Fig. 5.11 Genetic linkage map of wheat showing mapping QTLs with additive effects, epistatic effects, and QEs for RVA profile parameters. ■ locus involved in additive effects; ★ locus involved in epistasis effects

grain yield and its components. This indicated that many loci for epistatic effects might not have significant effects on RVA profile parameters alone, but might affect its expression by epistatic effects with the other loci. The results also suggest that some of the additive QTLs might be detected with effects confounded by epistatic effects, if the epistatic effects were ignored in QTL mapping. Thus, breeders must take into account such complexity and test for the effects of individual loci in the targeted genetic background in order to obtain the expected phenotypes of the interested genes.

5.3.4.2.4.2 Comparison of This Result with Previous Studies

Thirteen QTLs with additive effects (76.5 %) were located on the A genome. Sixteen QTLs with epistatic effects (88.9 %) were located on the A and D genomes. The A and D genomes seemed to play predominant roles in RVA profile parameters.

Thirty-five QTLs detected for RVA profile parameters in this study seemed to have similar or the same chromosomal locations with different genetic materials, as identified in the previous reports. We found two QTL clusters in this study. They were located on chromosomes 2A and 4A. The cluster on 2A was in the region of *Xwmc582–Xbarc264–Xwmc522–Xgwm448–Xwmc455*. This cluster included six QTLs for six RVA profile parameters. The cluster on 4A was in the region of *Xwmc718–Xwmc262–Xbarc343–Xwmc313–Xwmc497*. This cluster included six QTLs for TV, BD, FV, and SB. Chromosome 4A seemed to play a predominant role in defining RVA profile parameters. Batey et al. (2001) also reported that a predominant QTL for viscosity was mapped to chromosome 4A, centered on the *Wx-B1* locus.

We found nine QTLs on chromosomes 2B, 7A, and 7D, and Kuchel et al. (2006) reported three QTLs on chromosomes 2B, 7A, and 7D each, which accounted for between 7 and 10 % of the phenotypic variation in RVA. Sun et al. (2008) identified thirteen QTLs with additive effects for PV, TV, BD, FV, SB, and PT on chromosomes 1D, 2D, 3D, 6A, 6B, and 7B. A highly significant ($P < 0.001$) QTL was identified on the long arm of chromosome 7B (Batey et al. 2001). On chromosomes 2D and 7B, we did not found any QTL, but we did find some QTLs that were not reported in the literature. For example, a QTL cluster was located on chromosome 2A.

These differences might be attributed to the following possibilities: First, differences in mapping populations and experimental environments were studied in various QTL mappings. Second, different QTL mapping approaches were used in various studies. Third, some of the genes involved in the wheat RVA profile pathway did not show any allelic variation between the two parents (Huapei 3 and Yumai 57). Consequently, several loci could not be identified on some chromosomes.

In summary, a total of 17 additive QTLs and 18 pairs of epistatic QTLs which were distributed on 15 chromosomes were detected for RVA profile parameters in

168 DH lines derived from a cross Huapei 3 × Yumai 57 tested in three one-replicate environments. One major QTL, *QBd-4A*, was closely linked to *Xwmc262* 1.5 cM and could account for 21.34 % of the phenotypic variation of BD without any influence from the environment. Therefore, the *QBd-4A* could be used in MAS in wheat breeding programs. The results showed that both additive and epistatic effects were important as a genetic basis for RVA profile parameters and were also sometimes subject to environmental modifications.

5.3.4.3 QTL Mapping for RVA Characters Using Nuomai 1 × Gaocheng 8901 RIL Population

5.3.4.3.1 Phenotypic Variation and Correlation of RVA Parameters

Seven parameters of the parent and offspring pasting properties of phenotypic data are shown in Table 5.47. Pasting temperature and breakdown of Nuomai 1 were higher than those of Gaocheng 8901 (GC8901), and the remaining five RVA parameter values were less than those of GC8901. Except for peak viscosity and pasting temperature in Tai'an 2009, the rest parameters were among the average RVA parameters of those of parent. Variation range of seven parameters in population was large, and there were obvious bitransgressive segregation, representing continuous normal distribution. RVA parameters of the population are correlated highly (Table 5.48).

5.3.4.3.2 QTL and Effect Analysis of RVA

QTL analyzed for wheat starch paste of characteristics using phenotypic data in 2 years and its average. A total of 47 additive QTLs controlled 7 paste parameters were detected (Table 5.49). Among them, 15 QTLs were detected in two environments, 7 QTLs in all environments contributed more than 10 %, and they were considered as major QTL.

Seven additive QTLs controlling peak viscosity located on 2A, 2D, 6A, 7A, and 7D chromosomes. Of which, *QPv-7D* closely linked with *Wx-D1* was detected in two environments and the average P, explaining 4.73, 6.92 and 8.72 % of PVE, respectively. Its positive allele was from GC8901. But the major *QPv-6A* in E1 was an environment-specific QTL.

Eight additive QTLs for trough viscosity located on 1A, 2A, 4A, 7A, and 7D chromosomes. Of which, *QTrv-7A* and *QTrv-7D.1* were identified in two environments and P environment, whose positive alleles were derived from Gaocheng8901. These two QTLs were to the *Wx-A1* and *Wx-D1* loci, respectively. The major QTL, *QTrv-7D.1*, showed stable in two environments, but the rest QTL was only detected in a single environment.

Table 5.47 Phenotypic values for RVA parameters of two parents and the RIL population in two environments in wheat

Trait	Environment	Parents		RIL population			
		Nuomai 1	Gaocheng 8901	Range	Mean ± SD	Skewness	Kurtosis
Peak viscosity	Tai'an, 2009	219.83	228.58	120.22–329.08	252.23 ± 41.05	−0.89	0.89
	Anhui, 2011	202.25	207.08	60.50–297.83	206.80 ± 42.48	−0.74	0.30
Trough viscosity	Tai'an, 2009	86.50	152.58	57.84–251.33	141.53 ± 33.45	−0.16	0.69
	Anhui, 2011	52.83	139.17	33.25–198.25	126.47 ± 35.48	−0.59	−0.21
Breakdown	Tai'an, 2009	133.30	76.00	50.25–181.67	110.70 ± 24.21	−0.14	−0.00
	Anhui, 2011	150.42	67.92	27.25–140.08	80.33 ± 18.13	0.21	0.67
Final viscosity	Tai'an, 2009	127.50	274.00	80.00–361.00	243.15 ± 55.72	−0.89	0.58
	Anhui, 2011	69.50	222.17	49.50–291.50	189.39 ± 49.49	−0.87	0.16
Setback	Tai'an, 2009	41.00	121.42	22.16–194.67	101.52 ± 29.15	−0.28	0.11
	Anhui, 2011	17.67	83.00	14.17–101.83	62.92 ± 16.15	−0.92	0.83
Pasting time	Tai'an, 2009	3.46	6.40	3.40–7.00	6.04 ± 0.88	−2.22	3.40
	Anhui, 2011	3.80	6.53	3.67–6.93	6.18 ± 0.88	−2.18	3.15
Pasting temperature	Tai'an, 2009	63.60	62.90	60.75–82.35	67.01 ± 5.37	1.33	0.74
	Anhui, 2011	67.00	65.20	55.05–83.10	66.96 ± 3.72	1.78	1.69

Table 5.48 Coefficients of pairwise correlations of the mean values of RVA parameters

RVA parameters	Peak viscosity	Trough viscosity	Breakdown	Final viscosity	Setback	Pasting time
Trough viscosity	0.839**					
Breakdown	0.683**	0.176**				
Final viscosity	0.854**	0.915**	0.317**			
Setback	0.687**	0.622**	0.409**	0.885**		
Pasting time	0.461**	0.713**	−0.122**	0.685**	0.509**	
Pasting temperature	−0.2	−0.006	−0.028	0.069	0.142**	0.219**

**Significant correlations were seen among some RVA parameters

Eleven additive QTLs controlling breakdown value are distributed on 1B, 2A, 2D, 3A, 3B, 4A, and 6A chromosomes. Of which, only *QBd-2D.1* with flanking markers *wPt-6687* and *wPt-731336* was identified in the two environments and the average P, whose positive allele was from Nuomai1. It explained 30.02, 36.33 and 45.24 % of PVE, respectively. Although many QTL were detected, but fewer QTL was stably found in different environments.

Six additive QTLs controlling the final viscosity located on 2A, the 2D, 3A, 4A, 7A, and 7D chromosomes. Of which, *QFv-4A*, *QFv-7A*, and *QFv-7D* expressed stably in different environments, which were detected in two and average P environment, and the increasing effect QTL comes from paternal alleles |901; these three QTLs were close to *Wx-B1*, *Wx-A1*, and *Wx-D1*, whose contribution rate of *QFv-7D* in various environments is greater than 10 %, which was the major QTL.

QSb-4A, *QSb-7A*, and *QSb-7D* controlling setback can be detected repeatedly in two and average P environments, and *QSb-4A* and *QSb-7D* in each environment explain more than 10 % of the phenotypic variation, which was the major QTL.

Eight additive QTLs controlled pasting time was detected on 1A, 1D, 4A, 7A and 7D chromosomes. *QPt-4A.1*, *QPt-7A*, *QPt-7D.1* and *QPt-7D.2* were identified in two environments and the average environment. The major QTL, *QPt-7D.2*, explained 14.46, 13.00 and 15.64 % of PVE in three environments, respectively.

Four additive QTLs controlling pasting time temperature located on 1A, the 1D, 2A, and 7A chromosomes. Among them, *QPtem-1D* (*cfd-183/wPt-729773*) was detected repeatedly by three times, which was major QTL.

5.3.4.4 QTL Mapping for RVA Characters Using "Shannong 01-35x Gaocheng 9411" and RIL Population

5.3.4.4.1 Phenotypic Performance for RVA Profile Parameters

Average value of pasting parameters of Gaocheng 9411 was higher than that of Shannong 01-35 (Table 5.50). Average value of RIL population was among the value of that of parent. Seven parameters of RVA distributed continuously with

Table 5.49 QTL with significant additive effects for RVA parameters detected in different environments

Trait	Environment	QTL	Chromosome	Site (cM)	Interval markers	EstAdd	Phenotypic variance explained %
Peak viscosity	E1	*QPv-6A*	6A	133.00	*wPt-729920—wPt-664792*	−13.15	10.85
		QPv-7D	7D	0.00	*Wx-D1—wPt-664368*	−8.68	4.73
	E2	*QPv-2A*	2A	0.00	*wPt-0408—wPt-2838*	−7.81	3.61
		QPv-2D.1	2D	167.00	*wPt-730613—wPt-9749*	8.33	4.03
		QPv-7A	7A	162.00	*wPt-731311—Wx-A1*	−8.17	4.00
		QPv-7D	7D	0.00	*Wx-D1—wPt-664368*	−10.75	6.92
	PD	*QPv-2D.2*	2D	124.00	*wPt-6687—wPt-731336*	15.23	19.73
		QPv-2D.3	2D	130.00	*wPt-3757—wPt-1301*	−11.29	10.85
		QPv-7D	7D	0.00	*Wx-D1—wPt-664368*	−10.10	8.72
Trough viscosity	E1	*QTrv-1A*	1A	88.00	*wPt-2992—wPt-663934*	−8.07	3.11
		QTrv-4A.1	4A	85.00	*wPt-7924—wPt-664948*	−8.66	7.09
		QTrv-7A	7A	162.00	*wPt-731311—Wx-A1*	−8.02	6.08
		QTrv-7D.1	7D	0.00	*Wx-D1—wPt-664368*	−12.19	14.04
	E2	*QTrv-2A*	2A	0.00	*wPt-0408—wPt-2838*	−6.62	3.71
		QTrv-2D	2D	152.00	*wPt-8319—wPt-731130*	8.06	5.57
		QTrv-4A.2	4A	116.00	*Wx-B1—wPt-0105*	−8.20	5.71
		QTrv-7A	7A	162.00	*wPt-731311—Wx-A1*	−10.35	9.20
		QTrv-7D.1	7D	0.00	*Wx-D1—wPt-664368*	−11.58	11.49
	PD	*QTrv-4A.1*	4A	85.00	*wPt-7924—wPt-664948*	−8.02	7.64
		QTrv-7A	7A	162.00	*wPt-731311—Wx-A1*	−7.31	6.35
		QTrv-7D.2	7D	84.00	*wPt-3923—wPt-2592*	5.52	3.56
		QTrv-7D.1	7D	0.00	*Wx-D1—wPt-664368*	−11.22	14.95

(continued)

Table 5.49 (continued)

Trait	Environment	QTL	Chromosome	Site (cM)	Interval markers	EstAdd	Phenotypic variance explained %
Breakdown	E1	*QBd-1B*	1B	168.00	*wPt-6442—wPt-3824*	6.73	8.13
		QBd-2A	2A	283.00	*wPt-6775—wPt-9793*	−4.81	3.23
		QBd-2D.1	2D	124.00	*wPt-6687—wPt-731336*	12.93	30.02
		QBd-2D.2	2D	129.00	*wPt-665836—wPt-3757*	−8.95	14.39
		QBd-3A	3A	0.00	*wPt-4407—wPt-0714*	−4.56	3.74
		QBd-4A.1	4A	105.00	*Wx-B1—wPt-0105*	6.14	6.80
		QBd-6A	6A	122.00	*wPt-729920—wPt-664792*	−12.50	28.14
	E2	*QBd-2D.1*	2D	124.00	*wPt-6687—wPt-731336*	10.54	36.33
		QBd-2D.4	2D	131.00	*wPt-731220—wPt-730677*	−7.88	20.38
		QBd-3B	3B	119.00	*wPt-2757—wPt-666738*	3.82	4.78
	PD	*QBd-2D.1*	2D	124.00	*wPt-6687—wPt-731336*	11.30	45.24
		QBd-2D.3	2D	130.00	*wPt-3757—wPt-1301*	−8.88	27.91
		QBd-4A.2	4A	98.00	*wPt-664948—Wx-B1*	3.80	5.11
		QBd-6A	6A	129.00	*wPt-729920—wPt-664792*	−6.61	15.56
Final viscosity	E1	*QFv-4A*	4A	105.00	*Wx-B1—wPt-0105*	−15.60	8.29
		QFv-7A	7A	162.00	*wPt-731311—Wx-A1*	−14.41	7.08
		QFv-7D	7D	0.00	*Wx-D1—wPt-664368*	−22.64	17.45
	E2	*QFv-2A*	2A	0.00	*wPt-0408—wPt-2838*	−8.57	3.20
		QFv-2D	2D	154.00	*wPt-8319—wPt-731130*	10.99	5.33
		QFv-4A	4A	114.00	*Wx-B1—wPt-0105*	−15.42	10.42
		QFv-7A	7A	161.00	*wPt-731311—Wx-A1*	−16.06	11.38
		QFv-7D	7D	0.00	*Wx-D1—wPt-664368*	−19.11	16.10
	PD	*QFv-3A*	3A	0.00	*wPt-4407—wPt-0714*	−7.60	2.80
		QFv-4A	4A	108.00	*Wx-B1—wPt-0105*	−19.21	17.92
		QFv-7A	7A	162.00	*wPt-731311—Wx-A1*	−14.86	10.72
		QFv-7D	7D	0.00	*Wx-D1—wPt-664368*	−20.58	20.56

(continued)

Table 5.49 (continued)

Trait	Environment	QTL	Chromosome	Site (cM)	Interval markers	EstAdd	Phenotypic variance explained %
Setback	E1	*QSd-4A*	4A	103.00	*Wx-B1—wPt-0105*	−11.93	17.68
		QSd-7A	7A	162.00	*wPt-731311—Wx-A1*	−8.00	7.97
		QSd-7D	7D	0.00	*Wx-D1—wPt-664368*	−11.00	15.04
	E2	*QSd-4A*	4A	108.00	*Wx-B1—wPt-0105*	−7.36	22.44
		QSd-7A	7A	162.00	*wPt-731311—Wx-A1*	−4.08	6.89
		QSd-7D	7D	1.00	*Wx-D1—wPt-664368*	−7.90	25.87
	PD	*QSd-4A*	4A	103.00	*Wx-B1—wPt-0105*	−8.70	19.51
		QSd-7A	7A	162.00	*wPt-731311—Wx-A1*	−6.32	10.32
		QSd-7D	7D	2.00	*Wx-D1—wPt-664368*	−9.97	25.62
Pasting time	E1	*QPt-4A.1*	4A	86.00	*wPt-7924—wPt-664948*	−0.17	4.18
		QPt-4A.2	4A	103.00	*Wx-B1—wPt-0105*	−0.22	6.61
		QPt-7A	7A	162.00	*wPt-731311—Wx-A1*	−0.22	6.87
		QPt-7D.1	7D	79.00	*wPt-3923—wPt-2592*	0.17	4.03
		QPt-7D.2	7D	0.00	*Wx-D1—wPt-664368*	−0.32	14.46
	E2	*QPt-1A*	1A	155.00	*wPt-3904—wPt-5776*	−0.14	2.76
		QPt-1D.1	1D	165.00	*wPt-665814—wPt-666414*	0.20	5.62
		QPt-4A.1	4A	85.00	*wPt-7924—wPt-664948*	−0.28	10.87
		QPt-7A	7A	162.00	*wPt-731311—Wx-A1*	−0.27	10.06
		QPt-7D.1	7D	84.00	*wPt-3923—wPt-2592*	0.14	2.90
		QPt-7D.2	7D	0.00	*Wx-D1—wPt-664368*	−0.30	13.00
	PD	*QPt-1D.2*	1D	162.00	*wPt-9380—wPt-665814*	0.15	3.19
		QPt-4A.1	4A	85.00	*wPt-7924—wPt-664948*	−0.18	5.25
		QPt-4A.2	4A	103.00	*Wx-B1—wPt-0105*	−0.17	4.69
		QPt-7A	7A	162.00	*wPt-731311—Wx-A1*	−0.24	8.68
		QPt-7D.1	7D	83.00	*wPt-3923—wPt-2592*	0.16	4.06
		QPt-7D.2	7D	0.00	*Wx-D1—wPt-664368*	−0.32	15.64

(continued)

Table 5.49 (continued)

Trait	Environment	QTL	Chromosome	Site (cM)	Interval markers	EstAdd	Phenotypic variance explained %
Pasting temperature	E1	*Qptem-1D*	1D	0.00	*cfd-183—wPt-729773*	1.89	13.04
		Qptem-2A	2A	84.00	*wPt-668181—wPt-1853*	−1.43	6.74
	E2	*Qptem-1D*	1D	0.00	*cfd-183—wPt-729773*	1.13	9.95
	PD	*Qptem-1A*	1A	77.00	*wPt-664968—wPt-664972*	0.73	3.84
		Qptem-1D	1D	0.00	*cfd-183—wPt-729773*	1.43	14.66
		Qptem-7A	7A	158.00	*wPt-731311—Wx-A1*	−0.81	4.68

E1: Tai'an, 2008; E2′: Tai'an, 2009; E3: Suzhou, 2011; PD: pool data; positive values indicate that Nuomai 1 alleles increase corresponding trait value; negative values indicate that Gaocheng 8901 alleles increase corresponding trait value

Table 5.50 Phenotypic values for RVA profile parameters of the RIL population and the parents in different environments

Traits	Environment	Parents		RIL population						
		Shannong 01-35	Gaocheng 9411	Mean	SD	Min.	Max.	CV (%)	Skewness	Kurtosis
PV (RVU)	E1	139.92	156.33	152.93	28.83	74.42	355.50	18.85	1.62	13.14
	E2	92.75	195	186.83	36.03	64.75	367.00	19.28	−0.33	4.14
	E1	106.92	115.42	109.01	24.52	27.33	194.42	22.50	−0.25	0.77
TV (RVU)	E2	32.75	132.58	123.18	34.09	3.42	202.08	27.67	−1.06	1.58
	E1	177.58	192.58	189.06	32.87	74.42	350.00	17.39	0.16	3.14
FV (RVU)	E2	84.75	228.92	209.89	46.52	28.92	360.92	22.16	−1.15	2.87
	E1	33	40.92	43.93	11.81	24.17	161.08	26.88	5.73	54.91
BD (RVU)	E2	60	62.42	63.65	10.85	45.58	164.92	17.05	4.81	41.56
	E1	70.67	77.17	80.05	10.43	47.08	155.58	13.02	1.95	15.10
SB (RVU)	E2	52	96.33	86.71	13.74	25.33	158.83	15.84	−0.72	8.18
	E1	6.4	6.47	6.32	0.22	5.40	6.80	3.51	−0.86	1.80
PET (min)	E2	5.33	6.4	6.32	0.33	4.47	6.87	5.18	−2.43	8.25
	E1	64.45	65.4	64.76	1.72	59.00	74.25	2.66	1.90	8.74
PAT (°C)	E2	63.55	64.45	64.15	1.48	55.20	71.85	2.31	0.02	13.63

E1: 2008–2009 growing season at Tai'an site; E2: 2009–2010 growing season at Tai'an site

Table 5.51 Additive QTLs for RVA profile parameters in different environments

Trait	Environment	QTL	Left marker	Right marker	EstAdd	PVE (%)
PV	E1	*QPv3A-221*	*wPt-729826*	*wPt-666853*	−36.16	15.95
		QPv7A-224	*Xgpw2139*	*wPt-7151*	7.28	6.61
	E2	*QPv3A-191*	*wPt-667139*	*wPt-666832*	12.78	7.99
		QPv3A-223	*wPt-666853*	*wPt-4933*	−25.38	5.51
		QPv3B.1-64	*wPt-5295*	*wPt-667746*	7.58	4.42
		QPv4B.1-99	*wPt-7569*	*wPt-3908*	−11.34	9.85
		QPv6D-135	*wPt-664719*	*wPt-666615*	−7.75	4.74
		QPv6D-140	*wPt-666615*	*wPt-666008*	15.68	19.22
	PD	*QPv1D-38*	*wPt-9181*	*wPt-3738*	6.20	4.79
		QPv1D-105	*wPt-6963*	*wPt-667287*	−8.65	9.27
		QPv3A-190	*wPt-667139*	*wPt-666832*	9.24	6.74
		QPv3A-223	*wPt-666853*	*wPt-4933*	−18.84	4.74
		QPv4B.1-99	*wPt-7569*	*wPt-3908*	−8.48	8.59
		QPv5D-92	*wPt-0419*	*xgwm190*	−8.27	8.48
		QPv5D-149	*wPt-6429*	*wPt-671956*	8.09	7.95
		QPv6D-135	*wPt-664719*	*wPt-666615*	−7.11	6.22
		QPv7A-224	*Xgpw2139*	*wPt-7151*	6.53	5.31
TV	E1	*QTv3A-223*	*wPt-666853*	*wPt-4933*	−25.33	11.85
		QTv4B.1-99	*wPt-7569*	*wPt-3908*	−5.78	5.29
	E2	*QTv3A-190*	*wPt-667139*	*wPt-666832*	10.47	5.86
		QTv3A-223	*wPt-666853*	*wPt-4933*	−28.89	7.43
		QTv4B.1-99	*wPt-7569*	*wPt-3908*	−13.14	13.70
		QTv6D-135	*wPt-664719*	*wPt-666615*	−10.23	8.60
		QTv6D-142	*wPt-666615*	*wPt-666008*	16.10	21.23
	PD	*QTv3A-189*	*wPt-667139*	*wPt-666832*	7.70	5.59
		QTv3A-223	*wPt-666853*	*wPt-4933*	−31.36	14.83
		QTv4B.1-99	*wPt-7569*	*wPt-3908*	−9.13	11.24
		QTv6D-134	*xcfd49*	*wPt-672044*	−16.05	35.64
		QTv6D-144	*wPt-666615*	*wPt-666008*	14.67	29.94
FV	E1	*QFv3A-222*	*wPt-729826*	*wPt-666853*	−36.10	19.49
		QFv4B.1-99	*wPt-7569*	*wPt-3908*	−8.29	6.33
	E2	*QFv1D-40*	*wPt-9181*	*wPt-3738*	10.10	4.40
		QFv3A-189	*wPt-667139*	*wPt-666832*	14.39	5.99
		QFv3A-223	*wPt-666853*	*wPt-4933*	−37.92	6.63
		QFv4B.1-99	*wPt-7569*	*wPt-3908*	−17.94	13.23
		QFv6D-135	*wPt-664719*	*wPt-666615*	−14.53	8.98
		QFv6D-142	*wPt-666615*	*wPt-666008*	26.22	29.26
	PD	*QFv1D-11*	*wPt-665480*	*wPt-671415*	7.54	4.37
		QFv3A-191	*wPt-667139*	*wPt-666832*	9.49	4.24
		QFv3A-223	*wPt-666853*	*wPt-4933*	−42.71	15.05
		QFv4B.1-99	*wPt-7569*	*wPt-3908*	−13.13	12.70
		QFv6D-134	*xcfd49*	*wPt-672044*	−24.19	44.34
		QFv6D-143	*wPt-666615*	*wPt-666008*	21.65	35.67

(continued)

Table 5.51 (continued)

Trait	Environment	QTL	Left marker	Right marker	EstAdd	PVE (%)
BD	E1	*QBd3A-222*	*wPt-729826*	*wPt-666853*	−36.10	19.49
		QBd4B.1-99	*wPt-7569*	*wPt-3908*	−8.29	6.33
	E2	*QBd1B.1-9*	*wPt-731490*	*wPt-4555*	5.18	11.22
	PD	*QBd1A.2-33*	*Xgpw2180*	*wPt-671483*	2.18	5.79
		QBd5B.1-30	*wPt-9814*	*wPt-5737*	−2.67	8.90
		QBd5B.1-80	*Xgpw8015*	*wPt-0103*	2.29	6.21
SB	E1	*QSb3A-222*	*wPt-729826*	*wPt-666853*	−21.00	27.23
		QSb4B.1-99	*wPt-7569*	*wPt-3908*	−3.02	8.38
	E2	*QSb1D-40*	*wPt-9181*	*wPt-3738*	3.68	6.05
		QSb4B.1-99	*wPt-7569*	*wPt-3908*	−5.18	11.50
		QSb6D-135	*wPt-664719*	*wPt-666615*	−4.83	10.31
		QSb6D-141	*wPt-666615*	*wPt-666008*	7.94	27.75
	PD	*QSb4B.1-99*	*wPt-7569*	*wPt-3908*	−3.52	10.74
PET	E1	*QPet3A-223*	*wPt-666853*	*wPt-4933*	−0.16	5.54
	E2	*QPet2D-118*	*wPt-667294*	*wPt-4144*	−0.11	8.51
		QPet4B.1-99	*wPt-7569*	*wPt-3908*	−0.11	9.10
		QPet6D-135	*wPt-664719*	*wPt-666615*	−0.10	7.33
	PD	*QPet1A.2-33*	*Xgpw2180*	*wPt-671483*	−0.07	6.67
		QPet2B-73	*wPt-1140*	*wPt-4199*	0.07	6.97
		QPet3A-188	*wPt-667139*	*wPt-666832*	0.11	13.67
		QPet5D-146	*xgwm190*	*wPt-6429*	0.08	9.09
		QPet7B-173	*wPt-1266*	*wPt-0276*	0.10	14.88
PAT	E1	*QPat2D -159*	*Xgpw4085*	*wPt-2360*	−0.66	11.17
		QPat3A-222	*wPt-729826*	*wPt-666853*	−1.75	14.83
		QPat4A-160	*wPt-4064*	*wPt-6603*	−0.64	14.21
		QPat4A-164	*wPt-669526*	*wPt-668307*	0.88	27.05
		QPat4B.2-35	*wPt-8756*	*CFE149*	0.40	5.30
		QPat5D-164	*xgwm272*	*wPt-1197*	−0.36	4.65
		QPat6D-144	*wPt-666615*	*wPt-666008*	0.45	7.05
	E2	*QPat3A-223*	*wPt-666853*	*wPt-4933*	−1.78	15.41
		QPat3B.1-82	*wPt-5704*	*wPt-667891*	0.43	7.51
	PD	*QPat3A-222*	*wPt-729826*	*wPt-666853*	−2.11	31.82
		QPat3B.1-82	*wPt-5704*	*wPt-667891*	0.33	5.53
		QPat5B.2-68	*wPt-666268*	*wPt-3922*	0.31	5.28
		QPat5B.2-94	*wPt-8449*	*wPt-664746*	−0.41	9.54

E1: 2008–2009 growing season at Tai'an site; E2: 2009–2010 growing season at Tai'an site. Positive and negative values of additive effect (EstAdd) indicate that alleles to increase thousand-grain weight are inherited from Shannong 01-35 and Gaocheng 9411, respectively

large variation range and showed obviously bitransgressive segregation, representing continuous normal distribution. Part of the absolute value of skewness and kurtosis was very large.

5.3.4.4.2 QTL Mapping Analysis for RVA Parameters

Thirteen additional QTLs for peak viscosity were detected on 1D, 3A, 3B, 4B, 5D, 6D, and 7A chromosomes, explaining the phenotypic variation of 4.42–19.22 %, respectively (Table 5.51). Among them, the *QPv3A-221* (E1) and *QPv6D-140* (E2) 2 QTL explained phenotypic variation of 15.95 and 19.22 %, respectively, which was major QTL; its additive effect was from the Gaocheng 9411 and Shannong 01-35, respectively, which increased peak viscosity 36.16 and 15.68 RVU. *QPv3A-223*, *QPv4B.1-99*, and *QPv6D-135* were detected in E2 and detected *QPv3A-221* and *QPv3A-223* in E1 environment, having 2 cM distance; *QPv7A-224* was detected in the E1 and average.

Eight additive QTLs controlling trough viscosity distributed on 3A, 4B, and 6D chromosomes. The PVE of the five QTLs was more than 10 %, which were major QTLs. *QTv6D-134* detected in the average, explaining 35.64 % of the phenotypic variation. *QTv3A-223* was detected in the E1, E2, and average. The PVE was 11.85 and 14.83 %, respectively; *QTv4B.1-99* was detected in the E1, E2, and average, explaining 13.70 and 11.24 % of the phenotypic variation in E1 and E2, respectively. So, *QTv3A-223* and *QTv4B.1-99* were major QTLs with stable expression, and its additive effects were from Gaocheng 9411. *QTv3A-18* (PD) and *QTv3A-190* (E2) shared the common region with the contribution of 5.59 and 5.86 %, respectively. *QTv6D-142* (E2) and *QTv6D-144* (PD) were in the same marker interval with the contribution rate of 21.23 and 29.94 %, respectively.

Eleven additive QTLs for final viscosity were detected, which located on 1D, 3A, 4B, and 6D chromosomes, explaining 4.24–44.34 % of the phenotypic variation. Of which, five major QTLs explained 12.70–44.34 % of the phenotypic variation. Additive effect of *QFv6D-143* (PD, 35.67 %) was from Shannong 01-35, and additive effects of the other four major QTLs were from Gaocheng 9411; *QFv6D-134* explained 44.34 % of PVE. There was 1cM distance between *QFv3A-223* and *QFv3A-222*, which indicated they can be considered as one QTL. *QFv4B.1-99* identified in all environments was a stable QTL.

Six additive QTLs controlled were detected breakdown, located on 1A, on 1B, 3A, 4B, and 5B chromosomes, explaining 5.79–19.49 % of the phenotypic variation. Six additive QTLs were environment-specific expression. The PVE of *QFv3A-222* (E1) and *QFv1B.1-9* (E2) was over 10 %. The additive effects from Gaocheng 9411 and Shannong 01-35, respectively, can improve breakdown values with 36.10 and 5.18 RVU.

Five additive QTLs for setback located in chromosomes 1D, 3A, 4B, and 6D and explained 6.05–27.75 % of the phenotypic variation. Among them, the PVE of *QSb3A-222*, *QSb4B.1-99*, *QSb6D-135*, and *QSb6D-141* was more than 10 %, which were major QTLs. The additive effect of *QSb6D-141* (E2) accounting for 27.75 % was from maternal Shannong 01-35, which has increased setback value of 7.94 RVU; while the PVE of *QSb3A-222* (E1) was 27.75 %, which were from Gaocheng 9411, and its additive effect which increased the setback value by 21.00 RVU. *QSb4B.1-99* was detected in E1, E2, and average with explaining 8.38, 11.50 and 10.74 % of the phenotypic variation, respectively. It was major QTL with stable inheritance, and its additive effect were from Gaocheng 9411, which increased the setback values by 3.02–5.18 RVU under different environments.

Nine QTLs for peak time were detected, located on chromosomes 1A, 2B, 2D, 3A, 4B, 5D, 6D, and 7B, explaining 5.54–14.88 % of the phenotypic variation. Of which, *QPet7B-173* (PD) has the highest PVE with 14.88 %, and its additive effect was from Shannong 01-35.

Eleven additive QTLs controlled pasting temperature located on chromosomes 2D, 3A, 3B, 4A, 4B, 5B, 5D, and 6D and explained 4.65–31.82 % of the phenotypic variation. Five major QTLs were detected, explaining 11.17–31.82 % of the phenotypic variation. Among them, *QPat3A-222* was detected in the E1, with the PVE of 14.83 %, and the distance of *QPat3A-223* (E2, 15.41 %) and *QPat3A-222* was only 1 cM.

5.3.4.5 Research Progress of RVA Parameter QTL Mapping and Comparison with Previous Studies

5.3.4.5.1 Research Progress of RVA Parameter QTL Mapping

Wu et al. (2008) detected four QTLs for peak viscosity using 240 lines derived from RIL population PH82-2/Neixiang 188, which located on 1A, on 1B, 3A, and 7B chromosomes and detected five QTLs for breakdown, located on 1B, 4A, 5B, 6B, and 7A chromosomes. Among them, there was one QTL cluster on 1B chromosome, which related to Zeleny sedimentation value, mixing time, 8-min width, peak viscosity and breakdown; and so did on 1D chromosome, relating to Zeleny sedimentation, and mixing time and eight minutes width. In addition, Sun et al. (2008) and Mccartney et al. (2006) conducted a QTL mapping RVA parameters also (Table 5.52). At present, QTL for RVA parameters located on 17 chromosomes mainly 1A, 1B, 1D, 2A, 2D, 3A, 3B, 3D, 4A, 4B, 4D, 5B, 6A, 6B, 7A, 7B, and 7D. The highest PVE was 31.36 %.

5.3.4.5.2 Comparison of This Result with Previous Studies

QTL for RVA parameters using three groups was detected on chromosomes 4A and 7A all in three populations and was also detected on chromosomes 2A, 6A, 7D, and 2D in DH groups and Nuomai RIL population. QTL for RVA parameters was detected on 3A, 3B, 4B, 5B, and 6D chromosomes in 9411 × 01-35 RIL population. Because key enzyme of synthesis starch was on 4A, 7A, and 7D, QTLs associated with starch pasting parameters were detected on chromosome 4A in different populations.

QTLs for RVA parameters were detected in this study mainly on A and D genomes. Some QTLs located on same region (Table 5.52), indicating that the two

Table 5.52 Summary of QTL results of wheat flour RVA characteristics (PVE > 5 %)

Trait	QTL	Flanking marker	PVE (%)	Mapping population	References
Peak viscosity	*QPv.sdau-3D*	*Xwmc529–Xsrap8*	27.27	RIL	Sun et al. (2008)
	QPv.sdau-6B	*Xgwm644–Xgwm193*	15.83	RIL	Sun et al. (2008)
	QPv.sdau-7B	*Xubc857c–Xswes94*	9.54	RIL	Sun et al. (2008)
	Qpv.crc-4A	*Xgwm162*	12.5	DH	McCartney et al.(2006)
Tough viscosity	*QTv.sdau-3D*	*Xwmc529–Xsrap8*	31.36	RIL	Sun et al.(2008)
	QTv.sdau-6B	*Xgwm193–Xgwm608b*	18.94	RIL	Sun et al.(2008)
Final viscosity	*QFv.sdau-1D*	*Xwmc432a–Xwmc336c*	8.31	RIL	Sun et al.(2008)
	QFv.sdau-3D	*Xwmc529–Xsrap8*	24.48	RIL	Sun et al. (2008)
	QFv.sdau-6B	*Xgwm644–Xgwm193*	16.88	RIL	Sun et al. (2008)
	QRfv.crc-4B	*Xwmc826*	6.3	DH	McCartney et al. (2006)
	QRfv.crc-4D	*Xwmc617–Xwmc48*	7.4	DH	McCartney et al. (2006)
	QRfv.crc-7A	*Xwmc422*	5.5	DH	McCartney et al. (2006)
Setback	*QBd.sdau-1D*	*Xswes652–Xgwm458*	12.63	RIL	Sun et al. (2008)
	QBd.sdau-2D	*Xubc859a–Xswes624e*	19.62	RIL	Sun et al. (2008)
	QBd.sdau-3B	*Xubc823a–Xissr25a*	19.76	RIL	Sun et al. (2008)
	QRsb.crc-2A	*Xgwm294–Xwmc658*	5.5	DH	McCartney et al. (2006)
	QRsb.crc-4A	*Xwmc232*	10.4	DH	McCartney et al.(2006)
	QRsb.crc-4D	*Xwmc617–Xwmc48*	21.2	DH	McCartney et al. (2006)
Pasting temperature	*QPet.sdau-3D*	*Xwmc529–Xsrap8*	24.02	RIL	Sun et al. (2008)
	QPet.sdau-6B	*Xgwm644–Xgwm193*	19.63	RIL	Sun et al. (2008)
	QPet.sdau-7B	*Xubc857c–Xswes94*	9.22	RIL	Sun et al. (2008)
Pasting time	*QPat.sdau-6A*	*Xwmc163–Xswes119b*	29.92	RIL	Sun et al. (2008)
	QRpt.crc-7D	*Xgwm130*	11.2	DH	McCartney et al. (2006)
	QRfp.crc-7D	*Xwmc702*	14.1	DH	McCartney et al. (2006)

genomes captured major genes controlling starch pasting properties; QTL clusters were detected on 2A, 4A, 7A, 7D, 3A, 4B, and 6D chromosomes. QTL for starch pasting was detected on 2A, 2D, 3B, and 6D.

From this, in addition to the gene-controlled GBSSI enzyme, there are some important regions of chromosomes/genes that influence starch pasting properties. Therefore, QTL/gene cluster found in the present study has important reference value in the wheat improvement of starch pasting properties and special wheat quality breeding.

5.3.5 QTL Mapping of Falling Number

Wheat flour or other grains' powder suspension can be quickly pasted in a boiling water bath. Gelatinized starch was degraded in various degrees because of the different activity of α-amylase, which results in different starch viscosity and different falling speed of blender in paste. Falling Number Instrument was designed according to this principle. Given quantity of wheat or other grains powder, flour/water mixture was placed in a special tube and immersed in a bath of boiling water, then stir the mixture, and then the agitator in pasting was dropped. The time needed from a certain height to the bottom for agitator was recorded as falling number. Therefore, the falling number reflects the corresponding differences of α-amylase activity. The higher value suggested lower activity of α-amylase, conversely, showed highly α-amylase activity.

5.3.5.1 Determination

Falling number was measured according to GB/T10361 methods using FN1500 instrument.

5.3.5.2 Results and Analysis

5.3.5.2.1 Phenotypic Variation of Falling Number

In the three environments, falling number of Huapei 3 was lower than that of Yumai 57 (Fig. 5.12). Variation range of falling number of DH population was very large, showing a continuous distribution. And there was obvious bitransgressive segregation. The absolute values of skewness and kurtosis are less than 1.0 in normality test, indicating that the trait was quantitative trait controlled by multigene, which was suitable for QTL mapping analysis (Cao et al. 2001).

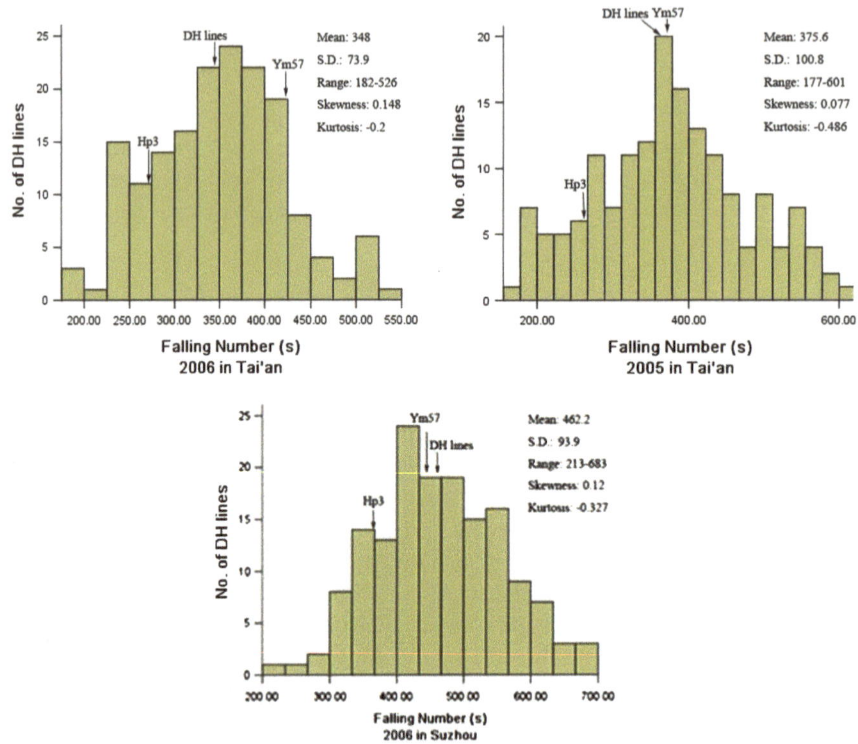

Fig. 5.12 Frequency distribution of falling number

5.3.5.2.2 Additive QTL and Its Interaction with the Environment

Two additive QTLs for falling number were detected (Table 5.53; Fig. 5.13), explaining 9.05 and 10.65 % of the phenotypic variation, respectively. Additive effect derived from the paternal Yumai. *Qfn-6A* located Xbarc1055-Xwmc553 interval explained the PVE of 10.65 % with increasing 30.07s. The QTL can be used in marker-assisted breeding. No additive QTL interacted with the environment was detected.

Table 5.53 Estimated additive (A) and additive × environment (AE) interactions of QTL for falling number

QTL	Flanking marker	Position (cM)	EstAdd A	H^2 (A, %)
Qfn-4A	*Xwmc262–Xbarc343*	6.5	−27.72	9.05
Qfn-6A	*Xbarc1055–Xwmc553*	47.0	−30.07	10.65

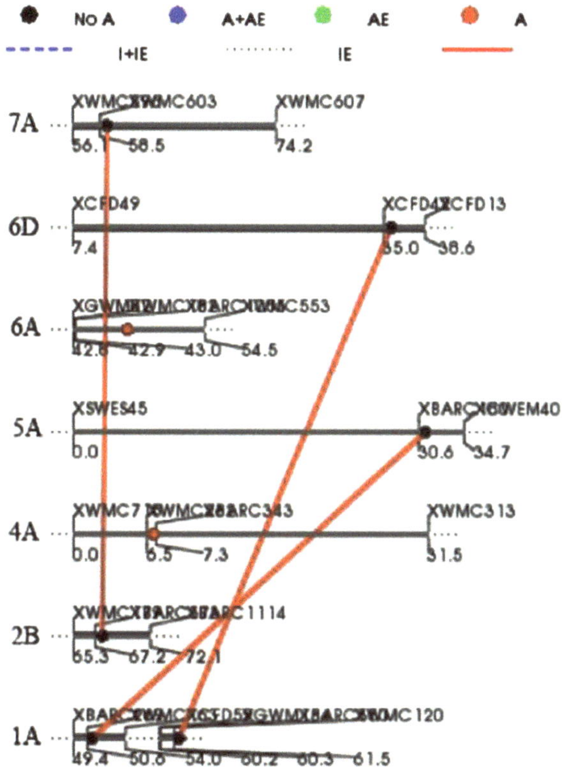

Fig. 5.13 Graphic presentation of the genetic architecture QTL with additive effects, epistatic effects, and QTL-by-environment interactions for FN

5.3.5.2.3 Epistatic QTL and Its Interactions with the Environment

Three epistatic QTLs for falling number were detected (Table 5.54; Fig. 5.13), explaining 4.82, 6.29, and 2.83 % of the phenotypic variation, respectively. Of which, the epistatic QTL *Qfn-1A.2/Qfn-6D* explained 6.29 % of the phenotypic variation. The total PVE of epistatic QTLs was 13.94 %. The epistatic QTL interactions with the environment were not detected.

Table 5.54 Estimated epistatic and epistasis × environment interaction (AAE) effects of QTL for falling number

QTL	Flanking marker	Position (cM)	QTL	Flanking marker	Position (cM)	A	H^2 (AA, %)
Qfn-1A.1	*Xbarc269–Xwmc163*	50.4	*Qfn-5A*	*Xbarc180–Xcwem40*	30.6	20.22	4.82
Qfn-1A.2	*Xbarc350–Xwmc120*	61.3	*Qfn-6D*	*Xcfd42–Xcfd13*	35.0	23.11	6.29
Qfn-2B	*Xbarc373–Xbarc1114*	67.2	*Qfn-7A*	*Xwmc603–Xwmc607*	58.5	5.50	2.83

5.3.5.2.4 Research Progress of Falling Number QTL Mapping and Comparison with Previous Studies

5.3.5.2.4.1 Research Progress of Falling Number QTL Mapping

Sun et al. (2008) detected one QTL for falling number, which explained 17.9 % of the phenotypic variation, which was located on chromosome 6B. Rasul et al. (2009) identified one QTL on chromosomes 4A and 4B, explaining 13.7 and 14.9 % of the phenotypic variation, respectively (Table 5.55).

5.3.5.2.4.2 Comparison of This Result with Previous Studies

Two additive QTLs and three epistatic QTLs for falling number were detected in this study. Of these, *Qfn-6A* was major QTL, explaining 10.65 % of the phenotypic variation. Two QTLs for falling number were detected on 4A and 6B, which were also reported in the previous study. The QTL might be related to the key starch synthesis enzymes on 4A; QTLs for falling number detected on 4B and 6B were closely related to preharvest sprouting traits.

5.3.6 QTL Mapping of Starch Content and Components

Starch is a polysaccharide polymer, different to other plant polysaccharides (e.g., cellulose and pectin), and there is no structural function of starch in plants. After processing, the starch solution will be thickening and gelation, so it can be widely used as thickening and gelling agents in food processing. The rheological properties of the starch varied widely, which can be expanded from the simple fluid gel to the gelatin, therefore the starch have wide industrial process utilization.

Starch molecule has an asymmetric carbon atom, possessing optically active; therefore, the size of optical rotation is proportional to the concentration of starch. The starch content can be determined according to the principle. Determination of total starch content using polarimetry is easy to operate, but the result was interfered by other optically active substances, resulting in higher results, so it is called “crude starch.” QTL analysis of starch content and components was less studied in the previous research, so this study could provide a reference for molecular marker-assisted breeding for the starch content and component.

Table 5.55 Summary of QTL results of wheat flour falling number

QTL	Flanking marker	PVE (%)	Mapping population	References
QFn.sdau-6B	*Xgwm132b–Xwmc487*	17.9	RIL	Sun et al.(2008)
QFN-4A.1	*Xwmc48*	13.7	DH	Rasul et al. (2009)
QFN-4B	*Xwmc349*	14.9	DH	Rasul et al. (2009)

5.3.6.1 Determination

The starch content was determined using polarimeter (WZZ-2B) method according to GB5006-85.

5.3.6.2 Results and Analysis

5.3.6.2.1 Phenotypic Variation of Crude Starch Content

The average starch content of Yumai 57 was significantly lower than that of Huapei 3 in the three environments (Fig. 5.14). The variation of crude starch content of DH population was very large, showing a continuous distribution, and there was obvious bitransgressive segregation. The absolute value of skewness and kurtosis is less than 1.0, indicating that it was a quantitative trait controlled by multigene and it suits for QTL mapping analysis (Cao et al. 2001).

5.3.6.2.2 Additive QTL of Crude Starch and Its Interaction with the Environment

Three additive QTLs for crude starch content were detected (Table 5.56; Fig. 5.15), explaining 3.04, 3.21, and 2.25 % of the phenotypic variation, respectively. One of the QTLs was from the Huapei 3, and two of them were from the paternal Yumai 57. It showed that the genes of starch content are distributed in the parents. *Qcs-5A* located between Xbarc358.2 and Xgwm186, explaining 3.21 % of the phenotypic variation. This locus increased crude starch content by 0.284. The total additive effects explained 8.5 % of the phenotypic variation. Interaction between additive QTL and the environment was detected.

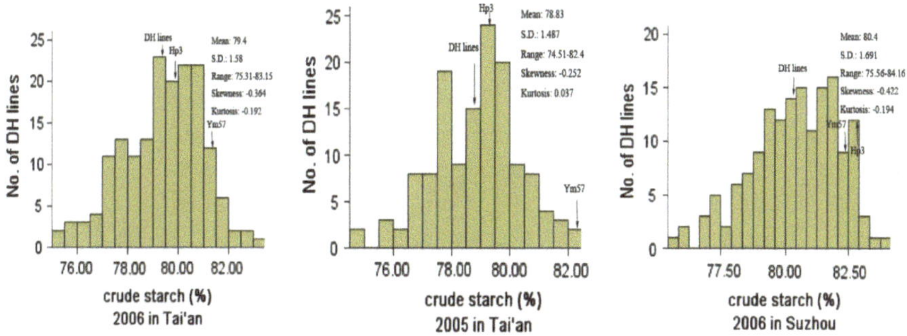

Fig. 5.14 Frequency distribution of crude starch

Table 5.56 Estimated additive (A) and additive × environment (AE) interactions of QTL for crude starch

QTL	Flanking marker	Position (cM)	EstAdd A	H^2 (A, %)
Qcs-3A	*Xbarc356–Xwmc489.2*	91.1	0.279	3.04
Qcs-5A	*Xbarc358.2–Xgwm186*	37.1	−0.287	3.21
Qcs-6D	*Xcfd42–Xcfd13*	35.0	−0.240	2.25

E1: Tai'an, 2005; E2: Tai'an, 2006; E3: Suzhou, 2006

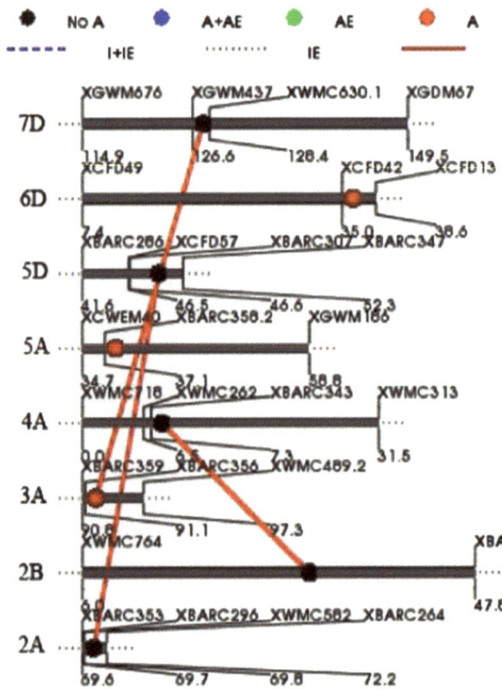

Fig. 5.15 Graphic presentation of the genetic architecture QTL with additive effects, epistatic effects, and QTL-by-environment interactions for crude starch

5.3.6.2.3 Epistatic QTL for Crude Starch and Its Interactions with the Environment

Three epistatic QTLs for crude starch content were detected (Table 5.57; Fig. 5.15), accounting for 6.37, 16.12, and 4.31 % of the phenotypic variation, respectively. *Qcs-2B/Qcs-4A* located Xwmc764-Xbrac200/Xbarc343-Xwmc313, explaining 16.12 % of the phenotypic variance. This major QTL can be used in marker-assisted breeding. Total epistatic effects explained 26.8 % of the phenotypic variance. No epistatic QTL was detected to be interacted with environment.

Table 5.57 Estimated epistatic and epistasis × environment interaction (AAE) effects of QTL for crude starch

QTL	Flanking marker	Position (cM)	QTL	Flanking marker	Position (cM)	AA	H^2 (AA, %)
Qcs-2A	*Xbarc296–Xwmc582*	69.7	*Qcs-5D*	*Xbarc307–Xbarc347*	48.6	−0.40	6.37
Qcs-2B	*Xwmc764–Xbrac200*	29.0	*Qcs-4A*	*Xbarc343–Xwmc313*	7.3	−0.64	16.12
Qcs-3A	*Xbarc356–Xwmc489.2*	91.1	*Qcs-7D*	*Xgwm437–Xwmc630.1*	58.5	−0.33	4.31

E1: Tai'an, 2005; E2: Tai'an, 2006; E3: Suzhou, 2006

5.3.6.3 Research Progress of Crude Starch Content QTL mapping and Comparison with Previous Studies

5.3.6.3.1 Research Progress of Crude Starch Content and Comparison of QTL Mapping

Sun et al. (2008) detected three QTLs controlling starch content distributed on 7B, 2A, and 2D chromosomes, explaining 21.95, 12.57, and 13.2 % of the phenotypic variation, respectively. McCartney et al. (2006) detected five QTLs on 1A, 1D, 7A, and 7D chromosomes, explaining 4.1 to 21.7 % of the phenotypic variation. Total phenotypic variance of five QTLs explained 59.1 % located on A and D genomes.

5.3.6.3.2 Comparison of This Result with Previous Studies

As shown in Table 5.58, QTLs for the crude starch content were mainly located on 1A, 1D, 2A, 2D, 7A, 7B, and 7D chromosomes, which indicated that these chromosomes contained the key loci controlling amylase synthesis. While in the present study, additive QTLs were detected only on 3A, 5A, and 6D chromosomes, but the epistatic QTLs were detected on 2A, 2B, 3A, 4A, 5D, and 7D chromosomes. These indicated that 2A and 7D chromosomes were also important for crude starch content.

5.4 QTL Mapping of Wheat Dough Rheological Characteristics

Dough quality was determined mainly by the rheological properties of dough which can largely reflect the inherent quality of the flour and the dough. It is very important for special flour production. At present, dough rheological properties were tested mainly by international testing instruments, such as farinograph,

Table 5.58 Summary of QTL results of wheat starch characteristics

Trait	QTL	Flanking marker	PVE (%)	Mapping population	References
Starch content	*QSp.sdau-7B*	*Xubc857c–Xswes94*	21.95	RIL	Sun et al. (2008)
	QTst.crc-1A	*Xwmc59–Xbarc158*	8.7	DH	McCartney et al. (2006)
	QTst.crc-1D	*Xbarc169–Xgdm126*	17.7	DH	McCartney et al. (2006)
	QAlc.sdau-2A	*Xubc840c–Xsrap29a*	12.57	RIL	Sun et al. (2008)
Amylose content	*QAlc.sdau-2D*	*Xissr23a–Xwmc181b*	13.2	RIL	Sun et al. (2008)
	QStd.crc-1D	*Xgdm33*	4.1	DH	McCartney et al. (2006)
Starch damage	*QStd.crc-7A*	*Xwmc139*	6.9	DH	McCartney et al. (2006)
	QStd.crc-7D	*Xgwm130–Xwmc405*	21.7	DH	McCartney et al. (2006)

extensograph, mixograph, and alveograph. So in this study, QTL mapping for these important dough quality traits was carried out.

5.4.1 *QTL Mapping for Farinograph Traits*

Farinograph is a common device for evaluating flour quality. Water absorption, development time, stability time, and tolerance index can be obtained from the farinograph curve. According to this information, we can make a judgment on the quality and range of application of flour.

Currently, farinograph parameters are the most important indicators to determine wheat gluten strength in breeding and food processing. It was widely used not only for the determination and evaluation of wheat and flour quality and functionality, but also to measure the effect of the different components on dough rheological properties and to predict the impact on the final product. Farinograph parameters are quantitative trait, so QTL mapping for these traits can help clarify the genetic mechanism at a single gene level and provide a reference for molecular marker-assisted breeding in selecting farinograph parameters.

5.4.1.1 Determination Methods

Farinograph parameters were tested by farinograph system (Brabender German), according to the AACC 54-21 method using 50 g flour sample, and the parameters include flour water.

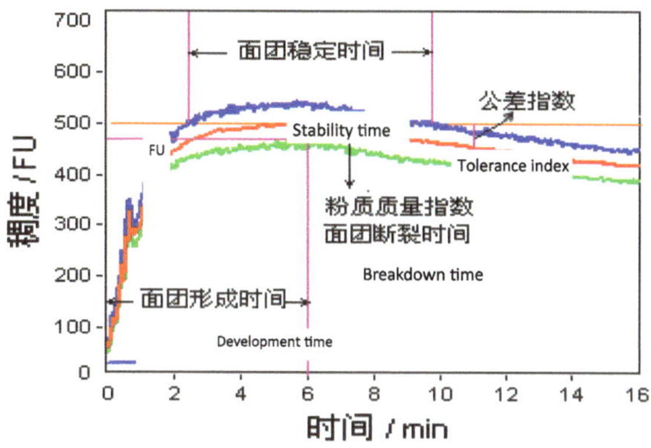

Fig. 5.16 Parameters commonly used in a farinograph

Absorption (FWA), dough development time (DDT), dough stability (DST), breakdown time (BDT), mixing tolerance index (MTI), and farinograph quality number (FQN) are parameters. Farinograph curve and some of parameters are shown in Fig. 5.16.

5.4.1.2 Results and Analysis

5.4.1.2.1 Phenotypic Data of Farinograph Parameters

In addition to DDT, most of the average parameters (Table 5.59) of lines were between Huapei 3 and Yumai 57. According to the DST, MTI, and BDT parameters, the flour quality of Yumai 57 was better than that of Huapei 3. There was a large variation in the DH population. Continuous distribution and transgressive segregation were found in this population. Except for DST, the absolute values of skewness and kurtosis were less than 1.0, indicating that the parameters were quantitative traits.

5.4.1.2.2 QTL Mapping for Farinograph Parameters

Forty-one QTLs for farinograph parameters and four QTLs interacted with environment were detected to be distributed on 19 chromosomes (Fig. 5.17). The QTLs explained 1.0 to 26.56 % of the phenotypic variation (Tables 5.60 and 5.61). All QTLs for each parameter explained 36.44 to 57.14 % of the phenotypic variation. Most of the additive QTLs were from Yumai 57, and the others were from Huapei 3, which indicated that the additive effect genes were dispersed in the two parents.

Table 5.59 Phenotypic data of farinograph parameters

Trait	Environment	Parents		DH population					
		Huapei 3	Yumai 57	Mean	Min.	Max.	SD	Skewness	Kurtosis
Flour water absorption (%)	Tai'an, 2005	62.4	56.6	60.79	53.2	70.9	3.44	0.108	−0.166
	Tai'an, 2006	67.0	57.3	62.42	55.4	69.1	3.15	−0.044	−0.960
	Suzhou, 2006	66.3	57.1	62.56	54.2	71.1	3.62	−0.075	−0.640
Dough development Time (min)	Tai'an, 2005	1.7	1.9	2.48	1.2	5.2	0.71	1.006	1.269
	Tai'an, 2006	2.0	2.0	2.69	1.4	4.9	0.75	0.736	0.132
	Suzhou, 2006	2.3	1.8	2.93	0.9	5.8	0.81	0.501	0.743
Dough stability time (min)	Tai'an, 2005	0.8	4.3	1.83	0.6	5.1	0.88	1.319	1.625
	Tai'an, 2006	1.1	6.5	2.37	0.8	6.5	1.00	1.352	2.439
	Suzhou, 2006	1.2	7.6	2.95	0.9	8.1	1.37	1.287	1.674
Mixing tolerance index (FU)	Tai'an, 2005	193	52	132.05	52.0	235.0	37.55	0.129	−0.310
	Tai'an, 2006	156	31	107.51	31.0	198.0	31.53	0.198	−0.047
	Suzhou, 2006	139	29	93.20	8.9	164.0	30.86	−0.066	−0.327
Breakdown time (min)	Tai'an, 2005	2.2	5.4	3.48	1.6	7.7	1.12	1.026	1.086
	Tai'an, 2006	2.7	7.0	3.96	2.1	8.2	1.19	1.000	1.015
	Suzhou, 2006	2.9	8.1	4.66	2.1	9.7	1.46	0.832	0.553

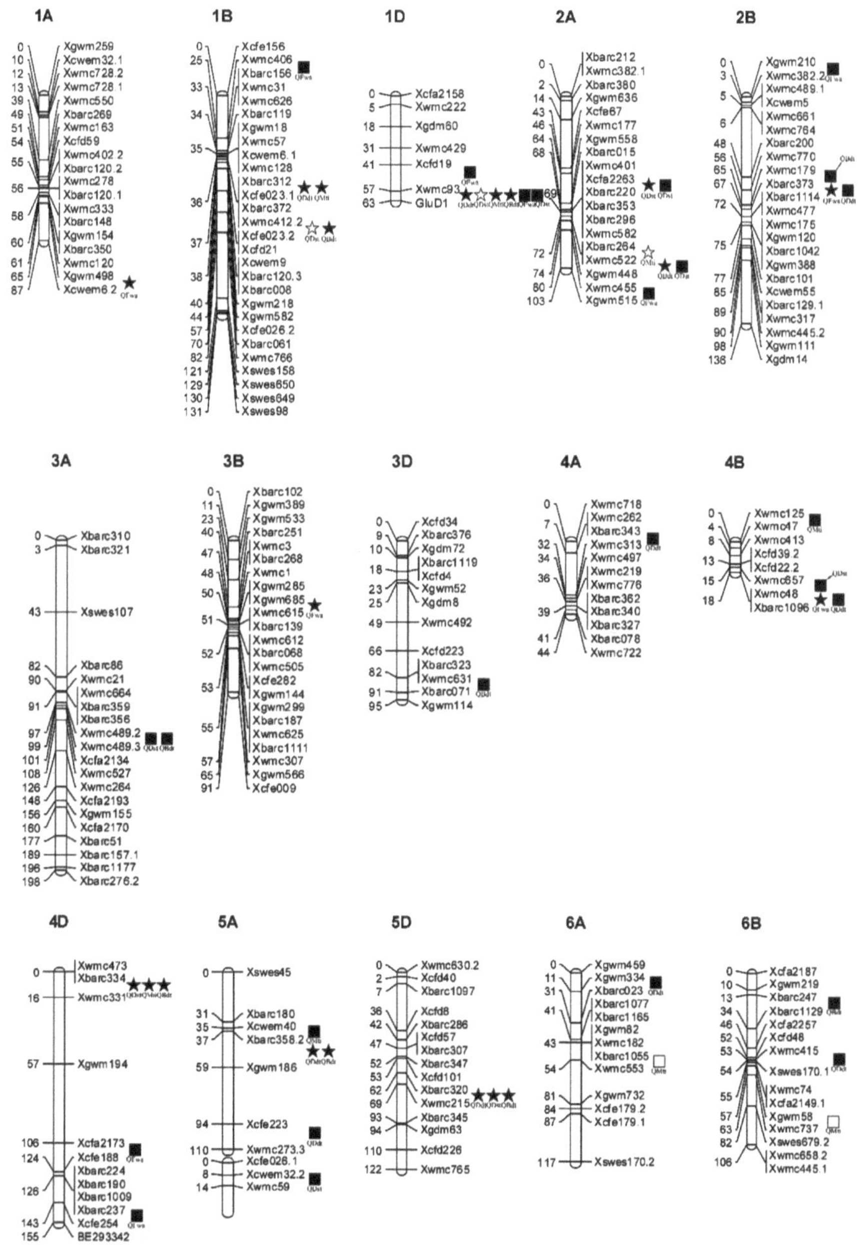

Fig. 5.17 Genetic linkage map of wheat showing mapping QTL for farinograph parameters. ★ locus involved in additive effects, ☆ locus involved in additive effects and AE interactions, ■ locus involved in epistasis effects, □ locus involved in epistasis effects and AAE interactions

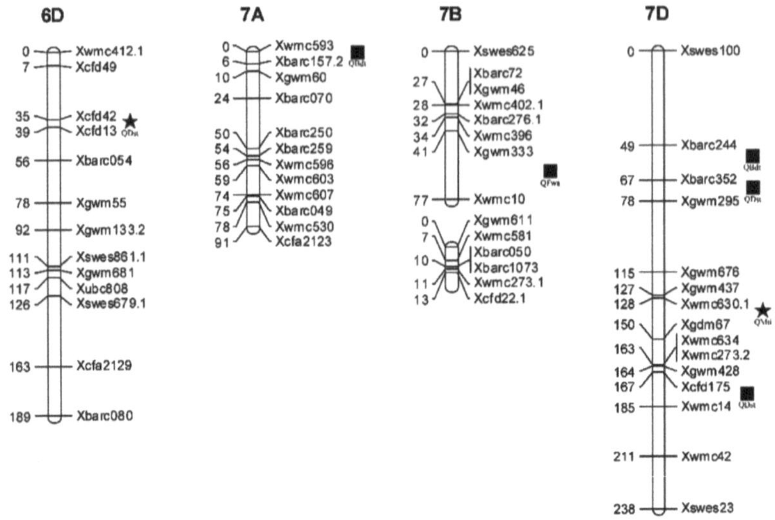

Fig. 5.17 (continued)

5.4.1.2.2.1 Additive QTL and Additive × Environment QTL

Four, four, six, five and six QTLs for FWA, DDT, DST, MTI and BDT were detected, respectively (Table 5.60; Fig. 5.17). For FWA, the QTLs were mainly distributed on 1A, 2B, 3B and 4B chromosomes with explaining from 4.21 to 12.36 % of the phenotypic variation. Of which, the QTL on 4B accounted for 12.36 % with increasing 1.25 ml water absorption. The QTLs controlling DDT were mainly on 1B, 1D, 5A and 5D chromosomes, which explained from 1.94 to 9.89 % of the phenotypic variation with increasing 0.11-0.24 min DDT. Among them, *QDdt-1D* had the highest PVE with right flanking marker *GluD1*. An additive QTL of DST in the interval of *Xwmc93* and *GluD1* had the highest additive effect, and explained 26.56 % of the phenotypic variation with increasing the DST by 0.53 min. This locus was the same as those loci controlling DDT, MTI and BDT, which indicated this locus was pleiotropic. For MTI, there were five QTLs distributing on 1B, 1D, 2A, 4D and 7D chromosomes with the PVE from 1.11 to 15.66 %. Six QTLs for BDT were on 1B. 1D, 2A, 4D, 5A and 5D chromosomes with explaining from 2.65 to 19.63 % of the phenotypic variation, which increased the BDT by 0.2–0.55 min. Therefore, there were some important gene loci on 1B and 1D chromosomes.

There were three additive QTLs interacting with environment, which accounted for 1.5–2.33 % of the phenotypic variation. Because of interaction with environment, *QDst-1B* and *QDst-1D* decreased the DST by 0.16 and 0.13 min, respectively, while *QMti-2A* increased the MTI by 5.26 FU, explaining 1.5 % of the phenotypic variation. It is interesting that there were four additive QTLs and one QTL × E was found close to *GluD1* on 1D chromosome.

Table 5.60 Estimated additive (A) and additive × environment (AE) interactions of QTL for farinograph parameters

Trait	QTL	Flanking marker	Position (cM)	A	H^2 (A,%)	AE1	H^2 (AE1, %)	AE3	H^2 (AE3, %)
FWA	*QFwa-1A*	*Xgwm498–Xcwem6.2*	86.1	−0.73	4.21				
	QFwa-2B.1	*Xbarc373–Xbarc1114*	67.2	0.68	3.63				
	QFwa-3B	*Xgwm685–Xwmc615*	50.3	−0.94	7.11				
	QFwa-4B	*Xwmc48–Xbarc1096*	18.4	−1.25	12.36				
DDT	*QDdt-1B*	*Xbarc312–Xcfe023.1*	36.1	−0.22	8.87				
	QDdt-1D	*Xwmc93-GluD1*	61.9	−0.24	9.89				
	QDdt-5A.1	*Xbarc358.2–Xgwm186*	38.1	0.11	1.94				
	QDdt-5D	*Xbarc320–Xwmc215*	62.3	−0.14	3.44				
DST	*QDst-1B*	*Xwmc412.2–Xcfe023.2*	36.4	−0.27	7.01			−0.16	2.33
	QDst-1D	*Xwmc93-GluD1*	61.9	−0.53	26.56			−0.13	1.73
	QDst-2A.1	*Xcfa2263–Xbarc220*	69.4	−0.13	1.49				
	QDst-4D	*Xbarc334–Xwmc331*	0.1	−0.14	1.87				
	QDst-5D	*Xbarc320–Xwmc215*	62.3	−0.18	3.24				
	QDst-6D	*Xcfd42–Xcfd13*	35.0	−0.14	1.95				
MTI	*QMti-1B*	*Xbarc312–Xcfe023.1*	36.1	13.74	15.66				
	QMti-1D	*Xwmc93-GluD1*	61.9	13.23	14.52				
	QMti-2A	*Xbarc264–Xwmc522*	72.2	10.07	8.41	4.25	1.50		
	QMti-4D	*Xbarc334–Xwmc331*	0.1	3.66	1.11				
	QMti-7D	*Xwmc630.1–Xgdm67*	134.4	7.31	4.43				
BDT	*QBdt-1B*	*Xwmc412.2–Xcfe023.2*	36.4	−0.37	8.66				
	QBdt-1D	*Xwmc93-GluD1*	61.9	−0.55	19.63				
	QBdt-2A	*Xwmc522–Xgwm448*	72.3	−0.21	2.93				
	QBdt-4D	*Xbarc334–Xwmc331*	4.1	−0.22	3.02				
	QBdt-5A	*Xbarc358.2–Xgwm186*	37.1	0.20	2.65				
	QBdt-5D	*Xbarc320–Xwmc215*	62.3	-0.23	3.26				

E1: Tai'an, 2005; E2: Tai'an, 2006; E3: Suzhou, 2006

Table 5.61 Estimated epistatic and epistasis × environment interaction (AAE) effects of QTL for farinograph parameters

Trait	QTL	Flanking marker	Position (cM)	QTL	Flanking marker	Position (cM)	AA	H^2 (AA, %)	AAE3	H^2 (AAE3, %)
FWA	*QFwa-1B*	*Xwmc406–Xbarc156*	27.7	*QFwa-2B.2*	*Xgwm210–Xwmc382.2*	1.0	0.88	6.15		
	QFwa-1D.1	*Xcfd19–Xwmc93*	49.9	*QFwa-4D.1*	*Xcfa2173–Xcfe188*	157.8	0.35	1.00		
	QFwa-1D.2	*Xwmc93-GluD1*	59.9	*QFwa-4D.2*	*Xbarc237–Xcfe254*	174.7	0.88	6.19		
	QFwa-2A	*Xwmc455–Xgwm515*	80.7	*QFwa-7B*	*Xgwm333–Xwmc10*	41.2	0.85	5.70		
DDT	*QDdt-2B*	*Xbarc373–Xbarc1114*	67.2	*QDdt-4A*	*Xbarc343–Xwmc313*	18.3	−0.15	4.21		
	QDdt-4B	*Xwmc48–Xbarc1096*	18.4	*QDdt-6B*	*Xwmc415–Xswes170.1*	53.2	−0.12	2.58		
	QDdt-5A.2	*Xcfe223–Xwmc273.3*	103.0	*QDdt-6A*	*Xgwm334–Xbarc023*	10.5	0.18	5.51		
DST	*QDst-1D*	*Xwmc93-GluD1*	61.9	*QDst-2A.1*	*Xcfa2263–Xbarc220*	69.4	0.17	2.65		
	QDst-2A.2	*Xwmc522–Xgwm448*	72.3	*QDst-3A*	*Xwmc489.2–Xwmc489.3*	98.3	−0.17	2.89		
	QDst-4B	*Xwmc657–Xwmc48*	16.7	*QDst-7D.1*	*Xcfd175–Xwmc14*	166.5	−0.19	3.30		
	QDst-5A	*Xcwem32.2–Xwmc59*	12.6	*QDst-7D.2*	*Xbarc352–Xgwm295*	68.8	0.15	2.12		
MTI	*QMti-4B*	*Xwmc125–Xwmc47*	0.0	*QMti-5A*	*Xcwem40–Xbarc358.2*	34.7	−5.00	2.08		
	QMti-6A	*Xbarc1055–Xwmc553*	48.0	*QMti-6B*	*Xgwm58–Xwmc737*	59.3	6.13	3.11	−3.96	1.31
BDT	*QBdt-2B*	*Xwmc179–Xbarc373*	65.3	*QBdt-6B*	*Xbarc247–Xbarc1129*	21.5	0.33	6.97		
	QBdt-3A	*Xwmc489.2–Xwmc489.3*	98.3	*QBdt-3D*	*Xwmc631–Xbarc071*	82.1	−0.24	3.70		
	QBdt-7A	*Xwmc593–Xbarc157.2*	5.0	*QBdt-7D*	*Xbarc244–Xbarc352*	58.6	−0.29	5.48		

E1: Tai'an, 2005; E2: Tai'an, 2006; E3: Suzhou, 2006

5.4.1.2.2.2 Epistatic QTL and Epistasis × Environment Interaction QTL

Sixteen epistatic QTLs for farinograph parameters were detected. These QTLs explained 1.00 to 6.97 % of the phenotypic variation. The total PVE for each parameter explained 5.19 to 19.04 % of the phenotypic variation.

One epistatic QTL (*QMti-6A/QMti-6B*) interacted with environment (Table 5.61; Fig. 5.17), which explained 1.31 % of the phenotypic variation with increasing tolerance index by 3.96FU.

In addition, it is worth noting that two-loci epistatic QTL interactions with the environment were detected close to the glutenin subunit *GluD1* on 1D chromosome.

5.4.1.3 Research Progress of Farinograph Parameter QTL Mapping and Comparison with Previous Studies

5.4.1.3.1 Research Progress of Farinograph Parameter QTL Mapping

Previous studies showed that chromosomes 1A, 1B, and 1D glutenin locus were important for controlling the changes in dough rheology traits. Perretant et al. (2000), Groos et al. (2004), Kuchel et al. (2006), and Sun et al. (2008) found that the wheat gluten loci were related to dough extensibility and elasticity. In addition, the impact of high molecular weight glutenin on the final quality has been confirmed (MacRitchie 1999), and the influence of *Glu-B3* and *Glu-A3* loci on dough extensibility has been also confirmed. Zhang et al. (2011) detected nine QTLs for 5 farinograph parameters, explaining 5.81–16.91 % of the phenotypic variation. Nine QTLs located on A and B genomes. Li et al. (2009) detected 10 QTLs for flour water absorption, which explained 3.1–38.7 % of the phenotypic variation distributed on the A, B, and D genomes. Thus, the chromosomes related to farinograph parameters were mainly on 1B, 2A, 2B, 2D, 3D, 4B, 5A, 5B, 5D, 6A, and 7D chromosomes (Table 5.62). Of which, chromosome 1B captured multiple QTLs for farinograph parameters, and its contribution rate was up to 38.7 %.

5.4.1.3.2 Comparison of This Result with Previous Studies

In the previous studies, QTLs for water absorption were located on 2A, 2B, 2D, 3D, 4B, 5A, 5B, 5D, 6A, and 7D chromosomes, and in this study, they were located on 1A, 2B, 3B, and 4B chromosomes. By comparison, the chromosomes 2B and 4B might be important for water absorption. Previous studies located QTL for development time on chromosomes 4B and 1B, while in this study, QTL for development time was located on 1B, 1D, 5A, and 5D. These indicated that 1B chromosome had captured the important loci, which may be related to HMW-GS subunit genes on 1B. In the previous study, QTLs for stability time were on

Table 5.62 Summary of QTL results of wheat dough farinograph parameters (PVE > 5 %)

Traits	QTL	Flanking marker	PVE (%)	Mapping population	References
Absorption (%)	*Qwa-2A*	*WPT-1480-WPT-9797*	8.03	RIL	Zhang et al. (2011)
	Qwa-5A	*WPT-8226-WPT-5467*	15.86	RIL	Zhang et al. (2011)
	QAbs.caas-4B	*wms6-wmc413*	5.7	RIL	Li et al. (2009)
	QAbs.caas-5D.1	*cfd18-cfd189*	31.3/38.7	RIL	Li et al. (2009)
	QAbs.caas-6A	*wms617-wms427*	8.8/7.3	RIL	Li et al. (2009)
	QAbs.caas-2D	*wms210-wmc111*	8.1	RIL	Li et al. (2009)
	QAbs.caas-5B	*vrnB-wmc75*	5.9	RIL	Li et al. (2009)
Stability time	*Qst-4B*	*WPT-7569–GWM0495*	6.74	RIL	Zhang et al. (2011)
	Qst-1B	*WPT-8832-WPT-1675*	16.91	RIL	Zhang et al. (2011)
Breakdown time	*Qbt-1B*	*WPT-8832-WPT-1675*	14.84	RIL	Zhang et al. (2011)
Development time	*Qda-1B*	*WPT-8832-WPT-1675*	12.23	RIL	Zhang et al. (2011)
Bandwidth	*Qbw-2B*	*WPT-6519-WPT-8760*	5.81	RIL	Zhang et al. (2011)
	Qbw-2B	*WPT-4559-WPT-7322*	8.62	RIL	Zhang et al. (2011)
Evaluation	*Qev-1B*	*WPT-8832-WPT-1675*	10.41	RIL	Zhang et al. (2011)

chromosome 1B, but in the present study, they were located on 1B, 1D, 2A, 4D, 5D, and 6D chromosomes; QTLs for breakdown time were located on the chromosome 1B in the previous study, while it was located on 1B, 1D, 2A, 4D, 3A, and 3D in this study. In all, the chromosomes 1B and 1D greatly contributed to farinograph parameters, while other loci slightly influenced.

5.4.2 QTL Mapping of Mixograph Traits

Traditionally, the mixograph is used to determine the rheological characteristics and bread-making quality in the milling and baking industries (Kunerth and D'Appolonia 1985). In general, the dough rheological properties are believed to be controlled by multigenes and thus cannot be fully explained by storage protein loci. Prior to this study, limited information on QTL mapping for mixographic characteristics is available. QTLs of mixing peak time, eight minutes, and width have been mapped on chromosomes 1B and 1D (Wu et al. 2008) using 240 RIL lines derived from the cross of PH82-2 and Neixiang 188. Zhang et al. (2009a, b, c, d) reported the mixing peak time, peak width, and eight-minute width on chromosomes 1A, 1B, and 1D, respectively, while studies by several other researchers have suggested that QTLs of mixograph traits are located on different chromosomes, such as 1A, 2A, 3A, 1B, 2B, 3B, 1D, 4D, 5B, 5D, 6B, 6D, 7A, and 7D (Huang et al. 2006; Nelson et al. 2006; Elangovan et al. 2008; Sun et al. 2008; Tsilo et al. 2011). These differences may be caused by using different genetic populations and different genetic maps.

Therefore, the objective of this study was to evaluate the genetic effects of QTL on mixograph traits by using the genetic map generated from RIL population. The results will provide the important information for improving the dough mixograph parameters, and some of the major QTLs would be used in MAS breeding programs.

5.4.2.1 Materials and Methods

5.4.2.1.1 Milling Flour

Seed samples of the RIL population and the parents, Gaocheng 8901 and Nuomai 1, obtained from the harvested population were normally stored for about 1 month and then milled using Buhler experimental mill (Buhler, Buhler-Miag Co., Germany) with a flour extraction yield of approximately 70 %.

5.4.2.1.2 Mixograph Parameters

Mixograph analysis was carried out using a 10-g mixograph system (National Manufacturing Co., USA) according to AACC-approved method 54-40A. Parameters were recorded at midline peak time (MPT), midline peak value (MPV), midline peak width (MPW), midline peak integral (MPI), and 8-min width (MTxW).

5.4.2.1.3 Statistic Analysis

Analysis of variance was carried out using SPSS version 13.0 (SPSS, Chicago, USA). QTLs with additive effects and epistatic effects expressed in RIL population under three environments were detected by the software QTL Network version 2.0 (Yang and Zhu 2005) based on the mixed linear model (Wang et al. 1999). QTL was abbreviated with every parameter followed by its relevant chromosome number. If there were more than one QTL on the same chromosome, the serial number was added after the chromosomal number separated by a dot.

5.4.2.2 Result

5.4.2.2.1 Correlation Analysis Between the Parameters of Mixing Parameters

The correlation between the mixing parameters is shown in Table 5.63. Midline peak time (MPT) showed significant negative correlation with peak value (MPV) and peak width (MPW), but had highly significant positive correlation with the area of peak integral (MPI) with the highest correlation coefficient by $r = 0.966^{**}$. Significantly positive correlation was seen between peak height (MPV) and peak width (MPW), with correlation $r = 0.613^{**}$; other parameters were also correlated with each other, but no significant correlation was found.

Table 5.63 Coefficients of pairwise correlations of the mean values of mixograph parameters

Mixograph parameters	MPT	MPV	MPW	MPI
MPV	−0.189**			
MPW	−0.160**	0.613**		
MPI	0.966**	0.003	−0.048	
MTxW	−0.03	0.018	0.037	−0.027

**Significant correlations were seen among some mixograph parameters

5.4.2.2.2 Phenotypic Variation for Dough Mixing Characteristics

Phenotypic variation for RIL lines and parents' data (Table 5.64) indicated that midline peak time (MPT), midline peak width (MPW), midline peak integral (MPI), and 8-min width (MT × W) of GC8901 were higher than those of WN1 under all environmental conditions, and the phenotypic variation among the RIL lines were observed. Continuous segregations were seen in MPT, and MPV, MPW, and MPI of RIL population and normal distribution were found because both absolute values of skewness and kurtosis are less than 1.0. Therefore, the data can be used for QTL analysis.

5.4.2.2.3 QTL for Dough Mixing Characteristics

Eighteen additive QTLs and three epistatic QTLs were detected (Tables 5.65 and 5.66). Seven additive QTLs for MPT parameter were identified on chromosomes 1A, 1B, 1D, and 6A with positive alleles from GC8901 except for *QMPT-1B*. The *QMPT-1D.1* was only detected in two environments (Tai'an 2008, 2009), and it was close to the *QMPT-1D.2*, accounting for 35.2, 22.22, and 36.57 % of the phenotypic variance in the three environments, respectively. The genetic distance from Glu-D1 marker was 4-5.9 cM.

Five additive QTLs for MPV parameter were mapped on 1A, 1D, 4A, and 7D chromosomes. The positive additive effect alleles of *QMPV-4A* and *QMPV-7D* came from WN1 in each environment. *QMPV-1D* and *QMPV-4A* were detected

Table 5.64 Phenotypic values for mixograph parameters of two parents and the RIL population in the three environments

Environment	Trait	Parents			RIL population		
		Nuomai 1	Gaocheng 8901	Range	Mean ± SD	Skewness	Kurtosis
Tai'an 2008	MPT	1.82	4.87	1.50–5.53	2.91 ± 0.05	0.428	−0.059
	MPV	67.24	58.64	48.50–77.92	61.90 ± 0.32	0.155	0.093
	MPW	22.74	23.92	11.43–32.81	23.55 ± 0.21	−0.103	0.405
	MPI	102.91	238.3	68.14–300.94	146.32 ± 2.57	0.526	0.255
	MTxW	4.86	13.9	2.64–20.72	6.06 ± 0.19	2.378	6.292
Tai'an 2009	MPT	3.00	4.51	1.50–9.01	3.90 ± 0.09	0.79	0.713
	MPV	62.48	57.63	47.26–73.89	60.87 ± 0.34	0.038	−0.58
	MPW	22.9	25.88	12.95–37.99	23.19 ± 0.31	0.368	0.246
	MPI	160.67	221.46	74.68–426.63	192.38 ± 4.20	0.709	0.571
	MTxW	7.65	13.55	3.43–43.98	9.27 ± 0.33	2.972	12.604
Anhui 2011	MPT	2.04	3.92	1.50–5.15	2.66 ± 0.04	0.717	0.527
	MPV	69.72	61.64	46.98–86.97	63.63 ± 0.38	0.588	0.885
	MPW	29.73	32.53	11.22–43.40	25.75 ± 0.30	0.526	0.899
	MPI	120.02	192.78	56.90–283.84	135.38 ± 2.49	0.779	0.91
	MTxW	7.37	16.61	4.21–33.64	9.85 ± 0.31	1.595	3.031

Table 5.65 Additive effects for mixograph of RIL population in different environments

Environment	Trait	QTL	Chromosome	Interval marker	Position (cM)	A[a]	H^2 (A) (%)[b]
Tai'an 2008	MPT	*QMPT-1A.1*	1A	wPt-9757-Glu-A1	87.7	−0.19	4.91
		QMPT-1D.1	1D	Glu-D1-wPt-3743	111.4	−0.51	35.29
		QMPT-6A.1	6A	wPt-729920-wPt-664792	132.7	−0.26	9.05
	MPV	*QMPV-1A.1*	1A	wPt-664666-wPt-9757	82.3	−1.29	6.19
		QMPV-1D	1D	cfd-183-wPt-729773	0	−1.77	11.63
		QMPV-4A	4A	wPt-664948-Wx-B1	100.4	1.35	6.78
		QMPV-7D	7D	Wx-D1-wPt-664368	0	1.06	4.23
	MPI	*QMPI-1A*	1A	wPt-9757-Glu-A1	87.7	−11.3	6.67
		QMPI-1B	1B	wPt-6642-wPt-3824	169.5	7.64	3.05
		QMPI-1D.1	1D	Glu-D1-wPt-3743	111.4	−24.32	30.94
		QMPI-6A	6A	wPt-729920-wPt-664792	130.7	−14.49	10.98
Tai'an 2009	MPT	*QMPT-1B*	1B	wPt-6442-wPt-3824	168.5	0.39	7.37
		QMPT-1D.2	1D	wPt-3743-wPt-666719	113.3	−0.68	22.22
		QMPT-6A.2	6A	wPt-664792-wPt-730772	143.8	−0.33	5.26
	MPV	*QMPV-1A.2*	1A	wPt-9757-Glu-A1	84.7	−1.38	6.74
		QMPV-1D	1D	cfd-183-wPt-729773	0	−1.72	10.57
		QMPV-4A	4A	wPt-664948-Wx-B1	99.4	1.33	6.3
	MPW	*QMPW-4A*	4A	Wx-B1-wPt-0105	107.8	1.45	8.92
	MPI	*QMPI-1A*	1A	wPt-9757-Glu-A1	87.7	−15.98	5.31
		QMPI-1B	1B	wPt-6442-wPt-3824	167.5	17.84	6.61
		QMPI-1D.2	1D	wPt-3743-wPt-666719	113.3	−31.73	20.92
		QMPI-6A	6A	wPt-729920-wPt-664792	131.7	−22.32	11.3

(continued)

Table 5.65 (continued)

Environment	Trait	QTL	Chromosome	Interval marker	Position (cM)	A[a]	H^2 (A) (%)[b]
Suzhou 2011	MPT	*QMPT-1A.2*	1A	Glu-A1-wPt-665259	88.6	−0.16	5.38
		QMPT-1D.1	1D	Glu-D1-wPt-3743	112.4	−0.42	36.57
	MPV	*QMPV-1A.1*	1A	wPt-664666-wPt-9757	81.3	−1.64	7.28
		QMPV-1D	1D	cfd-183-wPt-729773	0	−1.52	6.2
		QMPV-4A	4A	wPt-664948-Wx-B1	99.4	1.6	6.93
	MPI	*QMPI-1A*	1A	wPt-9757-Glu-A1	87.7	−9.77	6.24
		QMPI-1D.1	1D	Glu-D1-wPt-3743	112.4	−21.09	29.08

[a]Additive effects: Positive values indicate that positive effect alleles are derived from WN1, whereas negative values indicate that the negative effect alleles are contributed by Gc8901

[b]Percentage of phenotypic variation explained by QTL with additive effect

Table 5.66 Epistatic QTLs for mixograph of RIL population in different environments

Environment	Trait	QTL	Interval marker	Position (cM)	QTL	Interval marker	Position (cM)	AA[a]	H^2 (AA) (%)[b]
Tai'an 2008	MPW	*QMPW-2D.1*	wPt-7901-wPt-6687	120.3	*QMPW-4A.1*	wPt-671707–wPt-730913	127.3	−1.13	10.69
		QMWP-2D.1	wPt-7901-wPt-6687	120.3	*QMPW-4A.2*	wPt-5003–wPt-6440	136.2	0.47	1.86
		QMPW-2D.2	wPt-1301-wPt-6343	130.4	*QMPW-4A.2*	wPt-5003–wPt-6440	136.2	−0.61	3.1

Table 5.67 Summary of QTL results of wheat dough mixograph parameters (PVE > 10 %)

Trait	QTL	Flanking marker	(PVE)/%	Mapping population	References
MT	*QMt.caas-1B*	wmc128	19.2	IL	Li et al. (2012)
	QMt.caas-2D	barc159	13.7	Same as above	Same as above
	QMt.caas-4B	wms495	18.2	Same as above	Same as above
	QMt.caas-4A	barc1047	32.6	Same as above	Same as above
	QMt.caas-5A	wms293	13.1	Same as above	Same as above
	QMt.caas-6A	barc104	22.7	Same as above	Same as above
WS	*QWs.caas-4A*	barc1047	−14.5	Same as above	Same as above
	QWs.caas-6A	barc104	−14.2	Same as above	Same as above
8 min (MT × W)	*QMtxw.caas-1B*	wmc128	35.0	Same as above	Same as above
	QMtxw.caas-1D	wmc222	26.5	Same as above	Same as above
	QMtxw.caas-2D	barc159	21.9	Same as above	Same as above
	QMtxw.caas-4B	wms495	27.3	Same as above	Same as above
	QMtxw.caas-5D	wms272	17.6	Same as above	Same as above
	QMtxw.caas-3A	wms155	28.2	Same as above	Same as above
	QMtxw.caas-4A	barc1047	48.6	Same as above	Same as above
	QMtxw.caas-5A	wms293	22.1	Same as above	Same as above
MPW	*QMpw.caas-1A.1*	barc148	35.3	Same as above	Same as above
	QMpw.caas-1A.2	wms136	9.6	Same as above	Same as above
	QMpw.caas-1B	wmc128	18.9	Same as above	Same as above
	QMpw.caas-3A	wms155	10.7	Same as above	Same as above
	QMpw.caas-5A	wms304	13.9	Same as above	Same as above
MPTi	*QMpt.caas-4A*	barc1047	11.5	Same as above	Same as above

(continued)

Table 5.67 (continued)

Trait	QTL	Flanking marker	(PVE)/%	Mapping population	References
8 min (MT × V)	*QMtxv.caas-1A.1*	wms136	17.5	Same as above	Same as above
	QMtxv.caas-1B	wmc128	12.5	Same as above	Same as above
	QMtxv.caas-1D	wmc222	10.7	Same as above	Same as above
MPV	*QMpkh.crc-4D*	Xwmc617–Xwmc48	34.9	DH	McCartneyet al. (2006)
MPT	*QMmdt.crc-1B*	Xgwm403–Xgwm274	14.7	Same as above	Same as above
	QMmdt.crc-4D	Xwmc617–Xwmc48	26	Same as above	Same as above
PA	*QMetp.crc-1B*	Xgwm131–Glu-B1	16.4	Same as above	Same as above
	QMetp.crc-4D	Xwmc617–Xwmc48	19.4	Same as above	Same as above
MPW	*QMpbw.crc-4D*	Xwmc617–Xwmc48	21.4	Same as above	Same as above
	QMpbw.crc-7D	Xgwm130–Xwmc405	17.8	Same as above	Same as above
A	*QMteg.crc-2B*	Xgwm210–Xwmc25	10.3	Same as above	Same as above
	QMteg.crc-4D	Xgwm608	15	Same as above	Same as above
WS	*QMsap.crc-4D*	Xwmc617–Xwmc48	19.7	Same as above	Same as above

with stable effects, explaining 6.20 to 11.63 % and 6.30 to 6.93 % of the phenotypic variance, respectively. *QMPV-4A* was close to the *Wx-B1* marker with genetic distance of 2.4–3.4 cM. *QMPV-1A.1* was detected in Tai'an (2008) and Suzhou (2011), accounting for 6.67 and 6.24 % of the phenotypic variance, respectively. This QTL was close to *Glu-A1* marker. One additive QTL with increasing MPW by 1.45 from WN1 allele was detected on the 4A chromosome only expressed in one environment (Tai'an, 2009).

Five additive QTLs for MPI parameter were identified on 1A, 1B, 1D, and 6A chromosomes. Of which, the positive alleles were from GC8901 with the exception of *QMPI-1B*. *QMPI-1A* was detected across three environments with the genetic distances of 0.9 cM from the nearest marker *Glu-A1*, explaining 5.31 to 6.67 % of the phenotypic variance. *QMPI-1D.1* (Tai'an 2008; Suzhou 2011), *QMPI-6A* (Tai'an 2008, 2009), and *QMPI-1B* (Tai'an 2008, 2009) were detected in two environments. *QMPI-1D.2* was close to *QMPI-1D.1*, explaining 30.94, 20.92, and 29.08 % of the phenotypic variance, respectively. The genetic distance from *Glu-D1* marker was 4–5.9 cM.

Three epistatic QTLs were identified for MPW in Tai'an 2008, and they were located on chromosomes 2D and 4A (Table 5.66). *QMP W-2D.1/QMP W-4A. 1* had the largest effect. These QTLs explained from 1.86 to 10.96 % of the phenotypic variance.

5.4.2.3 Research Progress of QTL Mapping for Mixograph Parameters and Comparison with Previous Studies

5.4.2.3.1 Progress in QTL for Mixograph Parameters

Ma et al. (2005) studied the QTL for the dough maximum extensibility resistance and found that the QTLs were distributed on 2A, 5A, 1B, and 1D chromosomes. Meanwhile, the effects of additive and epistatic and interaction with environment were also analyzed, which indicated that the epistatic effects should be considered in molecular breeding besides the additive effects. Rousset et al. (2001) identified the QTLs for SDS sedimentation volume, mixing time, and bread volume on the Glu-A1 and Gli-B1/Glu-B3 loci, and there was another QTL detected on the distal of 1AL for mixing time and bread volume. In addition, each one of QTL was located on 1BL for high protein content and mixing time, respectively. Gras et al. (2001) testified the major QTL for mixograph on *Glu-B1*. Six QTLs were detected on 1D, 1BS, 2B, 5A, 2D, and 7AS chromosomes, explaining 20.3–63.8 %, 14.6–49.0 %, 9.8–46.5 %, and 12.7–34.1 % of the phenotypic variance, respectively (Zhang 2003). Wu et al. (2008) found three QTLs for mixing time on chromosomes 1B and 1D, explaining 7.9 to 55.3 % of the phenotypic variance and five QTLs for 8-min bandwidth. On 1B chromosome, there was one pleiotropic QTL simultaneously controlling Zeleny sedimentation volume, mixing time, 8-min bandwidth, peak viscosity, and setback; there was also one pleiotropic QTL on 1D chromosome for Zeleny sedimentation volume, mixing time, and 8-min bandwidth. Nelson et al. (2006) identified the genes on 1AS, 1BS, and 6DS for mixograph parameters, which were close to gliadin and LMW-GS loci. Huang et al. (2006) detected the QTLs for mixing time on 1B, 1D, and 3B using SSR markers.

In all, the QTLs involved 1A, 1B, 1D, 2B, 2D, 3A, 4A, 4B, 4D, 5A, 5D, 6A, 7D, and 7B chromosomes (Table 5.67). Of which, 1A, 1B, 1D, 2D, 2B, 4B, 4A, 5A, 6A, and 4D chromosomes had more than one QTL controlling several mixograph parameters, and the highest PVE was 48.9 %.

5.4.2.3.2 Comparison with Previous Research

In our study, the QTLs for mixograph parameters were mapped on chromosomes 1A, 1B, 1D, 6A, 2A, 3B, 4A, 6B, 6A, 7A, and 7D. Of which, there were more than one QTL on chromosomes 1A, 1B, 1D, and 6A with some major regions for mixograph parameters. These were perhaps caused by Glu-A1, Glu-B1, and

Glu-D1 loci on 1A, 1B, and 1D. However, the relationships between these loci and HMW-GS should be further studied.

5.4.3 QTL Mapping of Alveograph Traits

Alveograph is the instrument that reproduced the dough deformation in the stage of fermentation and baking, which was well applied in biscuit production. At present, the method of testing alveograph has been recognized by International Organization for Standardization (ISO), the American Association of Cereal Chemistry (AACC), the International Association for Cereal Chemistry (ICC), and the French Standards Association (AFNORV) and other international organizations as the standard method.

Different baked foods have different requirements in different gluten contents (Wieser et al. 2001). Bread requires the flour with good elasticity and extensibility; the ratio of P/L should be 0.8–1.4, and W is between 0.175 and 0.21 mJ/g. Too strong gluten will extend the fermentation time, which is difficult to control. Most cakes and biscuits require the flour with low elasticity, tenacity, and extensibility but require good plasticity. Flour used to produce biscuits and cakes is advised to choose wet gluten content in the range of 22–25 %, P/L 0.15–0.7, and W 0.084 mJ/g. Too strong gluten strength easily makes biscuits to be stiffened and deformed. Few QTLs for alveograph parameters were reported. Therefore, in this study, major QTLs obtained could be used in marker-assisted selection and breeding.

5.4.3.1 Determination of Alveograph Parameters

The alveograph parameters were measured with AACC method using alveograph made by Chopin Company in France.

5.4.3.2 Results and Analysis

5.4.3.2.1 Phenotypic Data Analysis

Phenotypic values of alveograph parameters of the DH population in the three environments are listed in Table 5.68. The alveograph parameters of Yumai 57 were higher than those of female Huapei 3. There was great variation range in the DH population, except for P/L. Normal distribution and transgressive segregation were found in the population. These indicated that they were quantitative traits controlled by multigenes.

Table 5.68 Phenotypic analysis of alveograph characters in the DH population

Environment	Trait	Parent		DH population					
		Huapei 3	Yumai 57	Mean	Maximum	Minimum	SD	Skewness	Kurtosis
E1	*P* (mm)	44	51	53.27	112.00	26.00	17.72	0.99	0.82
	L (mm)	79	92	90.04	159.00	40.00	23.47	0.39	0.25
	G	19.8	21.4	20.95	28.10	14.10	2.77	0.002	−0.02
	W (J)	71	174	112.53	263.00	40.00	47.89	0.89	0.35
	Ie	23.2	61.8	33.03	61.80	8.63	9.40	−0.08	−0.01
E2	*P* (mm)	58	59	69.43	135.00	33.00	20.24	0.96	0.66
	L (mm)	53	135	78.04	192.00	31.00	25.54	0.73	1.37
	G	16.2	25.9	19.41	30.80	12.40	3.19	0.22	0.08
	W (J)	83	231	136.08	295.00	52.00	48.96	0.90	0.23
	Ie	21.3	55	32.80	55.00	10.90	8.89	−0.13	−0.32
E3	*P* (mm)	38	50	69.28	197.00	38.00	20.71	1.88	1.58
	L (mm)	77	88	78.19	152.00	40.00	23.69	0.50	0.004
	G	19.5	20.9	19.46	27.40	14.10	2.97	0.14	0.37
	W (J)	65	136	126.36	308.00	54.00	47.42	1.12	1.13
	Ie	15.6	35.3	28.16	43.80	4.42	8.02	−0.29	−0.20

P: dough tenacity; L: dough extensibility; G: swelling index; W: dough strength; Ie: elasticity index

5.4.3.2.2 QTL for Alveograph Parameters

Based on mixed linear model (Wang et al. 1999), QTL analysis of alveograph parameters was carried out using QTL Network 2.0 (Yang and Zhu 2005). Seventeen additive QTLs and seven epistatic QTLs were detected for alveograph-related traits when $P < 0.005$ (Tables 5.69 and 5.70).

5.4.3.2.2.1 Additive Effect Analysis

Seventeen additive QTLs were detected for alveograph (Table 5.69). Of which, 4 additive QTLs were found for wheat dough tenacity (Table 5.69; Fig. 5.18) explaining 16.36 % of the total phenotypic variation. QTLs on 1B and 4B chromosome explained 6.40 and 1.64 % of variations, respectively, whose positive alleles came from Yumai 57; while the other two located on 2B and 7D chromosomes were from Huapei 3, accounting for 4.5 and 3.82 % of the phenotypic variation, respectively.

Three additive loci were detected for wheat dough extensibility (L), which were located on 1B, 3B, and 4B chromosomes and explained 23.93 % of the total phenotypic variation (Table 5.69; Fig. 5.18). The major QTL (*QDext1B*) had the

Table 5.69 Additive effects for alveograph characters of DH population in the three environments

Trait	QTL	Marker interval	Position (cM)	A[a]	H^2 (A)[b] (%)
P	*QDten1B*	*Xwmc626–Xbarc119*	33.3	−4.91	6.40
	QDten2B	*Xbarc373–Xbarc1114*	68.2	4.12	4.50
	QDten4B	*Xwmc48–Xbarc1096*	18.4	−2.49	1.64
	QDten7D	*Xgwm676–Xgwm437*	121.9	3.79	3.82
L	*QDext1B*	*Xbarc061–Xwmc766*	77.7	−9.25	13.82
	QDext3B	*Xgwm389–Xgwm533*	17.6	5.28	4.50
	QDext4B	*Xwmc48–Xbarc1096*	18.4	5.89	5.61
G	*QSin1B*	*Xbarc061–Xwmc766*	77.7	−1.04	11.66
	QSin1D	*Xwmc93–GluD1*	61.9	−0.54	3.19
	QSin3B	*Xgwm389–Xgwm533*	17.6	0.71	5.38
	QSin4B	*Xwmc48–Xbarc1096*	18.4	0.75	6.06
W	*QDstren1B*	*Xwmc626–Xbarc119*	33.3	−17.68	14.13
	QDstren1D	*Xwmc93–GluD1*	61.9	−19.81	17.74
	QDstren5A	*Xbarc358.2–Xgwm186*	37.1	8.46	3.23
Ie	*QEin1B*	*Xbarc312–Xcfe023.1*	36.1	−2.49	7.40
	QEin1D	*Xwmc93–GluD1*	61.9	−4.86	28.28
	QEin7D	*Xwmc630.1–Xgdm67*	136.4	−2.05	5.05

P: dough tenacity; L: dough extensibility; G: swelling index; W: dough strength; Ie: elasticity index

[a]Additive effects: Positive value indicates that allele from Huapei 3 enhances the dough alveograph characters, and negative value indicates that allele from Yumai 57 enhances the dough alveograph characters; [b]percentage of phenotypic variation explained by QTL with additive effect

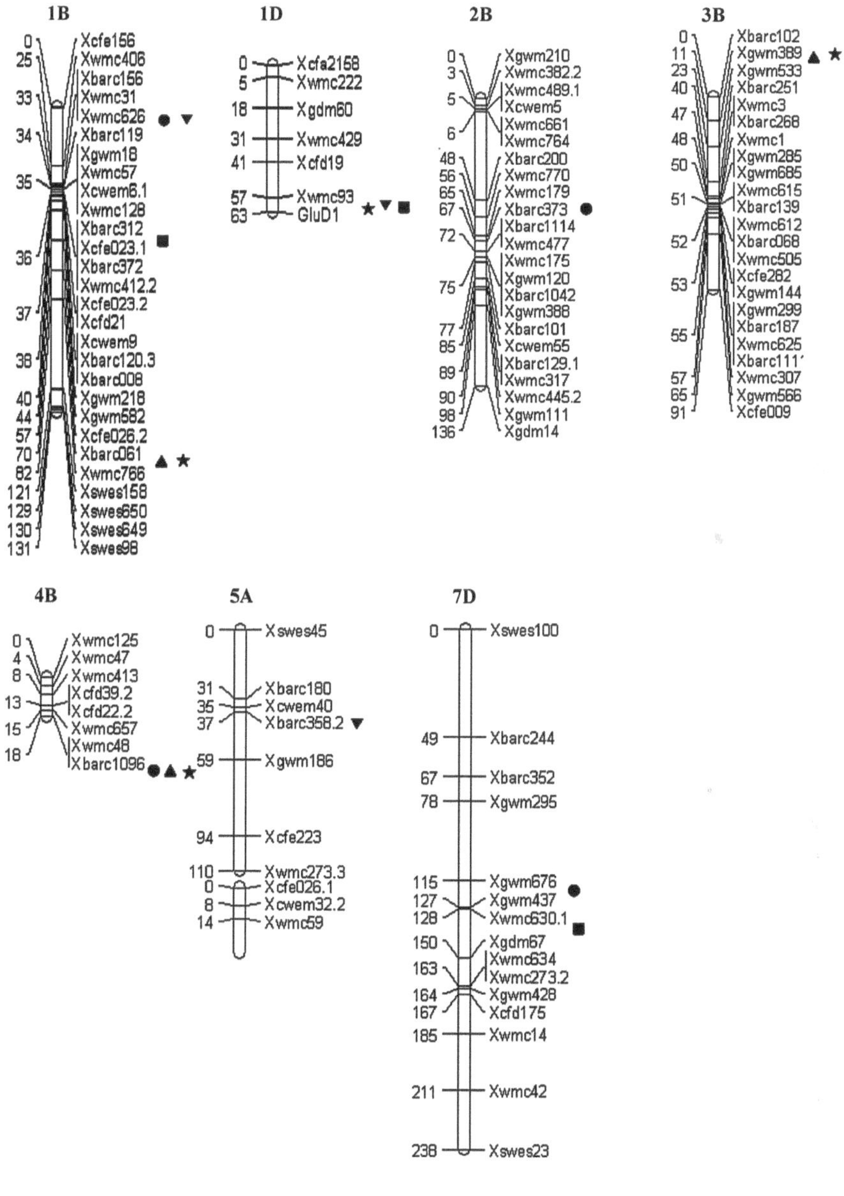

Fig. 5.18 Positions of additive QTLs conferring alveograph characters in the DH population. ● QTL for dough tenacity; ▲ QTL for dough extensibility; ▼ QTL for dough strength; ★ QTL for swelling index; ■ QTL for elasticity index

Table 5.70 Epistatic effects of QTLs for alveograph characters of DH population in the three environments

Trait	QTL	Marker interval	Position (cM)	QTL	Marker interval	Position (cM)	AA[a]	H^2 (AA)[b] (%)
P	*QDten4B*	*Xwmc47–Xwmc413*	5.2	*QDten6B*	*Xwmc415–GluB*	53.2	−4.26	4.82
L	*QDext1B*	*Xwmc412.2–Xcfe023.2*	36.4	*QDext1D*	*Xcfd19–Xwmc93*	47.9	−4.16	2.80
	QDext1B	*Xgwm582–Xcfe026.2*	55.6	*QDext1D*	*Xwmc93–GluD1*	60.9	−2.97	1.43
G	*QSin1B*	*Xbarc312–Xcfe023.1*	36.1	*QSin1D*	*Xwmc93–GluD1*	61.9	−0.33	1.15
W	*QDstren1B*	*Xswes158–Xswes650*	121.6	*QDstren3A*	*Xbarc356–Xwmc489.2*	96.1	−8.96	3.62
	QDstren2A	*Xgwm448–Xwmc455*	77.2	*QDstren4A*	*Xwmc219–Xwmc776*	35.6	−9.57	4.14
	QDstren4B	*Xwmc48–Xbarc1096*	18.4	*QDstren6B*	*Xwmc415–GluB*	53.2	−7.64	2.64

P: dough tenacity; L: dough extensibility; G: swelling index; W: dough strength

[a]Epistatic effects: Positive value represents that parent-type effect is bigger than recombinant-type effect, and negative value represents the opposite; [b]percentage of variation explained by epistatic QTL

maximum PVE with 13.82 %, which increased extensibility by 9.25. *QDext3B* and *QDext4B* alleles are derived from Huapei 3, which explained 4.50 and 5.61 % of the phenotypic variation, respectively.

Four additive QTLs on 1B, 1D, 3B, and 4B chromosomes for swelling index (G) explained 3.19–11.66 % of the phenotypic variation (Table 5.69; Fig. 5.18). *QSin1B* explained 11.66 % of the phenotypic variation with maximum PVE, *QSin1B* and *QSin1D* alleles were from Yumai 5, and the other two alleles were from Huapei 3.

Three additive QTLs for dough strength (W) were detected, which were located on 1B, 1D, and 5A chromosomes and explained 14.13, 17.74, and 3.23 % of the phenotypic variation, respectively (Table 5.69; Fig. 5.18). The allele of *QDstren5A* locus is derived from Huapei 3 and explained 3.23 % of the phenotypic variation, but the remaining two major gene alleles of *QDstren1B* and *QDstren1D* were from Yumai 57.

There were three QTLs for elasticity index (Ie) detected on 1B, 1D, and 7D chromosomes, which explained 7.4, 28.28, and 5.05 % of the phenotypic variation, respectively (Table 5.69; Fig. 5.18), and the three alleles were from Yumai 57. Of which, *QEin1D* (*Xwmc93/GluD1*) has the maximum effect, explaining 28.28 % of the phenotypic variation. Additive × environment interaction was not detected in the three environments.

5.4.3.2.2.2 Epistatic Effect Analysis

Seven epistatic QTLs were detected for alveograph traits (Table 5.70). One epistatic QTL for dough tenacity (P) on chromosome 4B-6B explained 4.82 % of the phenotypic variation. Two epistatic QTLs for dough extensibility (L) on chromosome 1B-1D explained 2.80 and 1.43 % of the phenotypic variation, respectively. One epistatic QTL controlled swelling index (G), located on chromosome 1B-1D, explaining 1.15 % of the phenotypic variation. Three epistatic QTLs controlled dough strength (W), located on chromosomes 1B-3A, 2A-4A, and 4B-6B, which explained 3.62, 4.14, and 2.64 % of the phenotypic variation, respectively. No epistatic QTLs controlling P/L ratio and elasticity index (Ie) of epistatic × environment interaction were found.

5.4.3.3 Research Progress of QTL Mapping for Alveograph Parameters and Comparison with Previous Studies

Previous studies (Xiao et al. 2003; Zhang et al. 2006) analyzed the relationship between alveograph parameters and the baking quality to understand the strength of wheat flour, elasticity, and extensibility using different wheat flour varieties. Wang (1998) measured 12 kinds of wheat flour samples using alveograph (including Chinese wheat varieties and France wheat varieties) and then analyze the relationship between alveograph parameters and the baking quality. The results indicated that utilization of alveograph parameters to predict wheat quality was a good scientific method, which can be used to speculate and determine the scope of application of wheat. The parameters were practically valuable to accurate and evaluate of the flour quality of wheat. Kang et al. (2005) studied the relationship between Ramen noodle score and alveograph parameters. The results showed that among the three alveograph parameters, maximum extensibility affected smallest on Ramen score, while the dough deformation indicator had an important effect on Ramen score. Dough extensibility indirectly influenced Ramen quality by affecting deformation energy. Therefore, the Ramen noodle flour could be selected mainly through the selection of dough extensibility and deformation, especially the latter. However, QTLs for alveograph parameters were less studied in domestic and international researches.

5.5 QTL Mapping of Wheat Processing Quality Traits

Processing quality of wheat includes processing quality, secondary processing quality, and eating quality. First processing quality refers to the requirement of milling process for the structure and physicochemical properties of wheat; secondary processing quality means requirement of baking or cooking food for flour biochemical characteristics. Eating quality refers to the adaptability of wheat flour for processing different foods. So in this chapter, QTL for quality of noodle and steamed bread was analyzed, which laid a foundation for genetic dissection of the complexed trait controlled by multigene and for marker-assisted selection in breeding.

5.5.1 QTL Mapping for Noodle Cooking Quality Traits

Noodle are traditional food in China, and it has more than two thousand years of history. According to the color and ingredients added in the process of noodle making, it can be divided into three categories: white water noodle (WWN), white salted noodle (WSN), and yellow alkaline noodle (YAN). The significant difference between them is the yellow color because of the presence of alkali and salt ingredients in yellow alkaline noodle (Morris and Rose 1996; Nagao 1996). In addition, fresh wet noodle (FWN), instant noodle, and all kinds of pure starch vermicelli also possessed a certain percentage of alkali and salt (Wei 2002).

For many years, people evaluated noodle quality using a sensory method whether at home or in abroad, but the sensory evaluation methods cannot meet the requirements of industrialization of food production in aspects of information exchange, quantitative expression, and scientific repeat. Determination of noodle quality using instruments makes up for the lack of sensory evaluation methods. Therefore, combination of sensory evaluation and instrument measurement has become an important method to evaluate noodle eating quality (Ren et al. 2007). So in this study, some traits of noodle quality were selected for QTL mapping.

5.5.1.1 Methods

5.5.1.1.1 Noodle Making and Testing

Noodle was made using a laboratory noodle machine (FuDe brand, JMT2-14 type) according to the procedures described by Deng et al. (2008). Water for 100 g samples (14 % moisture basis) was added according to the mixograph absorption values multiplied by 44 %. The prepared dough sheet was passed through cutting rollers producing noodle strands of about 2 mm width. About 20 g dry noodle were cooked in 1 L of boiling distilled water until the white core inside the noodle disappeared; the noodle were then poured into a wire sieve and rinsed with 25 °C distilled water for about 10 s.

5.5.1.1.2 Determination of the Optimal Cooking Time

Twenty grams of dry noodle was cooked in 1 l of boiling distilled water. Optimal cooking time was evaluated by observing the time of disappearance of the white core of the DWCN strand during cooking (every 15 s) by squeezing the DWCN between two transparent glass slides. Three determinations were performed to obtain the mean values.

5.5.1.1.3 Evaluation of the Quality of Cooked Noodle

The eating quality of the cooked noodle was subjectively evaluated by five trained panelists according to Chinese standard method SB/T10137-93. Cooked noodle parameters and evaluation criteria were determined according to Liu et al. (2003). Noodle scoring criteria is given in Table 5.71.

5.5.1.2 Results and Analysis

5.5.1.2.1 Phenotypic Variation of Noodle Cooking Quality Parameters

Seven noodle quality parameters in the DH population showed large variation and continuous distribution. Bitransgressive segregation was seen in this population (Table 5.72). Most of the absolute values of skewness and kurtosis of noodle parameters were less than 1.0, which indicated that seven noodle quality parameters were quantitative traits controlled by multiple genes, suitable for QTL mapping analysis.

Table 5.71 System for evaluating the quality of cooked dry Chinese noodle

Parameters	Full score	Criterion
Color	10	Evaluation under white light. Desirable: bright, white noodle surface, 8.5–10.0; medium: 6–8.4; undesirable: 1.0–6.0
Appearance	10	Smooth or deformed skin, the number, and size of specks or pinhole on noodle surface, evaluation under red light. Desirable: 8.5–10.0; medium: 6–8.4; undesirable: 1.0–6.0
Smoothness	5	Desirable: little resistance to the lips when drawn into mouth, 4.3–5; medium: 3–4.3; undesirable: 1–3
Palate	20	Firmness, the force required to cut through the noodle using front teeth. Desirable: medium firmness, 17–20; medium: 12–17; undesirable: too hard or too soft firmness, 1–12
Elasticity	25	Elastic and cohesive when chewed; desirable: 21–25; medium: 15–21; undesirable: 1–15
Stickiness	25	Desirable: not to stick to teeth when chewed, 21–25; medium: 15–21; undesirable: 10–15
Taste	5	Desirable: wheat flavor and aromatic taste when smelted and eaten, 4.3–5; medium: 3–4.3; undesirable: 1–3
Total score	100	Good quality, like much > 85; 85 > medium, acceptable > 70; poor quality, unacceptable < 70

5.5.1.2.2 Additive and Additive × Environment Interactions of QTL for Noodle Cooking Parameters

Each one of additive QTL for noodle color was detected on chromosomes 5D and 1D, respectively (Table 5.73). Additive effect is derived from maternal and paternal characteristics, indicating synergistic gene of noodle parameter distributed in the parents.

One additive QTL (*Qco-5D*) interacted with environment was detected (Table 5.73), explaining 7.2 % of phenotypic variation of noodle color, which increased the color by 0.32 score.

In addition, it is worth noting that one additive QTL for noodle color was located near the glutenin subunit *GluD1* on chromosome 1D.

5.5.1.2.3 Epistatic and Epistasis × Environment Interaction of QTL for Noodle Tasting Parameters

Five epistatic QTLs were detected for seven noodle taste parameters (Table 5.74). These QTLs explained 1.44 to 16.39 % of the phenotypic variation. Of which, *Qse-3A*/*Qse-5D* explained the largest phenotypic variance by 16.39 % (Table 5.74; Fig. 5.19), increased by tasting score 1.49. No epistatic QTL interacted with the environment was detected.

Table 5.72 Phenotypic performance for white water noodle score parameters

Trait	Environment	Parents		DH population					
		Huapei 3	Yumai 57	Mean	Min.	Max.	SD	Skewness	Kurtosis
Color	Tai'an, 2007	17.75	17.50	16.82	12.50	19.25	1.26	−0.971	1.110
	Tai'an, 2006	18.00	12.75	15.59	12.25	18.25	1.37	−0.279	−0.343
	Suzhou, 2006	17.75	16.50	16.46	12.75	18.00	0.81	−1.138	3.121
Appearance	Tai'an, 2007	16.00	16.00	16.24	14.00	18.00	0.95	−0.519	−0.297
	Tai'an, 2006	14.75	14.25	15.85	13.50	17.00	0.67	−0.553	0.158
	Suzhou, 2006	17.00	16.50	16.68	15.00	17.75	0.49	−0.442	0.395
Firmness	Tai'an, 2007	6.50	8.00	7.33	5.75	8.50	0.57	−0.142	−0.276
	Tai'an, 2006	7.75	7.25	7.21	5.25	8.25	0.61	−1.092	1.028
	Suzhou, 2006	7.75	7.25	7.59	6.25	8.50	0.85	−0.392	−0.287
Stickiness	Tai'an, 2007	24.75	27.25	24.86	22.25	27.25	1.06	−0.123	−0.407
	Tai'an, 2006	26.00	23.00	24.69	20.00	27.00	1.43	−1.031	1.065
	Suzhou, 2006	26.25	25.50	25.71	23.75	27.00	0.79	−0.572	−0.196
Smoothness	Tai'an, 2007	7.75	8.25	7.96	6.75	9.00	0.44	−0.126	−0.209
	Tai'an, 2006	8.00	7.25	7.87	6.75	8.75	0.39	−0.133	−0.313
	Suzhou, 2006	8.50	8.25	8.15	6.50	9.00	0.42	−0.892	1.975
Taste	Tai'an, 2007	7.50	9.25	7.94	5.50	9.50	0.80	−0.473	0.160
	Tai'an, 2006	9.25	6.25	7.70	5.50	9.25	0.74	−0.518	0.249
	Suzhou, 2006	8.75	8.00	7.87	6.00	8.75	0.43	−1.120	2.758
Total score	Tai'an, 2007	79.75	86.25	81.15	72.50	87.75	3.54	−0.168	−0.479
	Tai'an, 2006	83.75	70.75	78.91	67.50	86.75	4.13	−0.588	−0.065
	Suzhou, 2006	86.00	82.00	82.46	73.00	87.25	2.66	−0.786	0.958

Table 5.73 Estimated additive (A) and additive × environment (AE) interactions of QTL for white water noodle score parameters

Trait	QTL	Flanking marker	Position (cM)	A	H^2 (A, %)	AE3	H^2 (AE3, %)
Color	*Qco-5D*	*Xwm215–Xbarc345*	87.4	0.15	1.47	0.32	7.2
Appearance	*Qap-1D*	*Xwmc93-GluD1*	61.9	−0.06	2.45		

E1: Suzhou, 2006; E2: Tai'an, 2006; E3: Tai'an, 2007

Table 5.74 Estimated epistatic and epistasis × environment interaction (AAE) effects of QTL for white water noodle score parameters

Trait	QTL	Flanking marker	Position (cM)	QTL	Flanking marker	Position (cM)	AA	H^2 (AA, %)
Color	*Qco-3A*	*Xbarc157.1–Xbarc1177*	192.8	*Qco-5D.1*	*Xwmc630.2–Xcfd40*	2.0	0.14	1.44
	Qco-3A	*Xbarc157.1–Xbarc1177*	192.8	*Qco-5D.2*	*Xbarc1097–Xcfd8*	23.4	0.33	7.70
Appearance	*Qap-1B*	*Xbarc061–XWMC766*	72.7	*Qap-4A*	*Xwmc313–Xwmc497*	31.5	−0.16	4.91
Stickiness	*Qst-3A*	*Xbarc157.1–Xbarc1177*	191.8	*Qst-5D*	*Xbarc1097–Xcfd8*	19.4	0.41	12.02
Total score	*Qse-3A*	*Xbarc157.1–Xbarc1177*	191.8	*Qse-5D*	*Xbarc1097–Xcfd8*	19.4	1.49	16.39

E1: Suzhou, 2006; E2: Tai'an, 2006; E3: Tai'an, 2007

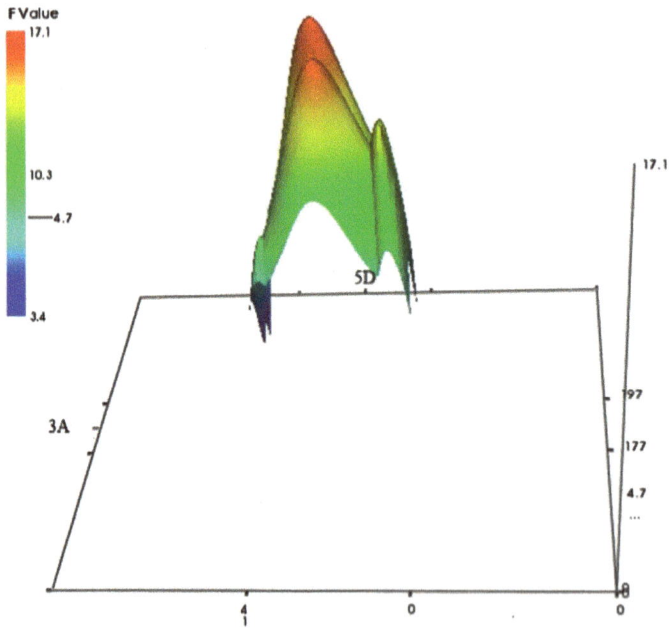

Fig. 5.19 3D visualization for the test statistics of genome scan for epistatic QTL associated with sensory evaluation (F value is taken as height)

5.5.1.3 Research Progress of QTL Mapping for Noodle Cooking Quality and Comparison with Previous Studies

5.5.1.3.1 Research Progress of QTL Mapping for Noodle Cooking Quality

Previous studies showed that glutenin sites on chromosomes 1A, 1B, and 1D controlled dough rheology. Perretant et al. (2000), Groos et al. (2004), Kuchel et al. (2006), and Sun et al. (2008) detected QTL of wheat gluten related to dough extensibility and toughness. In addition, the impact of HMW protein on the final quality has been demonstrated (MacRitchie 1999). Loci of Glu-A3 and Glu-B3 also have a great impact on dough extensibility.

5.5.1.3.2 Comparison of This Result with Previous Studies

Seven taste quality parameters of noodle are the main indicators of noodle quality. They are important parameters to determine food quality and processing technology and equipment. Few previous studies were found about QTL analysis of noodle scoring. In this study, an additive QTL on *GluD1* locus can be used in marker-assisted breeding, which proved the relationship between glutenin subunits and noodle taste at the genetic level.

In all, we detected two additive QTLs, 4 epistatic QTLs, and an interaction QTL with the environment for the noodle taste parameter. Of which, there were two major QTLs. QTLs closely linked with noodle taste test parameters could be used to improve the noodle quality through marker-assisted breeding.

5.5.2 QTL Mapping of Noodle Texture Property Analyzer (TPA) Parameters

Texture profile analysis (TPA) is an instrument to simulate human chewing action. It can record and draw the relationship between the force and the time, and give us six parameters (hardness, adhesiveness, springiness, cohesiveness, gumminess and resilience) (Fig. 5.20). These parameters could reflect human sensory evaluation on noodle such as hardness, brittleness, adhesion, elasticity, cohesiveness, gumminess and chewiness. Among them the hardness (Har) refers to the necessary force of the noodle to a certain denaturation, i.e., the maximum peak force of the first compression of the noodle. Adhesiveness (Adh) is the power to overcome the attractiveness when the probe run away from the sample surface at the first compression process (Area 3), indicated by the negative peak area between the two compression. Springiness (Spr) means the ration of gel height before and after the remove of pressure,which was calculated by Length 2/Length 1 (Fig. 5.20). Cohesiveness (Coh) indicates that the internal bonding force of the sample, that is the cohesion that pulling sample together. Gumminess (Gum) used to simulate the required

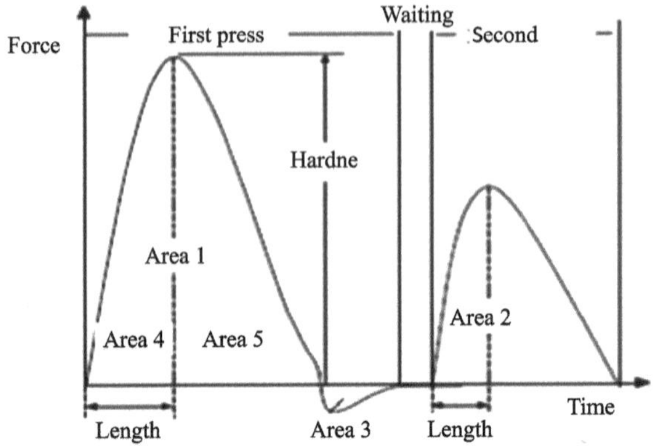

Fig. 5.20 Typical instrumental texture profile analysis curve

energy that rupture semi-solid sample into in a stable state for swallow, which was calculated by hardness multiplied by cohesiveness. Resilience (Res) indicated the degree of recovery for deformed gel at the same speed and the pressure conditions, which was calculated by Area5/Area 4.

5.5.2.1 Method of Texture Analysis

Dry noodle breaking strength (DNBS), dry noodle flexibility (DNF), and firmness of the cooked noodle were tested by the texture analyzer (TA.XT Plus; Stable Micro Systems Ltd., Godalming, Surrey, UK) equipped with the Windows version of Texture Expert software (Stable Micro Systems Ltd). Dry noodle of 10 cm length were prepared, and the average force and extension distance to break were measured. The test speed and distance were 2.5 mm/s and 10 mm. A set of six strands of cooked DWCN were placed parallel on a flat metal plate and compressed twice to 75 % of their original thickness, using the 3.175-mm-wide light knife blade. The testing parameters of the TPA test were carried out according to AACC method 16-50 (AACC, 2000) (Fig. 5.21). All texture analyses were measured in triplicate, and the coefficient of variation of the parameters was less than 5 %.

5.5.2.2 Results and Analysis

5.5.2.2.1 Phenotypic Analysis for TPA Parameters

Large variation was seen for the TPA parameters of DH population. They were continuous distribution and obviously transgression in DH population (Table 5.75). In addition, the absolute value of skewness and kurtosis in Suzhou and Tai'an in 2006 and 2007 is less than 1.0, which indicated that they were quantitative traits controlled by multigenes.

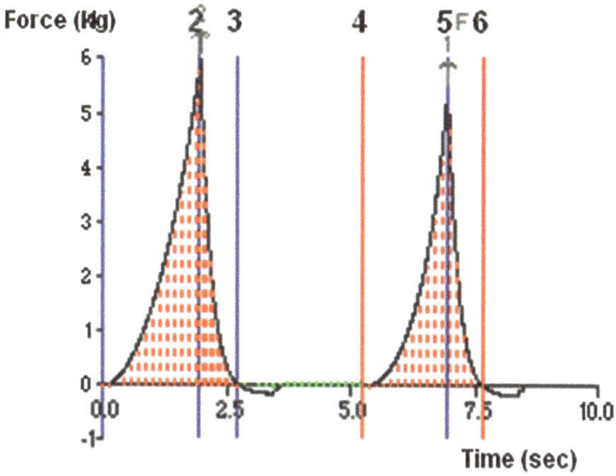

Fig. 5.21 Texture profile analysis of cooked noodle using the TA.XT plus

5.5.2.2.2 Additive and Additive × Environment Interactions of QTL for TPA Parameters

The numbers of QTLs for adhesiveness, springiness, cohesiveness, gumminess, chewiness, and resilience were 1, 1, 2, 1, 2, and 3, respectively (Table 5.76). The additive effects of most of the QTLs were from Yumai 57 (6 QTL, 60 %), which indicated the synergistic genes for TPA parameters distributed in the parents. These explained the obvious bitransgressive segregation phenomenon. There was one additive QTL for chewiness with the strongest additive effects, explaining 11.61 % of the phenotypic variance, which was located in the region of marker Xwmc412.2 and Xcfe023.2. No additive QTL interaction with environment was detected.

In addition, it is worth noting that one additive QTL located near glutenin subunit *GluD1* on chromosome 1D was detected.

5.5.2.2.3 Epistatic and Epistasis × Environment Interaction Effects of QTL for Noodle TPA Parameters

Three epistatic QTLs for 7 TPA parameters were identified (Table 5.77; Figs. 5.22 and 5.23), explaining 6.17–8.43 % of the phenotypic variance.

One epistatic QTL (*Qgum-1A/Qgum-3D*) interacted with the environment was detected (Table 5.77) and explained 2.13 % of the phenotypic variation with increasing gumminess of 5.83.

Table 5.75 Phenotypic performance for TPA parameters

Trait	Environment	Parents		DH population					
		Huapei 3	Yumai 57	Mean	Min.	Max.	SD	Skewness	Kurtosis
Hardness	Tai'an, 2007	238.59	400.41	340.20	152.76	460.36	53.14	−0.192	0.162
	Tai'an, 2006	332.75	477.53	421.86	267.50	889.10	92.54	2.331	8.111
	Suzhou, 2006	419.77	398.67	420.51	266.80	682.53	84.29	0.739	0.588
Adhesiveness	Tai'an, 2007	2.71	3.55	4.88	1.82	9.99	1.53	0.458	0.104
	Tai'an, 2006	4.00	7.06	6.82	3.05	13.64	2.12	1.034	1.124
	Suzhou, 2006	8.52	5.47	7.70	2.96	13.15	1.94	0.246	0.049
Springiness	Tai'an, 2007	0.92	0.94	0.89	0.74	0.97	0.04	−1.209	1.719
	Tai'an, 2006	0.94	0.89	0.87	0.65	0.94	0.05	−2.031	5.725
	Suzhou, 2006	0.76	0.88	0.83	0.67	0.95	0.06	−0.270	−0.873
Cohesiveness	Tai'an, 2007	0.57	0.62	0.57	0.45	0.63	0.03	−1.065	1.412
	Tai'an, 2006	0.60	0.56	0.57	0.40	0.63	0.03	−1.711	6.805
	Suzhou, 2006	0.54	0.56	0.55	0.45	0.65	0.03	−0.275	0.493
Gumminess	Tai'an, 2007	135.15	246.81	194.00	89.46	268.29	29.79	−0.257	0.391
	Tai'an, 2006	198.11	265.47	239.85	152.22	437.80	42.48	1.331	3.744
	Suzhou, 2006	227.91	224.58	237.72	136.00	374.80	43.46	0.542	0.264
Chewiness	Tai'an, 2007	124.76	230.62	172.60	82.42	241.55	27.28	−0.250	0.369
	Tai'an, 2006	186.05	234.99	207.96	135.76	313.30	29.28	0.355	0.447
	Suzhou, 2006	173.07	197.59	190.37	110.26	279.10	30.83	0.142	−0.235
Resilience	Tai'an, 2007	0.27	0.31	0.26	0.18	0.33	0.03	−0.115	−0.022
	Tai'an, 2006	0.31	0.24	0.26	0.16	0.33	0.03	−0.426	2.273
	Suzhou, 2006	0.24	0.27	0.25	0.18	0.36	0.03	0.462	1.260

Table 5.76 Estimated additive (A) and additive × environment (AE) interactions of QTL for TPA parameters

Trait	QTL	Flanking marker	Position (cM)	A	H^2 (A, %)
Adhesiveness	*Qadh-1B*	*Xcfe023.1–Xbarc372*	36.2	0.42	5.03
Springiness	*Qspr-1B*	*Xbarc372–Xwmc412.2*	36.3	−0.01	5.55
Cohesiveness	*Qcoh-1B*	*XbarcC372–Xwmc412.2*	36.3	−0.01	9.48
	Qcoh-4A	*Xwmc262–Xbarc343*	6.5	0.01	4.45
Gumminess	*Qgum-1B*	*Xwmc412.2–Xcfe023.2*	36.4	−11.56	8.39
Chewiness	*Qche-1B*	*Xwmc412.2–Xcfe023.2*	36.4	−10.26	11.61
	Qche-1D	*Xwmc93-GluD1*	61.9	−9.65	10.28
Resilience	*Qres-1B*	*Xcfe023.1–Xbarc372*	36.2	−0.0058	4.29
	Qres-4A	*Xwmc718–Xwmc262*	5.0	0.0077	7.46
	Qres-5D	*Xwmc215–Xbarc345*	84.4	0.0057	4.13

E1: Suzhou, 2006; E2: Tai'an, 2006; E3: Tai'an, 2007

5.5.2.3 Research Progress of QTL Mapping for Noodle TPA Parameters and Comparison with Previous Studies

5.5.2.3.1 Research Progress of Noodle TPA Parameters QTL Mapping

Szczeniak established "texture curve analytical method (TPA)" to comprehensively describe the physical properties of food in 1963, and the method was further improved by Bourne in 1978. At present, texture analyzer has become the main instruments to objectivity evaluate food quality, and the result has a high sensitivity and objectivity, which can be accurately and quantitatively treated, with the quantitative indicator to objectively and comprehensively evaluate food quality. This method would avoid the subjective influence on the result of food quality evaluation (Edwards et al. 1998; Lau et al. 2000; Zaidul et al. 2002; Epstein et al. 2002; Baxtera et al. 2004; Kealy 2006; Otegbayo et al. 2007).

QTL for noodle quality parameters was mainly related to the 1A, 1D, 3A, 3D, 4A, 4D, 5B, 6A, 6D, and 7A chromosomes. Among them, QTL located on 1B, 1D, 5D and 4A chromosomes was greater than others, and the contribution rate was up to 36 % (Table 5.78).

5.5.2.3.2 Comparison of This Result with Previous Studies

Previous studies have found that QTLs for noodle quality parameters mainly were located on 1B, 1D, 4A, and 5D chromosomes (Table 5.78). Of which, among them, there were many QTLs on chromosome 4A for controlling multinoodle TPA quality parameters, which might be related to W_x protein genes on 4A chromosome. In this study, the major QTLs were detected on chromosomes 1D and 5D for noodle quality, which might be associated with HMW-GS on 1D chromosome. These

Table 5.77 Epistatic and epistasis × environment interaction (AAE) effects of QTL for white water noodle score parameters

Trait	QTL	Flanking marker	Position (cM)	QTL	Flanking marker	Position (cM)	AA	H^2 (AA, %)	AAE2	H^2 (AAE2, %)
Gumminess	*Qgum-1A*	*Xcfd59–Xwmc402.2*	54.0	*Qgum-3D*	*Xgwm52–Xgdm8*	24.0	−10.21	6.54	−5.83	2.13
Chewiness	*Qche-1A*	*Xgwm259–Xcwem32.1*	1.0	*Qche-6B*	*Xcfa2187–Xgwm219*	10.0	−0.15	8.43		
	Qche-1A	*Xcfd59–Xwmc402.2*	54.0	*Qche-3D*	*Xgwm52–Xgdm8*	25.0	−0.12	6.17		

E1: Suzhou, 2006; E2: Tai'an, 2006; E3: Tai'an, 2007

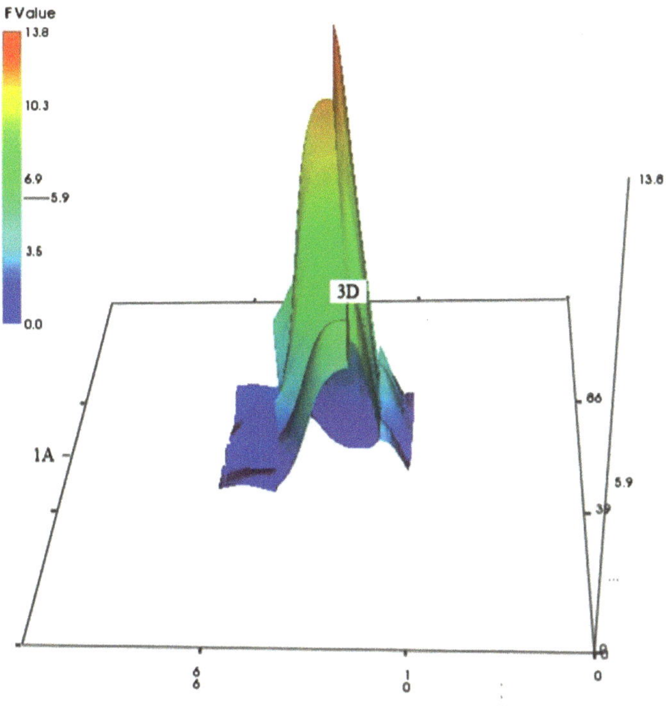

Fig. 5.22 3D visualization for the test statistics of genome scan for epistatic QTL associated with gumminess (F value is taken as height)

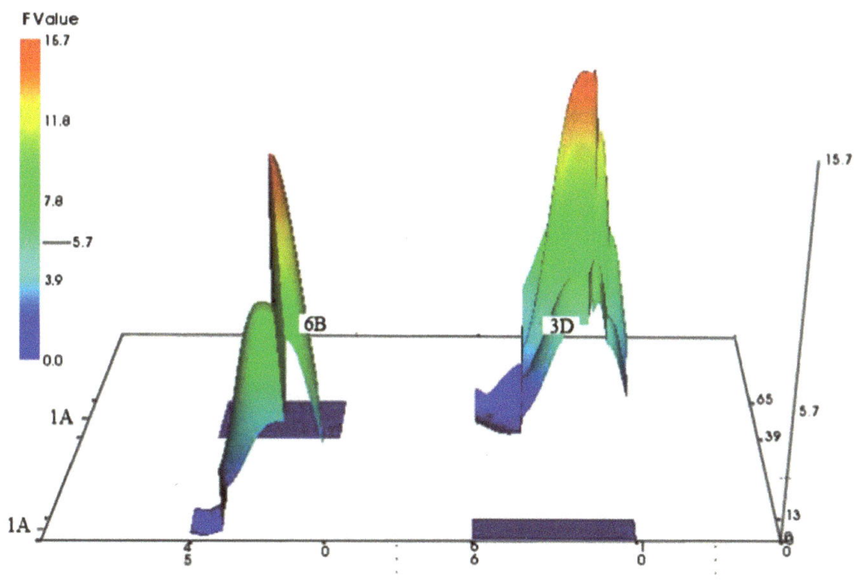

Fig. 5.23 3D visualization for the test statistics of genome scan for epistatic QTL associated with chewiness (F value is taken as height)

Table 5.78 Summary of QTL results of wheat noodle quality

Traits	QTLs	Flanking marker	PVE (%)	Population	References
Sensory quality					
Palate	*QSpal.sdau-1D*	*Glu-D1-Xsrap19*	3.28	RIL	Zhao et al. (2009)
Elasticity	*QSela.sdau-6D*	*Xswes426b–Xubc807d*	3.47	RIL	Zhao et al. (2009)
Stickiness	*QSsti.sdau-1D*	*Glu-D1-Xsrap19*	4.57	RIL	Zhao et al. (2009)
	QSsti.sdau-4A	*Xswes620–Xswes615*	6.78	RIL	Zhao et al. (2009)
Smoothness	*QSsmo.sdau-1D*	*Glu-D1-Xsrap19*	3.66	RIL	Zhao et al. (2009)
Taste	*QStas.sdau-4A.1*	*Xswes624c–Xissr25b*	3.07	RIL	Zhao et al. (2009)
	QStas.sdau-4A.2	*Xubc827d–Xissr22a*	2.75	RIL	Zhao et al. (2009)
Total	*QSts.sdau-4A*	*Xswes620–Xswes615*	2.98	RIL	Zhao et al. (2009)
Textural property					
Springiness	*QTspr.sdau-1A*	*Xwmc93d–Xgwm135*	3.55	RIL	Zhao et al. (2009)
Hardness	*QThd.sdau-3D*	*Xwmc529–Xsrap8*	2.58	RIL	Zhao et al. (2009)
Noodle color	*QY0a.crc-1A*	*Xwmc826–Xwmc11*	8.7	DH	McCartney et al. (2006)
	QY0a.crc-6A	*Xgwm334–Xbarc23*	13.3	DH	McCartney et al., 2006
	QY0b.crc-1A	*Xbarc119–Xwmc611*	7.7	DH	McCartney et al. (2006)
	QY0b.crc-4D	*Xwmc617–Xwmc48*	13.8	DH	McCartney et al. (2006)
	QY0b.crc-7A	*Xwmc809*	36	DH	McCartney et al. (2006)
	QY24 l.crc-5B	*Xgwm443–Xwmc740*	12.7	DH	McCartney et al. (2006)
	QY24b.crc-4D	*Xwmc617–Xwmc48*	11	DH	McCartney et al. (2006)
	QY24b.crc-7A	*Xwmc633–Xwmc809*	28.5	DH	McCartney et al. (2006)
	QYdl.crc-5D	*Xgwm16*	9.4	DH	McCartney et al. (2006)
	QYdb.crc-1D	*Xgwm337*	8.2	DH	McCartney et al. (2006)
	QYdb.crc-3A	*Xwmc264*	8	DH	McCartney et al. (2006)

results confirmed that the noodle quality was controlled not only by glutenin subunits but also by Wx protein loci.

5.5.3 QTL Mapping of Steamed Bread's Texture Property Analyzer (TPA) Parameters

Chinese steamed bread (CSB) is the most popular wheat product in China, and about 40 % of wheat consumption is particularly in northern China (He et al. 2003). CSB is a staple food for Chinese people, especially in northern China. It has been consumed for at least 2000 years in China and is gaining popularity in Korea, Japan, and some Southeast Asian countries. In China, there are several types of steamed bread, such as hard and soft northern-style steamed bread and soft southern-style steamed bread with respect to the differences in ingredients and quality parameters. Among them, hard northern-style steamed bread, with harder texture and firmer chewing properties, is usually prepared from strong gluten flour. It is preferred for eating by the people of northern China, such as Shandong, Shanxi, and Hebei provinces. The soft northern-style steamed bread is the mainly staple food for the people living in the central plain areas such as Henan, Shanxi, Anhui, and Jiangsu provinces. It is also liked by the people of the south of China. The steamed bread volume, texture, ratio of weight to volume, and color are important parameters for the steamed bread quality. So in this study, QTL mapping for steamed bread quality was analyzed, and the results would provide the reference to select the new cultivar with good steamed bread quality using MAS.

5.5.3.1 Materials and Methods

5.5.3.1.1 Plant Materials and Growing Conditions

Materials were same as ones of Sect. 5.2.3.1.1 in this chapter.

The DH lines and parents were planted in 2007 and 2008 in Tai'an (36.18°N, 117.13°E), Shandong Province, and in 2007 in Suzhou (31.32°N, 120.62°E), Anhui Province. In Tai'an, there were large differences in temperature and soil conditions between the years 2007 and 2008.

In autumn 2007, all lines and parents were sown in 2-m-long three-row plots (rows 25 cm apart); in autumn 2008, the lines were planted in 2-m-long four-row plots. Crop management followed local practices. The lines were harvested individually at maturity to prevent losses from shattering. Harvested grain samples from the three replicates at each environment were mixed and cleaned.

Hp3 and Ym67 are hard- and soft-grained, respectively. Prior to milling, hard, medium hard (between hard and soft), and soft wheat samples were tempered to around 16, 15, and 14 % moisture contents, respectively. Milling was performed using a Quadrumat Senior (Brabender, Germany) laboratory mill with a flour extraction rate of around 70 %.

5.5.3.1.2 Steamed Bread Preparation

Northern-style steamed bread was prepared according to the Chinese Business Standard (procedure 10139-93, Appendix A 1993). Milled flour samples (100 g) were mixed with yeast solution (1 g dry yeast dispersed in 48 ml water at 38 °C) and then fermented for 60 min in a fermentation cabinet (38 °C, 85 % RH) after 3 min of mixing. The fermented dough was divided into two pieces, and each was molded by hand for 3 min into a round dough piece with a smooth surface. After keeping in air for 15 min, the doughs were steamed for 20 min in a steam chamber (100 °C) followed by cooling at room temperature for 40 min.

5.5.3.1.3 Textural Property Testing

The steamed bread's textural property traits were determined using a texture analyzer (Texture Analyzer, Stable Micro Systems Company, England) according to the standard method 74-09 of the AACC (2000).

Steamed bread was placed in a covered bamboo container at room temperature immediately after steaming. A 25-mm slice was taken horizontally through the center of the steamed bread sample by placing the sample in a custom-built cutting machine, thus enabling a consistent slice thickness and slicing location. Compression tests were carried out on the slices 40 min after steaming. They were subjected to a constant deformation of 40 % with a BRDI/P35 indenter with 35 mm diameter and a traveling speed of 1 mm/s. Several textural parameters were recorded. Two compression cycles were conducted, measuring the proportion of height recovery of steamed bread during the time that elapsed between the end of the first cycle and the start of the second compression cycle. The initial compression was stopped for 4 s after 40 % compression and then continued until a two-cycle compression was completed (AACC 2000). All determinations were made at least three times.

5.5.3.1.4 Statistical Analysis

Analysis of variance (ANOVA) was carried out using the SPSS version 13.0 (SPSS, Chicago, USA) program. QTLs with additive effects and epistatic effects as well as QEs in the DH population were mapped by the software QTL Network version 2.0 (Yang and Zhu 2005) and IciMapping v2.2 (Wang et al. 2009) based on the mixed linear model. Composite interval analysis was undertaken using forward–backward stepwise multiple linear regression with a probability into and out of the model of 0.05 and a window size set at 10 cM. Significant thresholds for QTL detection were calculated for each data set using 1000 permutations and genome-wide error rates of 0.10 (suggestive) and 0.05 (significant). The final genetic model incorporated significant additive effects and epistatic effects as well as their environmental interactions.

A QTL was named for steamed bread's textural properties by the first two letters with the relevant chromosomal number, such as "Qad." for adhesiveness followed by a laboratory code (sau). If there were more than one QTL on a chromosome, a serial number was added after the chromosomal number, separated by a hyphen. The positions of these QTLs were indicated by the marker interval bracketing the QTL with the estimated distance (cM) from the left marker.

5.5.3.2 Results and Analysis

5.5.3.2.1 Phenotypic Assessments

The DH population had a wide range of variation for most traits. Mean values for steamed bread's textural property traits for the DH population and the parents grown under three environments are shown in Table 5.79. The DH means for hardness, gumminess, and chewiness were higher than those for Hp3 and Ym57, indicating transgressive segregation. Huapei 3 had higher values than Yumai 57 for all quality traits. Transgressive segregation occurred in all environments. Six of the seven traits, except adhesiveness, segregated continuously and followed normal distributions. The values of skewness and kurtosis were less than 1.0, except kurtosis for hardness at 1.37 (Table 5.79), indicating polygenic inheritance and suitability of the data for QTL analysis (Cao et al. 2001).

5.5.3.2.2 Coefficients of Steamed Bread's Textural Properties

The correlations among the steamed bread's textural property trait values are shown in Table 5.80. The most significant correlation was detected between chewiness and gumminess ($r = 0.99$), with a lower correlation between resilience and cohesiveness ($r = 0.95$). Significant negative correlations were found between springiness and adhesiveness ($r = -0.33$) and between resilience and hardness ($r = -0.31$).

5.5.3.2.3 QTLs for Hardness

Four QTLs for hardness with additive effects were mapped to chromosomes 2B, 6B, and 7B (Table 5.81; Fig. 5.24) and explained 6–35 % of the phenotypic variance. *Qha.sau-7B.1* (*Xwmc581–Xbarc050*) had the most significant effect. The Hp3 allele at *Qha.sau-7B.2* (*Xwmc273.1–Xcfd22.1*) increased hardness by 1002 g due to additive effects. The Ym57 allele increased hardness at *Qha.sau-2B* (*Xgwm111–Xgdm14-6D*), *Qha.sau-6B* (*Xcfd48–Xwmc415*), and *Qha sau-7B.1*

Table 5.79 Phenotypic performance of wheat steamed bread's textural properties under three environments

Trait	Parent		Mean	SE	Max.	Min.	SD	Skew.	Kurt.
	Huapei 3	Yumai 57							
Hardness/g	6644.74	6539.41	6935.01	103.40	11474.71	4146.6	1342.63	0.96	1.37
Adhesiveness/g s	−18.34	−232.33	−40.28	5.21	−2.19	−577.20	67.49	−4.86	36.79
Springiness	0.87	0.83	0.86	0.00	0.91	0.76	0.03	−0.86	0.81
Cohesiveness	0.72	0.67	0.71	0.00	0.80	0.58	0.05	−0.29	−0.54
Gumminess/g	4558.74	4297.45	4816.06	54.39	7543.61	3067.87	811.26	0.77	1.00
Chewiness/g	3931.48	3510.53	4095.10	58.32	6614.77	2714.43	705.09	0.67	0.93
Resilience	0.33	0.27	0.31	0.00	0.40	0.18	0.05	−0.15	−0.78

Mean, SE, Max., Min., SD, Skew., and Kurt. are the average, standard error, maximum, minimum, standard deviation, skewness, and kurtosis of all observations for the DH population in one environment

Table 5.80 Coefficients of pairwise correlations of the mean values among steamed bread's textural properties traits in the three environments

Trait	Hardness	Adhesiveness	Springiness	Cohesiveness	Gumminess	Chewiness
Hardness/g	−0.07					
Adhesiveness/g s	0.02	0.33**				
Springiness	−0.25**	0.42**	0.83**			
Cohesiveness	0.94**	0.09	0.18*	0.02		
Gumminess/g	0.89**	0.16*	0.29**	0.13	0.99**	
Chewiness/g	−0.31**	0.40**	0.70**	0.95**	−0.03	0.09

*$P < 0.005$ and **$P < 0.001$, respectively

(*Xwmc581–Xbarc050*) by 5.25 g, accounting for 6, 13, and 35 % of the phenotypic variance, respectively. This suggested that alleles for increased hardness were dispersed between the parents, resulting in small differences in phenotypic values between the parents, but permitting transgressive segregation among the DH lines. The total additive QTLs detected for hardness accounted for 78.8 % of the phenotypic variation.

An epistatic pair, *Qha.sau-3D/Qha.sau-6B*, for hardness explained 8 % of the phenotypic variance (Table 5.82; Fig. 5.24) and had a negative effect of −147 g. The QTL pair *Qha.sau -3D/Qha.sau-6B* was involved in AAE interactions in all three environments, accounting for 33.6 % of the phenotypic variance.

5.5.3.2.4 QTLs for Adhesiveness

Six additive QTLs for adhesiveness on five chromosomes, 1B, 2A, 4A, 3B, and 6B, increased adhesiveness from 11.1 to 116.6 gs and explained 8–298 % of the phenotypic variance. *Qad.sau-4A* had the highest PVE. Three QTLs (Qad.sau-1B, Qad.sau-2A, and Qad.sau-4A,) had negative effects on adhesiveness and were contributed by Yumai 57, while another three QTLs (Qad.sau-6B; Qad.sau-3B; and Qad.sau-1B) had positive effects on adhesiveness and were contributed by Huapei 3 alleles. These suggested that alleles increasing adhesiveness were dispersed between the parents. All six QTLs accounted for 84 % of the phenotypic variance (Table 5.81; Fig. 5.24).

Three pairs of epistatic effects for adhesiveness were located on chromosomes 1B/2A, 1B/3B, and 2A/3B, explaining 0.59 to 1.33 % of the phenotypic variance. *Qad.sau-1B/Qad.sau-2A* had a negative effect of −13.1 gs, whereas the other two pairs showed positive effects. The general contribution of the three pairs of epistatic QTLs was 2.8 %, and three main-effect QTLs were detected in epistatic effects. All of them were involved in AAE interactions, accounting for 5.7 % of the phenotypic variance (Table 5.82; Fig. 5.24).

Table 5.81 Estimated additive and additive × environment interaction of QTLs for steamed bread's textural properties based on averaged phenotypic data from 3 environments

Trait	QTL	Position (cM)	Flanking marker	Additive effects[a]	PVE[b] (%)
Hardness[1]	*Qha6B*	53	*Xcfd48–Xwmc415*	−1616.74	18.085
	Qha7B$_2$	8	*Xwmc581–Xbarc050*	−1361.02	35.1694
	Qha7B$_2$	12	*Xwmc273.1–Xcfd22.1*	1001.673	19.1932
	Qha2B	136	*Xgwm111–Xgdm14-6D*	−615.781	6.3582
Adhesiveness[1]	*Qad6B*	56	*Xwmc74–Xgwm58*	28.8054	13.3475
Adhesiveness[2]	*Qad3B*	23	*Xgwm533–Xbarc251*	60.4118	7.794
	Qad4A	44	*Xbarc078–Xwmc722*	−116.625	29.349
Adhesiveness[3]	*Qad1B*	33	*Xwmc31–Xwmc626*	11.1388	7.3432
	Qad1B	70	*Xcfe026.2–Xbarc061*	−20.7657	26.0446
	Qad2A	73	*Xbarc264–Xgwm448*	−17.2204	18.0936
Springiness[1]	*QSp2B*	2	*Xgwm210–Xwmc382.2*	−0.0125	8.7296
	QSp2B	77	*Xbarc101–Xcwem55*	0.0095	4.9431
	QSp3A	159	*Xgwm155–Xcfa2170*	0.0111	7.0213
	QSp6D	77	*Xbarc054–Xgwm55*	0.0108	6.3975
Cohesiveness[1]	*Qco2B*	1	*Xgwm210–Xwmc382.2*	−0.0275	8.9728
	Qco3A	167	*Xcfa2170–Xbarc51*	0.027	8.8641
	Qco3B	53	*Xcfe282–Xgwm299*	−0.0263	7.029
	Qco3B	58	*Xwmc307–Xgwm566*	0.0414	19.0557
	Qco6D	78	*Xgwm55–Xgwm133.2*	0.0193	4.358
Cohesiveness[3]	*Qco1A*	53	*Xwmc163–Xcfd59*	−0.0086	4.8898
	Qco3B	52	*Xbarc139–Xwmc612*	0.0106	6.3585
	Qco4A	37	*Xwmc776–Xbarc362*	−0.0091	5.5342
	Qco7D	185	*Xwmc14–Xwmc42*	−0.0092	4.5719

(continued)

Table 5.81 (continued)

Trait	QTL	Position (cM)	Flanking marker	Additive effects[a]	PVE[b] (%)
Gumminess[1]	*Qgu6A*	0	*Xgwm459–Xgwm334*	335.3752	5.6133
Gumminess[2]	*Qgu5D*	0	*Xwmc630.2–Xcfd40*	−284.643	5.7021
Gumminess[3]	*Qgu2B*	136	*Xgwm111–Xgdm14-6D*	−351.251	6.4099
Chewiness[1]	*QCh6A*	0	*Xgwm459–Xgwm334*	318.1299	6.2364
Chewiness[2]	*QCh5D*	0	*Xwmc630.2–Xcfd40*	−245.238	5.8545
	QCh6D	7	*Xwmc412.1–Xcfd49*	257.9171	6.459
Chewiness[3]	*QCh2B*	135	*Xgwm111–Xgdm14-6D*	−295.983	6.8447
Resilience[1]	*Qre2B*	2	*Xgwm210–Xwmc382.2*	−0.0336	11.8366
	Qre3A	163	*Xcfa2170–Xbarc51*	0.0293	9.1653
	Qre3B	53	*Xcfe282–Xgwm299*	−0.0258	5.9698
	Qre3B	58	*Xwmc307–Xgwm566*	0.0451	19.9542
	QRe5B$_2$	19	*Xbarc36–Xbarc140*	0.0191	3.9106
	Qre6D	78	*Xgwm55–Xgwm133.2*	0.0211	4.5736
Resilience[3]	*Qre3B*	52	*Xbarc139–Xwmc612*	0.0125	7.6372
	Qre4A	34	*Xwmc313–Xwmc497*	−0.0114	7.5644
	Qre7D	188	*Xwmc14–Xwmc42*	−0.0116	6.1283

[a]Additive effects: A positive value indicates that the Huapei 3 allele has a positive effect on the adhesiveness, and a negative value indicates that allele from Ym57 has a negative effect on the adhesiveness. [b]Contribution explained by putative main-effect QTL. [1]2008 in Tai'an; [2]2007 in Tai'an; [3]2007 in Suzhou

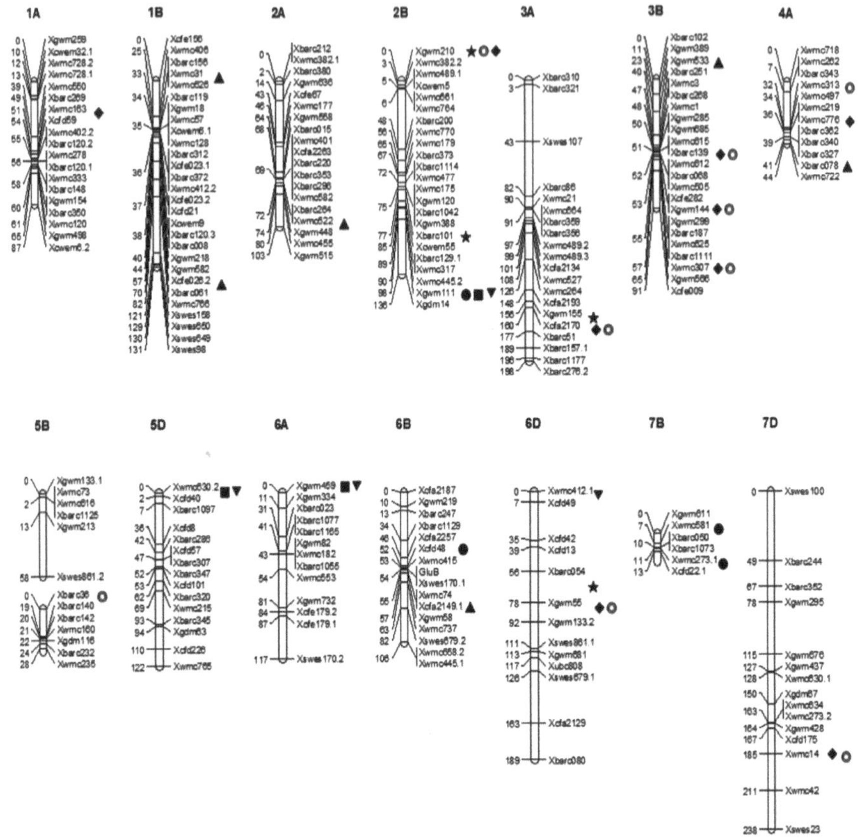

Fig. 5.24 Genetic linkage map of wheat showing mapping QTLs with additive effects for steamed bread's textural properties. ● Additive QTL for hardness, ▲ additive QTL for adhesiveness, ★ additive QTL for springiness, ◆ additive QTL for cohesiveness, ■ additive QTL for gumminess, ▼ additive QTL for chewiness, ◎ additive QTL for resilience

5.5.3.2.5 QTLs for Springiness

Four QTLs (*QSp.sau-2B.1*, *QSp.sau-2B.2*, *QSp.sau-3A*, and *QSp.sau-6D*) were detected for springiness, explaining 4.9–8.7 % of the phenotypic variance. The most important QTL was *QSp.sau-2B.1*, which explained 8.7 % of the phenotypic variance, had a negative effect on springiness, and was contributed by Yumai 57; the others were from Huapei 3.

5.5.3.2.6 QTLs for Cohesiveness

Nine additive QTLs on seven chromosomes (1A, 3A, 4A, 2B, 3B, 6D, and 7D) for cohesiveness explained phenotypic variance ranging from 4.36 to 19.06 %. *Qco.*

Table 5.82 Estimated epistasis (AA) and epistasis × environment interactions of QTLs for steamed bread's textural properties based on averaged phenotypic data from 3 environments

Trait	QTL	Flanking marker	Position (cM)	QTL_j	Trait	Flanking marker	Position (cM)	AA[a]	H^2 (AA%)[b]	AAE1	h^2 (AAE1 %)	AAE2	h^2 (AAE2 %)	AAE3	H^2 (AAE3 %)
Adhesiveness	*Qad1B*	*XCFE156–XWMC406*	0	*Qad2A*	4-3	*XBARC380–XGWM636*	1.6	−13.129**	1.33			−20.04*	2.32		
	Qad1B	*XCFE156–XWMC406*	0	*Qad3B*	8-3	*XGWM533–XBARC251*	23	9.786*	0.91			19.8673*	2.3		
	Qad2A	*XBARC380–XGWM636*	1.6	*Qad3B*	8-3	*XGWM533–XBARC251*	23	10.5699*	0.59			16.2597*	1.12		
Cohesiveness	*Qco1B-1*	*XCFE023.1–XBARC372*	36.1	*Qco3A-1*	7-16	*XCFA2170–XBARC51*	176.2	0.0121**	0.07	0.0144**	4.51				
	Qco1B-2	*XGWM582–XCFE026.2*	53.1	*Qco3A-2*	7-17	*XBARC51–XBARC157.1*	176.5	−0.0151**	0.27			−0.0167**	4.94		
Resilience	*Qre1B-1*	*XCFE023.1–XBARC372*	36.1	*Qre3A-1*	7-16	*XCFA2170–XBARC51*	175.2	0.0131**	0.06	0.0164**	5.31				
	Qre1B-2	*XGWM582–XCFE026.2*	52.1	*Qre3A-2*	7-17	*XCFA2170–XBARC51*	176.2	−0.0163**	0.25			−0.0197**	5.34		
hardness	*Qha3D*	*XGDM72–XBARC1119*	10	*Qha6B*	19-11	*XGWM58–XWMC737*	61.9	−147.3078**	8.3	−539.7731	11.974	124.1640**	14.2	415.6196	7.437

[a]Epistatic effects: A positive value indicates that the parental two-locus genotypes have a positive effect and that the recombinants have a negative effect

[b]Contribution explained by epistatic QTL. E1: 2008 in Tai'an; E2: 2007 in Tai'an; E3: 2007 in Suzhou

*$P < 0.005$ and **$P < 0.001$, respectively

sau-3B.2 had the highest contribution and explained 19.06 % of the phenotypic variance. Five QTLs (*Qco.sau-4A*, *Qco.sau-7D*, *Qco.sau-1A*, *Qco.sau-3B.1*, and *Qco.sau-2B*) had negative effects and were contributed by Yumai 57 alleles. The other four QTLs (*Qco.sau-3A*, *Qco.sau-3B.2*, *Qco6.sau-D*, and *Qco. sau-3B.3*) had positive effects and came from Huapei 3. These suggested that alleles increasing cohesiveness were dispersed between the parents, allowing for small differences in phenotype between the parents and transgressive segregation among the DH lines. All nine QTLs accounted for 69.63 % of the phenotypic variance (Table 5.81; Fig. 5.24).

There were two pairs of epistatic effects for cohesiveness, explaining 0.07 and 0.27 % of the phenotypic variance, respectively. The *Qco.sau-1B.1/3A.1* pair had positive effects, whereas the *Qco.sau-1B.2/3A.2* pair showed negative effects. The total contribution of the two pairs was 0.34 %, and no-main-effect QTLs were detected among them. Two pairs of epistatic QTLs were also involved in AAE interactions, accounting for 9.45 % of the phenotypic variance (Table 5.82; Fig. 5.24).

5.5.3.2.7 QTLs for Gumminess

Two additive QTLs (*Qgu.sau-5D* and *Qgu.sau-2B*) increased gumminess from 284.6 to 351.3 g, explaining phenotypic variance of 5.7 and 6.4 %. Both accounted for 12.1 % of the phenotypic variance and were from Yumai 57 alleles, in accordance with Yumai 57 having a much higher gumminess value than Huapei 3.

5.5.3.2.8 QTLs for Chewiness

Four QTLs, *Qch.sau-6A*, *Qch.sau-5D*, *Qch.sau-6D*, and *Qch.sau-2B* (Table 5.81; Fig. 5.24), for chewiness had significant additive effects. The phenotypic variance from 5.8 to 6.84 % is explained. The strongest, *Qch.sau-2B*, came from Ym57, which also contributed the favorable *Qch.sau-5D* allele. *Qch.sau-6A* and *Qch.sau-6D* from the Hp3 alleles also increased flour protein content and accounted for 6.24 and 6.46 % of the phenotypic variance, respectively. The additive QTLs for chewiness accounted for 25.39 % of the phenotypic variance.

5.5.3.2.9 QTLs for Resilience

Eight main-effect QTLs were identified for resilience. These QTLs increased the resilience by 1.25–4.51 % and accounted for phenotypic variance ranging from 3.91 to 19.95 %. The total contribution of the main-effect QTLs explained 76.74 % of the phenotypic variance. Four alleles (*Qre.sau-2B*, *Qre.sau-3B*, *Qre.sau-4A*, and *Qre.sau-7D*) came from Huapei 3, and the others were from Yumai 57.

Table 5.83 Summary of QTL results of wheat steamed bread quality

Traits	QTL	Flanking marker	PVE (%)	Mapping population	References
Volume (ml)	*QVol.sdau-6B*	*Xgwm193–Xgwm608b*	11.73	RIL	Fan et al. (2009)
Specific volume	*QSv.sdau-6B*	*Xgwm193–Xgwm608b*	11.19	RIL	Fan et al. (2009)
Appearance	*QApp.sdau-2D*	*Xgwm29b–Xgwm132a*	40.34	RIL	Fan et al. (2009)
Color	*QCol.sdau-5D*	*Xswes342b–Xsrap6b*	9.88	RIL	Fan et al. (2009)
Elasticity	*QEla.sdau-7B*	*Xgwm333–Xgwm297*	11.13	RIL	Fan et al. (2009)
Stickiness	*QSti.sdau-2B*	*Xsrap1a–Xgwm120*	32.46	RIL	Fan et al. (2009)
Smell	*QSme.sdau-4A*	*Xwmc308–Xsrap7c*	64.87	RIL	Fan et al. (2009)
Total score	*QTs.sdau-5B*	*Xgwm261a–Xgwm234*	29.8	RIL	Fan et al. (2009)

There were two pairs of epistatic QTLs, *Qco.sau-1B.1/Qco.sau-3A.1* and *Qco.sau-1B.2/Qre.sau-3A.2*; *Qco.sau-1B.1/Qco.sau-3A.1* had a small positive effect of 0.0131, and *Qco.sau-1B.2/Qre.sau-3A.2* had a negative effect. The contribution of the two pairs of epistatic QTLs was 0.31 %. One pair of epistatic effects was also involved in an AAE interaction, accounting for 5.34 % of the variation (Tables 5.82 and 5.83; Fig. 5.24).

5.5.4 QTL Mapping of Steamed Bread Specific Volume

5.5.4.1 Plant Materials and Growing Conditions

Plant materials and growing conditions were same as ones of Sect. 5.5.3.1.1 in this chapter.

5.5.4.1.1 Method of Specific Volume Testing

Steamed bread was placed in a covered bamboo container at room temperature immediately after steaming. Then, the steamed breads were cooled in the air for 40 min. Samples were measured using an electric balance for weight and a substitution method with rapeseed for volume. All scores were taken three times per sample. The mean values of CNSB specific volume in each environment were used for statistical analyses.

All determinations were made at least three times and were expressed on a 14–16 % moisture basis.

5.5.4.1.2 Statistical Analysis of the Phenotypic Assessments

A large variation for steamed bread specific volume was seen in the DH population. Mean values for CNSB specific volume traits for the DH population and the parents in the three environments are shown in Table 5.84. Transgressive segregation was observed in all the environments, with some lines higher or lower than the parents. A normal distribution was observed for the specific volume traits in this population (Fig. 5.25). The absolute values of skewness and kurtosis were less than 1.0 (Table 5.84), which indicated it had polygenic inheritance.

5.5.4.2 QTL Analyses of Steamed Bread Specific Volume

Five additive QTLs and fourteen pairs of epistatic effects were identified for specific volume of CNSB in three years (Tables 5.85; Fig. 5.26) by using the software QTL Network version 2.0 and IciMapping v2.2, based on the mixed linear model.

Five additive QTLs were detected for specific volume on five chromosomes 6A, 7B2, 7D, 1A, and 3D (Fig. 5.26). They explained phenotypic variance from 5. 11 to 9.75 %. All of the five QTLs had negative effects on specific volume and were contributed by Yumai 57 alleles which have the higher specific volume. The total additive QTLs detected for specific volume accounted for 38.454 % of the phenotypic variance.

Table 5.84 Phenotypic performance of wheat CNSB specific volume in the three environments

Trait	Parents		DH population					
Specific volume	Huapei 3	Yumai 57	Mean	Max.	Min.	SD	Skew.	Kurt.
	1.89	2.09	1.98	2.28	1.66	0.11	0.2	0.06

Table 5.85 Estimated additive (A) of QTLs for steamed bread specific volume in the three environments

Trait	Chr.	Site (cM)	Left marker	Right marker	Additive effects	LOD	Contributions (%)
Specific volume 1	*Qsv-6A*	68	*Xwmc553*	*Xgwm732*	−0.06	2.31	9.37
Specific volume 2	*Qsv-7B2*	13	*Xwmc273.1*	*Xcfd22.1*	−0.06	3.12	7.38
Specific volume 2	*Qsv-7D*	75	*Xbarc352*	*Xgwm295*	−0.05	2.461	6.84
Specific volume 3	*Qsv-1A*	54	*Xcfd59*	*Xwmc402.2*	−0.04	2.191	5.11
Specific volume 3	*Qsv-3D*	94	*Xbarc071*	*Xgwm114*	−0.06	3.711	9.75

(1) 2007 Tai'an; (2) 2008 Tai'an; (3) 2007 Suzhou

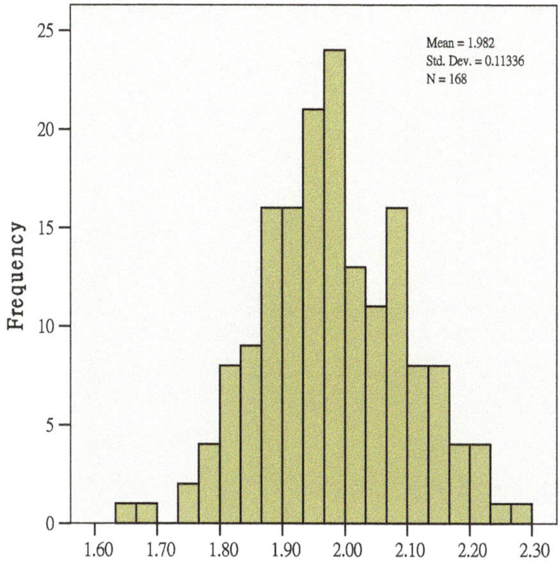

Fig. 5.25 Frequency distribution of steamed bread specific volume-related traits in 168 doubled haploid lines derived from the cross of Huapei 3 × Yumai 57 evaluated at three environments

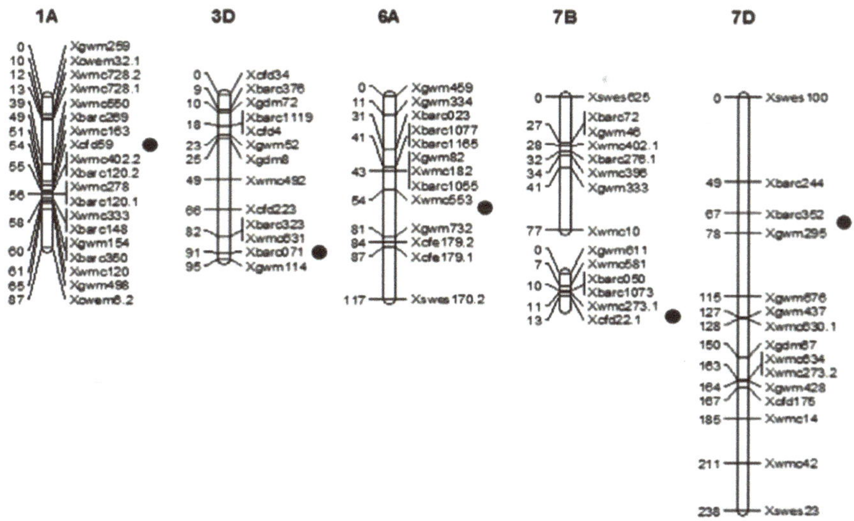

Fig. 5.26 Positions of additive QTLs conferring steamed bread specific volume-related traits in 168 doubled haploid lines derived from the cross of Huapei 3 × Yumai 57 evaluated in the three environments

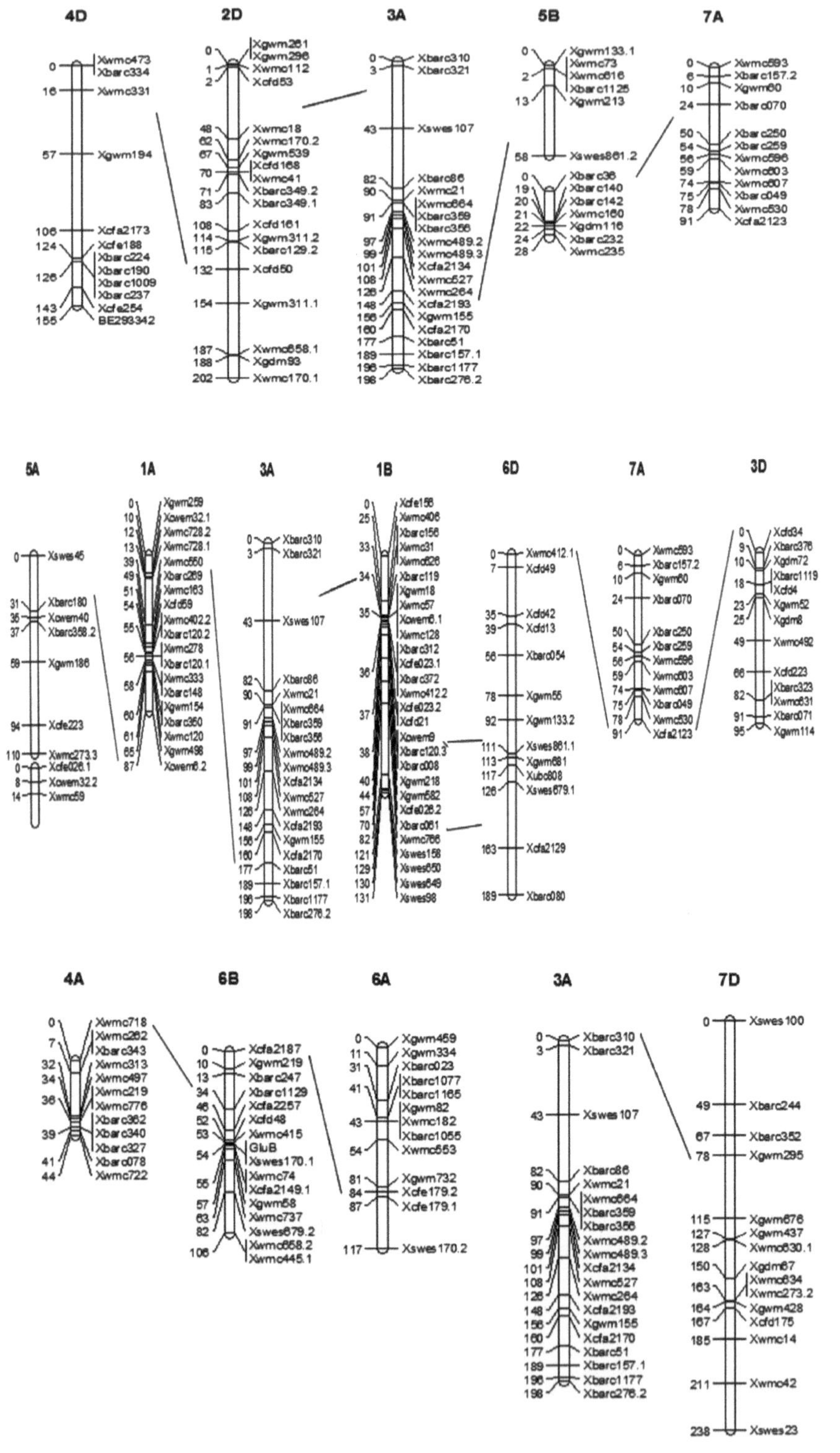
4D
Xwmc473
Xbarc334
Xwmc331
Xgwm194
Xcfa2173
Xcfe188
Xbarc224
Xbarc190
Xbarc1009
Xbarc237
Xcfa254
BE293342
2D
Xgwm261
Xgwm296
Xwmc112
Xcfd53
Xwmc18
Xwmc170.2
Xgwm539
Xcfd168
Xwmc41
Xbarc349.2
Xbarc349.1
Xcfd161
Xgwm311.2
Xbarc129.2
Xcfd50
Xgwm311.1
Xwmc658.1
Xgdm93
Xwmc170.1
3A
Xbarc310
Xbarc321
Xswes107
Xbarc86
Xwmc21
Xwmc664
Xbarc359
Xbarc356
Xwmc489.2
Xwmc489.3
Xcfa2134
Xwmc527
Xwmc264
Xcfa2193
Xgwm155
Xcfa2170
Xbarc51
Xbarc157.1
Xbarc1177
Xbarc276.2
5B
Xgwm133.1
Xwmc73
Xwmc616
Xbarc1125
Xgwm213
Xswes861.2
Xbarc36
Xbarc140
Xbarc142
Xwmc160
Xgdm116
Xbarc232
Xwmc235
7A
Xwmc593
Xbarc157.2
Xgwm60
Xbarc070
Xbarc250
Xbarc259
Xwmc596
Xwmc603
Xwmc607
Xbarc049
Xwmc530
Xcfa2123
5A
Xswes45
Xbarc180
Xcwem40
Xbarc358.2
Xgwm186
Xcfe223
Xwmc273.3
Xcfe026.1
Xcwem32.2
Xwmc59
1A
Xgwm259
Xcwem32.1
Xwmc728.2
Xwmc728.1
Xwmc550
Xbarc269
Xwmc163
Xcfd59
Xwmc402.2
Xbarc120.2
Xwmc278
Xbarc120.1
Xwmc333
Xbarc148
Xgwm154
Xbarc350
Xwmc120
Xgwm498
Xcwem6.2
3A
Xbarc310
Xbarc321
Xswes107
Xbarc86
Xwmc21
Xwmc664
Xbarc359
Xbarc356
Xwmc489.2
Xwmc489.3
Xcfa2134
Xwmc527
Xwmc264
Xcfa2193
Xgwm155
Xcfa2170
Xbarc51
Xbarc157.1
Xbarc1177
Xbarc276.2
1B
Xcfe156
Xwmc406
Xbarc156
Xwmc31
Xwmc626
Xbarc119
Xgwm18
Xwmc57
Xcwem6.1
Xwmc128
Xbarc312
Xcfe023.1
Xbarc372
Xwmc412.2
Xcfe023.2
Xcfd21
Xcwem9
Xbarc120.3
Xbarc008
Xgwm218
Xgwm582
Xcfe026.2
Xbarc061
Xwmc766
Xswes158
Xswes650
Xswes649
Xswes98
6D
Xwmc412.1
Xcfd49
Xcfd42
Xcfd13
Xbarc054
Xgwm55
Xgwm133.2
Xswes861.1
Xgwm681
Xubc808
Xswes679.1
Xcfa2129
Xbarc080
7A
Xwmc593
Xbarc157.2
Xgwm60
Xbarc070
Xbarc250
Xbarc259
Xwmc596
Xwmc603
Xwmc607
Xbarc049
Xwmc530
Xcfa2123
3D
Xcfd34
Xbarc376
Xgdm72
Xbarc1119
Xcfd4
Xgwm52
Xgdm8
Xwmc492
Xcfd223
Xbarc323
Xwmc631
Xbarc071
Xgwm114
4A
Xwmc718
Xwmc262
Xbarc343
Xwmc313
Xwmc497
Xwmc219
Xwmc776
Xbarc362
Xbarc340
Xbarc327
Xbarc078
Xwmc722
6B
Xcfa2187
Xgwm219
Xbarc247
Xbarc1129
Xcfa2257
Xcfd48
Xwmc415
GluB
Xswes170.1
Xwmc74
Xcfa2149.1
Xgwm58
Xwmc737
Xswes679.2
Xwmc658.2
Xwmc445.1
6A
Xgwm459
Xgwm334
Xbarc023
Xbarc1077
Xbarc1165
Xgwm82
Xwmc182
Xbarc1055
Xwmc553
Xgwm732
Xcfe179.2
Xcfe179.1
Xswes170.2
3A
Xbarc310
Xbarc321
Xswes107
Xbarc86
Xwmc21
Xwmc664
Xbarc359
Xbarc356
Xwmc489.2
Xwmc489.3
Xcfa2134
Xwmc527
Xwmc264
Xcfa2193
Xgwm155
Xcfa2170
Xbarc51
Xbarc157.1
Xbarc1177
Xbarc276.2
7D
Xswes100
Xbarc244
Xbarc352
Xgwm295
Xgwm676
Xgwm437
Xwmc630.1
Xgdm67
Xwmc634
Xwmc273.2
Xgwm428
Xcfd175
Xwmc14
Xwmc42
Xswes23

◀ **Fig. 5.27** Positions of estimated epistasis QTLs conferring steamed bread specific volume-related traits in 168 doubled haploid lines derived from the cross of Huapei 3 × Yumai 57 evaluated in the three environments

Fourteen pairs of epistatic effects were identified for specific volume and located on chromosomes 1B-3A, 2D-3A, 3A-5B1, 6D-7A, 1A-3A, 1B-6D, 2D-4D, 4A-6B, 6A-6B, 1A-5A, 1B-6D, 3A-7D, 3D-7A, and 5B2-7A (Fig. 5.27). They explained the phenotypic variance ranging from 4.83 to 20.39 %. *Qsa1B/Qsa3D-2* and *Qsa2A/Qsa3A* had negative effects of −0.1525 and −0.1516. The general contribution of three pairs of epistatic QTLs was 164.95 %.

Among them, seven QTLs (*Qsv-1B/Qsv-3A*, *Qsv-2D/Qsv-3A*, *Qsv-3A/Qsv-5B1*, *Qsv-1B/Qsv-6D*, *Qsv-2D/Qsv-4D*, *Qsv-4A/Qsv-6B*, and *Qsv-3A/Qsv-7D*) explained 13.88, 20.39, 18.88, 12.31, 18.78, 11.98, and 17.05 % of the phenotypic variance, respectively (Table 5.86). All of them were major QTLs and could be used in the molecular marker-assisted selection (MAS) in wheat breeding programs.

The *QTL* (*Qsv-1B*) was detected in both environment 1 and environment 3, which explained 13.88 and 4.83 % of the phenotypic variance.

In summary, a total of five QTLs and fourteen pairs of QTLs with epistatic effects were detected for specific volume in 168 DH lines derived from a cross Huapei 3 × Yumai 57. Five putative QTLs for CNSB specific volume were detected on 5 chromosomes with explain from 5.11 % to 9.75 % of the phenotypic variation. All of them had negative effects on specific volume and were contributed by Yumai 57 alleles. *Qsv-1B* was detected in both environment 1 and 3 with 13.88 and 4.83 % of the phenotypic variation. Fourteen pairs of QTLs with epistatic effects were detected for specific volume. Seven major QTLs, *Qsv-1B/Qsv-3A*, *Qsv-2D/Qsv-3A*, *Qsv-3A/Qsv-5B1*, *Qsv-1B/Qsv-6D*, *Qsv-4A/Qsv-6B*, and *Qsv-3A/Qsv-7D* could account for 13.88, 20.39, 18.88, 12.31, 18.78, 11.98 and 17.05 % of the phenotypic variation of specific volume. The information obtained in this study will be useful for manipulating the QTLs for CNSB specific volume property by molecular marker-assisted selection (MAS).

5.5.5 QTL Mapping of Steamed Bread Color

5.5.5.1 Materials and Methods

5.5.5.1.1 Plant Materials and Growing Conditions

Plant materials and growing conditions were same as ones of Sect. 5.5.3.1.1 in this chapter.

Table 5.86 Estimated epistasis (AA) of QTLs for steamed bread specific volume in the three environments

Trait	QTL1	Site 1 (cM)	Flanking marker	QTL2	Site 2 (cM)	Flanking marker	Epistatic effects	LOD	Contributions (%)
Specific volume 1	*Qsv-1B*	34	*Xbarc119–Xgwm18*	*Qsv-3A*	26	*Xbarc321–Xswes107*	0.07	3.13	13.88
Specific volume 1	*Qsv-2D*	26	*Xcfd53–Xwmc18*	*Qsv-3A*	16	*Xbarc321–Xswes107*	0.08	4.08	20.39
Specific volume 1	*Qsv-3A*	140	*Xwmc264–Xcfa2193*	*Qsv-5B1*	42	*Xgwm213–Xswes861.2*	−0.08	3.58	18.88
Specific volume 1	*Qsv-6D*	0	*Xwmc412.1–Xcfd49*	*Qsv-7A*	86	*Xwmc530–Xcfa2123*	0.05	3.39	8.58
Specific volume 2	*Qsv-1A*	48	*Xwmc550–Xbarc269*	*Qsv-3A*	174	*Xcfa2170–Xbarc51*	0.06	3.41	7.83
Specific volume 2	*Qsv-1B*	76	*Xbarc061–Xwmc766*	*Qsv-6D*	144	*Xswes679.1–Xcfa2129*	0.08	3.14	12.31
Specific volume 2	*Qsv-2D*	130	*Xbarc129.2–Xcfd50*	*Qsv-4D*	28	*Xwmc331–Xgwm194*	0.1	3.15	18.78
Specific volume 2	*Qsv-4A*	2	*Xwmc718–Xwmc262*	*Qsv-6B*	26	*Xbarc247–Xbarc1129*	−0.09	3.03	11.98
Specific volume 2	*Qsv-6A*	84	*Xcfe179.2–Xcfe179.1*	*Qsv-6B*	8	*Xcfa2187–Xgwm219*	−0.07	3.84	8.54
Specific volume 3	*Qsv-1A*	86	*Xgwm498–Xcwem6.2*	*Qsv-5A*	30	*Xswes45–Xbarc180*	−0.05	7.48	7.37
Specific volume 3	*Qsv-1B*	38	*Xcwem9–Xbarc120.3*	*Qsv-6D*	108	*Xgwm133.2–Xswes861.1*	0.07	3.42	4.83
Specific volume 3	*Qsv-3A*	0	*Xbarc310–Xbarc321*	*Qsv-7D*	78	*Xbarc352–Xgwm295*	−0.08	15.89	17.05
Specific volume 3	*Qsv-3D*	0	*Xcfd34–Xbarc376*	*Qsv-7A*	90	*Xwmc530–Xcfa2123*	−0.04	5.18	5.14
Specific volume 3	*Qsv-5B2*	2	*Xbarc36–Xbarc140*	*Qsv-7A*	22	*Xgwm60–Xbarc070*	−0.05	7.56	9.39

(1) 2007 Tai'an; (2) 2008 Tai'an; (3) 2007 Suzhou

5.5.5.1.2 Methods of Steamed Bread Color Testing

Steamed bread was placed in a covered bamboo container at room temperature immediately after steaming. Samples were carried out using a Minolta Color Meter 310 (Minolta Camera Co, Ltd, Japan) on the basis of L*, a*, and b* values (CIE, 1976). Inside color L* value indicates the lightness, 0–100 representing darkness to lightness. The a* value gives the degree of the red–green color, with a higher positive a* value indicating more red. The b* value indicates the degree of the yellow–blue color, with a higher positive b* value indicating more yellow (Hutchings, 1994). Colorimeter scores were taken three times per sample. The mean values of CNSB color in each environment were used for statistical analyses.

All determinations were made at least three times and were expressed on a 14 % moisture basis.

5.5.5.2 Results and Analysis

5.5.5.2.1 Statistical Analysis of the Phenotypic Assessments

A large variation for CNSB color was seen in the DH population, and the mean values under three environments are shown in Table 5.87. The means of CNSB surface and inside color of Hp3 were lower than those of Ym57. Transgressive segregants were observed in all the environments, with some lines higher or lower than the parents. All of the six color traits among the DH population segregated continuously and followed a normal distribution (Fig. 5.28), and both absolute values of skewness and kurtosis were less than 1.0 (Table 5.87), indicating that they were polygenic inheritance and suitable for QTL analysis (Cao et al. 2001).

The correlations among the CNSB color trait values are shown in Table 5.88. The most significant correlations were detected between inside color a* and surface color a* ($r = 0.89$). The lower ones were between inside color b* and surface color b* ($r = 0.62$) which indicated the similar correlation between surface color and inside color in yellow–blue color. Significantly negative correlations were seen between surface color a* and inside color L* ($r = -0.23$), which indicated that the inside color was lower than the surface color.

5.5.5.2.2 QTL Mapping of Steamed Bread Color

*5.5.5.2.2.1 QTLs for Surface Color L**

Only one additive QTL was identified on chromosome 3A for surface color L* (Table 5.89; Fig. 5.29), explaining 6.2614 % of the phenotypic variance. QTL increased the surface color L* by 0.686 whose additive effect was from Hp3.

No epistatic effects were identified for surface color L*.

5.5.5.2.2.2 QTLs for Surface Color a*

Four additive QTLs were detected for surface color a*, which were located on four chromosomes 4A, 6A, 7B1, and 2B. They increased surface color a* from 0.08 to 0.11, explaining phenotypic variance from 5.01 to 8.89 %. *Qsa7B1* with the highest PVE explained 8.89 % of the phenotypic variance. Two QTLs (*Qsa7B1* and *Qsa 2B*) had negative effects on surface color a* and were contributed by Yumai 57 alleles, while the other two QTLs (*Qsa 4A* and *Qsa 6A*) had positive effects on surface color a* and from Huapei 3 alleles. These indicated that the alleles that increased surface color a* were dispersed within the two parents. All four QTLs accounted for 27.53 % of the phenotypic variance (Table 5.89; Fig. 5.29).

Three pairs of epistatic effects were identified for surface color a* on chromosomes 1B-3D-1, 1B-3D-2, and 2A-3A. They explained the phenotypic variance ranging from 1.81 % to 2.58 %. *Qsa1B/Qsa3D-2* and *Qsa2A/Qsa3A* had negative effects of −0.15 and −0.15. The general contribution of three pairs of epistatic QTLs was 4.39 %. All of them were involved in AAE interactions, accounting for 14.53 % of the phenotypic variance (Table 5.90). This suggested that the alleles, which increased surface color a*, were dispersed within the two parents.

5.5.5.2.2.3 QTLs for Surface Color b*

Five QTLs on 5 different chromosomes (*Qsb3A*, *Qsb6B*, *Qsb6A*, *Qsb6D*, and *Qsb7A*) were detected for surface color b*, explaining 4.55 to 9.41 % of the phenotypic variance. The most important one Qsb7A was on chromosome 7A close to the locus *Xgwm60–Xbarc070,* which explained 9.4069 % of the phenotypic variance. The Hp3 alleles at two loci, *Xbarc321–Xswes107* and *Xbarc247–Xbarc1129,* increased the surface color b* by 0.40 and 0.48 owing to additive effects. The Ym57 allele increased the surface color b* at the *Qsb6A Xcfe179.1–Xswes170.2*, *Qsb6D Xswes861.1–Xgwm681*, and *Qsb7A Xgwm60–Xbarc070*,

Table 5.87 Phenotypic performance of wheat CNSB color under three environments

Trait	Parents		DH population						
	Huapei 3	Yumai 57	Mean	SE	Max.	Min.	SD	Skew.	Kurt.
Surface color L*	85.16333	86.43	84.1704	0.20	86.93	55.28	2.57	−0.582	0.568
Surface color a*	−1.62	−2.03	−1.4541	0.04	0	−2.95	0.47	−0.317	0.685
Surface color b*	16.42667	15.62667	16.3352	0.08	19.67	11.14	1.07	−0.171	0.184
Inside color L*	81.14	81.22667	78.979	0.19	81.78	52.12	2.43	−0.213	0.14
Inside color a*	−1.37333	−1.87333	−1.2454	0.04	0.6	−2.62	0.48	0.354	0.31
Inside color b*	15.88333	15.86	15.3665	0.13	34.59	10.31	1.76	0.61	0.513

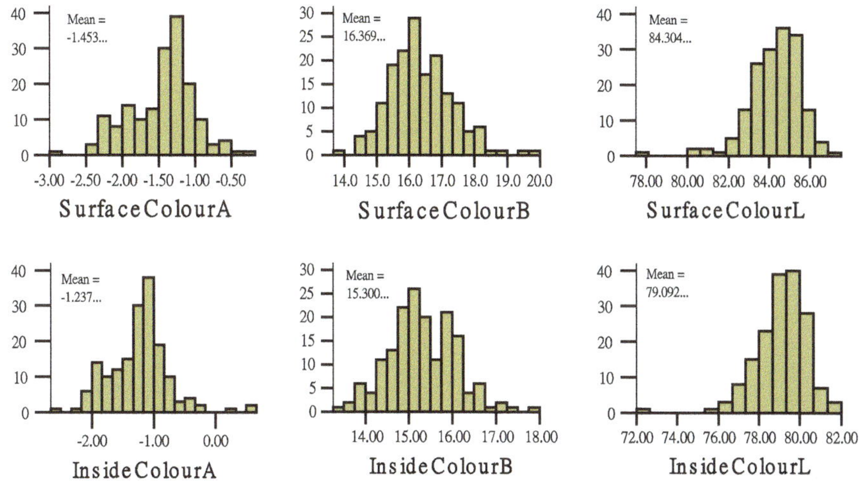

Fig. 5.28 Frequency distribution of steamed bread color-related traits in 168 doubled haploid lines derived from the cross of Huapei 3 × Yumai 57 evaluated at 3 environments

Table 5.88 Coefficients of pairwise correlations of the mean values among steamed bread's textural properties traits in the three environments

Trait	Surface color L*	Surface color a*	Surface color b*	Inside color L*	Inside color a*	Inside color b*
Surface color a*	0.35**					
Surface color b*	0.37**	0.03				
Inside color L*	0.30**	−0.23**	0.09			
Inside color a*	0.32**	0.89**	0.03	−0.23**		
Inside color b*	0.03	−0.06	0.62**	0.07	−0.06	
Scores	0.40**	−0.24**	−0.31**	0.1	−0.2*	−0.19*

*$P < 0.005$ and **$P < 0.001$, respectively

accounting for 4.55, 5.02, and 9.41 % of the phenotypic variance, respectively. The total additive QTLs detected for hardness accounted for 29.59 % of the phenotypic variance.

*5.5.2.2.2.4 QTLs for Inside Color L**

Four additive QTLs were detected for inside color L* on four chromosomes 3A, 4A, 6D, and 7D. They explained phenotypic variance. The highest PVE of *QiL7D* was 7.61 %. Two QTLs (*QiL4A Xbarc362–Xbarc* and *QiL 6D Xbarc054–Xgwm55*)

Table 5.89 Estimated additive and additive × environment interaction of QTLs for steamed bread's textural properties based on averaged phenotypic data from three environments

Trait	Environment	Chr.	Site (cM)	Left marker	Right marker	Additive effects[a]	LOD	Contributions[b] (%)
Surface color L*	3	3A	160	*Xgwm155*	*Xcfa2170*	0.686	2.6489	6.2614
Surface color a*	2	4A	39	*Xbarc362*	*Xbarc078*	0.0814	2.0462	5.0139
Surface color a*	2	6A	46	*Xgwm82*	*Xwmc553*	0.1035	2.8874	8.3724
Surface color a*	2	7B1	48	*Xgwm333*	*Xwmc10*	−0.1123	3.0236	8.8887
Surface color a*	3	2B	90	*Xwmc445.2*	*Xgwm111*	−0.0853	2.0059	5.2591
Surface color b*	1	3A	43	*Xbarc321*	*Xswes107*	0.3971	2.2533	5.1782
Surface color b*	1	6B	13	*Xbarc247*	*Xbarc1129*	0.4829	2.2686	5.4404
Surface color b*	3	6A	117	*Xcfe179.1*	*Xswes170.2*	−0.2958	2.0131	4.5464
Surface color b*	3	6D	111	*Xswes861.1*	*Xgwm681*	−0.6921	2.1663	5.0229
Surface color b*	3	7A	23	*Xgwm60*	*Xbarc070*	−0.4238	3.8771	9.4069
Inside color L*	2	3A	93	*Xbarc356*	*Xwmc489.2*	0.4697	2.0912	6.0903
Inside color L*	3	4A	40	*Xbarc362*	*Xbarc078*	−0.5641	2.6807	6.4456
Inside color L*	3	6D	76	*Xbarc054*	*Xgwm55*	−0.5638	2.595	6.4391
Inside color L*	3	7D	9	*Xswes100*	*Xbarc244*	0.643	2.1571	7.6086
Inside color a*	2	6A	43	*Xgwm82*	*Xwmc553*	0.1397	2.2192	5.2298
Inside color a*	2	6D	112	*Xswes861.1*	*Xgwm681*	2.1266	11.4843	28.9737
Inside color a*	2	7B1	41	*Xwmc396*	*Xgwm333*	−0.1749	3.0987	7.3677
Inside color a*	3	1B	42	*Xgwm218*	*Xgwm582*	−0.0983	2.5034	6.6931
Inside color a*	3	2B	90	*Xwmc445.2*	*Xgwm111*	−0.1093	3.7458	8.5806
Inside color a*	3	3D	90	*Xbarc323*	*Xbarc071*	0.0965	2.8059	6.574
Inside color a*	3	6D	77	*Xbarc054*	*Xgwm55*	0.0895	2.449	5.54
Inside color a*	3	7D	185	*Xwmc14*	*Xwmc42*	0.0938	2.2495	5.0927
Inside color b*	1	1B	37	*Xcfd21*	*Xcwem9*	0.39	4.4845	11.1449

(continued)

Table 5.89 (continued)

Trait	Environment	Chr.	Site (cM)	Left marker	Right marker	Additive effects[a]	LOD	Contributions[b] (%)
Inside color b*	2	1B	35	*Xcwem6.1*	*Xwmc128*	0.2898	3.1515	8.3607
Inside color b*	3	1B	26	*Xwmc406*	*Xbarc156*	0.2489	2.7869	5.4702
Inside color b*	3	2D	1	*Xwmc112*	*Xcfd53*	0.5489	12.5355	26.1864
Inside color b*	3	3A	146	*Xwmc264*	*Xcfa2193*	−0.3231	4.5832	9.2048
Inside color b*	3	6D	92	*Xgwm133.2*	*Xswes861.1*	0.2802	2.9039	5.2032
Inside color b*	3	7D	36	*Xswes100*	*Xbarc244*	−0.3668	4.0124	11.2281
Inside color b*	3	7D	130	*Xwmc630.1*	*Xgdm67*	−0.3774	5.6375	11.8396

[a]Additive effects: A positive value indicates that the Huapei 3 allele has a positive effect on the adhesiveness, and a negative value indicates that allele from Ym57 has a negative effect on the adhesiveness

[b]Contribution explained by putative main-effect QTL

Environment 1: 2008 in Tai'an; 2: 2007 in Tai'an; 3: 2007 in Suzhou

L*, a*, b* name of traits

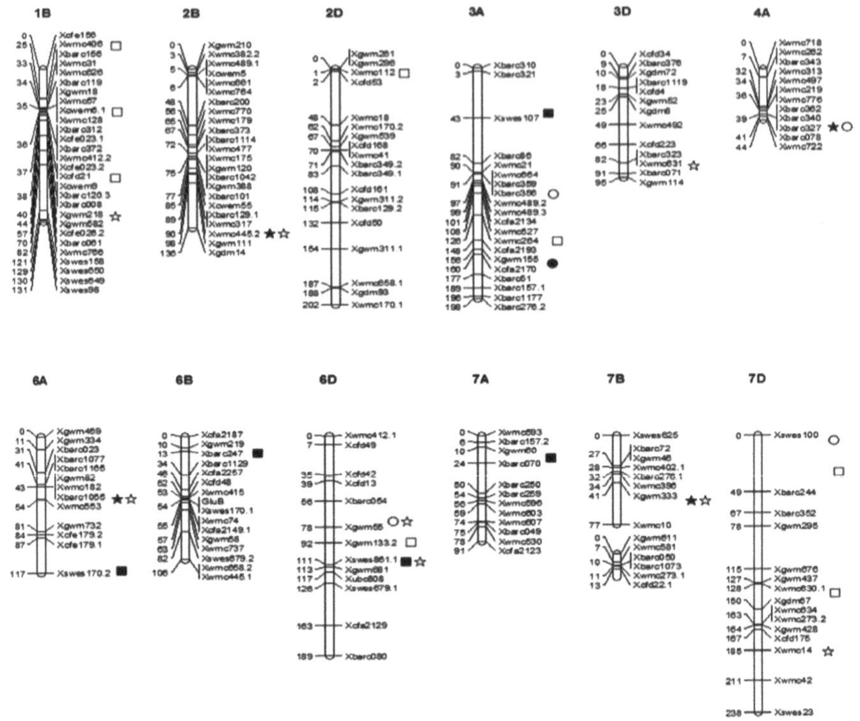

Fig. 5.29 Genetic linkage map of wheat showing mapping QTLs with additive effects for CNSB color. ●, ★, and ■ indicate surface color L* a* b*, respectively; ○, ☆, and □ indicate inside color L* a* b*, respectively

had negative effects on inside color L* from Yumai 57 alleles, while the other two QTLs (*QiL 3A Xbarc356–Xwmc489.2* and *QiL7D Xswes100–Xbarc244*) had positive effects on inside color L* and was transmitted by Huapei 3 alleles. These suggested that the alleles increasing inside color L* were dispersed within the two parents. All four QTLs could account for 26.58 % of the phenotypic variance (Table 5.89; Fig. 5.29).

*5.5.5.2.2.5 QTLs for Inside Color a**

Eight additive QTLs were detected for inside color a* on seven chromosomes *Qia6A*, *Qia6D*, *Qia7B1*, *Qia1B*, *Qia2B*, *Qia3D*, and *Qia7D*. They increased inside color a* from 0.09 to 2.17, explaining phenotypic variance from 5.09 % to 28.97 %. *Qia 6D* made the highest contribution and explained 28.97 % of the phenotypic variance. Three QTLs (*Qia7B1*, *Qia1B*, and *Qia2B*) had negative effects on inside color a* and were contributed by Yumai 57 alleles, while the other five QTLs (*Qia6A*, *Qia6D*, *Qia3D*, *Qia6D,* and *Qia7D*) had positive effects on adhesiveness and were transmitted by Huapei 3 alleles. The total additive QTL detected for inside color a* accounted for 74.0516 % of the phenotypic variance.

Table 5.90 Estimated epistasis (AA) and epistasis × environment interactions of QTLs for steamed bread's textural properties based on averaged phenotypic data from three environments

Trait	QTL	Flanking marker	Site (cM)	QTL	Flanking marker	Site (cM)	AA[a]	H^2 (AA%)[b]	AAE1	H^2 (AAE1 %)	AAE2	h^2 (AAE2 %)	AAE3	H^2 (AAE3 %)
Surface color a*	Qsa1B-1	XWMC31-XWMC626	33	Qsc3D-1	XCFD34-XBARC376	3			−0.1708**	3.58				
	Qsa1B-2	XGWM218-XGWM582	39.7	Qsc3D-2	XGDM8-XWMC492	36.4	−0.1525**	2.58	−0.1906**	2.7				
	Qsa2A	XWMC455-XGWM515	88.1	Qsc3A	XBARC310-XBARC321	0	−0.1516**	1.81	−0.3403**	5.47	0.1637**	1.36	0.1646**	1.42

[a]Epistatic effects: A positive value indicates that the parental two-locus genotypes have a positive effect and that the recombinants have a negative effect

[b]Contribution explained by epistatic QTL

E1: 2008 in Tai'an; E2: 2007 in Tai'an; E3: 2007 in Suzhou

$*P < 0.005$ and $**P < 0.001$, respectively

*5.5.5.2.2.6 QTLs for Inside Color b**

Eight QTLs were detected for inside color b* on 5 chromosomes *Qib1B*, *Qib2D*, *Qib3A*, *Qib6D*, and *Qib7D* (Table 5.89; Fig. 5.29). All the eight QTLs were identified with significant additive effects. The part of the phenotypic variance explained by these QTLs ranged from 5.2 % to 26.19 %. *Qib1B* was detected in all three environments on 1B with the flanking markers *Xcfd21–Xcwem9*, *Xcwem6.1–Xwmc128*, and *Xwmc406–Xbarc156* for inside color b*, respectively. The *Qib2D* explained up to 26.19 % of the variation, and the positive allele of this QTL was from Hp3. For the *Qib1B*, *Qib2D*, and *Qib6D*, the favorable allele also came from Hp3. The Ym57 allele increased the inside color b* at the *Qib7D* and *Qib3A*, accounting for 23.07 % and 9.20 % of the phenotypic variance. The total additive QTL detected for inside color b* accounted for 88.64 % of the phenotypic variance.

In general, thirty QTLs with additive effects were detected for six traits in the three environments (Table 5.89; Fig. 5.29), ranging from one to eight QTLs for each trait, and were distributed on 12 of the 21 chromosomes.

Three pairs of QTLs with epistatic effects and/or epistasis × environment (AAE) effects were detected for surface color a* in the three environments (Table 5.90). No epistatic effects were found in other color traits for samples across the three environments. Two major QTLs, *Qia6D* and *Qib2D*, closely linked to *Xswes 861.1* and *Xwmc112*, could account for 28.9737 and 26.1864 % of the phenotypic variation of inside color a* and inside color b*, respectively. So the *Qia6D* and *Qib 2D* could be used in the molecular marker-assisted selection (MAS) in wheat breeding programs.

5.5.5.3 Progress in QTLs for Steamed Bread Quality and Comparison with Previous Studies

5.5.2.4.1 Progress in QTLs for Steamed Bread Quality

Fan et al. (2009) detected eight QTLs for TPA parameters, explaining 9.88 to 64.87 % of the phenotypic variance. They were located on A, B, and D genomes (Table 5.91). Li et al. (2009) identified ten QTLs for flour absorption accounting for 3.1–38.7 % of the phenotypic variance on A, B, and D genomes.

For the specific volume of steamed bread, Fan et al. (2009) found the QTL on 6B chromosome with the flanking marker Xgwm193-Xgwm608b, which was different from our results. This perhaps was caused by different populations, growing conditions, and different analysis methods. But till now, there was no report on the QTLs for the steamed bread color.

Table 5.91 Summary of QTL results of wheat steamed bread quality

Traits	QTL	Flanking marker	PVE/%	Mapping population	References
Volume (ml)	QVol.sdau-6B	Xgwm193–Xgwm608b	11.73	RIL	Fan et al. (2009)
Specific volume	QSv.sdau-6B	Xgwm193–Xgwm608b	11.19	Same as above	Same as above
Appearance	QApp.sdau-2D	Xgwm29b–Xgwm132a	40.34	Same as above	Same as above
Color	QCol.sdau-5D	Xswes342b–Xsrap6b	9.88	Same as above	Same as above
Elasticity	QEla.sdau-7B	Xgwm333–Xgwm297	11.13	Same as above	Same as above
Stickiness	QSti.sdau-2B	Xsrap1a–Xgwm120	32.46	Same as above	Same as above
Smell	QSme.sdau-4A	Xwmc308–Xsrap7c	64.87	Same as above	Same as above
Total score	QTs.sdau-5B	Xgwm261a–Xgwm234	29.8	Same as above	Same as above

5.5.2.4.2 Comparison with Previous Studies

The QTLs for protein content, mixograph parameters, bread volume, and steamed bread adhesiveness have been reported on 2B chromosome in the previous researches (Campbell et al. 2001; Fan et al. 2009). But in our study, there were two QTL clusters for TPA parameters on 2B chromosome, of which there was one QTL cluster located between Xgwm111 and Xgdm14-6D having 6 QTLs controlling chewiness, gumminess, and hardness, respectively. Between Xgwm210 and Xwmc382.2, there was one cluster containing three QTLs, which controlled adhesiveness, cohesiveness, and resilience, respectively. So the 2B chromosome is important for steamed bread quality.

In our study, thirty additive QTLs and three epistatic QTLs for steamed bread color were detected on 11 chromosomes (3A, 4A, 6A, 7A, 1B, 2B, 7B1, 2D, 3D, 6D, and 7D).

Two major QTLs, *Qia6D* and *Qib2D*, could account for 28.9737 and 26.1864 % of the phenotypic variation of inside color a* and inside color b*, which could be used in MAS in wheat breeding programs. Three epistatic QTLs for surface color a* were located on 1B and 2A chromosomes, but no other epistatic QTLs for other color traits in the three environments were detected.

Five additive QTLs and fourteen epistatic QTLs were identified in our study. Of which, five additive QTLs had the larger PVE (5.11–9.75 %) with the same genetic effect from Ym57. The epistatic *Qsv-1B* for specific volume was identified in E1 and E3 with explaining 13.88 and 4.83 % of the phenotypic variance, respectively, which indicated that this locus is important for the specific volume of steamed bread.

The results showed that both additive and epistatic effects were important in the genetic basis of CNSB color and were also sometimes subject to environmental modifications.

References

Abdel-Aal E-SM, Huclw P. Amino acid composition and in vitro protein digestibility of selected ancient wheats and their end products. J Food Compos Anal. 2002;15:737–747.

Avivi L. High grain protein content in wild tetraploid wheat Triticum dicoccoides Korn. In: Proceedings of 5th international wheat genetics symposium, New Delhi. 1978, vol. 1, pp. 372–380.

Bao JS, Sun M, Corke H. Analysis of genetic behavior of some starch properties in indicia rice (Oryza sativa L.): thermal properties, gel texture, swelling volume. Theor Appl Genet. 2002;104:408–13.

Batey IL, Hayden MJ, Cai S, Sharp PJ, Cornish GB, Morell MK, Appels R. Genetic mapping of commercially significant starch characteristics in wheat crosses. Aust J Agric Res. 2001;52:1287–96.

Baxtera G, Blancharda C, Zhao J. Effects of prolamin on the textural and pasting properties of rice flour and starch. J Cereal Sci. 2004;40:205–11.

Blanco A, Smeone R, Gadaleta A. Detection of QTLs for grain protein content in durum wheat. Theor Appl Genet. 2006;112:1195–204.

Campbell BT, Baenziger PS, Gill KS, Eskridge KM, Budak H, Erayman M, Dweikat I, Yen Y. Identification of QTLs and environmental interactions associated with agronomic traits on chromosome 3A of wheat. Crop Sci. 2003;43:1493–505.

Campbell KG, Bergman CJ, Gualberto DG, Anderson JA, Giroux MJ, Hareland G, Fulcher RG, Sorrels ME, Finney PL. Quantitative trait loci associated with kernel traits in a soft × hard wheat cross. Crop Sci. 1999;39:1184–95.

Campbell KG, Finney L, Bergman CJ, Gualberto DG, Anderson JA, Giroux M, Siritunga D, Zhu J, Gendre F, Roue C, Vérel A, Sorrells ME. Quantitative trait loci associated with milling and baking quality in a soft 9 hard wheat cross. Crop Sci. 2001;41:1275–85.

Cao G, Zhu J, He C, et al. Impact of epistasis and QTL × environment interaction on the developmental behavior of plant height in rice (*Oryza sativa* L.). Theor Appl Genet. 2001;103 (1):153–60.

Cavanagh CR, Taylor JL, Larroque O, Coombes N, Verbyla AP, Nath Z, Kutty I, Rampling L, Butow B, Ral J-P, Tomoskozi S, Balazs G, Békés F, Mann G, Quail KJ, Southan M, Morell MK, Newberry M. Sponge and dough bread making: genetic and phenotypic relationships with wheat quality traits. Theor Appl Genet. 2010;121:815–28.

Charmet G, Robert N, Branlard G, et al. Genetic analysis of dry matter and nitrogen accumulation and protein composition in wheat kernels. Theor Appl Genet. 2005;111(3):540–50.

Chen F, Li GY, Geng HW, Xia LQ, Xia XC, He ZH. Review and prospect of wheat kernel hardness and its molecular genetics basis. Sci Agric Sin. 2005;38(6):1088–94 (in Chinese with English abstract).

Crawford AC, Shaw K, Stefanova K, et al. A molecular toolbox for xanthophyll genes in wheat. http://ses.library.usyd.edu.au/bitstream.

Demeke T, Morris CF, Campbell KG, et al. Wheat polyphenol oxidase. Crop Sci. 2001;41 (6):1750–7.

Deng ZY, Tian JC, Zhao L, Zhang YX, Sun CL. High temperature-induced changes in high molecular weight glutenin subunits of Chinese winter wheat and its influences on the texture of Chinese noodles. J Agron Crop Sci. 2008;194:262–9.

Distelfeld A, Uauy C, Olmos S, Schlatter AR, Dubcovsky J, Fahima T. Microcolinearity between a 2-cM region encompassing the grain protein content locus *Gpc-6B1* on wheat chromosome 6B and a 350-kb region on rice chromosome 2. Funct Integr Genomics. 2004;4:59–66.

Eberly LE. Methods in molecular biology, topics in biostatistics. Humana Press, Totowa, 2007; pp 165–87.

Edwards RH, Berrios JDJ, Mossman AP, Takeoka GR, Wood DF, Mackey BE. Texture of jet cooked, high amylose corn starch-sucrose gels. Lebensmittel-Wissenschaft Und-Technologie. 1998;31:432–8.

Elangovan M, Rai R, Oak M, Dholakia BB, Lagu MD, Tiwari R, Gupta RK, Tamhankar S, Röder MS, Gupta VS. Revealing the genetic relationship of dough mechanical properties with loaf volume using QTL analysis of mixograph traits in wheat. J Cereal Sci. 2008;47:587–98.

Elouafi I, Nachit MM, Martin LM. Identification of a microsatellite on chromosome 7B showing a strong linkage with yellow pigment in durum wheat (*Triticum turgidum* L. var. durum). Hereditas. 2001;135:255–61.

Epstein J, Morris CF, Huber KC. Instrumental texture of white salted noodles prepared from recombinant inbred lines of wheat differing in the three granule bound starch synthase waxy genes. J Cereal Sci. 2002;35:51–63.

Fan YD, Sun HY, Zhao JL, Ma YM, Li RJ, Li SS. QTL mapping for quality traits of northern-style hand-made Chinese steamed bread. J Cereal. 2009;49:225–9.

Finney KF, Barmore MA. Loaf volume and protein content of hard winter and spring wheats cereal. Chemistry. 1948;25(5):291–312.

Francki M, Carter M, Ryan K, et al. Comparative organization of wheat homoeologous group 3S and 7L using wheat-rice synteny and identification of potential markers for genes controlling xanthophyll content in wheat. Funct Integr Genomics. 2004;4(2):118–30.

Gonzalez-Hernandez JL, EliasEM, KianianSF. Mapping genes for grain protein concentration and grain yield on chromosome 5B of *Triticum turgidum*(L.) var. dicoccoides. Euphytica. 2004;139:217–25.

Grama A, Gerechter-Amitai ZK, Blum A, et al. Breeding bread wheat cultivars for high protein content by transfer of protein genes from Triticum dicoccoides. Cereal grain protein improvement. 1984, pp. 145–153.

Groos C, Bervas E, Charmet G. Genetic analysis of grain protein content, grain hardness and dough rheology in a hard × hard bread wheat progeny. J Cereal Sci. 2004;40:93–100.

Groos C, Robert N, Bervas E, Charmet G. Genetic analysis of grain protein-content, grain yield and thousand-kernel weight in bread wheat. Theor Appl Genet. 2003;106:1032–40.

Groos C, Bervas E, Chanliaud E, Charmet G. Genetic analysis of breadmaking quality scores in bread wheat using a recombinant inbred line population. Theor Appl Genet. 2007;115:313–23.

Guo CQ, Bai ZA, LiaoPA, Jin WK. New high quality and yield wheat variety Yumai 57. China Seed Ind. 2004; 4:54.

Hai Y, Kang MH. Breeding of a new wheat variety Huapei 3 with high yield and early maturing. Henan Agri Sci. 2007; 5:36–37.

He XY, He ZH, Zhang LP, et al. Allelic variation of polyphenol oxidase (PPO) genes located on chromosomes 2A and 2D and development of functional markers for the PPO genes in common wheat. Theor Appl Genet. 2007;115(1):47–58.

He ZH, Liu AH, Peña RJ, Rajaram S. Suitability of Chinese wheat cultivars for production of northern style Chinese steamed bread. Euphytica. 2003;131:155–63.

He ZH, Yang J, Zhang Y, Quail KJ, Peña RJ. Pan bread and dry white Chinese noodle quality in Chinese winter wheats. Euphytica. 2004;139:257–67.

Hristov N, Mladenov N, Djuric V, Kondic-Spika A, Marjanovic-Jeromela A, Simic D. Genotype by environment interactions in wheat quality breeding programs in southeast Europe. Euphytica. 2010;174:315–24.

Hu R, Tian J. Relationship between main quality characteristics and wheat flour color. J Triticeae Crops. 2006;26(3):96–101.

Hua JP, Xing YZ, Wu WR, Xu CG, Sun XL, Yu SB, Zhang QF. Single-locus heterotic effects and dominance by dominance interaction can adequately explain the genetic basis of heterosis in an elite hybrid. Proc Natl Acad Sci USA, 2003;100:2574–9.

Huang XQ, Cloutier S, Lycar L, Radovanovic N, Humphreys DG, Noll JS, Somers DJ, Brown PD. Molecular detection of QTL for agronomic and quality traits in a doubled haploid population derived from two Canadian wheats (*Triticum aestivum* L.). Theor Appl Genet. 2006;113:753–66.

Huang XQ, Kempf H, Ganal MW, Röder MS. Advanced backcross QTL analysis in progenies derived from a cross between a German elite winter wheat variety and a synthetic wheat (*Triticum aestivum* L.). Theor Appl Genet. 2004;109:933–43.

Jiang X, Hao Z, Tian J. Variations in amino acid and protein contents of wheat during milling and northern-styled breadmaking. Cereal Chem. 2008;85(4):502–6.

Kang ZY. Relationship between score of hand-extended noodle and quantity and quality of gluten. J Triticeae Crops. 2003;23(2):3–6 (in Chinese with English abstract).

Kang ZY, Wang JJ, Shang XW. Relationship between Score of hand-extended noodle and alveogram. Acta Tritical Crops. 2005;25(3):81–84.

Kealy T. Application of liquid and solid rheological technologies to the textural characterization of semi-solid foods. Food Research International. 2006;39(3):265–76.

Khlestkina EK, Giura A, Roder MS, Borner A. A new gene controlling the flowering response to photoperiod in wheat. Euphytica. 2009;165:578–85.

Knežević D, Šurlan-Momirović G, Ćirić D. Allelic variation at Glu-1 loci in some Yugoslav wheat cultivars. Euphytica. 1993;69:89–94.

Knott DR. The genetic nature of mutations of a gene for yellow pigment linked to Lr19 in 'Agatha' wheat. Can J Genet Cytol. 1984;26:392–3.

Kruger JE, Anderson MH, Dexter JE. Effect of flour refinement on raw cantonese noodle color and texture. Cereal Chem. 1994;71:177–182.

Kruger JE, Morgan B, Preston KR, MaltsuoR R. Evaluation of some characteristics of Chinese steamed buns prepared from Canadian wheat flours. Can J Plant Sci. 1992;72:369–75.

Kuchel H, Langridge P, Mosionek L, Williams K, Jefferies SP. The genetic control of milling yield, dough rheology and baking quality of wheat. Theor Appl Genet. 2006;112:1487–95.

Kulwal P, Kumar N, Kumar A, Balyan HS, Gupta PK. Gene networks in hexaploid wheat: interacting quantitative trait loci for grain protein content. Funct Integr Genomics. 2005;5:254–9.

Kunert A, Naz AA, Dedeck O, Pillen K, Léon J. AB-QTL analysis in winter wheat: I. Synthetic hexaploid wheat (*T. turgidum* ssp. dicoccoides × *T. tauschii*) as a source of favourable alleles for milling and baking quality traits. Theor Appl Genet. 2007;115:683–95.

Kunerth WH, D'Appolonia BL. Use of the mixograph and farinograph in wheat quality evaluation. In: Hamed F, editor, Rheology of Wheat Products. USA:American Association of Cereal Chemists; 1985. pp. 27–49.

Lau MH, Tang J, Paulson AT. Texture profile and turbidity of gellan/gelatin mixed gels. Food Res Int. 2000;33:665–71.

Li S, Jia J, Wei X, et al. A intervarietal genetic map and QTL analysis for yield traits in wheat. Mol Breeding. 2007a;20(2):167–78.

Li Y, Song Y, Zhou R, Branlard G, Jia J. Detection of QTLs for bread-making quality in wheat using a recombinant inbred line population. Plant Breeding. 2009;128:235–43.

Li SP, Wang CS, Chang XP, Jing RL. Genetic dissection of developmental behavior of grain weightin wheat under diverse temperature and water regimes.Genetica. 2012;140:393–405.

Li H, Ye G, Wang J. A modified algorithem for the improvement of composite interval mapping. Genetics. 2007b;175:361–74.

Li X, Yang W, Li Y, Liu D, Yan H, Meng Q, Zhang T. A SSR marker for leaf rust resistance gene Lr19 in wheat. Sci Agric Sini. 2005;38:1156–9.

Li WH, Zhang DH. The balance analysis of the amino acid content in seed filling period of wheat. 2000, vol. 2, pp. 21–23 (in Chinese)

Lin ZJ, Miskelly DM, Moss JH. Suitability of various Australian wheats for Chinese-style steamed bread. J Sci Food Agric. 1990;53:203–13.

Liu JJ, He ZH, Zhao ZD, Peña RJ, Rajaram S. Wheat quality traits and quality parameters of cooked dry white Chinese noodles. Euphytica. 2003;131:147–154.

Liu YP, Quan SY, Li XP, Lan SQ, Liu YH, Li JP. Protein content and amino acid composition and qualities of different blue and purple grain wheat (in Chinese). Acta Agric Boreali-Sin. 2002;17:103–7.

Liu GF, Yang J, Xu HM, Zhu J. Influence of epistasis and QTL 9 environment interaction on heading date of rice (*Oryza sativa* L.). J Genet Genomics. 2007;34:608–15.

Ma W, Appels R, Bekes F, Larroque Q, Morell MK, Gale KR. Genetic characterization of dough rheological properties in a wheat doubled haploid population: additive genetic effects and epistatic interactions. Theor Appl Genet. 2005;111(3):410–422.

Ma XQ, Tang JH, Teng WT, Yan JB, Meng YJ, Li JS. Epistatic interaction is an important genetic basis of grain yield and its components in maize. Mol Breed. 2007;20:41–51.

MacRitchie F. Wheat proteins: characterization and role in flour functionality. Cereal Foods World. 1999;44(4):188–93.

Marais GF. Gamma irradiation induced deletions in an alien chromosome segment of the wheat 'Indis' and their use in gene mapping. Genome. 1992;35:225–9.

Marchylo BA, Dexter JE, Clarke FR, et al. Relationships among bread-making quality, gluten strength, physical dough properties, and pasta cooking quality for some Canadian durum wheat genotypes. Can J Plant Sci. 2001;81(4):611–20.

Mares DJ, Campbell AW. Mapping components of flour and noodle colour in Australian wheat. Aus J Agric Res. 2001;52:1297–309.

Mccartney CA, Somers DJ, Lukow O, Ames N, Noll J, Cloutier S, Humphreys DG, Mccallum BD. QTL analysis of quality in traits in the spring wheat wheat cross RL4452 × AC domain. Plant Breeding. 2006;125:565–75.

Mesdag J. Variations in the protein content of wheat and its influence on the sedimentation value and the baking quality. Euphytica. 1964;13:250–61.

Miskelly DM. Flour components affecting paste and noodle color. J Sci Food Agric. 1984;35:463–71.

Morris CF, Rose SP. Quality requirements of cereal users, wheat. In: Henry RJ, Kettlewell PS, editors. Cereal grain quality. London: Chapman and Hall; 1996, pp. 3–54.

Nagao S. Processing technology of noodle products in Japan. In: Kruger JE, Matsuo RB, Dick JW, editors. Pasta and noodle technology. American Association Cereal Chemistry. St. Paul, MN. 1996, pp. 169–194.

Nelson JC, Andreescu C, Breseghello F, Finney PL, Gualberto DG, Bergman CJ, Peña RJ, Perretant MR, Leroy P, Qualset CO, Sorrells ME. Quantitative trait locus analysis of wheat quality traits. Euphytica. 2006;149:145–59.

Otegbayo B, Aina J, Abbey L, Sakyi-Dawson E, Bokanga M, Asiedu R. Texture profile analysis applied to pounded yam. J Texture Stud. 2007;38:355–72.

Özberk I, Kılıç H, Atlı A, Özberk F, Karlı B. Selection of wheat based on economic returns per unit area. Euphytica. 2006;152:235–45.

Ozturk S, Kahraman K, Tiftik B, Koksel H. Predicting the cookie quality of flurs by using Mixolab. Euro Food Res Technol. 2008;227:1549–54.

Panthee DR, Pantalone VR, Sams CE, Saxton AM, West DR, Orf JH, Killam AS. Quantitative trait loci controlling sulfur containing amino acids, methionine and cysteine, in soybean seeds. Theor Appl Genet. 2006a;112:546–53.

Panthee DR, Pantalone VR, Saxton AM, West DR, Sams CE. Genomic regions associated with amino acid composition in soybean. Mol Breeding. 2006b;17:79–89.

Parker GD, Chalmers KJ, Rathjen AJ, Langridge P. Mapping loci associated with flour colour in wheat (*Triticum aestivum* L.). Theor Appl Genet. 1998;97:238–45.

Parker GD, Langridge P. Development of a STS marker linked to a major locus controlling flour color in wheat (*Triticumaestivum* L.). Mol Breed 2000;6:169–74.

Patil RM, Oak MD, Tamhankar SA, et al. Mapping and validation of a major QTL for yellow pigment content on 7AL in durum wheat (*Triticum turgidum* L. ssp. durum). Mol Breeding. 2008; 21(4):485–496.

Peña E, Bernardo A, Soler C, Jouve N. Do tyrosine crosslinks contribute to the formation of the gluten network in common wheat (*Triticum aestivum* L.). J Cereal Sci. 2006;44:144–53.

Perretant MR, Cadalen T, Charmet G, Sourdille P, Nicolas P, Boeuf C, Tixier MH, Branlard G, Bernard S, Bernard M. QTL analysis of bread-making quality in wheat using a doubled haploid population. Theor Appl Genet. 2000;100:1167–75.

Pozniak CJ, Knox RE, Clarke FR, Clarke JM. Identification of QTL and association of a phytoene synthase gene with endosperm colour in durum wheat. Theor Appl Genet. 2007;114:525–37.

Prasad M, Kumar N, Kulwal PL, Röder M, Balyan HS, Dhaliwal HS, Gupta PK. QTL analysis for grain protein content using SSR markers and validation studies using NILs in bread wheat. Theor Appl Genet. 2003;106:659–67.

Prasad M, Varshney RK, Kumar A, Balyan HS, Sharma PC, Edwards KJ, Singh H, Dhaliwal HS, Roy JK, Gupta PK. A microsatellite marker associated with a QTL for grain protein content on chromosome arm 2DL of bread wheat. Theor Appl Genet. 1999;99:341–5.

Rasul G, Humphreys DG, Brûlé-Babel A, McCartney CA, DePauw RM, Knox RE, Somers DJ. Mapping QTLs for pre-harvest sprouting traits in the spring wheat cross 'RL4452/AC Domain'. Euphytica. 2009;168(3):363–78.

Ren JS, Chen J, Wang C. Influence factors of noodle texture outcome tested witht exture analyzer. Food Sci Technol. 2007;5:228–31 (in Chinese with English abstract).

Roncallo PF, Cervigni GL, Jensen C, Miranda R, Carrera AD, Helguera M, Echenique V. QTL analysis of main and epistatic effects for flour color traits in durum wheat. Euphytica. 2012;185 (1):77–92.

Rousset M, Brabant P, Kota RS, Dubcovsky J, Dvorak J. Use of recombinant substitution lines for gene mapping and QTL analysis of bread making quality in wheat. Euphytica. 2001;119:81–7.

Sax K. The association of size differences with seed coat pattern and pigmentation in Phaseolusvulgaris. Genetics 1923;8:552–60.

Snape JW, Quarrie SA, Laurie DA. Comparative QTL mapping and its application in cereal breeding. In: Proceedings of FAO/IAEA international symposium on the use of induced mutations and molecular techniques for crop improvement. 1995, pp. 39–49.

Sourdille P, Perretant MR, Charmet G, Leroy P, Gautier MF, Joudrier P, Nelson JC, Sorrels ME, Bernard M. Linkage between RFLP markers and genes affecting kernel hardness in wheat. Theor Appl Genet. 1996;93:580–6.

Su D, Ding C, Li L, Su D, Zheng X. Effect of endoxylanases on dough properties and making performance of Chinese steamed bread. Eur Food Res Technol. 2005;220:540–5.

Su ZQ, Hao CY, Wang LF, Dong YC, Zhang XY. Identification and development of a functional marker of TaGW2 associated with grain weight in bread wheat (*Triticum aestivum* L.). Theor Appl Genet. 2011;122:211–23.

Sun HY, Lu JH, Fan YD, Zhao Y, Kong F, Li RJ, Wang HG, Li SS. Quantitative trait loci (QTLs) for quality traits related to protein and starch in wheat. Prog Nat Sci. 2008;18:825–31.

Sun XD, Wang LK, Ren HB, Lan J. The application of tristimulus colorimeter in the determination of flour color. Technol Oil Food. 2002;10:31–3.

Sun XC, Marza F, Ma HX, Carver BF, Bai GH. Mapping quantitative trait loci for quality factors in an inter-class cross of US and Chinese wheat. Theor Appl Genet. 2010;120:1041–51.

Tang SQ. QTL mapping for cooking and nutrient quality traits of rice. MAE Dissertation of Zhejiang University, 2007 (in Chinese with English abstract).

Tang JW, Yin GH, Wang LN, Han YL, Huang F, Yu HF, Yang GY, Li XP. Inheritance of wet gluten content and gluten index in wheat. Acta Agron Sin. 2011;37(9):1701–6 (in Chinese with English abstract).

Tsilo TJ, Simsek S, Ohm JB, Hareland GA, Chao S, Anderson JA. Quantitative trait loci influencing endosperm texture, dough-mixing strength, and bread-making properties of the hard red spring wheat breeding lines. Genome. 2011; 54:460–470.

Uauy C, Distelfeld A, Fahima T, Blechl A, Dubcovsky J. A NAC gene regulating senescence improves grain protein, zinc, and iron content in wheat. Science. 2006;314:1298–301.

Wang S. Analysis on the quality of Wheat by Alveograph instrument. J Zhengzhou Grain College. 1998;19(4):80–4 (in Chinese).

Wang RX, Hai L, Zhang XY, You GX, Yan CS, Xiao SH. QTL maping for grain filling rate and yield-related traits in RILs of the Chinese winter wheat population Heshangmai X Yu8679. Theor Appl Genet. 2009;118:313–25.

Wang SZ, Li CX, Luo YR, Jiang LN. Investigation on effects for genotypes and region distribution to the grain amino acid contents of winter wheat. Acta Bot Boreal Occident Sin. 2001;21 (3):437–45 (in Chinese with English abstract).

Wang J, Li H, Wan X, Pfeiffer W, Crouch J, Wan J. Application of identified QTL-marker associations in rice quality improvement through a design breeding approach. Theor Appl Genet. 2007;115:87–1.

Wang RX, Zhang XY, Wu L, Wang R, Hai L, You GX, Yan CS, Xiao SH. QTL analysis of grain size and related traits in winter wheat under different ecological environments. Sci Agric Sin. 2009; 42(2):398–407 (in Chinese with English abstract).

Wang LQ, Zhong M, Li XH, Yuan DJ, Xu YB, Liu HF, He YQ, Luo LJ, Zhang QF. The QTL controlling amino acid content in grains of rice (*Oryza sativa*) are co-localized with the regions involved in the amino acid metabolism pathway. Mol Breeding. 2008;21:127–37.

Wang DL, Zhu J, Li ZK, Paterson AH. Mapping QTLs with epistatic effects and QTL × environment interactions by mixed linear model approaches. Theor Appl Genet. 1999;99:1255–64.

Wei YM. Grain quality and food quality. Xian. Shanxi People's Publishing House, 2002 (in Chinese).

Wieser H, Kieffer R. Correlations of the amount of gluten protein types to the technological properties of wheat flours determined on a micro-scale. J Cereal Sci. 2001;34(1):19–27.

Witkowski E, Waga J, Witkowska K, Rapacz M, Gut M, Bielawska A, Luber H, Lukaszewski AJ. Association between frost tolerance and the alleles of high molecular weight glutenin subunits present in Polish winter wheats. Euphytica. 2008;159:377–84.

Wu YP, Zhang YL, Xiao YG, Yan J, ZhangY, Zhang XK, Zhang LM, Xia XC, He ZH. QTL mapping for important quality traits in common wheat. Sci Agric Sin. 2008; 41(2):331–9 (in Chinese with English abstract).

Xiao CC. The application of blow bubble instrucment in controlling wheat of biscuits special powder and the property of flour quality. Food Grease. 2003;8:20–1 (in Chinese with English abstract).

Yan J, Zhang LL, Wang XM, Xue WT, Yong RZ, Tizion F, Cheng JP. QTL mapping of yield related traits in durum wheat wild emmer wheat RIL population. J Shandong Agric Univ (Nat Sci). 2011; 42(2):163–171 (in Chinese with English abstract).

Yang J, Zhu J. Predicting superior genotypes in multiple environments based on QTL effects. Theor Appl Genet. 2005;110:1268–74.

Zaidul ISM, Karim AA, Manan DMA, Ariffin A, Norulaini NNA, Omar AK. Study of rheological profile analysis related to texture for mixtures of sago-wheat gel. Int J Food Prop. 2002;5:585–98.

Zanetti S, Winzeler M, Feuillet C, Keller B, Messmer M. Genetic analysis of bread-making quality in wheat and spelt. Plant Breeding. 2001;120:13–9.

Zhang LP. Genetic Analysis and QTL mapping of quality traits in common wheat. Chinese Academy of Agricultural Sciences, Ph.D. 2003.

Zhang Y, He ZH, Guo YY, Zhang AM, Maarten VG. Effect of environment and genotype on bread-making quality of spring-sown spring wheat cultivars in China. Euphytica. 2004;139:75–83.

Zhang QC, Shao LG, Wang Y. Analysis and evaluation of spring wheat quality with alveograph NG consistograph. Heilongjiang Agric Sci. 2006;5:74–7 (in Chinese with English abstract).

Zhang Y, Tang J, Zhang Y, Yan J, Xiao Y, Zhang Y, Xia X, He Z. QTL mapping for quantities of protein fractions in bread wheat (*Triticumaestivum* L.). Theor Appl Genet. 2011;122:971–87.

Zhang X, Tian JC. The color advantage of Chinese wheat with high whiteness and analysis of factors affecting color formation. Sci Agri Sini. 2008;41:347–53.

Zhang KP, Tian JC, Zhao L. Molecular genetic analysis of flour color using a doubled haploid population in bread wheat (*Triticum aestivum* L.). Euphytica. 2009c;165(3):471–84.

Zhang K, Tian J, Zhao L, Liu B, Chen G. Detection of quantitative trait loci for heading date based on the doubled haploid progeny of two elite Chinese wheat cultivars. Genetica. 2009b;135 (3):257–65.

Zhang KP, Tian JC, Zhao L, Wang SS. Mapping QTLs with epistatic effects and QTL × environment interactions for plant height using a doubled haploid population in cultivated wheat. J Genet Genomic. 2008; 35:119–127.

Zhang YL, Wu YP, Xiao YG, Yan J, Zhang Y, Zhang Y, Ma CX, Xia XC, Ma CX, He ZH. QTL mapping for milling, gluten qualityand flour pasting properties in a recombinant inbred line population derived from a Chinese soft × hard wheat cross. Crop Pasture Sci. 2009d;60:587–97.

Zhang LP, Yan J, Xia XC, He ZH, Sutherland MW. QTL mapping for kernel yellow pigment content in common wheat. Acta Agron Sin. 2006; 32(1):41–45 (in Chinese with English abstract).

Zhang K, Zhang Y, Chen G, Tian J. Genetic analysis of grain yield and leaf chlorophyll in common wheat. Cereal Res Commun. 2009a;37(4):499–511.

Zhang Y, Zhang Y, He ZH, Ye GY. Milling quality and protein properties of autumn-sown Chinese wheats evaluated through multi-location trials. Euphytica. 2005;143:209–22.

Zhao YL. QTLs mapping and cloning of micronutrient-related genes in hexaploid wheat. Ph.d. dissertation of Chinese Academy of Agricultural Sciences, 2005 (in Chinese with English abstract).

Zhao L, Liu B, Zhang KP, Tian JC, Deng ZY. Detection of QTLs with additive effects, epistatic effects, and QTL × environment interactions for zeleny sedimentation value using a doubled haploid population in cultivated wheat. Agric Sci China. 2009;8:1039–45.

Zhao L, Zhang KP, Liu B, Deng ZY, Qu HL, Tian JC. A comparison of grain protein content QTL and flour protein content QTL across environments in cultivated wheat. Euphytica. 2010;1:11–22.

Zheng X, Wu JG, Lou XY, Xu HM, Shi CH. Mapping and analysis of QTLs on maternal and endosperm genomes for histidine and arginine in rice (*Oryza sativa* L.) across environments. Acta Agron Sin. 2008;34(3):369–375 (in Chinese with English abstract).

Chapter 6
Genetic Analysis of Main Physiological and Morphological Traits

Abstract Wheat physiological and morphological traits are the most important traits for wheat (*Triticum aestivum* L.) yield. In this chapter, quantitative trait loci (QTL) mapping for physiological traits including photosynthetic Characters, microdissection characteristics of Stem, heading date and cell membrane permeability of leaf, and for morphological traits of containing root-related traits and leaf-related traits were analyzed in different environments using the DH population, RIL population or natural population. Photosynthesis related traits of wheat were mapped under field and phytotron environments, respectively. Eight additive QTLs and three pairs of epistatic QTLs for chlorophyll were detected in field environments and 17 additive QTLs for conferring photosynthesis and its related traits were identified in phytotron environments. Furthermore, 18 additive loci for dry matter production (DMA) and Fv/Fm were detected. For microdissection characteristics of wheat stem, a total of 12 QTLs controlling anatomical traits of second basal internode on chromosomes 1B, 4D, 5B, 5D, 6A and 7D, and 20 additive QTLs for anatomical traits of the uppermost internode on chromosomes 1A, 1B, 2A, 2D, 3D, 4D, 5D, 6A, 6D and 7D were detected based on DH population. Two additive QTLs on chromosomes 1B and 5D in DH population, five additive QTLs on chromosomes 3B, 5B, 6A, 6B and 7D in RIL population derived from the cross of Nuomai 1 × Gaocheng 8901 and 12 additive QTLs on chromosomes 1A, 1B, 4B, 6A and 6B based on a RIL population derived from the cross of Shannong 01-35 × Gaocheng 9411 were identified for heading date. For cell membrane permeability of leaf, a total of 21 additive QTLs were detected on chromosomes 1B, 2A, 3A, 3B, 5B, 6A, 6B, 6D, 7B and 7D, respectively in three different environments based on a DH population. Seven additive QTLs and 12 pairs of epistatic QTLs for root-related traits were mapped on chromosomes 1A, 1D, 2A, 2B, 2D, 3A, 3B, 5D, 6D and 7D using IF_2 population derived from Huapei 3 × Yumai 57.31 additive QTLs and 22 pairs of epistatic QTLs conferring leaf morphology were detected based on a DH population. Finally, by genome-wide association analysis with a natural population derived from the founder parent Aimengniu and its progenies, 61 marker-trait associations (MTAs) involving 46 DArT markers distributed on 14 chromosomes (1B, 1D, 2A, 2B, 2D, 3A, 3B, 4A, 5B, 6A, 6B, 6D, 7A and 7B) for leaf-related traits were identified and the R^2 ranges from 0.1 to 16.4 %. These results provide a

J. Tian et al., *Genetic Analyses of Wheat and Molecular Marker-Assisted Breeding, Volume 1*, DOI 10.1007/978-94-017-7390-4_6

better understanding of the genetic factors for wheat physiological and morphological traits and facilitate marker-assisted selection strategy in wheat breeding.

Keywords Physiological traits · Morphological traits · Photosynthetic characters · Dry matter production · Microdissection characteristics · Heading date · Cell membrane permeability · Root traits · Leaf-related traits · QTL mapping

Wheat physiological and morphological traits are closely related to yield. From a physiological point of view, yield potential is the general performance for assimilates from unit photosynthesis furthest transfer to harvest organs. Final yield is formed by comprehensive coordination of source–sink translocation, that is to say the coordination among the accumulating rate of photosynthate, the distributing ability to grain, duration of distribution, and the turnover capacity of assimilates which stored in stem, leaf, and sheath. The production and transport of photosynthate product has direct relationships with aboveground plant type a leaf type and underground root. Therefore, for the improvement of wheat physiological trait, root, overground plant morphology, and plant anatomy features are considered in the first place, meanwhile several traits are related to photosynthetic characteristics, i.e., canopy structure, light-intercepting capability, photosynthetic capacity, and the storage and turnover capacity of carbohydrate. Hence, this chapter will connect physiological traits with morphological traits of wheat to discuss.

Most of the physiological traits are quantitative characters, which are controlled by multiple genes and easily affected by environmental conditions. So, genetic analyses of wheat physiological traits are started from QTL mapping and then discussed the number of genes, gene effect, and interaction effect. For example, for wheat root, researchers always focus on QTL analysis under abiotic stress; for photosynthetic characteristics, researchers always focus on QTL analysis of photoelectric energy conversion system, chlorophyll, fluorescence parameter, photosynthetic rate, stomatal conductance, flag leaf senescence, etc. Although QTL analysis of physiological traits has made good progress, but these results are difficult to be used for genetic improvement of wheat, because phenotypic determination of physiological traits has more difficulty in multi-year and multisite trails; moreover, mechanism of QTLs and those interaction effects are further complications when comparing to yield trait. These QTLs results have few direct applications in wheat genetic improvement. Hence, genetic analysis of physiological traits is needed to be deeper researched, in order to obtain molecular markers for improving wheat physiological traits, and then speed up the genetic improvement of physiological character and enhance yield and quality of wheat.

6.1 QTL Mapping of Photosynthetic Characters in Wheat

Photosynthesis is closely related to crop yield. The purpose of agricultural production is to enhance photosynthesis of crop, accumulate more organics, and then increase yield, according to various agricultural technical measures.

Hence, photosynthesis is the basis of enhancing crop yield, while breeding varieties with high photosynthetic efficiency is an important approach to improve crop yield. Researches related to QTLs analysis of physiological traits in rice (Nagata et al. 2002); soybean, sorghum (Ritter et al. 2008); barley (Guo et al. 2008); maize (Hund et al. 2005; Leipner et al. 2008; Pelleschi et al. 2006); cotton, and sunflower, etc., were conducted. However, similar researches for wheat (*Triticum aestivum* L.) are relatively few. The recent development of molecular markers and measuring technology related to photosynthesis, QTL analysis of wheat has started. However, it is difficult to precisely determine phenotype of photosynthetic property, especially photosynthetic property for population, because physiological traits are greatly influenced by environment and mechanism of photosynthesis is complex. Meanwhile, the determining methods have limitations. So far, most of the researches referenced QTL analyses of physiological traits were focused on chlorophyll content at seedling stage, dry matter accumulation, leaf photosynthetic rate, stomatal conductance, transpiration rate, inter-cellular CO_2 concentration, and leaf fluorescence parameters, etc. Further, QTL analysis of photosynthetic characters of population in field was few. Therefore, in this study, a set of double-haploid lines (DHLs) derived from a cross of two elite Chinese wheat cultivars were used to map QTLs for photosynthesis-related traits. And the purposes of this study were to obtain closely linked molecular markers that could be used for marker-assisted selection in wheat breeding programs.

6.1.1 QTL Mapping of Photosynthesis Characters of Wheat in Field

6.1.1.1 Materials and Methods

6.1.1.1.1 Materials

One hundred and sixty-eight DH lines derived from the cross of Huapei 3 (HP3)/ Yumai 57 (YM57) were used as materials.

6.1.1.1.2 Planting and Processing in Field Trails

The field trials were conducted on the experimental farm at Shandong Agricultural University (Tai'an, China, 36° 57′N, 116° 36′E) in 2005–2006 and 2006–2007, and in Suzhou Academy of Agricultural Sciences, (Anhui province) in 2006–2007, providing data for three environments. The experimental field consisted of a randomized block with two replications. In the autumn of 2005, all DH lines and parents were grown in a plot with three rows in 2-m length and 25 cm between rows. In the autumn of 2006, the lines were grown in a plot with four 2-m rows spaced 25 cm apart. Crop management was carried out following the local practice. The soil was brown earth, in which the available N, P, and K contents in the top 20 cm were 40.2, 51.3, and 70.8 mg/kg, respectively. Before planting, 37,500 kg/hectare (ha) of farmyard manure or barnyard manure (nitrogen content, 0.05–0.1 %), 375 kg/ha of urea, 300 kg/ha of phosphorus diamine fertilizer, 225 kg/ha of potassium chloride, 15 kg/ha of zinc sulfate were added as fertilizers. Plots were irrigated in winter (December 1, 2006), and at jointing (April 3, 2007), anthesis (May 4, 2007), and grain filling (May 15, 2007). Topdressings of 300 and 75 g/ha urea were applied with the irrigation water at jointing and anthesis, respectively. In 2007–2008, all DH lines and parents were grown in a plot with five rows in 2-m length and 25 cm between rows. And two environments were set including environment I (2008 (+N)) and environment II (2008 (−N)). Moreover, base fertilizer, additional fertilizer, and irrigation in environment I were the same as 2006–2007, while there was no additional fertilizer in environment II, but base fertilizer and irrigation were the same as 2006–2007. Crop management was carried out following the local yield comparison trial.

6.1.1.1.3 Determining Methods

6.1.1.1.3.1 Determination of Wheat Chlorophyll Content at Grain Filling Stage in Field

For leaf chlorophyll content analyses, flag leaves were taken from five plants per plot at the grain filling stage (around 12 May) and saved in −80 °C ultra-low-temperature freezer. Samples of approximately 0.2 g of leaf tissue (taken from the middle of the leaves) were placed to 20 mL tubes and 10 mL 80 % acetone were added. All tubes were placed in dark at 4 °C for 24 h, and oscillated regularly till leaf tissue turned pale. And then OD was measured at 662 nm and 645 nm with a spectrophotometer UV-4802 (Unico instrument Co., Ltd, Shanghai, China). Chlorophyll a and b contents were estimated, adapting the procedure described by Porra et al. (1989).

6.1.1.1.3.2 Determination of Wheat Chlorophyll Fluorescence Parameters in Field

At jointing, anthesis, and grain filling stages, five uppermost leaves (fully expanded) of each line and the parents were sampled. And chlorophyll fluorescence was measured on the leaf using a portable fluorometer (Handy PEA; Hansatech Instruments, King's Lynn, UK) at ambient temperature after 20-min adaptation of leaves to dark conditions on the day of sampling. The fast chlorophyll a fluorescence transient (OJIP) was induced by pulsed light with 3000 μmol m^{-2} s^{-1}, and changes in fluorescence were registered during irradiation of 10 μs to 1 s with the initial rate of 105 data per second. The meaning and formula of each parameter for OJIP was as follows:

Fo, initial fluorescence, fluorescence level when plastoquinone electron acceptor pool (Qa) is fully oxidized;
Fm, maximum fluorescence, fluorescence level when Qa is transiently fully reduced;
Fv, variable fluorescence, Fv = Fm−Fo, maximum variable fluorescence, reflecting the reduction of Qa; and
Fv/Fm, maximum quantum efficiency of PSII, reflecting the maximum efficiency of PSII reaction center converting luminous energy.

6.1.1.2 Result and Analysis

6.1.1.2.1 QTL Mapping of Chlorophyll Content

6.1.1.2.1.1 Variation of Chlorophyll Content

Mean values of chlorophyll contents for the parents Huapei 3 and Yumai 57, as well as the 168 DH lines under three different environments are shown in Table 6.1. Male parent Yumai 57 had larger values than Huapei 3 for chlorophyll a and b contents, and the differences were visible. The distribution of chlorophyll a and b contents was continuous in the DH lines, showing their quantitative nature. Meanwhile, a transgressive separation was found from the DH lines (Figs. 6.1 and 6.2). Therefore, the distributive character of phenotypic data was suitable for QTL analysis. Correlation analysis showed that there was a highly positive correction between chlorophyll a and chlorophyll b, and the coefficient of correlation was 0.823**.

6.1.1.2.1.2 QTL Mapping and Effect Analysis for Chlorophyll Content

For chlorophyll, eight additive QTLs and three pairs of epistatic QTLs were detected (Tables 6.2 and 6.3). Among them, four additive QTLs and one pair of epistatic QTL had QTL × environment interaction effects.

Table 6.1 Phenotypic data of leaf chlorophyll content (mg g^{-1} FW)

Trait	Parent		DH population					
	Huapei 3	Yumai 57	Mean	Maximum	Minimum	SD	Skewness	Kurtosis
chlorophyll a content (mg g^{-1} FW)								
Suzhou 2006	25.42	31.01	27.94	32.16	21.44	2.24	−0.55	0.41
Tai'an 2006	24.33	32.56	25.20	34.84	17.42	2.79	0.20	0.36
Tai'an 2005	22.69	27.51	23.86	28.17	18.76	2.03	−0.05	−0.44
chlorophyll b content (mg g^{-1} FW)								
Suzhou 2006	9.59	10.24	10.21	11.76	7.84	0.82	−0.54	0.46
Tai'an 2006	7.98	10.96	9.22	12.74	6.37	1.02	0.18	0.35
Tai'an 2005	7.24	10.43	8.95	11.73	4.67	1.23	−0.27	0.65

SD Standard deviation

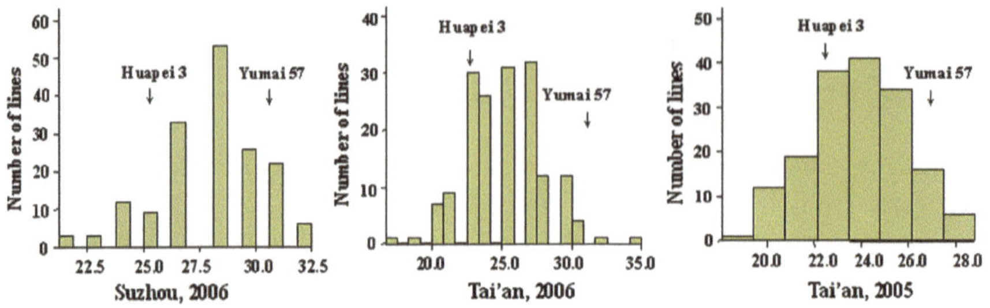

Fig. 6.1 Frequency distribution of chlorophyll a content

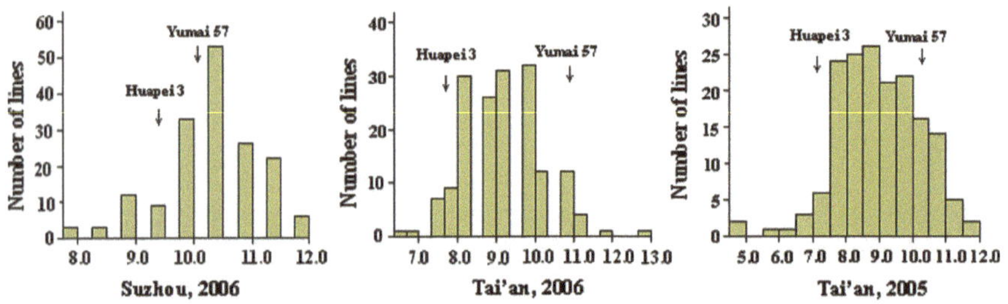

Fig. 6.2 Frequency distribution of chlorophyll b content

6.1.1.2.1.2.1 QTL Mapping and Effect Analysis for Chlorophyll a Content

Four additive QTLs controlled chlorophyll a content were detected on chromosomes 1B, 4A, 5D, and 7A, respectively. And the variance of chlorophyll a content explained by the QTLs ranged from 0.84 to 12.95 %. Among them, *qChla5D* had the highest phenotypic contribution, which could explain 12.95 % of total phenotypic variation, and its positive allele originated from Yumai 57. Environmental interaction effect was detected in *qChla5D*, explaining 21.27 % of total variation.

Three pairs of epistatic QTLs associated with chlorophyll a content were identified on chromosomes 2A-2B and 2A-3B(2), respectively. The pair of QTL (*qChla2Ab*/*qChla3B*) involved in environmental interaction and explained 1.62 % of total phenotypic variation.

6.1.1.2.1.2.2 QTL and Effect Analysis for Chlorophyll b Content

Four additive QTLs controlled chlorophyll a content were on chromosomes 2D, 4A, 5A, and 5D, respectively. And the variance of chlorophyll b content explained by the QTLs ranged from 1.37 to 23.29 %. Among them, *qChlb5D* had the highest phenotypic contribution, which could explain 23.29 % of total phenotypic variation, and its positive allele originated from Yumai 57. Further, *qChlb2D*, *qChlb4A,* and *qChlb5A* involved in environmental interaction, which explained 5.81 % of total variation. No pair of epistatic QTL for chlorophyll b content was detected in this study.

Table 6.2 Estimated additive (A) and additive × environment (AE) interactions of QTL for chlorophyll content

Trait	QTL	Flanking marker	Position (cM)	A	H^2 (%) d	A × E1		A × E2		A × E3	
						AE1	H^2 (%)	AE2	H^2 (%)	AE3	H^2 (%)
Chlorophyll a	*qChla1B*	Xbarc120.3–Xbarc008	38.6	0.34	1.89						
	qChla4A	Xwmc718–Xwmc262	3.0	−0.53	4.41						
	qChla5D	Xwmc215–Xgdm63	74.3	−0.90	12.95			−0.78	9.87	0.88	12.40
	qChla7A	Xwmc607–Xbarc049	74.6	0.23	0.84						
Chlorophyll b	*qChlb2D*	Xcfd53–Xwmc18	2.8	−0.14	1.70					−0.18	2.94
	qChlb4A	Xwmc718–Xwmc262	0.0	−0.26	6.15					−0.12	1.32
	qChlb5A	Xcfe026.1–Xcwem32.2	7.0	0.12	1.37					0.13	1.55
	qChlb5D	Xwmc215–Xgdm63	73.3	−0.51	23.29						

E1: Suzhou, 2006; E2: Tai'an, 2006; E3: Tai'an, 2005

Table 6.3 Estimated epistasis (AA) and epistasis × environment (AAE) interactions of QTL for chlorophyll content

Trait	QTL	Flanking marker	Position (cM)	QTL	Flanking marker	Position (cM)	AA	H^2 (%)	AA × E	
									AAE	H^2 (%)
Chl a	*qChla2Aa*	Xbarc296–Xcfa2263	69.0	qChla2B	Xbarc373–Xwmc477	77.6	0.57	3.97		
	qChla2Ab	Xbarc264–Xgwm448	75.1	qChla3B	Xcfe009–Xwmc3	51.8			−0.32	1.62
	qChla2Ac	Xwmc455–Xgwm515	104.9	qChla3B	Xcfe009–Xwmc3	51.8	−0.34	1.86		

6.1.1.2.2 QTL of Chlorophyll Fluorescence Parameters

6.1.1.2.2.1 Phenotypic Variations of Chlorophyll Fluorescence Parameters

Differences were found for chlorophyll fluorescence parameters between Huapei 3 and Yumai 57 (Table 6.4). The phenotypic value of PSII Fv/Fm for Huapei 3 was higher than Yumai 57 in all environments. In the environment of 2008 (−N), Fv/Fm for the two parents was higher than that in 2007 (+N) and 2008 (+N). The values of Chla/b were inconsistent in 2007 (+N) and 2008 (+N). No difference was found for Fo in nitrogen-deficiency environment and normal environment. The distribution of all parameters was continuous in the DH lines, and in accordance with normal distribution. Meanwhile, a transgressive separation was found from the DH lines.

6.1.1.2.2.2 QTL and Effect Analysis of Chlorophyll Fluorescence Parameters

A total of fourteen additive QTLs and five pairs of epistatic QTLs were identified for chlorophyll and fluorescence parameter, distributing on chromosomes 2A, 3A, 4A, 5A, 6A, 1B, 3B, 4B, 7B, 2D, 3D, 5B, 5D, and 6D, respectively (Table 6.5 and Fig. 6.3).

Five additive QTLs associated with Chl a, Chl b, and Chla/b were mapped on chromosomes 4A, 2D, and 5D, respectively. Among them, two major QTLs

Table 6.4 Phenotypic performance of chlorophyll content and chlorophyll a fluorescence of DH population in field test

Treatment	Trait	Parent			DH population					
		Ym57	Hp3	Mean	Max	Min	SD	Skewness	Kurtosis	CV (%)
2007 (+N)	Chl a	27.51	22.69	27.94	32.16	21.44	2.24	−0.55	0.41	0.08
	Chl b	10.43	7.24	10.21	11.76	7.84	0.82	−0.54	0.46	0.08
	Chla/b	2.64	3.13	2.73	2.76	2.71	0.01	0.5	0.77	0.01
	Fo	500	508	520	592	451	25.5	−0.14	0.36	0.05
	Fm	3103	2846	2809	3355	2124	233.0	−0.01	−0.23	0.08
	Fv	2603	2338	2286	2823	1587	229	−0.15	−0.13	0.1
	Fv/Fm	0.82	0.83	0.82	0.84	0.75	0.02	−0.9	1.22	0.25
2008 (+N)	Chl a	31.06	24.99	29.65	34.32	22.88	2.38	−0.43	0.36	0.08
	Chl b	10.98	9.06	8.54	13	5.46	1.26	0.36	0.43	0.15
	Chla/b	2.83	2.76	3.55	4.87	2.42	0.5	0.19	−0.5	0.14
	Fo	452	473	454	502	402	17.34	−0.12	0.52	0.04
	Fm	2458	2894	2710	3235	2378	161.2	0.16	−0.12	0.06
	Fv	2006	2421	2257	2747	1921	152.7	0.12	−0.16	0.07
	Fv/Fm	0.816	0.837	0.83	0.849	0.801	0.01	−0.82	0.95	0.01
2008 (−N)	Fo	452	448	453	503	315	44.1	−0.88	−0.55	0.01
	Fm	2382	2565	2552	3175	1761	296.3	−0.26	−0.49	3.47
	Fv	1930	2117	2007.3	2698	1414	263.1	−0.14	−0.44	0.13
	Fv/Fm	0.81	0.825	0.83	0.85	0.78	0.01	−0.58	0.57	0.01

Chl a chlorophyll a content; *Chl b* chlorophyll b content; *Chla/b* chlorophyll a/chlorophyll b; *Fo* initial fluorescence; *Fm* maximum fluorescence; *Fv* variable fluorescence; *Fv/Fm* maximum quantum efficiency of PSII

Table 6.5 Estimated additive (A) QTLs for wheat chlorophyll content and chlorophyll fluorescence of DH population in field test

Trait	QTL	Flanking marker	Site (cM)	A[a]	H^2 (A, %)[b]
Chl a	*qChla4A*	Xwmc718–Xwmc262	1.0	−0.70	8.24
	qChla5D	Xwmc215–Xbarc345	74.4	−0.97	16.12
					24.36
Chl b	*qChlb2D*	Xcfd53–Xwmc18	1.7	−0.44	11.59
	qChlb5D	Xbarc320–Xwmc215	67.3	−0.69	28.49
					40.08
Chl a/b	*qChla/b5D*	Xbarc320–Xwmc215	66.3	0.08	4.34
Fo	*qFo2A*	Xwmc455–Xgmw515	102.7	−9.00	11.08
	qFo5D	Xwmc215–Xbarc345	82.4	8.31	9.54
					20.62
Fm	*qFm3B*	Xgwm389–Xgwm533	15.6	−59.17	6.25
	qFm4B	Xwmc47–Xwmc413	4.2	48.78	4.25
					10.5
Fv	*qFv3B*	Xgmw389–Xgwm533	15.6	−57.77	6.18
	qFv4B	Xwmc47–Xwmc413	4.2	48.06	4.28
					10.46
Fv/Fm	*qFv/Fm5A*	Xgwm186–Xcfe223	58.8	0.0035	3.85
	qFv/Fm6A	Xcfe179.2–Xcfe179.1	84.1	−0.0031	4.23
	qFv/Fm6D	Xgwm55–Xgwm133.2	90.9	0.0047	9.38
					16.16

Note: [a]Additive effects, a positive value indicates that allele from Hp3 increases the trait, a negative value indicates that allele from YM57 increases the trait
[b]Contribution explained by additive QTL

(*qChla5D* and *qChlb5D*) flanked by Xwmc215 could explain 16.12 and 28.49 % of total variation, respectively. Other three additive QTLs (*qChla4A, qChlb2D,* and *qChla/b5D*) explained 8.24, 11.59, and 4.34 % of total variation, respectively.

Two additive QTLs controlling Fo were detected on chromosomes 2A and 5D, accounting for 20.62 % of total phenotypic variation. Further, the positive alleles of *qFo2A* and *qFo5D* came from Huapei 3 and Yumai 57, respectively, and which explained 9.54 and 11.08 % of phenotypic variation, respectively.

For Fm, two additive QTLs (*qFm3B* and *qFm4B*) were detected, whose positive alleles originated from Yumai 57, and could explain 7.86 and 7.38 % of phenotypic variation, respectively.

For Fv, two additive QTLs (*qFv3B* and *qFv4B*) were identified, jointly explaining 10.46 % of the total variation, whose location on chromosomes were as same as the two QTLs controlling Fm. However, their positive alleles came from Huapei 3.

Three additive QTLs (*qFv/Fm5A, qFv/Fm6A,* and *qFv/Fm6D*) associated with PSII Fv/Fm were detected, jointly explaining 16.16 % of phenotypic variation, and the positive alleles came from Huapei 3.

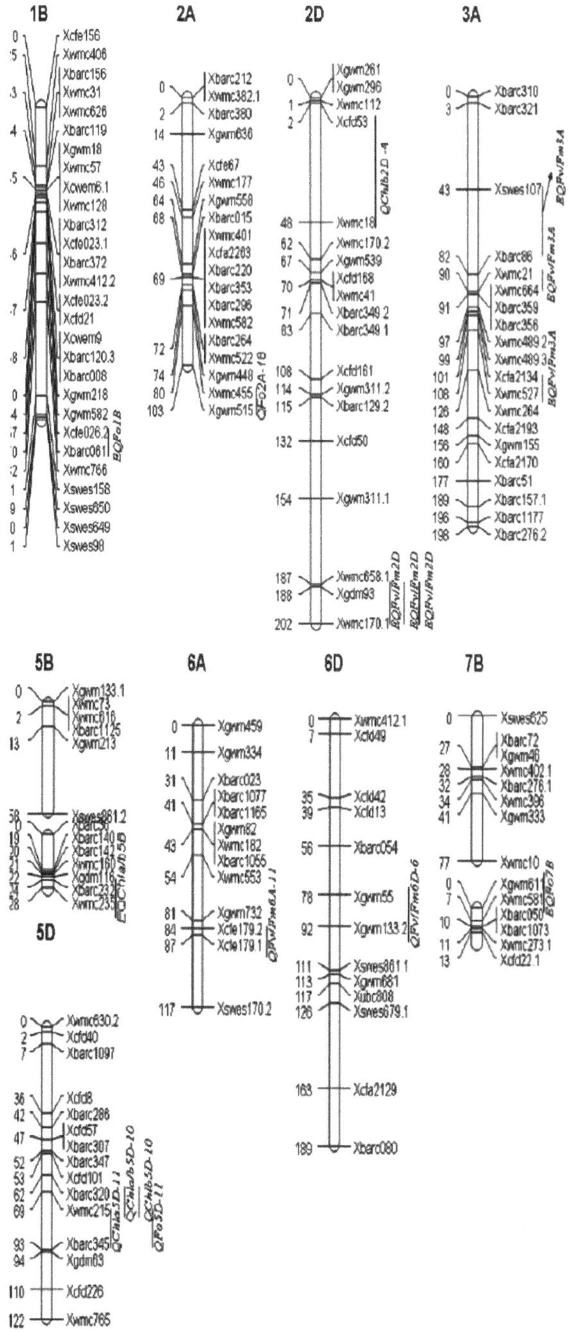

Fig. 6.3 The position of additive QTLs and epistatic QTLs conferring chlorophyll content and chlorophyll fluorescence of DH population in field test

Table 6.6 Estimated digenic epistatic (AA) effects of QTLs for wheat chlorophyll fluorescence of DH population in field test

Trait	QTL	Flanking marker	Site/cm	QTL	Flanking marker	Site/cM	AA[a]	H^2 (AA, %)[b]
Fo	*qFo1B*	Xcfe026.2–Xbarc061	68.3	qFo7B	Xgwm611–Xwmc581	5.0	9.36	12.1
Fv/Fm	*qFv/Fm2D*	Xgdm93–Xwmc170.1	201.8	qFv/Fm3A	Xcfa2134–Xwmc527	107	–	0.16
	qFv/Fm2D	Xgdm93–Xwmc170.1	201.8	qFv/Fm3A	Xwmc21–Xwmc664	90.3	−0.02	1.44
	qFv/Fm2D	Xgdm93–Xwmc170.1	201.8	qFv/Fm3A	Xswes107–Xbarc86	43.1	0.03	3.03
								4.63
Chla/b	*qChla/b3D*	Xbarc1119–Xcfd4	17.8	qChla/b5B	Xbarc232–Xwmc235	25.7	−0.01	3.54

Note: [a]Positive value indicates that the parental two-loci genotypes is greater than the recombinant-type effect, and the negative value means that the parent-type effect is less than the recombinant-type effect have a positive effect and that the recombinants have a negative effect
[b]Contribution explained by epistatic QTL

Five pairs of epistatic QTLs controlling Fo, Fv/Fm, and chl a/b were detected, distributing on chromosomes 1B-7B, 2D-3A, and 3D-5B, respectively (Table 6.6 and Fig. 6.3). And they could explain 12.1, 4.63, and 3.54 % of the phenotypic variation.

6.1.2 QTL Mapping of Photosynthesis of Wheat Seedlings in Phytotron

6.1.2.1 Planting and Determining Methods in Phytotron

6.1.2.1.1 Planting Trails

Two environment conditions including environment I (from September to October 2007) and environment II (from February to April 2008) were set in net room and phytotron in Shandong Agricultural University. A total of 168 lines and parents were planted in cultivate bowls (diameter for 10 cm and height for 8 cm) with homogeneous and fertile soils. Furthermore, each line and parent was planted for three bowls, and five plants were cultivated in a bowl. Under environment I, materials were sowed on September 5, 2007, while materials were sowed on February 28, 2008, under environment II. Materials management was carried out following the conventional potting trial and transforming the location of cultivate bowl once a week to reduce the difference in growing environment among lines and parents. After one month, all the materials were transferred to a phytotron (ACC-1, Hangzhou), and the upper two full extended leaves were sampled to determine the photosynthesis parameters after 7 days for adaptation. In phytotron, the day/might temperature was controlled in 24/18 °C, photon flux density 400 μmol m^{-2} s^{-1}, photoperiod 12 h/12 h, and relative humidity 60 %. In order to avoid the effect of circadian rhythms on determining of parameters, preliminary work was conducted, and multipoint photosynthesis and fluorescence parameters were determined on 5, 7, and 9 days after wheat in phytotron. It was found that photosynthesis and fluorescence parameters of leaf were basically stable in one day after 7 days.

6.1.2.1.2 Determining Methods

6.1.2.1.2.1 Determination of Leaf Gas Exchange Parameters at Seedling Stage

Net photosynthetic rate (Pn), stomatal conductance (Gs), inter-cellular CO_2 concentration (Ci) of the lines and parents were determined using portable photosynthesis system (CIRAS-2, PP Systems, UK) after 7 days stored in phytotron. Concentration of CO_2 was controlled in 380 μmol mol^{-1} by the system, and illumination intensity was controlled in 1000 μmol m^{-2} s^{-1} by LED red-white source.

6.1.2.1.2.2 Determination of Chlorophyll Fluorescence Parameters of Wheat Seedlings

After gas exchange parameters were determined, the same position of leaves were put in clip holders for a 20-min period of darkness adaptation and measuring the fast chlorophyll a fluorescence transient (OJIP) by using a Handy PEA (Hansatech instruments, Norfolk, UK) instrument, and the determination method was the same as that described above.

6.1.2.1.2.3 Determination of Chlorophyll Content of Wheat Seedlings

After determining photosynthetic character and fluorescence parameters, the samples of all lines and parents were taken according to the method described above. OD was measured at 662, 645, and 470 nm with a spectrophotometer UV-4802 (Unico instrument Co., Ltd, Shanghai, China). And then chlorophyll a, chlorophyll b, and carotenoid contents were estimated.

6.1.2.2 QTL Mapping and Effects Analysis of Photosynthetic Characters in Wheat Seedlings

QTL analyses were performed using QTL Network 2.0 software based on the mixed linear model approach. When $P < 0.005$, 17 additive QTLs and 20 pairs of epistatic QTLs conferring photosynthesis and its related traits were identified; furthermore, all additive QTLs and 16 pairs of epistatic QTLs involved in environmental interaction (Tables 6.7 and 6.8, Fig. 6.4).

Two additive QTLs (*QPn4D-11* and *QPn5D-11*) conferring Pn distributing on chromosomes 4D and 5D were detected, whose positive alleles came from Yumai 57 and Huapei 3, respectively, and could explain 2.47 and 7.15 % of phenotypic variation. Moreover, both the two additive QTLs involved in environmental interaction. Meanwhile, four pairs of epistatic QTLs, distributed on chromosomes 1B-3A, 1B-3A, 1B-3D, and 1D-5B, were also detected and could explain 2.17, 1.58, 1.09, and 3.22 % of phenotypic variation, respectively.

For Tr, one QTL (*QE4D-11*) accounting for 3.81 % of phenotypic variation was detected and involved in environmental interaction. Three pairs of epistatic QTLs, distributed on chromosomes 3A-4A, 3B-4D, and 3B-6D, were detected and could explain 2.59, 4.17, and 1.18 % of phenotypic variation. Moreover, all the three pairs of epistatic QTLs involved in environmental interaction, jointly accounting for 8.66 and 9.75 % of phenotypic variation in the two environments, respectively.

For Ci/Cr, two additive QTLs (*QGs4D-11* and *QGs5D-13*) were identified, accounting for 4.17 and 2.64 % of phenotypic variation. And both the two QTLs involved in environmental interaction. The phenotypic variations of *QGs4D-11* were larger in the two environments, which were 3.48 and 3.45 %, respectively. Five pairs of epistatic QTLs associated with Ci/Cr were also detected, distributing

Table 6.7 Additive effects of QTLs for photosynthesis and related physiological traits at seedling stage of wheat in two environments

Trait	Locus	Flanking marker	Site/cM	*A*	H^2 (%)	AE1	H^2 AE1 (%)	AE2	H^2 AE2 (%)
Chl a	*QCa5B-5*	Xgwm213–Xswes861.2	68.1	0.05	1.2	−0.03	0.6	0.20	0.6
	QCa5D-10	Xbarc320–Xwmc215	66.3	0.18	18.2	−0.15	12.5	0.17	12.3
Chl b	*QCb5B-5*	Xgwm213–Xswes861.2	68.1	0.02	1.8	−0.01	0.2	0.05	0.2
	QCb5D-10	Xbarc320–Xwmc215	66.3	0.04	10.4	−0.05	14.1	0.09	13.9
Car	*QCx5D-10*	Xbarc320–Xwmc215	66.3	0.04	27.3	−0.02	6.3	0.06	6.1
Pn	*QPn4D-11*	Xcfe254–Be293342	194.5	0.52	2.5	0.20	0.4	0.50	0.4
	QPn5D-11	Xwmc215–Xbarc345	79.4	0.89	7.2	−0.59	3.2	0.65	3.1
Tr	*QTr4D-11*	Xfe254–Be293342	194.5	0.12	3.8	0.05	0.6	0.30	0.6
Gs	*QGs4D-11*	Xcfe254–Be293342	194.5	13.24	4.2	12.12	3.5	12.38	3.5
	QGs5D-13	Xgdw63–Xcfd226	93.6	10.55	2.6	−5.81	0.8	6.74	0.8
Ci	*QCi5B-5*	Xgwm213–Xswes861.2	68.1	5.70	1.2	27.70	28.9	25.10	27.7
	QCi5D-9	Xcfd101–Xbarc320	58.6	2.71	0.3	–	–	2.05	0.2
Ci/Cr	*QCi/Cr4A-3*	Xbarc343–Xwmc313	16.3	0.01	0.6	−0.02	4.9	0.09	4.6
	QCi/Cr5B-5	Xgwm213–Xswes861.2	68.1	0.01	1.4	0.03	6.9	0.04	7.0
	QCi/Cr5D-9	Xcfd101–Xbarc320	58.6	0.02	5.1	0.02	2.8	0.07	2.6
Fm	*QFm1A-1*	Xgwm259–Xcwem32.1	4.0	140.90	1.4	−114.00	0.9	138.00	0.9
	QFm1A-17	Xwmc120–Xgwm498	64.5	126.44	1.2	53.60	0.2	55.80	0.2

A additive effect; H^2 contribution rate; E1: environment I (from September to October 2007); E2: environment II (from February to April 2008). P_n net photosynthetic rate; T_r transpiration rate; G_s stomatal conductance; C_i intercellular CO_2 concentration; C_i/C_r gas conductance; F_o initial fluorescence; F_m maximum yield of fluorescence in darkness; F_v variable fluorescence; F_v/F_m maximal photochemical efficiency of PSII; *Chl a* Chl a content; *Chl b* Chl b content; *Car* Carotinoid content. "–" data were not taken

Table 6.8 Epistatic effects of QTLs for photosynthesis and related physiological traits at seedling stage of wheat in two environments

Trait	QTL	Flanking markers	Site (cM)	QTL	Flanking markers	Site (cM)	AA	H^2 AA (%)	AAE1	H^2 AAE1 (%)	AAE2	H^2 AAE2 (%)
Chl a	*QCa3B-14*	Xwmc505–Xcfe282	52.6	*QCa5D-7*	Xbarc307–Xbarc347	48.6	0.04	0.9	−0.06	2.0	0.09	2.1
Chl b	*QCb3B-14*	Xwmc505–Xcfe282	52.6	*QCb5D-7*	Xbarc307–Xbarc347	46.6	0.01	0.2	−0.02	2.8	0.05	2.1
Car	*QCx1B-15*	Xcfe023.2–Xcfd21	37.3	*QCx4D-5*	Xcfa2173–Xcfe188	153.8	0.01	2.0	–	–	–	–
	QCx1B-21	Xgwm582–Xcfe026.2	44.6	*QCx4D-5*	Xcfa2173–Xcfe188	153.8	0.01	0.8	–	–	–	–
Pn	*QPn1B-24*	Xwmc766–Xswes158	92.9	*QPn3A-12*	Xwmc527–Xwmc264	120.8	0.49	2.2	−0.21	0.4	0.40	0.4
	QPn1B-27	Xswes649–Xswes98	130.1	*QPn3A-11*	Xcfa2134–Xwmc527	107.0	0.42	1.6	−0.16	0.2	0.24	0.2
	QPn1B-27	Xswes649–Xswes98	130.1	*QPn3D-6*	Xgwm52–Xgdm8	24.0	0.35	1.1	−0.47	2.0	0.50	2.0
	QPn1D-4	Xwmc429–Xcfd19	33.2	*QPn5B-1*	Xgwm133.1–Xwmc73	1.0	0.59	3.2	−0.86	6.7	1.04	8.5
Tr	*QTr3A-18*	Xbarc157.1–Xbarc1177	192.8	*QTr4A-10*	Xbarc327–Xbarc078	40.8	−0.10	2.6	0.11	3.0	−0.25	2.8
	QTr3B-1	Xbarc102–Xgwm389	6.0	*QTr6D-6*	Xgwm55–Xgwm133.2	77.9	0.13	4.2	−0.10	2.7	0.18	4.2
	QTr3B-3	Xgwm533–Xbarc251	23.0	*QTr6D-7*	Xgwm133.2–Xswes861.1	109.0	−0.07	1.2	0.11	3.0	−0.19	2.8
Gs	*QGs1A-11*	Xwmc278–Xbarc120.1	56.3	*QGs2B-18*	Xcwem55–Xbarc129.1	86.0	13.07	4.1	−5.55	0.7	4.99	0.6
	QGs1D-4	Xwmc429–Xcfd19	31.2	*QGs5B-2*	Xwmc73–Xwmc616	1.6	5.82	0.8	−5.99	0.9	0.62	5.1
	QGs2D-13	Xgwm311.2–Xbarc129.2	115.3	*QGs7D-12*	Xcfd175–Xwmc14	172.5	16.32	6.3	−18.59	8.2	18.83	8.4
	QGs4B-3	Xwmc413–Xcfd39.2	7.7	*QGs5B-2*	Xwmc73–Xwmc616	1.6	−9.74	2.3	–	–	–	–
	QGs4B-5	Xcfd22.2–Xwmc657	13.2	*QGs5B-5*	Xgwm213–Xswes861.2	68.1	−9.92	2.3	12.56	3.8	−12.27	3.6
Fm	*QFm1B-1*	Xcfe156–Xwmc406	0.0	*QFm2B-17*	Xbarc101–Xcwem55	77.4	120.10	1.0	61.24	0.3	−65.20	0.3
	QFm3B-21	Xwmc307–Xgwm566	63.0	*QFm7B-7*	Xgwm333–Xwmc10	68.2	118.80	1.0	45.80	0.2	−49.90	0.2
	QFm1B-9	Xcwem6.1–Xwmc128	34.9	*QFm1B-21*	Xgwm582–Xcfe026.2	47.6	168.00	2.0	–	–	–	–
	QFm5B-6	Xbarc232–Xwmc235	27.7	*QFm5D-10*	Xbarc320–Xwmc215	65.3	122.6	1.1	38.40	0.1	−39.40	0.1

AA epistatic effect. Other abbreviations are the same as in Table 6.7. "–" data are not taken

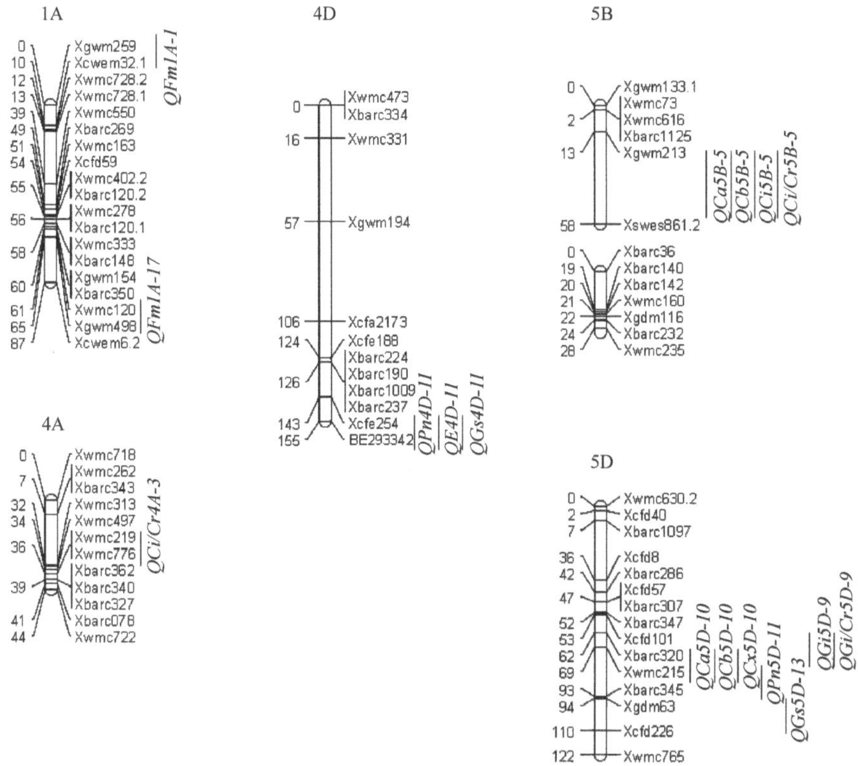

Fig. 6.4 Chromosome positions of additive QTLs for photosynthesis and related traits in 168 double-haploid lines derived from the cross of Huapei 3 × Yumai 57 at seedling stage of wheat

on chromosomes 1A-2B, 1D-5B, 2D-7D, 4B-5B, and 4B-5B, and explained 4.06, 0.81, 6.33, 2.25, and 2.33 % of phenotypic variation, respectively. In addition to *QGs4B-3/QGs5B-2*, other four pairs of QTLs involved in environmental interaction. Furthermore, *QGs2D-13/QGs7D-12* had the highest phenotypic contribution, accounting for 8.21 and 8.42 %, respectively, in the two environments.

Two additive QTLs conferring Ci, distributing on chromosomes 5B and 5D, were detected and explained 1.22 and 0.28 % of phenotypic variation. Among them, *QCi5B-5* had higher environmental interaction effect, accounting for 28.94 and 27.7 % of phenotypic variation in the two environments, respectively. In environment I, the parental effect was greater than recombinant effect, but that was opposite in environment II. No pair of epistatic QTL conferring Ci was detected.

Three additive QTLs for Ci/Cr (*QTL-QCi/Cr4A-3, QCi/Cr5B-5,* and *QCi/Cr5D-9*) were identified, accounting for 0.58, 1.37, and 5.07 %, respectively. The total variation of additive effect and environmental interaction effect were 14.69 and

14.21 %, respectively. No pair of epistatic QTL conferring Ci/Cr was detected (Tables 6.7 and 6.8, Fig. 6.4).

For chlorophyll a content, two additive QTLs (*QCa5B-5* and *QCa5D-10*) were detected, accounting for 1.2 and 18.23 % of phenotypic variation. And one pair of epistatic QTL on chromosome 3B-5D for chlorophyll a content involved in environmental interaction (Tables 6.7 and 6.8, Fig. 6.4).

Two additive QTLs (*QCb5B-5* and *QCb5D-10*) conferring chlorophyll b content were detected, accounting for 1.78 and 10.4 % of phenotypic variation, and their positive alleles came from Huapei 3, which was in accordance with Huapei 3 having the higher content of chlorophyll b. Both the two QTLs involved in environmental interaction, and *QCb5D-10* had the higher phenotypic contribution, explaining 14.12 and 13.9 % of phenotypic variation in two environments, respectively. One pair of epistatic QTL for chlorophyll b was detected, distributing on chromosomes 3B-5D, which involved in environmental interaction.

For carotenoid contents, only one additive QTL (*QCx5D-10*) was identified, accounting for 27.25 % of phenotypic variation, and whose positive alleles came from Huapei 3. Meanwhile, QE interaction could explain 6.3 and 6.12 % of phenotypic variation in the two environments, respectively. Two pairs of epistatic QTLs were also detected on chromosomes 1B-4D, accounting for 2.02 and 0.75 % of phenotypic variation; however, they did not involved in QE interaction (Tables 6.7 and 6.8, Fig. 6.4).

For Fm, two additive QTLs (*QFm1A-1* and *QFm1A-17*) were detected, accounting for 1.43 and 1.15 % of phenotypic variation, respectively. And both the two additive QTLs involved in QE interaction, but the contributions to phenotypic variation were small. Four pairs of epistatic QTLs on chromosomes 1B-2B, 3B-7B, 1B-1B, and 5B-5D, respectively, were detected, explaining 1.04, 1.02, 2.04, and 1.08 % of phenotypic variation, respectively. In addition to the pair of epistatic QTLs linked by Xcwem6.1–Xwmc128 and Xgwm582–Xcfe026.2 locating on chromosome 1B, other three pairs of epistatic QTLs all involved in QE interaction (Tables 6.7 and 6.8, Fig. 6.4).

6.1.3 QTL Mapping of Dry Matter Production (DMA) and Fv/Fm at Jointing and Anthesis Stage in Field

6.1.3.1 Materials and Methods

6.1.3.1.1 Planting Materials

Materials and Planting were same as one of the Sect. 6.1.1.1.1 in this chapter.

6.1.3.1.2 Determining Methods

6.1.3.1.2.1 Determining DMA at Jointing and Flowering Stage of Wheat in Field

Each genotype was tagged at jointing (first internode about 2 cm above the soil) and at flowering (anthers burst on more than 50 % of panicles). Five stems from each DHL were cut at the soil surface and then put in ice from both growth stages. Samples were treated at 105 °C for 30 min and further dried at 65 °C until reaching constant dry weight. The leaves were separated from the stem and the weights of each stem with the sheath and corresponding leaf were separately measured using a JA3003A electronic balance (Jingtian Instruments, Shanghai, China). The DM weight of each plant was the sum of the values of the stem and the leaf. The DMA of leaves, stems, and plants was calculated according to the difference in weight between the jointing and anthesis stages. The means of five replications from each plot were used for statistical analysis.

6.1.3.1.2.2 Determination of Fv/Fm at Jointing and Flowering Stage of Wheat in Field

The upper unfolded leaves at the jointing and anthesis stages were used to measure the maximum quantum efficiency (Fv/Fm), and the determining method as described above. Mean values of five replications per plot were taken for data analysis.

6.1.3.2 Result and Analysis

6.1.3.2.1 Phenotypic Variation Among DHLs

The phenotypic variation of DHLs and the parents for DMA of culms, leaves, total plants, and Fv/Fm at the jointing stage and anthesis stage in 2007 and 2008 are summarized in Table 6.9. HP3 and YM57 differed significantly in the measured traits and phenotypic values of HP3 for the majority of traits at both growth stages were much higher than those of YM57. However, the DMA of leaves for YM57 was higher than that of HP3 at the jointing stage. The mean values of DHLs were intermediate between the parents for most of the traits. Some lines had more extreme values than the parents, showing substantial transgress segregation. In addition, all target traits showed considerable phenotypic variation and continuous distributions, indicating their quantitative nature. The skewness and kurtosis of DMA were less than 1.0, implying polygenic inheritance and suitability of the data for QTL analysis, whereas the Fv/Fm values were often a little higher than 1.0, indicating the distribution of Fv/Fm was skewed to some extent.

Table 6.9 Phenotypic data for DMA and Fv/Fm in two developmental stages in the 2007 and 2008 crop seasons

Season growth stage	Trait	Parent		DH population					
		HP3	YM57	Mean	Max	Min	SD	Skew	Kurt
2007 Jointing	Culm (g·culm-1)	0.51	0.16	0.46	1	0.06	0.2	0.41	−0.27
	Leaves (g·culm-1)	0.11	0.14	0.07	0.4	−0.15	0.1	0.29	0.37
	Plant (g·culm-1)	0.62	0.30	0.54	1.34	0.1	0.23	0.73	0.64
	Fv/Fm	0.835	0.815	0.809	0.84	0.74	0.02	−1.1	2.26
2007 Anthesis	Culms (g·culm-1)	1.84	0.84	0.74	3.34	−1.53	0.79	0.45	0.91
	Leaves (g·culm-1)	0.08	−0.06	−0.02	0.17	−0.23	0.09	−0.1	−0.52
	Plant (g·culm-1)	1.92	0.78	0.87	3.46	−0.82	0.78	0.59	0.85
	Fv/Fm	0.82	0.82	0.82	0.84	0.75	0.02	−0.9	1.22
2008 Jointing	Culms (g·culm-1)	0.39	0.37	0.28	0.73	0.01	0.14	0.74	0.46
	Leaves (g·culms-1)	0.14	0.25	0.13	0.31	−0.08	0.07	−0.11	0.25
	Plants (g·culm-1)	0.53	0.62	0.4	0.87	−0.02	0.18	0.32	−0.14
	Fv/Fm	0.84	0.84	0.84	0.85	0.81	0.01	−1.15	2.6
2008 Anthesis	Culms (g·culm-1)	1.1	0.65	0.54	1.96	−0.9	0.49	−0.08	0.56
	Leaves (g·culm-1)	0.01	−0.07	−0.04	0.17	−0.22	0.08	−0.06	−0.3
	Plants (g·culm-1)	1.11	0.58	0.51	2.04	−1.09	0.55	0.04	0.43
	Fv/Fm	0.837	0.816	0.83	0.849	0.801	0.01	−0.82	0.95

Culms DMAs of culms; *Leaves* DMAs of leaves; *Plants* DMAs of plants; *Fv/Fm* maximum quantum efficiency of PSII; the same as below

6.1.3.2.2 Correlation Analysis for Identified Traits

Correlations among all the identified traits at the two growth stages in both years are given in Table 6.10. The correlations between DMA in culms and leaves at anthesis were much higher than those at the jointing stage in both years, with the exception of the highly significant correlations rA12 = 0.648**, rA22 = 0.737**, and rJ12 = 0.163, rJ22 = 0.378** (1, 2 represent the years 2007 and 2008, respectively). However, the DMAs of plants showed high positive correlations with those of both culms and leaves. In addition, the correlation coefficients between plants and culms (rG12 = 0.523**, rG22 = 0.996**, rJ12 = 0.943**, and rJ22 = 0.925**) were much higher than those between plants and leaves (rG12 = 0.344**, rG22 = 0.789**, rJ12 = 0.456**, and rJ22 = 0.699**) at the two growth stages in both years.

Table 6.10 Correlation coefficients for dry matter accumulation (DMA) and Fv/Fm at two developmental stages in the 2007 and 2008 crop seasons

Trait	Fv/Fm1 (n = 168)	Culms1 (n = 110)	Leaves1 (n = 110)	Plants1 (n = 110)	Fv/Fm2 (n = 168)	Culms2 (n = 168)	Leaves2 (n = 168)	Plants (n = 168)
Fv/Fm1 (n = 168)	–	0.077	0.015	0.03	0.129	0.141	0.115	0.142
Culms1 (n = 134)	−0.085	–	0.648**	0.523**	−0.006	−0.079	−0.063	−0.08
Leaves1 (n = 134)	−0.037	0.163	–	0.344**	0.081	−0.049	−0.095	−0.059
Plants1 (n = 134)	−0.082	0.943**	0.456**	–	−0.164	−0.098	−0.035	−0.093
Fv/Fm2 (n = 168)	0.032	−0.021	0.042	0	–	0.053	0.106	0.063
Culms2 (n-168)	−0.076	0.403**	−0.0164	0.312**	−0.201*	–	0.737**	0.996**
Leaves2 (n = 168)	−0.016	0.038	−0.092	−0.015	−0.083	0.378**	–	0.798**
Plants2 (n = 168)	−0.066	0.33**	−0.0162	0.239**	−0.19*	0.925**	0.699**	–

Numbers in the upper right segment apply to the anthesis stage; those at the lower left are for the jointing stage (*1* in 2007; *2* in 2008); significant at $*P = 0.05$ and $**P = 0.01$; other abbreviations are the same as in Table 6.9

This suggested that DMA in culms plays an important role in plant development. Fv/Fm was poorly correlated with the parameters for DMA.

6.1.3.2.3 QTL Mapping and Effect Analysis of DMA and Fv/Fm in Field

6.1.3.2.3.1 Additive QTLs and Additive QTL × Environment Interactions

A total of 18 additive loci affecting the measured traits were detected. Map locations and additive effects of the QTL and interaction effects between additive QTLs and environments are summarized in Table 6.11 and Fig. 6.5, respectively. It is interesting that all QTLs showing interacting effects with environments were identified at the jointing stage.

The three loci showing significant associations with DMA in culms explained from 7.02 to 14.02 % of the phenotypic variation. All loci derived their additive effects from favorable alleles of HP3. A major QTL, *Qculm5D-10*, was detected at the jointing stages, accounting for 14.02 % of the phenotype variation. The other two QTLs *Qculm1D-2* and *Qculm3B-21*, involved at the anthesis stage, explained 7.02 and 9.93 % of the phenotypic variation, respectively.

For DMA in leaves, seven additive QTLs, 4 at jointing and 3 at anthesis, were located on chromosomes 2A, 3A, 3B, 4A, 5A, 5B, and 5D. Five of these were conferred by favorable alleles from HP3. All QTLs with A-QEIs were identified at the jointing stage, explaining from 1.25 to 3.84 % of the phenotypic variation. No major loci were involved.

Five QTLs controlling DMA in plants were located on chromosomes 1D, 3B, 4B, 5D, and 6A, accounting for 0.37 to 9.34 % of the phenotypic variation. The favorable allele of *Qplant4B-7* came from YM57, and the other four favorable alleles were from HP3. Three QTLs with A-QEIs were identified at the jointing stage, explaining from 0.34 to 1.74 % of the phenotypic variation. No major loci were involved.

Three regions on chromosomes 5A, 6A, and 6D, associated with Fv/Fm, were detected at the anthesis stage. These loci accounted for 3.19–7.26 % of the phenotypic variation. Two of the favorable alleles were from HP3, and the other was from YM57. No loci were involved in additive and environmental interactions.

6.1.3.2.3.2 Epistatic QTL and Epistatic QTL × Environment Interactions

The 12 pairs of epistatic QTLs for DMA (Table 6.12 and Fig. 6.5) explained phenotypic variation ranging from 0.18 to 13.11 %. Among them, five pairs not only had epistatic effects, but also had E-QEI effects at jointing.

Three pairs of epistatic QTLs were detected for DMA in culms; one pair showed both epistatic effects and also E-QEI effects. Two epistatic pairs involved at the jointing stage had negative effects, which meant that recombinant types had higher effects than the parents. The single pair detected at anthesis showed positive effects, that is, parental effects were larger than recombinant effects.

Table 6.11 QTL detected in the HP3 × YM57 DH mapping population at two growth stages in 2007 and 2008

Trait	Locus	Flanking markers	Jointing site (cM)	A[a]	H^2 (A, %)	AE1	H^2 (AE1, %)	AE2	H^2 (AE2, %)	Anthesis site (cM)	A[a]	H^2 (A, %)[b]
Culmsc	*Qculm5D-10*	XBARC320–XWMC215	68.3	0.07	14.02							
	Qculm1D-2	XWMC222–XGDM60								11.2	0.13	7.02
	Qculm3B-21	XWMC307–XGWM566								64	0.16	9.93
					14.02							16.95
Leaves	*Qleaves3A-1*	XBARC310–XBARC321	1	0.0058	0.55	−0.0109	1.94	0.011	1.98			
	Qleaves3B-2	XGWM389–XGWM533	17.6	0.0073	0.86	−0.0094	1.45	0.0095	1.47			
	Qleaves5B-5	XGDM116–XBARC232	22	−0.00015	0.04	0.0153	3.84	−0.0153	3.81			
	Qleaves5D-9	XCFD101–XBARC320	53.6	−0.0151	3.74	−0.0088	1.26	0.0088	1.25			
	Qleaves2A-18	XWMC455–XGWM515								97.7	−0.0156	3.71
	Qleaves4A-10	XBARC327–XBARC078								38.8	−0.0043	0.28
	Qleaves5A-2	XCWEM32.2–XWMC59								12.6	0.014	2.97
					5.19							6.96
Plants	*Qplant4B-7*	XWMC48–XBARC1096	18.4	−0.0245	1.35	0.0141	0.45	−0.0137	0.43			
	Qplant5D-10	XBARC320–XWMC215	64.3	0.0303	2.07	−0.0124	0.34	0.124	0.34			
	Qplant6A-1	XGWM459–XGWM334	2	0.0127	0.37	0.0278	1.74	0.0278	1.74			
	Qplant1D-2	XWMC222–XGDM60								12.2	0.1179	4.37
	Qplant3B-21	XWMC307–XGWM566								65	0.1705	9.14
					3.79							13.51
Fv/Fm	*QFv/Fm5A*	XGWM186–XCFE223								58.8	0.0035	4.07
	QFv/Fm6A	XCFE179.2–XCFE179.1								84.1	−0.0031	3.19
	QFv/Fm6D	XGWM55–XGWM133.2								90.9	0.0047	7.26
												14.52

[a]Additive effects, a positive value indicates that the allele from HP3 increases the trait value; a negative value indicates that allele from YM57 increases the trait value
[b]Contribution explained by additive effect QTL; other abbreviations are the same as in Table 6.9

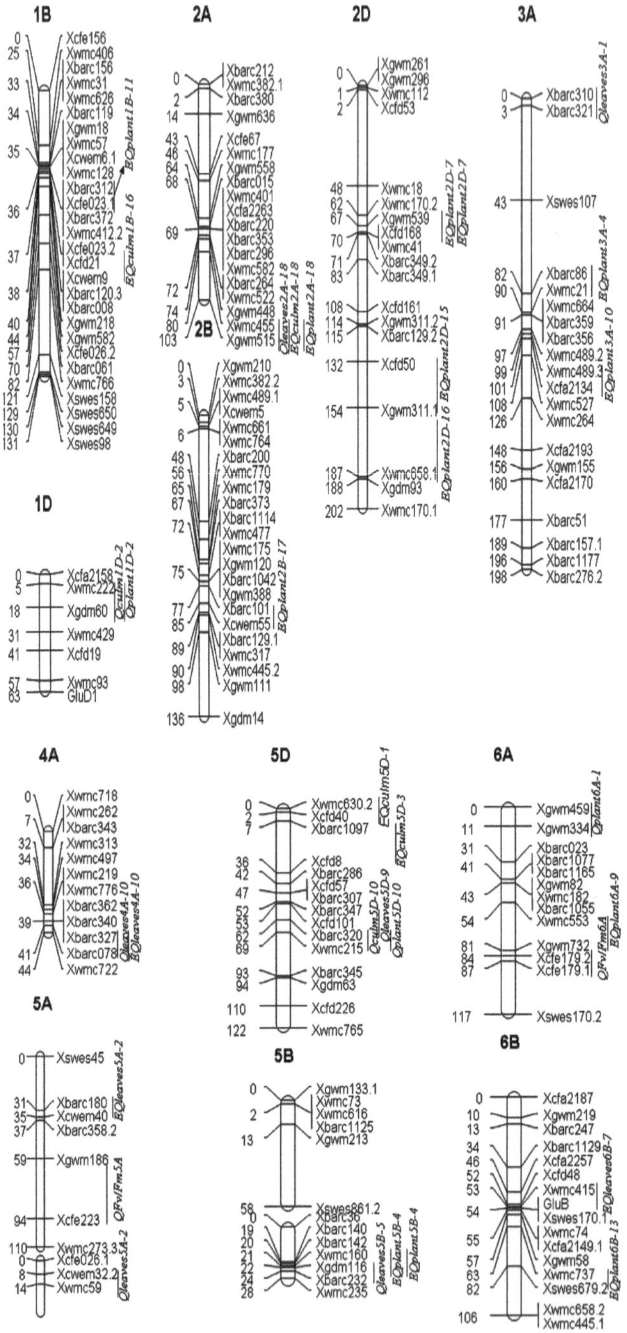

Fig. 6.5 The position of additive QTLs and epistatic QTLs conferring dry matter production and Fv/Fm at two developmental stages in 2007 and 2008

Table 6.12 Estimated digenic epistatic (AA) and epistasis × environment interaction (AAE) effects of QTLs for DMA at two developmental stages in 2007 and 2008

Trait	Locus						Jointing						Anthesis stage	
	Locus 1	Flanking markers	Site (cM)	Locus 2	Flanking markers	Site (cM)	AA[a]	H^2 (AA, %)[b]	AAE1	H^2 (AAE1, %)	AAE2	H^2 (AAE2)	AA[a]	H^2 (AA, %)
Culm	*Qculm5D-1*	*Xwmc630.2–Xcfd40*	0	*Qculm7B-2*	*Xwmc581–Xbarc050*	7.3	−0.01	0.52	−0.01	0.23	0.01	0.23		
	Qculm5D-3	*Xbarc1097–Xcfd8*	28.4	*Qculm7B-2*	*Xwmc581–Xbarc050*	7.3	−0.03	2.39						
	Qculm1B-16	*Xcfd21–Xcwem9*	37.4	*Qculm2A-18*	*Xwmc455–Xgwm515*	80.7							0.15	8.89
Leaves	*Qleaves5A-2*	*Xbarc180–Xcwem40*	30.6	*Qleaves7B-5*	*Xwmc273.1–Xcfd22.1*	12.7	0.01	0.53	−0.01	2.36	0.01	2.44		
	Qleaves4A-10	*Xbarc327–Xbarc078*	38.8	*Qleaves6B-7*	*Xwmc415-Glub*	53.2							0.03	13.11
Plant	*Qplant2D-7*	*Xgwm539–Xcfd168*	68.4	*Qplant3A-4*	*Xbarc86–Xwmc21*	86.5	0.01	0.18	−0.03	2.49	0.03	2.22		
	Qplant2D-7	*Xgwm539–Xcfd168*	68.4	*Qplant3A-10*	*Xwmc489.3–Xcfa2134*	98.7	−0.03	2.81	−0.01	0.28	0.01	0.28		
	Qplant2D-15	*Xcfd50–Xgwm311.1*	132.3	*Qplant5B-4*	*Xwmc160–Xgdm116*	21.4	−0.03	2.83						
	Qplant2D-16	*Xgwm311.1–Xwmc658.1*	186.5	*Qplant5B-4*	*Xwmc160–Xgdm116*	21.4	−0.02	0.59						
	Qplant6A-9	*Xwmc553–Xgwm732*	81.5	*Qplant6B-13*	*Xwmc737–Xswes679.2*	70.8	−0.02	1.05	−0.01	0.18	0.01	0.18		
	Qplant1B-11	*Xbarc312–Xcfe023.1*	36.1	*Qplant2A-18*	*Xwmc455–Xgwm515*	80.7							0.17	8.88
	Qplant2B-17	*Xbarc101–Xcwem55*	84.4	*Qplant7A-1*	*Xwmc593–Xbarc157.2*	4.0							0.17	9.27

[a]Positive value indicates that the effect of the parental two-loci genotype is greater than the recombinant, and a negative value means that the parental effect is less than the recombinant
[b]Contribution explained by epistatic QTL; other abbreviations are the same as in Table 6.9

Two pairs of epistatic QTLs affected DMA in leaves were detected (one at each growth stage). The *Qleaves4A-10/Qleaves6B-7* pair, with positive effects, explained 13.11 % of the phenotypic variation.

Seven pairs of epistatic QTLs affected DMA in plants. These included four pairs only for epistatic effects, and three pairs involved in both epistatic and E-QEI effects at jointing. Two pairs of epistatic QTL with positive effects explaining variation of 8.88 and 9.27 % were identified in the anthesis stage. No major loci were involved.

6.1.3.2.3.3 Distribution of the Additive and Epistatic QTLs

Overall, 16 chromosomes carried 18 additive QTLs for the four traits (Table 6.11, Fig. 6.5). An interesting feature was the highly concentrated distribution of additive QTLs in a few chromosomal regions, and the existence of QTL hot spots, namely chromosomal regions shared by multiple QTLs. For example, the additive QTLs involved in DMA in culms and plants, *Qculm1D-2* and *Qplant1D-2*, *Qculm3B-21* and *Qplant3B-2*, and *Qculm5D-10* and *Qplant5D-10*, were identified within the same chromosomal intervals, viz. XWMC222–XGDM60, XWMC307–XGWM566, and XBARC320–XWMC215, respectively. Some QTL clustering occurred in neighboring marker intervals, e.g., flanking markers XCFD101 to XWMC215 were shared by QTLs for DMA in culms, leaves, and plants on chromosome 5D. Similarly, clustered groups were also found for loci associated with the 12 pairs of epistatic QTLs (Table 6.12, Fig. 6.5), further increasing the locus densities in clustered regions.

6.1.4 Research Progress of Photosynthetic Characters QTL Mapping and Comparison of the Results with Previous Studies

6.1.4.1 Research Progress of Wheat Photosynthesis QTL Mapping

Cao et al. (2004) detected 16 QTLs for chlorophyll content under nitrogen (N) sufficient environment and N deficient environment. Yang et al. (2007) analyzed the QTL for chlorophyll fluorescence and related traits under conditions of rainfed and well-watered and reported that a total of 18 additive QTLs, including 11 QTLs detected under rainfed condition and seven QTLs detected under well-watered condition were located on eight chromosomes 1A, 5A, 6A, 7A, 1B, 3B, 4D, and 7D. The variance explained by the QTLs ranging from 7.27 to 72.72 % depended on the traits. Four QTLs controlling Chlorophyll b under two water regimes were located on chromosomes 1A, 5A, and 7A. Only one QTL for Fo was detected under rainfed condition and was located on chromosome 1B. One QTL of each water regime involved in Fm were identified and located separately on chromosomes 7A and 1B. Two QTLs for Fv under rainfed condition were detected and located on chromosomes 7A and 7D, respectively. No epistatic QTL was identified for Chlorophyll b under two water regimes, for Fm under rainfed condition, as well as for Fo and Fv/Fo under the well-watered condition. In this research, there was no QTL controlling one given trait to be mapped on the same marker interval under two water regimes. Therefore, the results imply that there were different QTL expression patterns under different water conditions. More QTLs were revealed in stress conditions than in non-stressed conditions, suggesting that environmental stress can induce the expression of genes originally keeping silent under non-stressed conditions to alleviate plant damages from environmental stress. Cao et al. (2004), Li et al. (2013), Czyczyło-Mysza et al. (2013), Vijayalakshmi et al. (2010) and Ali et al. (2013) analyzed QTLs for some traits, i.e., chlorophyll, fluorescence, PS parameters, carotenoid, flag leaf senescence in wheat.

So far, scholars at home and abroad have studied wheat photosynthesis and related traits at different growth stages and that under different environments using RIL, DH, and other populations. QTL for about 11 traits related to photosynthesis and physiology were analyzed, and 224 QTLs were obtained conferring different traits. Among them, 101 QTLs whose effect is greater than 10 % were detected, furthermore, the highest contribution to phenotypic variation was 49.59 % (Table 6.13). Those QTLs referred to 21 chromosomes, especially 24 QTLs were found on chromosome 6B, which had the largest number of QTLs, followed by chromosome 5B (23 QTLs were detected) and chromosome 2D (19 QTLs). It can be seen, chromosomes 2D, 5B, and 6B were very important to traits related to photosynthesis and physiology of wheat.

6.1.4.2 Comparison of the Results with Previous Studies

QTL conferring Fm distributed on chromosome 1A, which were detected in this study, was nearby *QRaw.ipk-1A*, which also controlled Fm, detected by Börne et al. (2002). The QTL controlling Ci/Cr on chromosome 4A was near by the QTL associated with cereal protein content (Cao et al. 2004). The QTLs (*qCHO-5B* and *qCHN-5B*) conferring chlorophyll content detected in this study, which nearby *QTgwg.cgb-5B* controlling thousand seeds weight at grain filling stage. Meanwhile, QTLs for thousand seeds weight, yield, and protein content were also detected on the similar loci (Groos et al. 2003). *QPn4D-11*, *QE4D-11*, and *QGs4D-11* detected on chromosome 4D were adjacent to the QTL for Fv/Fo (Yang et al. 2007). Su et al. (2006) mapped major QTLs controlling grain yield on chromosome 3B in winter wheat, and in this study, *QCulmc.sau-3B*, *QLeavesc.sau-3B* and *QPlantc.sau-3B*, were detected on the same chromosome. In addition, the loci *QLeavesc.sau-2A*, *QPlantc.sau-4B*, and *QFv/fmc.sau-5A* coincided with loci for grain weight per ear and post-anthesis DMA per culm (Su et al. 2006; Huang et al. 2003; Quarrie et al. 2005). These indicate that most of the QTLs associated with photosynthesis and related traits were in accordance with the previous results. Meanwhile, many QTLs for some traits, which were not determined before, and QTLs, which were not identified, were also detected in this study (see the previous paper and Table 6.13).

Table 6.13 Summary of QTL of wheat photosynthetic physiology (PVE > 10 %)

Environment	Trait	QTL	Flanking markers	PVE (%)	Population	Reference
Nitrogen supply with 4.0 mmol/l	Chl	*qCHO-2Ba*	Xfbb62–Xfba272	10.85	RIL	Cao et al. (2004)
		qCHO-7D	Xfba8–Xbcd1872	13.45		
Nitrogen supply with 0.4 mmol/l	Chl	*qCHN-6A*	Xpsr312–Xfbb145	11.73		
	SOD activity	*QSod.sdau-2D*	Xissr859a–Xswes624e	16.64	RIL	Wei et al. (2007)
	POD activity	*QPod.sdau-4A*	Xissr23b–Xwmc308	49.56		
Normal	Fv/Fm	*QFv/Fm.csdh-2A**	gwm339	12	DH	Czyczyło-Mysza et al. (2013)
		*QFv/Fm.csdh-6A**	csb112(Dhn5)	13.8		
		*QFv/Fm.csdh-7A.2**	barc108a	12.8		
		*QFv/Fm.csdh-2D.3**	gwm349	17.7		
	PI	*QPI.csdh-3A.1**	cfa2234	12.8		
		*QPI.csdh-4A**	gwm30b	11.2		
		*QPI.csdh-5A**	psp3003b	13.3		
		*QPI.csdh-4B.1**	Rht-B1	18.4		
		*QPI.csdh-6B.1**	wPt-2424	11.4		
		*QPI.csdh-4D**	wPt-5809	24.6		
	ABS/CSm	*QABS.csdh-1A.1*	m51p65.5	10.7		
		*QABS.csdh-1A.2**	wPt-731617	10.9		
		*QABS.csdh-2B**	wPt-8776	12.2		
		*QABS.csdh-5B.2**	psp3037	13		
		QABS.csdh-5B.3	wPt-1548	10.9		

(continued)

Table 6.13 (continued)

Environment	Trait	QTL	Flanking markers	PVE (%)	Population	Reference
		*QABS.csdh-6B.1**	wPt-2424	14.3		
		*QABS.csdh-6B.3**	wPt-2564	13.1		
		QABS.csdh-3D	wPt-732092	10.5		
	TRo/CSm	*QTRo.csdh-1A.1**	wPt-731617	11.6		
		*QTRo.csdh-2B**	m65p64.5	12.1		
		QTRo.csdh-5B.2	psp3037	12.2		
		*QTRo.csdh-5B.3**	psr806.2	16		
		*QTRo.csdh-5B.4**	wPt-1548	15		
		*QTRo.csdh-6B.3**	wPt-2564	13.1		
	ETo/CSm	*QETo.csdh-6A**	wPt-667844	10.9		
		*QETo.csdh-6B.1**	wPt-2424	20.5		
		*QETo.csdh-6B.2**	wg232.4	21.7		
		*QETo.csdh-6B.3**	wPt-2564	20.2		
		*QETo.csdh-4D**	wPt-5809	23.8		
		*QETo.csdh-7D.1**	wPt-744354	14.3		
		*QETo.csdh-7D.2**	gwm37	10		
	DIo/CSm	*QDIo.csdh-5A*	Vrn-A1	10.9		
		*QDIo.csdh-5B.1**	wPt-5346	12.3		
		*QDIo.csdh-5B.2**	wmc73	15		
		QDIo.csdh-5B.3	wPt-5737	10.3		
		*QDIo.csdh-1D.3**	wPt-4671	18.5		
		*QDIo.csdh-7D.1**	wPt-744354	15.1		
	RC/CSm	*QRC.csdh-3A**	cfa2234	14.4		
		*QRC.csdh-4B**	Rht-B1	20.7		

(continued)

Table 6.13 (continued)

Environment	Trait	QTL	Flanking markers	PVE (%)	Population	Reference
		*QRC.csdh-5B**	wPt-5346	15.2		
		*QRC.csdh-7D**	barc154	17.6		
	chla + b	*Qchla + b.csdh-3B**	wPt-1682	11.5		
		*Qchla + b.csdh-2D.1**	wPt-6574	22.4		
		*Qchla + b.csdh-2D.2**	wPt-730613	16.7		
		*Qchla + b.csdh-2D.3**	gwm349	13		
	SPAD	*QSPAD.csdh-1B.1**	wPt-2389	15		
		*QSPAD.csdh-1B.2**	wPt-3451	11.6		
		*QSPAD.csdh-4B.1**	Rht-B1	15.6		
		*QSPAD.csdh-6B.2**	gwm191	13.6		
		*QSPAD.csdh-6B.3**	wPt-2564	13.5		
		*QSPAD.csdh-2D**	wPt-6574	17.4		
	Car	*QCar.csdh-3A.1**	wPt-2478	11		
		*QCar.csdh-3A.2**	wPt-4569	14.2		
		QCar.csdh-3D.1	dupw173	10		
		QCar.csdh-3D.2	wPt-4569	32		
		*QCar.csdh-4D**	psr375.1	12.3		
		*QCar.csdh-6D.1**	wPt-665675	13.4		
		*QCar.csdh-6D.3**	wPt-732626	13.8		
	DWP	*QDWP.csdh-5A**	wPt-668257	11.3		
		*QDWP.csdh-4B.2**	psp3030b	14.7		
		*QDWP.csdh-3D**	dupw173	11.6		
	Chl	*QcChl2.2*	HVM54	12.8/13.6	RIL	Guo et al. (2008) (Barley)

(continued)

Table 6.13 (continued)

Environment	Trait	QTL	Flanking markers	PVE (%)	Population	Reference
		QcChl4.1	p71m88-02	10.2/9.6		
	CFL	*qFC2.2*		20.2	DH	Xue et al. (2008) (Barley)
	MRS	*Qmrso.ksu-5A*	GTG.AGC-254- CGA. CGCT-485	12	RIL	Vijayalakshmi et al. (2010)
		Qmrso.ksu-6A	CAG.AGC-101- AGG. CTT-212	26		
		Qmrsh.ksu-2A	Xgwm356- CGT. TGCG-349	19		
		Qmrsh.ksu-2A	CGT.TGCG-349- CTCG. ACC-242	21		
	TMRS	*Qtmrso.ksu-3B*	CGT.CTCG-146- GTG. AGCT-206	18		
		Qtmrso.ksu-7B	Xbarc340–Xgwm43	12		
		QTmrsh.ksu-2A	Xgwm356-CGT. TGCG-349	17		
		QTmrsh.ksu-6A	CGT.GTG-343- CGA. CGCT-406	30		
	PGMS	*Qpgmso.ksu-4B*	Xgwm368-Xksum62	17		
	Fv/Fm	*QFv/Fmh.ksu-7A*	CGA.CGCT-272- Xbarc121	11		
	Chl at 4 DPA	*QChlc.tamu-1B*	Xbarc128	22.5		Ali et al. (2013)
	Chl at 8 DPA	*QChlc.tamu-1B*	Xbarc128	45.3		

Note: *Fv/Fm* the maximum photochemical efficiency; *PI* overall performance index of PSII photochemistry; *ABS/CSm* light energy absorption; *TR/CSom* amount of excitation energy trapped in PSII reaction centers; *ET/CSom* amount of energy used for electron transport; *DI/CSom* energy amount dissipated from PSII; *RC/CSm* number of active reaction centers; chl a+b The concentrations of total chlorophyll a + b; *SPAD* chlorophyll meter readings; *Car* carotenoids; *DWP* dry weight per plant at harvest; *CFL* chlorophyll content of flag leaf; *MRS* maximum rate of senescence; *TMRS* time to maximum rate of senescence; *PGMS* percent greenness at maximum senescence; *DPA* days post-anthesis

6.2 QTL Conferring Microdissection Characteristics of Wheat Stem

The structure of stem was closely related to lodging resistance of wheat and consists of epidermis, mechanical tissue, elementary tissue, vascular bundle, and pith. Furthermore, the vascular bundle plays an important role in transportation of photosynthetic products, mineral nutrients, and water. The number, size, and capacity of the vascular bundle influence the transportation ability, especially for photosynthetic products. The number and area of the vascular bundles are the basis of large sink and free flow. The growth of vascular bundle is affected by both variety and growing environment, and very complex. In wheat breeding practice with high yield, the relationship between structure of vascular bundle in stem, and size, and plumpness of grain becomes one of the important tissues for research. The capacity of the vascular bundle system transporting assimilates from the source to the sink may be one of the limiting factors for crop yield. Therefore, there is important significance for improving lodging resistance and yield in wheat by studying the structure of vascular bundle in stem.

6.2.1 QTL Mapping for Anatomical Traits of Second Basal Internode

6.2.1.1 Materials and Methods

6.2.1.1.1 Materials

Materials and Planting were same as one of the Sect. 6.1.1.1.1 in this chapter.

6.2.1.1.2 Field Trails

The field trials were conducted on the experimental farm at Shandong Agricultural University (Tai'an, China, 36°9′N, 117°9′E) and Jiyuan Agricultural Science Institute (Jiyuan, Henan province, 35°5′N, 112°38′E) in 2008–2009 and 2009–2010. And nine different environmental conditions were set, as follows:

2008–2009, one normal environmental condition was set in Jiyuan, and four environmental conditions (normal, rainfed, well-watered, and late-sowing) were set in Tai'an. While, in 2009–2010, one normal environmental condition was also set in Jiyuan, and three environmental conditions (normal, rainfed, and well-watered) were set in Tai'an.

In the autumn of 2008, the test materials were sowed on October 6–8 in the normal, rainfed, and well-watered conditions, and they were sowed on November

22 in the late-sowing condition. While in the autumn of 2009, the sowing date was October 4–7 in the normal, rainfed, and well-watered conditions.

All DH lines and parents were grown in a plot with four rows in 2-m length, 26.7 cm between rows and 2.2 cm between plants. And the basic seedling number was about 120,000.

For normal condition, crop management was carried out following the local practice. At jointing and anthesis stages, 225 and 75 kg/ha of urea were added, respectively. Meanwhile, plots were irrigated before winter and at jointing and anthesis. For rainfed condition, crop management was carried out following the normal practice. At jointing and anthesis, 225 and 75 kg/ha of urea were added, respectively. However, there was no irrigation during the whole growth period. For well-watered condition, crop management was also carried out following the normal practice. And plots were irrigated before winter and at jointing and anthesis. However, no fertilizer was applied at jointing and anthesis. For late-sowing condition, crop management was also carried out following the local practice, but the sowing date was delayed.

6.2.1.1.3 Determining Methods

6.2.1.1.3.1 Hanging Tag

At the beginning of May, main stems (flowered on the same day) in the center of every plot were marked.

6.2.1.1.3.2 Sampling

The transverse hand section was made for 2 cm in the middle of the second basal internode at milk-spike stage.

6.2.1.1.3.3 Fixing and Saving

The materials were put into carnoy's fluid, immediately and were extracted until no air bubbles were appeared. Then, carnoy's fluid was changed and air was extracted. After that, the materials were stored at 0–4 °C for use.

6.2.1.1.3.4 Section

Settled segment of stem was sectioned to slices with about 20 μm thickness (less than two layers of cells) by using thin razor.

6.2.1.1.3.5 Microscopy

Better sections were selected under the microscope and then dyed.

6.2.1.1.3.6 Dyeing

The selected sections were stained for 3 min using safranin, rinsed for 1 min, and again stained for 15 s using Fast Green, and then rinsed.

6.2.1.1.3.7 Microscopy

A drop of distilled water was taken on the glass slide, and the dyed sections were put on the glass slide and then observed using low power lens (Nikon YS100). Finally, the poorly dyed sections were rejected.

6.2.1.1.3.8 Photographing

Cover glass was put on the selected sections and photographed in DP71 high resolution by using microscope (Olympus BX51).

6.2.1.1.3.9 Statistics

The stem diameter (SD) for basal internode was measured, and matching parameter was set, and then the stem anatomical structure-related traits such as the number of large and small vascular bundle (LVB, SVB), culm wall thickness (CWT), and the pith diameter (PD) were measured by using the graphic program Image-pro Plus 6.0.

6.2.1.2 Result and Analysis

6.2.1.2.1 Variation of Phenotype

Five traits for anatomical structure of the second basal internode including the number of LVB, the number of SVB, SD, CWT, and PD were analyzed by using the DH population. The variations of phenotypic data of five traits related to the second basal internode in four environments for two years were summarized in Table 6.14.

Huapei 3 had the higher values of anatomical traits than Yumai 57 in Jiyuan and Tai'an (late-sowing condition) in 2008–2009, and in Tai'an (normal condition) in 2009–2010. However, Yumai 57 had the higher values in Tai'an (normal condition) in 2008–2009. The ranges of variation of the test traits were large, which was in accordance with normal distribution and the distribution was continuous. Meanwhile, a transgressive separation was found from the DH lines. Therefore, the phenotypic data were suitable for QTL analysis.

Table 6.14 Phenotypic values of the anatomical traits in the DH population

Env.	Trait	Parent		DH population					
		HP3	YM57	Min	Max	Average	SD	Skew	Kurt
E1	LVB	44.75	42.5	34.3	47	40.4	2.59	0.2	−0.04
	SVB	21	25	12.3	31.8	20.79	3.94	0.37	−0.07
E2	LVB	41.25	40.25	30	48	39.78	3.03	0.13	0.69
	SVB	22.5	19.25	11.75	33	20.81	3.99	0.38	−0.09
E5	LVB	33	35.75	26	41	32.9	2.94	0.04	−0.1
	SVB	16	25.25	13	26	18.25	2.6	0.17	−0.11
E7	LVB	37	39.57	31	45	37.27	2.59	0.06	0.05
	SVB	18.86	22.29	13	30	20.82	3.57	0.17	−0.38
E1	SD	4.79	5.06	3.52	5.71	4.63	0.39	0.31	0.26
	CWT	0.54	0.57	0.37	0.66	0.5	0.05	0.5	0.22
	PD	3.72	3.92	2.71	4.44	3.62	0.35	0.05	−0.29
E2	SD	4.56	4.36	3.52	5	4.12	0.3	0.42	−0.18
	CWT	0.41	0.41	0.34	0.52	0.4	0.03	0.77	0.61
	PD	3.75	3.55	2.73	4.16	3.32	0.28	0.35	−0.13
E5	SD	3.68	3.75	2.99	4.26	3.63	0.25	0.12	−0.17
	CWT	0.39	0.41	0.33	0.56	0.43	0.05	0.29	−0.14
	PD	2.91	2.92	2.21	3.37	2.77	0.22	0.3	0.22
E6	SD	4.31	4.11	3.36	4.94	4.15	0.27	0.03	0.33
	CWT	0.8	0.63	0.39	1.3	0.63	0.14	1.19	3.26
	PD	2.71	2.85	1.82	3.79	2.88	0.38	−0.19	0.2
E7	SD	3.75	4.49	3.39	5.11	4.17	0.31	0.4	0.09
	CWT	0.86	0.67	0.52	1.46	0.83	0.17	0.81	0.78
	PD	2.03	3.16	1.18	4	2.51	0.48	0.04	0.01
E8	SD	4.06	4.01	3.26	4.59	3.88	0.28	0.35	−0.32
	CWT	0.66	0.49	0.34	1.02	0.54	0.09	1.3	4.25
	PD	2.75	3.04	1.44	3.72	2.79	0.35	−0.26	1.19

LVB large vascular bundles; *SVB* small vascular bundles; *SD* stem diameter; *CWT* culm wall thickness; *PD* pith diameter; E1, Jiyuan, Henan province in 2008–2009 under normal environment; E2–E5, Tai'an, Shandong province in 2008–2009 under normal, rainfed, irrigation, late-sowing environment, respectively; E6, Jiyuan, Henan province in 2009–2010 under normal environment; E7–E9, Tai'an, Shandong province, in 2009–2010 under normal, rainfed, irrigation environment, respectively. The same as below

6.2.1.2.2 QTL Mapping and Effect Analysis of Anatomical Structure of the Second Basal Internode

A total of five QTLs conferring LVB on chromosomes 5D and 4D were detected in two years in four different environmental conditions (Table 6.15). Among them, the QTL on chromosome 5D detected in Jiyuan in 2008–2009 had the highest contribution, accounting for 13.69 % of phenotypic variation. A total of seven QTLs for SVB on chromosomes 1B, 5B, 6A, and 7D were identified, and the QTL detected in

Table 6.15 The additive effects for vascular bundle number of the second basal internode

Env.	Trait	Chromosome	Flanking marker	Site	Range	A	*P* value	h^2(a) %
E1	LVB	5D-11	Xwmc215–Xbarc345	77.3	63.2–84.3	−1.1724	0.0000	13.69
	SVB	1B-6	Xbarc119–Xgwm18	33.8	27.7–34.5	−1.6968	0.0000	17.12
		6A-4	Xbarc1077–Xbarc1165	41.2	35.5–42.2	−0.8242	0.0019	3.48
E2	LVB	4D-8	Xbarc190–Xbarc1009	165.5	156.7–172.4	−0.9156	0.0000	8.27
		5D-10	Xbarc320–Xwmc215	66.2	60.5–84.3	−1.1546	0.0000	11.37
	SVB	5B2-1	Xbarc36–Xbarc140	14.0	6.0–20.1	−1.5074	0.0000	10.87
E5	LVB	5D-2	Xcfd40–Xbarc1097	2.4	0.0–6.4	−0.8399	0.0000	10.39
	SVB	1B-7	Xgwm18–Xwmc57	34.5	25.7–34.9	−0.6899	0.0000	8.54
		1B-25	Xswes158–Xswes650	126.1	110.4–130.6	−0.7139	0.0001	7.80
		6A-8	Xbarc1055–Xwmc553	50.7	43.7–63.2	−0.9192	0.0000	9.28
E6	LVB	4D-10	Xbarc237–Xcfe254	169.4	164.3–177.4	−0.8101	0.0001	8.25
	SVB	7D-5	Xgwm676–Xgwm437	117.9	104.3–124.9	1.3033	0.0000	10.58

From E1 to E6 as the same as Table 6.14

Jiyuan condition in 2008–2009 had the highest contribution, accounting for 17.12 % of phenotypic variation. Meanwhile, the locus had the highest value of additive effect, which could increase the number of SVB by 1.69. In addition to the QTL controlling SVB on chromosome 7D, the positive alleles all came from Yumai 57.

The QTLs conferring LVB on chromosomes 5D and 4D and the QTLs conferring SVB on chromosomes 6A and 1B were detected twice in the four environmental conditions, accounting for 10.39–11.36, 8.25–8.27, 3.48–9.28, and 8.54–17.12 % of phenotypic variation, respectively. Other QTLs for the number of vascular bundle detected in this study were only one time and had poor reproducibility.

A total of seven QTLs controlling SD of the second basal internode on chromosomes 1A, 1B, 2D, 4B, and 5D were detected; furthermore, the QTL on chromosome 5D, which was detected in Jiyuan condition in 2008–2009, had the highest contribution, accounting for 15.49 % of phenotypic variation, and its additive effect was also the highest, which could increase SD by 0.19 mm (Table 6.16). In addition to the QTL conferring SD on chromosome 1A, the positive alleles of other QTLs all came from Yumai 57.

For CWT, seven QTLs on chromosomes 1B, 3D, 5B, and 6A were detected, and the QTL on chromosome 1B, detected in Jiyuan condition in 2008–2009, had the highest contribution, accounting for 13.41 % of phenotypic variation. Except for the QTL on chromosome 6A, positive alleles of other QTLs all originated from Yumai 57.

For PD, eight QTLs on chromosomes 1B, 2A, 2D, 3D, 4D, and 5D were detected, and the QTL detected in Jiyuan condition in 2008–2009 had the highest contribution, explaining 20.95 % of phenotypic variation. Except for the QTLs conferring PD on chromosomes 1B, 2D, and 5D, the positive alleles of other QTLs all came from Huapei 3.

The QTLs for SD and SWT on chromosome 1B stably express in all conditions in 2008–2009, but were not identified in all conditions in 2009–2010.

Table 6.16 Estimated the additive effect for SD, CWT, and PD in the DH population

Env.	Trait	Chromosome	Flanking marker	Site	Range	A	*P* value	h^2 (a) %
E1	SD	1B-6	Xbarc119–Xgwm18	33.8	33.0–34.5	−0.1167	0.0000	4.95
		2D-4	Xcfd53–Xwmc18	16.6	0.9–29.6	−0.1505	0.0000	8.12
		5D-10	Xbarc320–Xwmc215	64.2	58.5–69.2	−0.1910	0.0000	15.49
	CWT	1B-13	Xbarc372–Xwmc412.2	36.1	35.2–36.1	−0.0200	0.0000	13.41
	PD	2D-4	Xcfd53–Xwmc18	18.6	0.9–31.6	−0.1300	0.0000	6.91
		3D-11	Xwmc631–Xbarc071	86.0	77.3–93.9	0.0914	0.0002	11.33
		5D-10	Xbarc320–Xwmc215	66.2	62.2–76.3	−0.1655	0.0000	20.95
E2	SD	1B-2	Xwmc406–Xbarc156	28.7	24.0–33.0	−0.1111	0.0000	11.08
	CWT	1B-12	Xcfe023.1–Xbarc372	36.1	36.1–37.0	−0.0104	0.0000	9.65
	PD	1B-2	Xwmc406–Xbarc156	27.7	9.0–33.0	−0.0711	0.0006	9.48
		2D-3	Xwmc112–Xcfd53	0.9	0.0–22.6	−0.0714	0.0002	8.62
E5	SD	1B-9	Xcwem6.1–Xwmc128	34.9	33.8–35.2	−0.0852	0.0000	11.37
	CWT	1B-4	Xwmc31–Xwmc626	33.0	30.7–34.9	−0.0129	0.0000	6.79
		5B1-5	Xgwm213–Xswes861.2	68.1	53.1–68.1	−0.0144	0.0000	8.21
		6A-9	Xwmc553–Xgwm732	70.2	59.2–83.4	0.0166	0.0000	9.20
	PD	7A-4	Xbarc070–Xbarc250	23.5	18.8–36.5	0.0632	0.0001	8.08
E6	SD	1A-5	Xwmc550–Xbarc269	44.3	29.4–53.6	0.1109	0.0000	13.47
E7	SD	4B-7	Xwmc48–Xbarc1096	18.4	14.7–18.4	−0.0863	0.0002	7.92
	CWT	6A-8	Xbarc1055–Xwmc553	46.7	40.5–54.2	0.0152	0.0000	9.62
E8	CWT	3D-10	Xbarc323–Xwmc631	82.0	75.3–88.0	−0.0333	0.0000	12.84
	PD	2A-17	Xgwm448–Xwmc455	74.6	71.7–79.6	0.1035	0.0001	9.71
		4D-1	Xwmc473–Xbarc334	0.0	0.0–9.0	0.0881	0.0004	7.99

From E1 to E6 as the same as Table 6.14

For the QTLs conferring PD detected in 2009–2010, only the QTL on chromosome 2D was detected in the two environments, while other QTLs were detected only one time in different environmental condition.

6.2.2 QTL Mapping for Anatomical Traits of the Uppermost Internode

6.2.2.1 Materials and Methods

6.2.2.1.1 Materials

The test materials were the same as the previous paper for the second basal internode.

6.2.2.1.2 Field Trails

Field trials were conducted in 2006–2007 and 2007–2008 in Tai'an, Shandong province, China. The experimental field design consisted of a randomized block design with two replications. One environment (environment I) was conducted in 2006. All lines and parents were grown in 2-m-long four-row plots (25 cm apart). Crop management was carried out following the local practices, which were irrigation in wintering, jointing, anthesis, and grain filling stages. An additional 225 and 75 kg/ha of urea were top-dressed at jointing and anthesis, with irrigation, respectively. The total water (millimeter) (rainfall 122.0 mm; irrigation 270.0 mm) and accumulated temperature days were 2605.1 °C day during the whole life of the wheat. Three environments (environment II–IV) were conducted in 2007 in the same soil conditions. The total water (mm) (rainfall 172.5 mm; irrigation 180.0 mm) and accumulated temperature days were 2362.5 °C day during the whole life of the wheat. The management of ground fertilizers, irrigation, and top-dressed fertilizers of environment II was the same as that of environment I; the management of ground fertilizers and irrigation of environment III was the same as that of environment I, but there was no top-dressed urea applied at the jointing and anthesis stages; the management of ground fertilizers and top-dressed urea of environment IV was the same as that of environment I, but there was no irrigation during the wheat's entire growing season.

6.2.2.1.3 Determining Methods

At anthesis stage, three main stems (flowered on the same day) in the center of every plot were selected. The lengths of uppermost internodes and the diameters at 2 cm below the neck of the spike were determining by using ruler and vernier caliper, respectively. Furthermore, 2 cm of uppermost internodes were put in stationary liquid (absolute ethyl alcohol/glacial acetic acid = 3/1) for 4–10 h, and then put in 70 % ethanol in the refrigerator. The numbers of large vascular bundles (LVB) located in the inner parenchyma of the stem (with diameters equal to or greater than 10 μm) were recorded, and the numbers of total vascular bundles (TVB) were counted with a Nikon YS100 (Nikon Co. Ltd, Nanjing, China) microscope (magnified 100×). According to TVB = LVB + SVB, the numbers of small vascular bundles (SVB) and the ratios of large to small vascular bundles (L/S) were calculated (Fig. 6.6). Mean values were used for the data analysis.

6.2.2.2 QTL Mapping for Anatomical Traits of the Uppermost Internode

6.2.2.2.1 Variation of Phenotypic Data of Uppermost Internode-Related Traits

Huapei 3 is a weak spring and precocity variety, while Yumai 57 is a semi-winterness and medium maturing variety. All lines and parents were planted in October 2006 and 2007 on experimental farm in Shandong Agricultural University. And four environmental conditions including environment I (normal

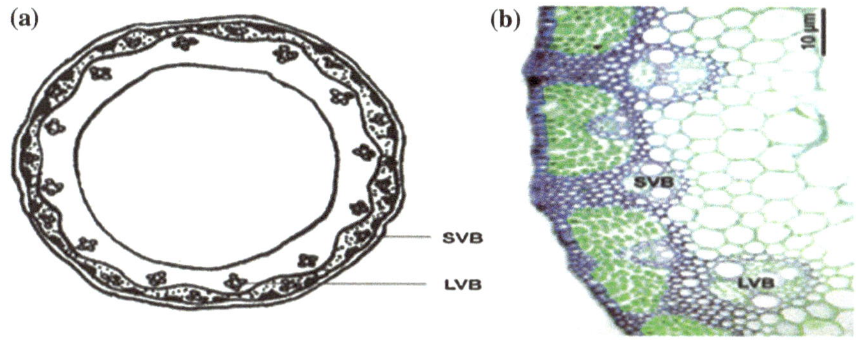

Fig. 6.6 The diagram and microscope structure of the internode in wheat **a** transverse section structure of wheat internode; **b** magnified structure of uppermost internode in wheat (magnified 200). *LVB* large vascular bundles; *SVB* small vascular bundles

irrigated and top-dressed urea in 2007), environment II (normal irrigated and top-dressed urea in 2008), environment III (normal irrigated but no top-dressed urea in 2008), and environment IV (no irrigated but top-dressed urea in 2008) were set. There were obvious differences in uppermost internode length (UIL), the number of total vascular bundles (TVB) and the number of small vascular bundles (SVB) between the parents. And the frequency distributions for eight traits (UIL, UID, CWT, CWA, TVB, LVB, SVB, and L/S) examined in the DH population of wheat exhibited continuous variations and more or less normal distributions with transgressive segregation, indicating that all the eight traits are quantitative traits controlled by polygenes (Table 6.17 and Fig. 6.7), and their quantitative nature which is suitable for QTL mapping.

6.2.2.2.2 QTL Mapping and Effect Analysis of Uppermost Internode-Related Traits

QTL conferring eight traits associated with morphological anatomy of uppermost internode were performed by using the software QTL Network 2.0 based on a mixed linear model. A total of 20 additive QTLs and one pair of epistatic QTL on chromosomes 1A, 1B, 2A, 2D, 3D, 4D, 5D, 6A, 6D, and 7D were detected (Table 6.18 and Fig. 6.7).

6.2.2.2.2.1 QTL Mapping for UIL

Three additive QTLs for UIL, located on chromosomes 1B, 4D, and 7D (Table 6.18 and Fig. 6.7) were detected and expressed differently in the four environments. In environment I, two QTLs on chromosomes 4D and 7D were detected; in environment

Table 6.17 Phenotypic performance of vascular bundle system under four environments

Trait	Env.	Hupei 3	Yumai 57	DH population					
				Mean	Max	Min	SD	Skewness	Kurtosis
UIL (cm)	EI	30.17	25.77	24.35	35.15	12.50	5.16	−0.11	−0.60
	E II	28.77	25.93	25.28	36.53	13.97	5.28	−0.00	−0.91
	E III	26.37	24.87	25.26	38.10	12.30	5.01	−0.14	−0.62
	E IV	27.97	25.30	23.96	33.83	12.50	4.88	0.00	−0.76
UID (mm)	E I	2.73	2.33	2.56	3.28	1.71	0.32	−0.07	−0.37
	E II	2.68	2.55	2.55	3.17	2.00	0.21	0.16	0.28
	E III	2.40	2.36	2. 42	3.35	1.85	0.25	0.60	0.90
	E IV	2.42	2.32	2. 24	3.18	1.67	0.23	0. 87	1.81
CWT (μm)	E I	390.0	368.3	406.8	560.0	290.0	50.05	0.27	−0.18
	E II	377.8	355.8	354.8	438.3	283.3	29.21	−0.00	−0.35
	E III	399.2	356.7	412.3	505.0	310.0	37.48	0.20	−0.31
	E IV	393.8	384.6	366.8	468.3	296.7	32.78	0.12	−0.15
CWA (mm^2)	E I	3.36	2.26	2.76	4.26	1.29	0.54	0.25	−0.06
	E II	2.74	2.45	2.44	3.36	1.57	0.33	0.15	−0.14
	E III	2.50	2.24	2.60	4.17	1.70	0.38	0.55	1.27
	EIV	2.50	2.00	2.16	3.45	1.41	0.32	0.71	1.33
TVB	E I	56	63	54	76	40	7.11	0.56	0.33
	E II	52	64	53	72	41	5.40	0.25	−0.04
	EIII	46	61	51	70	38	5.69	0.44	0.20
	E IV	46	60	53	73	38	5.74	0.68	0.75
LVB	E I	12	13	14	22	5	3.19	0.24	−0.05
	E II	12	14	13	20	6	2.49	0.35	0.50
	E III	11	12	12	18	7	2.28	0.30	0.08
	E IV	11	13	13	20	8	2.54	0.22	−0.20
SVB	EI	44	50	40	59	27	6.28	0.49	−0.03
	E II	40	50	41	59	29	4.90	0.23	0.28
	E III	35	49	39	57	26	5.50	0.49	0.24
	E IV	35	47	39	54	29	5.01	0.48	0.49
L/S	E I	0.27	0.26	0.36	0.67	0.11	0.10	0.50	0.12
	E II	0.30	0.28	0.31	0.53	0.16	0.07	0.50	0.19
	E III	0.31	0.24	0.31	0.59	0.13	0.08	0.54	0.40
	EIV	0.31	0.28	0.34	0.62	0.17	0.08	0.52	0.52

UIL uppermost internode length; *UID* uppermost internode diameter; *CWT* culm wall thickness; *CWA* culm wall area; *TVB* the number of total vascular bundles; *LVB* the number of large vascular bundles; *SVB* the number of small vascular bundles; *L/S* the ratio of large and small vascular bundles; *EI* normal irrigated and top-dressed urea in 2007; *EII* normal irrigated and top-dressed urea in 2008; *EIII* normal irrigated but no top-dressed urea in 2008; *EIV* no irrigated but top-dressed urea in 2008

II, three QTLs on chromosomes 1B, 4D and 7D were detected; in environment III, two QTLs on chromosomes 4D and 7D were detected; and in environment IV, the QTLs were detected on chromosomes 1B and 7D. The QTLs on chromosomes 4D and 7B were contributed by Huapei 3, and the contributions were 9.54 ~ 22.04 %; while the

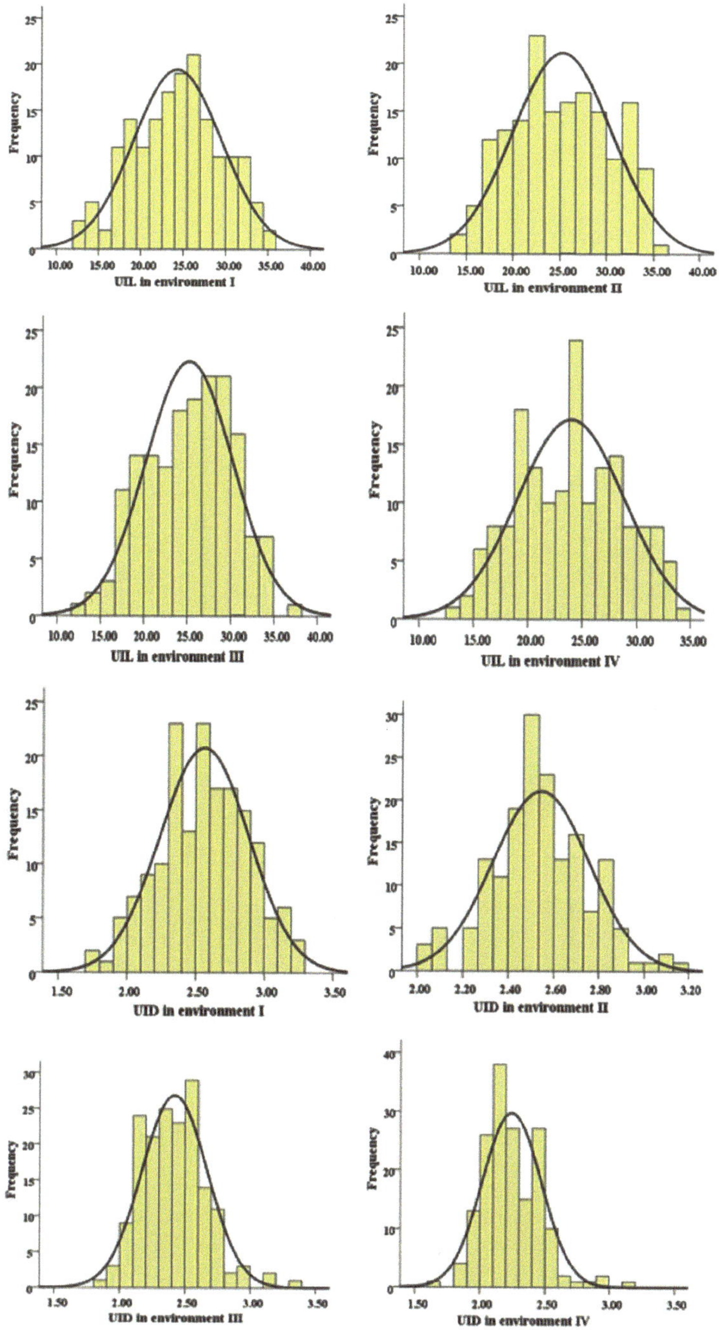

Fig. 6.7 Distribution of vascular bundle system and correlative traits of uppermost internode in DH population (environment and traits see Table 6.17)

Table 6.18 Main-effect QTL affecting TVB, LVB, SVB, and L/S in four environments

QTL Loci	Flanking marker	Position	A[a]	H^2 (A, %)[b]
QTL detected	Environment I			
q*UIL-4D*	XBARC334–XWMC331	2.1	2.21	17.19
q*UIL-7D*	XGWM676–XGWM437	120.9	1.84	11.97
qUID-5D	XWMC215–XBARC345	71.4	0.16	22.67
qCWA-5D	XWMC215–XBARC345	75.4	0.29	25.61
qTVB-5D	XWMC215–XBARC345	69.4	2.03	8.11
qLVB-5D	XWMC215–XBARC345	73.4	1.57	22.95
qSVB-5D-1	XCFD40–XBARC1097	2.4	−1.79	8.11
qL/S-5D	XWMC215–XBARC345	77.4	0.04	12.45
QTL detected	Environment II			
qUIL-1B	XWMC31–XWMC626	33.0	−1.43	7.08
qUIL-4D	XBARC334–XWMC331	5.1	1.66	9.54
qUIL-7D	XGWM676–XGWM437	118.9	2.05	14.49
qTVB-2D-1	XCFD53–XWMC18	1.7	−1.49	7.61
qTVB-7D	XGWM676–XGWM437	114.9	1.70	9.88
qSVB-7D	XGWM676–XGWM437	117.9	1.80	13.11
QTL detected	Environment III			
qUIL-4D	XBARC334–XWMC331	6.1	1.72	10.74
qUIL-7D	XGWM676–XGWM437	118.9	2.47	22.04
qCWT-3D	XWMC631–XBARC071	82.1	−10.97	8.52
qTVB-2D-2	XWMC112–XCFD53	1.0	−1.87	10.49
qSVB-2D	XWMC112–XCFD53	1.0	−1.52	8.96
qSVB-6A	XBARC1077–XBARC1165	41.2	−1.42	7.80
qSVB-7D	XGWM676–XGWM437	119.9	1.73	11.58
QTL detected	Environment IV			
qUIL-1B	XWMC31–XWMC626	33.0	−1.49	9.00
qUIL-7D	XGWM676–XGWM437	118.9	1.87	14.16
qTVB-1A	XWMC333–XBARC148	57.8	−2.04	12.76
qSVB-1A	XWMC333–XBARC148	57.8	−1.92	12.05
qSVB-5D-2	XBARC320–XWMC215	66.3	1.70	9.53
qSVB-7D	XGWM676–XGWM437	116.9	1.78	10.44
qL/S-7D	XGWM295–XGWM676	101.3	−0.03	15.17

Note: [a]Additive effects, a positive value indicates that allele from Hp3 increases the trait, a negative value indicates that allele from YM57 increases the trait

[b]Contribution explained by additive QTL

Environment I normal irrigated and top-dressed urea in 2007; *Environment II* normal irrigated and top-dressed urea in 2008; *Environment III* normal irrigated but no top-dressed urea in 2008; *Environment IV* no irrigated but top-dressed urea in 2008

QTLs on chromosome 1B were contributed by Yumai 57, the contributions were 7.08 ~ 9.00 %. The QTL located within XGWM676–XGWM437 on chromosome 7D had the most significant effect, explaining 22.04 % of phenotypic variation, and expressed stably in the four environments.

6.2.2.2.2.2 QTL Mapping for UID

Only one additive QTL conferring UID on chromosome 5D was detected in environment I, explaining 22.67 % of phenotypic variation. And its positive allele came from Huapei 3, increasing UID by 0.16 mm.

6.2.2.2.2.3 QTL Mapping for CWT

Only one additive QTL conferring CWT on chromosome 3D was detected in environment III, while its contribution to phenotypic variation was small. And its positive allele came from Yumai 57, increasing CWT by 10.97 μm.

6.2.2.2.2.4 QTL Mapping for CWA

Only one additive QTL for CWA on chromosome 5D was identified in environment I, explaining 25.61 % of phenotypic variation, and its additive effect was contributed by Huapei 3 alleles, increasing CWA by 0.29 mm^2.

6.2.2.2.2.5 QTL Mapping for TVB

A total of five additive QTLs for TVB were detected on chromosomes 1A, 2D (two regions), 5D, and 7D under four different environments (Table 6.18 and Fig. 6.8). Under environment I, one QTL on chromosome 5D was detected; under environment II, two QTLs on chromosome 2D and 7D were detected; under environment III, one QTL on chromosome 2D was detected; under environment IV, one QTL was detected on chromosome 1A. The additive effects of *qTVB-5D* and *qTVB-7D* were contributed by Huapei 3 alleles, and the rest were contributed by Yumai 57 alleles. The QTL detected under environment III and IV (*qTVB-1A* and *qTVB-2D-2*) had the most significant effects, explaining 12.76 and 10.49 % of the phenotypic variance, respectively.

One pair of epistatic QTL for TVB was detected on chromosome 2A and 6D (Fig. 6.9), accounting for 10.81 % of phenotypic variation and increasing by two TVBs. The epistatic QTL had no additive effect, was sensitive to environment, and only expressed in environment III.

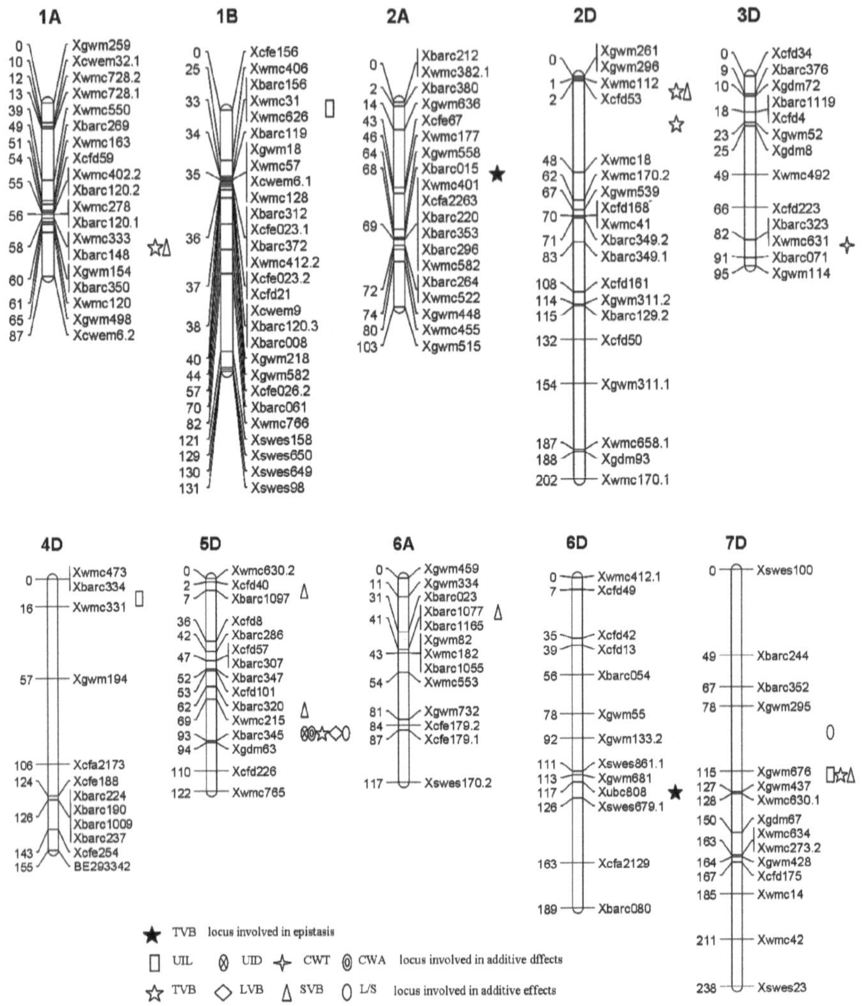

Fig. 6.8 QTL for the vascular bundle system and correlative traits of uppermost internode in SSR linkage map

6.2.2.2.2.6 QTL Mapping for LVB

Only one *qLVB-5D* for LVB was detected on chromosome 5D, contributed by Huapei 3 alleles. The *qLVB-5D* had a significant effect, accounting for 22.95 % of the phenotypic variance, and increased by two LVBs.

6.2.2.2.2.7 QTL Mapping for SVB

A total of six additive QTLs conferring SVB were detected on chromosomes 1A, 2D, 5D (2 regions), 6A, and 7D (Table 6.18 and Fig. 6.8). Under environment I,

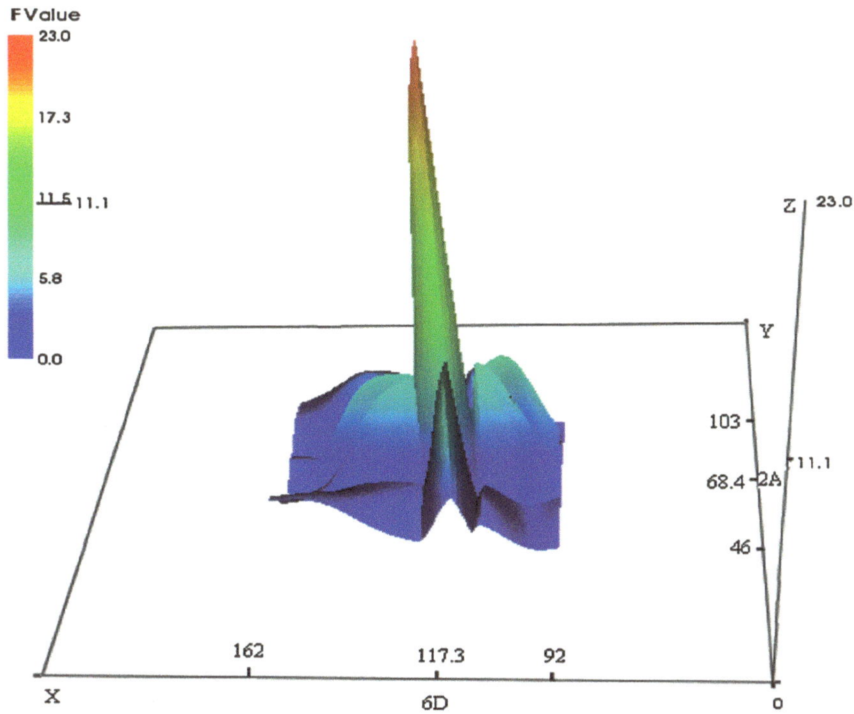

Fig. 6.9 3D visualization for the test statistics of genome scan for epistatic QTL associated with total vascular bundle under environment III (normal irrigated but no top-dressed urea in 2008) between 2A and 6D (F value is taken as height); AA: 1.9; H: 10.81 %

one QTL on chromosome 5D was detected; under environment II, one QTL on chromosome 7D were detected; under environment III, the QTLs on chromosome 2D, 6A, and 7D were detected; under environment IV, the QTLs on chromosomes 1A, 5D, and 7D were detected. Among them, the QTLs on chromosomes 1A, 2D, and 7D were main-effect QTLs.

6.2.2.2.2.8 QTL Mapping for L/S

Two QTLs affecting L/S were identified in environments I and IV and contributed by Huapei 3 and Yumai 57, explaining 12.45 and 15.17 % phenotypic variance, respectively.

In a word, after analyzing anatomical characteristics of the basal internode and uppermost internode, it was found that there were main-effect QTLs on chromosomes 5D and 7D, declaring that important gene and region confer the traits on these chromosomes. Meanwhile, some important genes were also found on chromosomes 1B, 4D, and 2D.

6.2.3 Research Progress of Anatomical Traits of Culm QTL Mapping and Comparison of the Results with Previous Studies

6.2.3.1 Research Progress of QTL Conferring Anatomical Traits of Culm and the Comparative Analysis with This Study

6.2.3.1.1 Research Progress of QTL Conferring Anatomical Traits of Culm

Because stem strength was closely related to lodging resistance of wheat, it will be of great significance for enhancing the lodging resistance of wheat by studying QTL for stem strength-related traits. Marza et al. (2006), Huang et al. (2006), and Zhang et al. (2008) conducted QTL analysis of lodging resistance of wheat by using DH population or RIL population. And a total of 16 QTLs, including four main-effect QTLs, were detected on chromosomes 1B, 1D, 2B, 4A, 4B, 4D, 5A, 6D, and 7D, and the highest contribution was 23 % of phenotypic variation.

Few researches related to QTL for anatomical traits of stem were conducted. Only Keller et al. (1999) in abroad and Guo et al. (2002) in domestic studied UIL, UID, CWT, culm wall strength, the length, and diameter of other internode; however, few studies associated with TVB, LVB, SVB, and L/S were conducted (Table 6.19).

Table 6.19 Summary of QTL of wheat stem microdissection traits (PVE > 10 %)

Trait	QTL	Flanking marker	PVE (%)	Population	Reference
Stem trait		Xpsr949–Xgwm18	12	RIL	Keller et al. (1999)
		Xpsr958–Xpsr566c	15		
		Xpsr933b–Xglk529a	15		
		Xpsr598–Xpsr570	21		
		Xgwm397–Xglk315	23		
		Xpsr918b–Xpsr1201a	31		
		Xpsr370–Xpsr580b	20		
Stem strength	*QSs-3A*	Xwmc527–Xwmc21	10.61	DH	Hai et al. (2005)
	QSs-3B	Xgwm108–Xwmc291	16.6		
PD	*QPd-1A*	Xgwm135–Xwmc84	10.72		
	QPd-2D	Xgwm311–Xgwm301	18.7		
LVB	*5A*	xgwml86–xgwm415	18	DH	Guo (2002)
	4B	xgwm368–xgwm276	38		
SVB	*2A*	xgwm294–xgwm356	14		
	5B	xgwm99–xgwml64	11		

Abbreviations are the same as in Table 6.14

6.2.3.2 Comparison of this Study with Previous Researchers

We researched QTL for anatomical traits of the second basal internode and the uppermost internode for the first time. A total of 62 QTLs, including 31 major QTLs (contribution is greater than 10 %), were detected on chromosomes 1B, 2D, 4D, 5D, and 7D, and the most significant QTL could explain 25.61 % of phenotypic variation. Further, some loci controlled multiple traits. In addition, it was found that LVB and SVB were controlled by different genes, and the locus (*Qlvb.sdau-5D*) on chromosome 5D, controlling LVB, was detected in several environments. Comparing with the results of spike yield, leaf morphology, and related traits studied by Zhang et al. (2008) using the same DH population, the QTLs for LVB and spike and leaf traits were on the same or near regions and tended to be co-located within the genome, which can be used as marker to polymerize multiple excellent traits in breeding program.

The main-effect QTLs conferring UID, CWA, TVB, LVB, and L/S were all located within the interval XBARC320–XWMC215–XBARC345 on chromosome 5D, and nearby the main-effect interval controlling grain yield and spike-correlated traits (kernels for spike, the total number of spikelets, density of spikelet) (Zhang et al. 2009). In addition, the QTLs conferring UIL, LVB, and SVB located within the interval of XGWM676–XGWM437 on chromosome 7D had high contribution and stably expressed in four different environments, which can be used in marker-assisted selection (MAS) to polymerize several traits and improve multiple traits simultaneously in wheat breeding.

In a word, the QTL conferring stem-correlated traits distributed on chromosomes 1B, 2A, 2D, 3D, 4D, 5D, 6A, 7A, and 7D. Comparing with previous researches, more loci were detected on genome D in this study; furthermore, QTL cluster controlling important physiology and yield-correlated traits was located on chromosome 5D.

6.3 QTL Mapping and Effect Analysis of Heading Date

Heading date is an important trait that is a major determinant of the regional and seasonal adaptation of wheat varieties. Appropriate heading date and anthesis are important target traits for breeding, which not only correlate with growth period, but also directly or indirectly affect some important agronomic traits such as yield, disease resistance, and stress resistance. According to the different signal response to the environment, there are three categories of genes influence heading date including the following: (1) vernalization response (Vrn), controlling winter wheat took on low temperature treatment for a certain time before ear differentiation; (2) photoperiod response (Ppd), decides the response to the length of sunlight; and (3) earliness per se (Eps), when vernalization and photoperiod are satisfied, the number of days for wheat to heading date is determined by Eps, which control developmental rate independently of the other two genes. Vrn-A1, Vrn-B1, and

Vrn-D1 were located on long arms of chromosomes 5A, 5B, and 5D, and Vrn-A1 had the highest effect, showing the vernalization insensitivity. Now, Vrn-A1 and Vrn-B1 have been successfully cloned (Yan et al. 2004). The genes Ppd-A1, Ppd-B1, and Ppd-D1 were located on chromosomes 2A, 2B, and 2D, respectively, and Ppd-D1 had the highest effect, followed by Ppd-B1. Furthermore, these genes were all insensitive to photoperiod. However, few research for Eps was conducted. While, Eps was located on chromosomes 2B, 3A, 4A, 4B, 6B, 6D, and 7B by using aneuploid of Chinese spring and chromosome substitution. Song et al. (2006) identified nine QTLs for wheat heading date on chromosomes 2D, 3B (2 regions), 3D, 4A, 5B, 6B, 6D, and 7D, explaining 3.97–22.91 % of phenotypic variation, by using two mapping populations (Hanxuan 10 × Lumai 14 and Wenmai 6 × Shanhongmai) in field and greenhouse. Since some researches regarding QTL analysis for wheat growth period were conducted, few could be used in MAS. Therefore, in this study, several populations were used to analyze the QTL for wheat growth period, in order to find reliable and stable markers that can be used in MAS in wheat breeding programs.

Heading date was recorded as the number of days from sowing to 50 % of spikes fully emerging in a plot. And heading was noted when 1/3 of spikes emerged from the flag leaves.

6.3.1 *QTL Analysis of Heading Date Based on a DH Population Derived from the Cross of Huapei 3 × Yumai 57*

6.3.1.1 Phenotypic Variation of Heading Date

The heading date for the DH population and the parents in three environments were described in Fig. 6.10. Huapei 3 headed significantly earlier than Yumai 57 in all

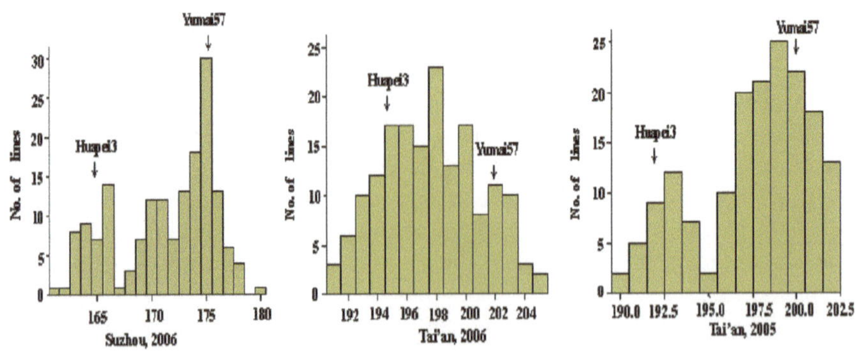

Fig. 6.10 Frequency distribution of heading date

three environments. Transgressive segregants were observed for heading date among DH lines in the three environments. The heading date of the DH population segregated continuously and followed a normal distribution, indicating its polygenic inheritance and suitability of the data for QTL analysis.

6.3.1.2 QTLs with Additive Effects and Additive × Environment (AE) Interactions

Two additive QTLs were detected for heading date on chromosomes 1B and 5D (Table 6.20). A highly significant QTL, designated as *Qhd5D*, was observed within the Xbarc320–Xwmc215 interval on the chromosome 5DL, accounting for 53.19 % of the phenotypic variance. The second QTL, *Qhd1B*, could explain 3.49 % of the phenotypic variance. The Huapei 3 alleles at the *Qhd5D* reduced days-to-heading by 2.77 days due to additive effects, but increased days-to-heading by 0.71 days at the *Qhd1B*. This suggested that alleles for reducing the heading date were dispersed within the two parents. This result was in accordance with the presence of a wide range of variation and transgressive segregations of wheat heading date in the DH population. The total additive QTLs for heading date accounted for 56.68 % of the phenotypic variance. The *Qhd5D* showed AE interactions in two environments, accounting for 3.81 and 1.51 % of the phenotypic variance, respectively. The general contribution of the two AE effects on wheat heading date was 5.32 %.

6.3.1.3 Epistasis and Epistasis × Environment (AAE) Interactions

Two pairs of digenic epistatic interactions were identified for heading date, located on chromosomes 2B–6D and 7A–7D (Table 6.21), explaining phenotypic variance from 2.45 to 3.44 %, respectively. The general contribution of digenic epistatic interactions to heading date was 5.90 %. The *Qhd2B/Qhd6D* was involved in AAE interactions in two environments, which explain 0.65 and 0.73 % of the phenotypic variance, respectively. The total contribution of AAE interactions was 1.38 %.

6.3.2 QTL Analysis of Heading Date Based on a RIL Population Derived from the Cross of Nuomai 1 × Gaocheng 8901

6.3.2.1 Phenotypic Variation of Heading Date

The heading date of the RIL population and the parents in three environments were described in Table 6.22 and Fig. 6.11. Nuomai 1 headed significantly earlier than Gaocheng 8901 in all three environments. Transgressive segregants were observed

Table 6.20 Estimated additive and additive × environment interaction of QTL for heading time

Trait	QTL	Flanking marker	Site (cM)	A	H^2 (A, %)	AE1	H^2 (AE1, %)	AE2	H^2 (AE2, %)	AE3	H^2 (AE3, %)
Heading date	*QHd1B*	Xwmc406–Xbarc156	26.7	0.71	3.49						
	QHd5D	Xbarc320–Xwmc215	67.2	−2.77	53.19	−0.74	3.81			0.47	1.51

E1: Suzhou, 2006; E2: Tai'an, 2006; E3: Tai'an 2005

Table 6.21 Estimated epistasis (AA) of QTL for heading time

Trait	QTL	Flanking marker	Site (cM)	QTL	Flanking marker	Site (cM)	AA	H^2 (AA, %)	AAE1	H^2 (AAE1, %)	AAE2	H^2 (AAE2, %)	AAE3	H^2 (AAE3, %)
Heading date	*qHd2B*	Xbarc129.1–Xgwm111	98.3	qHd6D	Xgwm133.2–Xgwm681	91.9	−0.59	2.45	−0.31	0.65	0.33	0.73		
	qHd7A	Xwmc603–Xwmc596	62.6	qHd7D	Xgwm295–Xgwm676	40.3	0.71	3.44						

E1: Suzhou, 2006; E2: Tai'an, 2006; E3: Tai'an 2005. AAE1: epistasis effect of Env.1 (Suzhou 2006); AAE1 H^2/%: H^2/% of Env.1 (Suzhou 2006); AAE2: epistasis effect of Env.2 (Tai'an 2006); AAE2 H^2/%: H^2/% of Env.2 (Tai'an 2006); AAE3: epistasis effect of Env.3 (Tai'an 2005); AAE3 H^2/%: H^2/% of Env.3 (Tai'an 2005)

Table 6.22 Phenotypic values for heading stage of two parents and the RIL population in three environments in wheat

Trait	Environment	Parent		RIL population			
		Nuomai1	Gaocheng8901	Range	Mean ± SD	Skewness	Kurtosis
Heading date (d)	E1	197	201	194 ~ 206	200.1 ± 2.22	0.38	0.44
	E2	199	202	197 ~ 207	202.4 ± 1.80	0.07	−0.66
	E3	205	208	203 ~ 211	206.4 ± 1.73	0.1	−0.80

E1: Tai'an, 2008; E2: Tai'an, 2011; E3: Tai'an, 2011

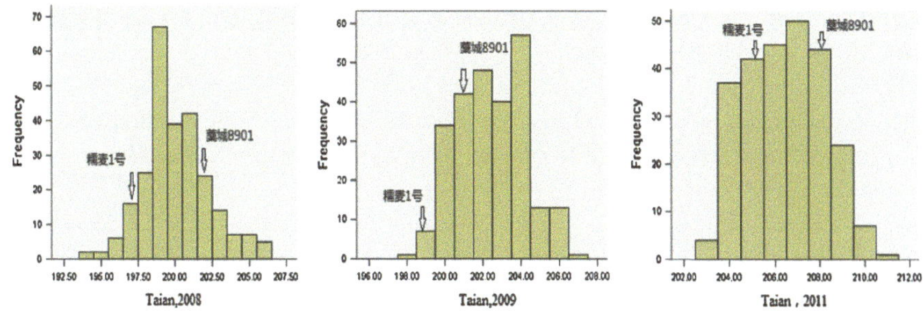

Fig. 6.11 The heading time distribution of the RIL population in three different environments

for heading date among RIL lines in the three environments. The heading date of the RIL population segregated continuously and followed a normal distribution, and both absolute values of skewness and kurtosis were less than 1.0, indicating its polygenic inheritance.

6.3.2.2 QTLs with Additive Effects and Additive × Environment (AE) Interactions

A total of five additive QTLs conferring heading date on chromosomes 3B, 5B, 6A, 6B, and 7D were identified (Table 6.23). *Qhs-6A* and *Qhs-6B* were detected four times in the three different environments and mixed environment and contributed by Gaocheng 8901, accounting for 16.16 ~ 25.64 % and 5.75 ~ 9.88 % of

Table 6.23 QTL with significant additive effects for heading stage detected in different environments

Trait	Env.	QTL	Chr.	Site (cM)	Interval marker	A	H^2 (A, %)
Heading date	E1	*Qhs-3B*	3B	58.00	wPt-9510-wPt-664393	0.53	5.65
		Qhs-6A	6A	41.00	xgpw-4085-wPt-667562	−0.89	16.16
		Qhs-6B	6B	17.00	xgpw-3241-wPt-0315	−0.55	5.75
	E2	*Qhs-6A*	6A	39.00	xgpw-4085-wPt-667562	−0.91	25.64
		Qhs-6B	6B	14.00	xgpw-3241-wPt-0315	−0.49	7.19
	E3	*Qhs-3B*	3B	57.00	wPt-9510-wPt-664393	0.44	6.55
		Qhs-6A	6A	44.00	xgpw-4085-wPt-667562	−0.87	25.44
		Qhs-6B	6B	12.00	xgpw-3241-wPt-0315	−0.53	9.22
	PD	*Qhs-3B*	3B	58.00	wPt-9510-wPt-664393	0.40	5.12
		Qhs-5B	5B	112.00	wPt-9454-wPt-3457	−0.38	4.57
		Qhs-6A	6A	43.00	xgpw-4085-wPt-667562	−0.75	18.03
		Qhs-6B	6B	15.00	xgpw-3241-wPt-0315	−0.57	9.88
		Qhs-7D	7D	90.00	wPt-730876-wPt-8343	0.85	22.67

E1: Tai'an, 2008; E2: Tai'an, 2011; E3: Suzhou, 2011; *PD* Pool data; Positive values indicate that Nuomai 1 alleles increase corresponding trait value; Negative values indicate that Gaocheng 8901 alleles increase corresponding trait value phenotypic variation, respectively. And the contributions of *Qhs-6A* were all greater than 10 % in each environment. *Qhs-3B* was detected in E1, E3, and pool data (PD) three times and contributed by Nuomai 1, explaining 5.65, 6.55, and 9.22 % of phenotypic variation. In PD, *Qhs-7D* with a LOD value 15.49, located on chromosome 7D, accounted for 22.67 % of phenotypic variation. *Qhs-5B* was only detected on PD.

6.3.3 QTL Analysis of Heading Date Based on a RIL Population Derived from the Cross of Shannong 01-35 × Gaocheng 9411

6.3.3.1 Phenotypic Variation of Heading Date

There were smaller differences between two parents and bigger differences among lines in heading date and anthesis under the three environments. Furthermore, heading date varied from 194 to 206 in E1 and anthesis varied from 204 to 213. The heading time and flowering time of the RIL population segregated continuously and followed a normal distribution, and both absolute values of skewness and kurtosis were less than 1.0 (Table 6.24).

6.3.3.2 QTLs with Additive Effects and Additive × Environment (AE) Interactions

A total of 12 additive QTLs for heading time and four additive QTLs for flowing time were identified on chromosomes 1A, 1B, 4B, 6A, and 6B, respectively, using phenotypic data from E1, E2, and E3 and the mean value of the three environments (Table 6.25). The QTLs distributing on chromosomes 1A, 4B, and 6B were contributed by Gaocheng 8901, the rest were contributed by Shannong 01-35.

Four major QTLs for heading time, *QHt1A.1-54* (PD), *QHt1A.2-132* (E2, PD), *QHt1B.1-87* (E1, PD), and *QHt1B.2-44* (E2, E3) were detected, accounting for 10.75 ~ 30.32 % of phenotypic variation. Furthermore, *QHt1A.2-132*, *QHt1B.1-87,* and *QHt1B.2-44* were detected by using both individual environment and average environment, which were stably major QTLs. In addition, *QHt1A.2-133* and *QHt1A.2-132* detected in E1 were located within the same interval.

The QTL for flowering time, designated as *QFt1B.1-87* (E2) and *QFt1B.1-105* (E3), were major QTLs, explaining 13.23 and 15.77 % of phenotypic variation, respectively. Meanwhile, *QFt1B.1-87* and *QHt1B.1-87*, controlling heading time, were located on the same locus; and *QFt6B.3-5* and *QHt6B.3-0* were within the same interval.

Table 6.24 Phenotypic values for heading time and flowering time of the RIL population and the parents in different environments

Trait	Environment	Parent		RIL population			
		Shannong 01-35	Gaocheng 9411	Mean ± SD	Range	Skewness	Kurtosis
Heading time (d)	E1	202	202	201.52 ± 1.96	194–206	−0.19	0.88
	E2	204	206	204.51 ± 1.66	201–208	−0.04	−0.90
	E3	205	206	205.49 ± 1.80	202–210	0.12	0.04
Flowering time (d)	E1	208	209	208.95 ± 1.30	204–213	0.21	1.31
	E2	209	210	210.3 ± 1.33	205–214	−0.20	0.79
	E3	209	210	210.84 ± 1.53	208–216	0.69	0.88

E1: 2008–2009 growing season at Tai'an site; E2: 2009–2010 growing season at Tai'an site; E3: 2010–2011 growing season at Tai'an site, the same as below

Table 6.25 Additive QTL for heading time and flowering time in different environments

Trait	Environment	QTL	Left marker	Right marker	A	LOD	PVE (%)
Heading time	E1	*QHt1A.2-133*	wPt-672089	wPt-730213	−0.60	5.03	9.32
		QHt1B.1-87	wPt-5562	wPt-8971	0.86	9.31	17.08
	E2	*QHt1A.2-132*	wPt-672089	wPt-730213	−0.51	5.52	9.46
		QHt1B.1-86	wPt-5562	wPt-8971	0.98	17.56	30.32
	E3	*QHt1B.1-104*	wPt-5363	wPt-1363	0.70	8.45	10.75
		QHt1B.2-44	wPt-4497	CFE026	0.56	7.45	9.54
		QHt4B.1-60	xcfd54-4D	Xgpw2172	−0.44	4.28	5.79
		QHt6A.1-61	wPt-5652	CFE041	0.48	5.67	6.95
	PD	*QHt1A.1-54*	wPt-6005	wPt-730172	−0.62	4.79	14.40
		QHt1A.2-132	wPt-672089	wPt-730213	−0.54	6.96	11.07
		QHt1B.1-87	wPt-5562	wPt-8971	0.90	17.24	26.58
		QHt1B.2-44	wPt-4497	CFE026	0.50	6.87	9.43
		QHt4B.1-66	Xgpw2172	wPt-1505	−0.47	4.96	8.09
		QHt6B.1-14	wPt-0259	wPt-2095	0.41	4.73	6.27
		QHt6B.3-0	wPt-1325	wPt-669607	−0.43	5.27	6.62
Flowering time	E2	*QFt1B.1-87*	wPt-5562	wPt-8971	0.51	6.18	13.23
	E3	*QFt1B.1-105*	wPt-1781	wPt-0974	0.71	9.35	15.77
		QFt6B.3-5	wPt-1325	wPt-669607	−0.49	5.11	9.92
	PD	*QFt1B.1-104*	wPt-5363	wPt-1363	0.36	4.73	6.71

Positive and negative values of additive effect (EstAdd) indicate that alleles to increase thousand-grain weight are inherited from Shannong 01-35 and Gaocheng 9411, respectively

6.3.4 *Research Progress of Growth Period QTL Mapping and Comparison of the Results with Previous Studies*

6.3.4.1 Research Progress of QTL Mapping for Growth Period of Wheat

Heading date is a quantitative trait controlled by multiple genes. The QTL expressed differently in different environment, because of the interaction between genotype and environment. Many researches regarding QTL for growth period of wheat have been conducted, and there were loci detected on each chromosome. However, there were differences in detected loci by using different test materials, linkage map, and environment. Song et al. (2005, 2006), Yao et al. (2010), Xu (2005), and Hanocq et al. (2004) analyzed QTL conferring heading date of wheat by using different DH and RIL population and identified 21 QTLs, including seven major QTLs, with the highest effect of 22.91 % (Table 6.26). Summary analysis showed that most of the researchers found QTLs for heading date on chromosomes 7B, 2D, and 3B.

Table 6.26 Summary of QTL of wheat heading date (PVE > 10 %)

Trait	QTL	Flanking marker	(PVE)/%	Population	Reference
Heading date	*QDH.CAAS-5D*	Vrn-D1-WMS212	24.40/49.80	RIL	Yao et al. (2010)
	QDH.CAAS-7B.2	wPt4230-wPt4660	19.53		
	3B	WMS299-M539.1	16.36	DH	Song (2005)
	5B	WMS371-WMS335	22.91		
	6B	A1142.1-A8166.1	11.09		
	4B	WMS265-WMC161	10.38		
	3B	A2478.1-WMC505.1	12.53		
	7D	WMS295-WMC346	12.68		
	5B	WMS371-WMS335		RIL	Hanocqet et al. (2004)
	Ht-2D1	xgwm261–xgwm349	20.77	RIL	Xu (2005)
	qHd 5D	Xbarc320–Xwmc215	53.19	DH	Zhang (2008)

6.3.4.2 Comparsion of the Results with One of the Previous Studies

The major QTL (*qHd5D*) for heading time detected in this DH population were located within Xbarc320–Xwmc215 interval on chromosome 5DL, explaining 53.19 % of phenotypic variation, and closely linked with Vrn-D1. And it was contributed by precocious parent Huapei 3. In addition, *qHd5D* was closely linked with Xwmc215, with the genetic distance of 2.1 cm. Therefore, it was more likely used in MAS and polymerizing breeding programs. Another additive QTL (*qHd1B*) located within the interval Xwmc406–Xbarc156 on chromosome 1BS explained 3.49 % of phenotypic variation and was not found in the previous studies. It may be allelic loci of *Ppd-H2* on chromosome 1B and that needed further research to prove.

QTLs conferring heading date were located on chromosomes 3B, 6A, and 6B in the RIL population derived from the cross of Nuomai 1 × Gaocheng 8901. And the most significant QTL (*Qhs-6A*) was located on chromosome 6A, which was not found in the previous studies, indicating that chromosome 6A was a main chromosome for controlling heading date. Song et al. (2006) also identified QTLs conferring heading date on chromosome 3B and 6B, but the loci were different with those detected in this study, which may be correlated with Eps.

QTL conferring growth period was identified on chromosomes 1A, 1B, 4B, 6A, and 6B in the RIL population derived from the cross of Shannong 01-35 × Gaocheng 9411. QTLs (*QHt1A.2-132*, *QHt1B.1-87*, and *QHt1B.2-44*) with stable expression were the new-found main-effect QTLs and could be used in MAS. In addition,

QTL identified on chromosome controlled heading time and flowering time simultaneously, that is *QFt1B.1-87* (controlling flowering time) and *QHt1B.1-87* (controlling heading time) was the same locus, which performed pleiotropic effects.

6.4 QTL Mapping of Cell Membrane Permeability of Wheat Leaf Treated by Low Temperature

Chilling injury and frost damage occur frequently in the most of the winter wheat growing areas. In northern winter wheat region of China, climate is cold, both frost damage in winter and late spring cold in spring cause large loss of yield. Therefore, chilling injury and frost damage is one of the highlights of researching stress resistance in wheat. Some physiological and biochemical changes will happen during cold resistance of wheat, and some physiological traits such as malondialdehyde (MDA) content, soluble protein content, and cell membrane permeability were all identification index for cold resistance. Too low temperature will damage the structure of cell membrane, and result in wheat tissue injury or death. Hence, cool tolerance of cell membrane closely correlated with cold resistance of wheat. Ju et al. (2012) determined cell membrane permeability of cold wheat leaf by using conductivity method, which was a relatively reliable method to determine cold resistance in wheat. Brube et al. (1988) showed that cold resistance of wheat was a quantitative trait, controlled by polygenes, and affected by environment easily. Furthermore, the genes those controlled cold resistance of wheat were a kind of modificator gene, which perform cold resistance only under low temperature and short day. Waldman et al. (1975) and Limin et al. (1997) located the gene for cold endurance of wheat on chromosomes 5A and 5D and deemed that wheat varieties with the gene from group D perform stronger cold resistance than that from group A. However, until now, no researches related to QTL for cold resistance of wheat were conducted. Therefore, in this study, the DH population derived from two parents with different cold resistance was used to analyze QTL for cold resistance by determining cell membrane permeability of leaf treated by low temperature. And the purpose was to identify molecular markers, closely linked with cold resistance, which were used in cold resistance breeding of wheat, furthermore, and lay a theoretical foundation for mining the genes controlling cold resistance in wheat.

6.4.1 QTL Mapping for Cell Membrane Permeability of Wheat Leaf

6.4.1.1 Test Materials

Materials and Planting were same as one of the Sect. 6.1.1.1.1 in this chapter.

6.4.1.2 Field Trails

All DH lines and parents were planted in Baoding (Hebei province, E1), Cangzhou (Hebei province, E2), and Handan (Hebei province, E3) on October 4, 2010. The experimental field consisted of a randomized block design with three replications. All DH lines and parents were grown in a plot with three rows in 2 m length, 26.7 cm between rows and 2.2 cm between plants. Crop management was carried out following the local practices.

6.4.1.3 Determining Method of Cell Membrane Permeability of Leaf

In late December 2010, five leaves (intermediate leaves of the plant) in the center of every plot were selected and washed by tap water and deionized water for three times successively, and then moisture was blotted on the surface of the leaves. Each sample of 0.2 g was cut into about 1 cm of small pieces, and put into two tubes, and then treated by room temperature (control) and low temperature (−18 °C), respectively. Cell membrane permeability was determined by using conductivity method.

A volume of 10 mL deionized water was added to each sample, including control and treatment, and then vacuumized for 15 min. After gently shaking, the tubes were put in room temperature for 10 min. Electrical conductivity of control (C) and treatment (R) was determined by using conductometer (DDS-11A) according to the method described by Shen et al. (modified slightly). And then, the tubes of treatment were put into the boiling water bath for 5 min, and the electrical conductivity (K) after cooling to room temperature was determined. The relative transuding rate of electrolyte (A, %) was used to show cell membrane permeability, whose value was calculated by using the formula: $A = (R - C)/(K - C) \times 100$.

6.4.1.4 Data Analysis

Analysis of phenotypic data was carried out using the SPSS program (version 17.0, SPSS, Chicago, USA). The inclusive composite interval mapping (ICIM) was applied by means of the QTL IciMapping 2.2 to identify QTLs for cell membrane permeability under three environments, based on the molecular genetic map constructed by Zhang et al. (2009). A logarithm of odds (LOD) of 2.5 and Sep of 1 cM were set to declare QTL as significant. QTL effects were estimated as the proportion of phenotypic variance (R^2) explained by the QTL. QTL was named referring to the method described by McIntosh et al.

Table 6.27 Variations of cell membrane permeability of leaf treated by low temperature (−18 °C) in parents and DH population in three environments

Environment	Parent		DH population				
	Huapei 3	Yumai 57	Mean ± SD	Range	CV (%)	Skewness	Kurtosis
E1	36.22	32.32*	30.11 ± 5.39	15.84–48.92	17.9	0.322	0.855
E2	29.45	20.34*	28.29 ± 4.07	18.04–40.60	14.4	0.106	−0.249
E3	33.73	31.09*	31.34 ± 4.37	20.42–42.48	13.9	0.092	−0.486

E1: Baoding site; E2: Cangzhou site; E3: Handan site
*Indicates significant difference between parents ($P < 0.05$) according to *t*-test

6.4.1.5 Results and Analysis

6.4.1.5.1 Analysis of Phenotypic Variation

In three different environments, significant difference was found in cell membrane permeability of leave treated by low temperature between parents and large range of variation was observed among DH lines. And the coefficients of variations were 17.9 % (E1), 14.4 % (E2), and 13.9 % (E3), respectively. The cell membrane permeability of the DH population segregated continuously and followed a normal distribution, and both absolute values of skewness and kurtosis were less than 1.0 (Table 6.27), indicating its polygenic inheritance and suitability of the data for QTL analysis.

6.4.1.5.2 QTL Analysis of Cell Membrane Permeability of Leaf in Wheat

A total of 21 additive QTLs conferring cell membrane permeability of leaf were detected on chromosomes 1B (three regions), 2A (two regions), 3A (three regions), 3B (three regions), 5B (five regions), 6A (one region), 6B (one region), 6D (one region), 7B (one region), and 7D (one region), respectively, in three different environments. Seven, nine, and five QTLs were found in E1, E2, and E3, respectively, and most of them were contributed by Huapei 3, which had stronger cold resistance (Table 6.28).

The QTLs located on chromosome 5B, including *qCMP-5B-1* (E1), *qCMP-5B-2* (E2), and *qCMP-5B-4* (E3), were located within the interval Xgwm213–Xswes861.2, were away from Xswes861.2 for 0.0 cM, and were detected in the three environments. The locus had most significant contribution in three environments, accounting for 17.5, 8.1, and 14.0 % of phenotypic variation.

In addition, *qCMP-1B-1*, *qCMP-3B-2*, *qCMP-5B-1,* and *qCMP-5B-4* were all main-effect QTLs, whose contributions were all greater than 10 %, accounting for 18.4, 17.7, 17.5, and 14.0 % of phenotypic variation. Except for *qCMP-3B-2*, their positive alleles were all came from Huapei 3. And other 17 additive QTLs were minor genes, whose contributions smaller than 10 %.

Table 6.28 Position, effect, and phenotypic contribution of additive QTL for cell membrane permeability of leaf treated by low temperature (−18 °C) in three environments

QTL	Site/cM	Marker interval	LOD	Additive effect	PVE (%)
Environment 1					
qCMP-2A-1	42	Xgwm636–Xcfe67	2.796	−2.395	7.4
qCMP-2A-2	102	Xwmc455–Xgwm515	2.652	10.113	4.3
qCMP-3A-1	188	Xbarc51–Xbarc157.1	2.691	2.313	6.6
qCMP-3B-1	90	Xgwm566–Xcfe009	2.636	−9.926	4.2
qCMP-5B-1	58	Xgwm213–Xswes861.2	10.046	21.176	17.5
qCMP-6B	83	Xswes679.2–Xwmc658.2	4.036	17.207	7.6
qCMP-7B	48	Xgwm333–Xwmc10	3.484	13.845	7.3
Environment 2					
qCMP-1B-1	23	Xcfe156–Xwmc406	8.073	2.846	18.4
qCMP-1B-2	39	Xbarc008–Xgwm218	2.828	−1.67	6.0
qCMP-3A-2	196	Xbarc157.1–Xbarc1177	2.828	−1.583	5.7
qCMP-3B-2	86	Xgwm566–Xcfe009	6.697	−3.450	17.7
qCMP-3B-3	50	Xgwm285–Xgwm685	3.817	2.658	9.8
qCMP-5B-2	58	Xgwm213–Xswes861.2	3.974	−1.966	8.1
qCMP-5B-3	1	Xgwm133.1–Xwmc73	2.580	1.519	5.2
qCMP-6A	19	Xgwm334–Xbarc023	2.726	1.834	7.6
qCMP-6D	118	Xubc808–Xswes679.1	2.651	−3.303	6.5
Environment 3					
qCMP-1B-3	22	Xcfe156–Xwmc406	3.420	−1.787	7.9
qCMP-3A-3	97	Xbarc356–Xwmc489.2	2.829	1.681	7.1
qCMP-5B-4	58	Xgwm213–Xswes861.2	6.452	2.477	14.0
qCMP-5B-5	1	Xbarc36–Xbarc140	2.700	−1.663	7.0
qCMP-7D	211	Xwmc14–Xwmc42	3.667	2.892	9.7

Positive and negative additive effects indicate that the positive alleles are from Huapei 3 and Yumai 57, respectively. *PVE* phenotypic variation explained

6.4.2 Research Progress of Cell Membrane Permeability QTL Mapping and Comparison of the Results with Previous Studies

6.4.2.1 Research Progress of QTL Conferring Cold Resistance of Wheat

Although Brube et al. (1988) and Waldman et al. (1975) found that genes related to cold resistance were on chromosomes 4D, 5A, 5D, and 7A, etc., but the specific locations were not clear. With the development of genetic map and QTL analysis, Båga et al. (2007), Galiba et al. (1995), Vágújfalvi et al. (2003), Sutka et al. (2001), Tóth et al. (2003), and Liu et al. (2005) studied the cold resistance and relative

Table 6.29 Summary of QTL of wheat resistance to cold (PVE > 10 %)

Trait	Site	Flanking marker	PVE (%)	Population	Reference
Cold resistance	1D	E37M60_(72); barc152_ (145)	P = 0.001	DH	Båga et al. (2007)
	1D	barc169_122	P = 0.0005		
	2A	gwm296_177	P = 0.005		
	5A	wmc206_224; cfd2_326	P = 0.0001		
	6D	cfd76_153	P = 0.005		
	5A	Vrn1	LOD > 3	SCRL	Galiba et al. (1995)
		Xpsr426,	LOD > 3		
		Xwg644	LOD > 3		
		Xcdo504	LOD > 3		
		Frl	LOD > 3		
	5A	Xbcd508	49 %	RIL	Vágújfalvi et al. (2003)
		CBF3	Transcription factor		
		Vrn-A1/Xpsr426/Xwg644			Sutka (2001)
		Fr1			
	5D	Vrn-D1			
		Fr2			
	5B	Vrn-B1			Tóth et al. (2003)
		Fr-B1			
		Xgwm639			
	2A	Xgwm372–Xgwm249	10.45/15.61/17.14	DH	Liu et al. (2005)
	4B	Xwroe48–DuPw043	16.97		
	2A	BARC208–Xgwin95	19.8		

DH double haploid; *SCRL* single chromosome recombinant lines; *RIL* recombinant inbred lines

transuding rate of electrolyte under low temperature using DH, RIL, and SCRL (single chromosome recombinant lines) populations and identified 24 loci and their linking molecular markers, among them 19 QTLs, were major QTLs, including one transcription factor, three vernalization genes and two cold-resistant genes. Precious results showed that important QTLs conferring cold resistance distributed on chromosomes 2A, 5A, and 5D (Table 6.29).

6.4.2.2 Comparison of this Study with Previous Researchers

In this study, a total of 21 QTLs conferring cell membrane permeability of leaf treated by low temperature (−18 °C) were detected, including four major QTLs (contribution greater than 10 %), which located on chromosomes 1B, 3B, and 5B, respectively. Furthermore, *qCMP-5B-1* (E1) and *qCMP-5B-4* (E3) were located within the interval Xgwm213–Xswes861.2, and a locus was also detected within this interval in environment 2, accounting for 8.1 % of phenotypic variation, and was away from Xswes861.2 for 0.0 cM. Hence, Xswes861.2 could be used in MAS in wheat breeding programs of cold resistance. In previous studies, QTLs conferring cold resistance were mainly located on chromosomes 1D, 2A, 2B, 5A, 6D, 7B, 4B, 5D, and 5B, while this study showed that the chromosomes 2A, 6B, 7B, 3A, 6A, 6D, and 7D were also related to cold resistance, except the chromosomes 1B, 3B, and 5B. Comparing with the previous results, it was found that the related QTLs on chromosomes 2A, 7B, 5A, 5B, and 5D were very important for cold resistance in wheat.

6.5 QTL Mapping of Root Traits in Wheat

Root is an important organ absorbing water and minerals, whose development directly affects the growth and development of overground parts and material production, and is the foundation of the high and stable yield for crop (Liu et al. 2002; Moudal and Kour 2004; Partha et al. 2004). Development of root is not only affected by environment and cultivating condition but also controlled by genes. Caradus (1995) indicated that the traits correlated with root size such as root weight, root volume, the number of root, root length, root surface area, and the ratio of root dry weight to shoot dry weight had higher heritability; furthermore, the traits correlated with root morphology such as root average diameter, root hair length, adventitious root grade, branch grade, root density, and density of root length also had higher heritability. These root traits were all quantitative traits. In addition, Jing et al. (1997) researched the heritability of root morphology and its relationship with drought resistance and showed that there was significant positive correlation between drought resistance of wheat seedling and root dry weight.

Now, most of the researches related to root traits in wheat were focused on mapping QTL under abiotic stress such as high temperature and drought, using efficiency of NPK, salt stress, water stress, and heavy metal stress; in addition, majority studied root traits at seedling stage. Zhang and Xu (2002) identified QTL and interaction QTL conferring the number of root, root diameter, root dry weight, the ratio of root dry weight to shoot dry weight, and growth rate of root using a RIL population in wheat. Zhou et al. (2005) analyzed QTL and interaction QTL for the number of root, maximum root length, root raw weight, root dry weight, the ratio of root raw weight to shoot raw weight and the ratio of root dry weight and shoot dry weight under two different environments including water stress and no stress conditions, by using a DH population containing 150 progeny lines derived from a cross between Hanxuan 10 and Lumai 14. Landjeva et al. (2010) identified QTL for vigor located on the wheat D genome at seedling stage. Ibrahim et al. (2012) researched QTL conferring root morphology at wheat seedling stage under drought environment. Bai et al. (2013) researched the QTL for root traits at seeding stage and its relationship with plant height. Because wheat root is closely related to final yield and that it is difficult to improve wheat root by using traditional breeding method, we can use MAS to speed up the further improvement of wheat root. Hence, in this study, immortalized F_2 (IF_2) population of wheat derived from a DH population was used to analyze QTL for root traits at seedling stage, in order to find the markers, closely linking with root traits, to conduct molecular-assisted breeding.

6.5.1 QTL Mapping and Effects' Analysis of Root Traits

6.5.1.1 Experiment Designing

A total of 30 seeds of 168 single crossed derived from a DH population (Huapei 3 × Yumai 57) and parents were sampled, and soaked with 1 % H_2O_2 for 24 h. After washing for 2 ~ 3 times by water, the samples were put in a light incubator (nighttime temperature was set as 12 ± 2 °C, while daytime temperature was 20 ± 4 °C) and cultured with deionized water until the first leaf emerged. Six excellent plants of each cross were sampled and cultured on foamed plastic with thickness of 0.5 cm (perforated with the diameter of 1 cm), and then fixed by disinfected sponge. At last, wheat seedlings were cultured with Hoagland's solution (An et al. 2006) in cultivating basins (with height of 30 cm) for three replicates. Furthermore, 1 L of Hoagland's solution consisted of 1 mmol $Ca(NO_3)2{\cdot}4H_2O$, 0.2 mmol KH_2PO_4, 0.5 mmol $MgSO_4{\cdot}7H_2O$, 1.5 mmol KCl, 1.5 mmol $CaCl_2$, 1 μmol H_3BO_3, 50 nmol $(NH4)6Mo_7O_{24}{\cdot}4H_2O$, 0.5 μmol $CuSO_4{\cdot}5H_2O$, 1 μmol $ZnSO_4{\cdot}7H_2O$, 1 μmol $MnSO_4{\cdot}H_2O$, and 0.1 mmol Fe^{3+}-EDTA, and the pH of the solution was 6.0. Meanwhile, cultivating basins were brushed by black paint to supply dark environment for the growth of root. Replacement of the nutrient solution was done every three days.

Three individuals with consistent growth of each cross were sampled when the fourth leaf emerged, washed with distilled water, and divided into stems and roots using scissor. The traits including root total length (RTL), root surface area (RSA), root average diameter (RAD), root volume (RV), root tip number (RT), and maximum root length (MRL) were measured using the WinRHIZO root analysis system. The fresh roots and shoots were killed out for 10 min under 105 °C and then dried to balance weight under 80 °C. Furthermore, shoot dry weight (SDW) and root dry weight (RDW) were weighed. Root–shoot ratio was the root dry weight to shoot dry weight.

6.5.1.2 Results and Analysis

6.5.1.2.1 Phenotypic Variation and Correlation of Root Traits in IF_2 Population

Big differences were found in nine root-related traits between parents. And the nine root traits of IF_2 population segregated continuously and followed a normal distribution, and both absolute values of skewness and kurtosis were less than 1.0, except for RDW/SDW (Table 6.30), indicating its polygenic inheritance and suitability of the data for QTL analysis.

In addition to RAD, RTL was significantly or extremely significantly positively correlated with other seven traits, and the correlation between RTL and RSA was the largest ($r = 0.981$, $P < 0.01$), while RAD was negatively correlated with RT and RTL (r = 0.417, 0.314, respectively, $P < 0.01$), and they both reached significant level (Table 6.31).

Table 6.30 Analysis of root traits at seedling stage in the IF_2 population derived from Huapei 3 × Yumai 57

Root trait	Parent		Immortalized F_2 population			
	Huapei 3	Yumai 57	Mean ± SD	Range	Skewness	Kurtosis
RTL (cm)	96.12	165.01	151.56 ± 4.47	35.19–365.00	0.65	1.86
RSA (cm^2)	9.48	13.29	12.05 ± 0.33	3.01–26.50	0.38	0.96
RAD (μm)	320.23	260.41	250.21 ± 0.00	210.12–290.31	0.17	−0.55
RV (mm3)	70.38	90.47	76.13 ± 0.00	20.31–0.15	0.14	0.26
RT	199	302	262.00 ± 7.43	61.00–528.00	0.49	0.37
MRL (cm)	16.71	17.48	19.88 ± 0.29	10.35–27.8	−0.36	0.14
SDW (mg)	19.77	24.60	18.13 ± 0.50	4.50–43.75	0.61	2.80
RDW (mg)	6.7	6.4	6.14 ± 0.15	1.60–11.95	0.09	0.52
RDW/SDW	0.34	0.26	0.34 ± 0.00	0.13–0.82	3.23	22.25

RTL root total length; *RSA* root surface area; *RAD* root average diameter; *RV* root volume; *RT* root tip number; *MRL* maximum root length; *SDW* shoot dry weight; *RDW* root dry weight

Table 6.31 Coefficients of pairwise correlations of mean values of root traits at seedling stage in the IF_2 population derived from Huapei 3 × Yumai 57

Traits	RTL	RSA	RAD	RV	RT	MRL	SDW	RDW
RSA	0.981**							
RAD	−0.314**	−0.136						
RV	0.916**	0.977**	0.067					
RT	0.831**	0.788**	−0.417**	0.708**				
MRL	0.846**	0.829**	−0.322**	0.773**	0.774**			
SDW	0.750**	0.786**	−0.033	0.791**	0.652**	0.683**		
RDW	0.870**	0.915**	0.015	0.924**	0.712**	0.768**	0.779**	
RDW/SDW	0.093	0.091	0.032	0.084	0.058	0.012	−0.344*	0.216*

*Significant at 0.05 probability level
**Significant at 0.01 probability level. Abbreviations are the same as in Table 6.30

6.5.1.2.2 QTL Mapping and Effect Analysis of Root Traits in the IF_2 Population

A total of seven additive QTLs (Table 6.32 and Fig. 6.12) and 12 pairs of epistatic QTLs (Table 6.33 and Fig. 6.13) for eight root traits were mapped on chromosomes 1A, 1D, 2A, 2B, 2D, 3A, 3B, 5D, 6D, and 7D. Additive (A), dominant (D) effects were observed across these QTLs, and the interactions between additive and

Table 6.32 Intervals, effects, and contributions of QTL for root traits at seedling stage in the IF_2 population derived from Huapei 3 × Yumai 57

Root trait	QTL	Flanking marker	Position (cM)	Additive		Dominance		Gene action
				A	H^2 (%)	D	H^2 (%)	
RTL	*QRtl2D*	XWMC41–XBARC349.2	69.5	−15.04	4.44	−19.07	8.88	OD
RSA	*QSa2D*	XWMC170.2–XGWM539	65.4			−2.50	8.18	
RAD	*QAd2A*	XBARC380–XGWM636	1.6	6.67	9.32			
RV	*QRv7D*	XGWM295–XGWM676	107.3	7.00	0.03	−20.00	11.91	OD
MRL	*QMrl6D*	XWMC412.1–XCFD49	0	−1.32	9.98	1.18	3.01	PD
RDW	*QRdw2D*	XWMC41–XBARC349.2	69.5	−0.63	3.53	−1.15	11.1	OD
	QRdw7D	XGWM295–XGWM676	101.3			−1.35	9.81	

For additive effect, a positive value indicates that the allele from Huapei 3 increases plant height. For dominant effect, a positive value indicates that the heterozygote has a higher phenotypic value than the homozygote. In the column of "Gene action," PD, D, and OD denote partial dominant (*D*/*A* < 1.00), dominant (*D*/*A* = 1.00), and overdominant (*D*/*A* > 1.00), respectively. Other abbreviations are the same as in Table 6.30

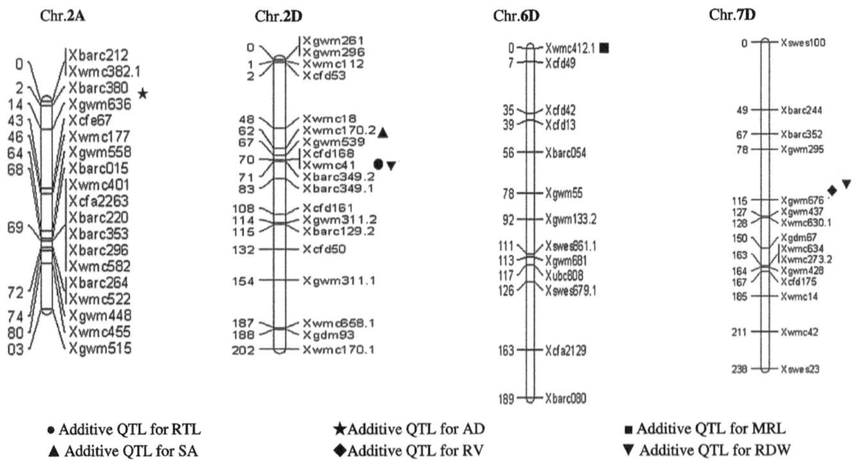

● Additive QTL for RTL ★Additive QTL for AD ■ Additive QTL for MRL
▲ Additive QTL for SA ◆Additive QTL for RV ▼ Additive QTL for RDW

Fig. 6.12 Positions of additive QTL associated with root traits at seedling stage in the IF_2 population derived from Huapei 3 × Yumai 57

Table 6.33 Estimated epistasis and contributions of QTL for root traits at seedling stage in the IF_2 population derived from Huapei 3 × Yumai 57

Trait	QTL	Marker interval	Position (cM)	QTL	Marker interval	Position (cM)	AA	H^2 (%)	AD	H^2 (%)	DA	H^2 (%)	DD	H^2 (%)
RTL	*QRtl2D*	XCFD50–XGWM311.1-2A	141.1	QRtl5D	XBARC307–XBARC347	50.5	25.75	14.05			39.91	2.78	76.57	3.40
	QRtl3B	XCFE282–XGWM144	53.0	QRtl5D	XBARC1097–XCFD8	18.4	41.75	6.38			29.20	0.17	46.98	1.34
RSA	*QSa2D*	XCFD53–XWMC18	1.6	QSa7A	XBARC070–XBARC250	30.5					3.31	17.40		
	QSa2D	XCFD53–XWMC18	1.6	QSa7A	XWMC593–XBARC157.2	0							3.86	9.13
RV	*QRv1A*	XCFD59–XWMC402.2	54.0	QRv2A	XGWM558–XBARC015	65.4			20.90	0.04				
	QRv1A	XCFD59–XWMC402.2	54.0	QRv2A	XGWM448–XWMC455	73.6							20.30	1.43
	QRv1A	XBARC148–XGWM154	58.9	QRv2A	XGWM448–XWMC455	73.6								
	QRv1A	XGWM498–XCWEM6.2	80.8	QRv2A	XGWM448–XWMC455	73.6	14.30	5.46			10.40	0.20	10.7	0.83
	QRv1A	XGWM498–XCWEM6.2	80.8	QRv2A	XWMC455–XGWM515	89.1	12.10	2.95						
RT	*QRt1D*	XCFD19–XWMC93	40.9	QRt2B	XBARC200–XWMC770	52.7	7.23	0.20	69.63	7.2	72.40	7.02		
RDW	*QRdw1A*	XWMC402.2–XBARC120.2	55.1	QRdw2A	XGWM448–XWMC455	73.6	0.69	8.49						
	QRdw2A	XWMC382.1–XBARC380	1.0	QRdw3A	XWMC527–XWMC264	115.7	1.12	8.68	1.38	3.58			2.13	3.36

For the epistatic effect, the positive value means that the parent-type effect is greater than the recombinant-type effect, and the negative value means that the parent-type effect is less than the recombinant-type effect. Abbreviations are the same as in Table 6.30

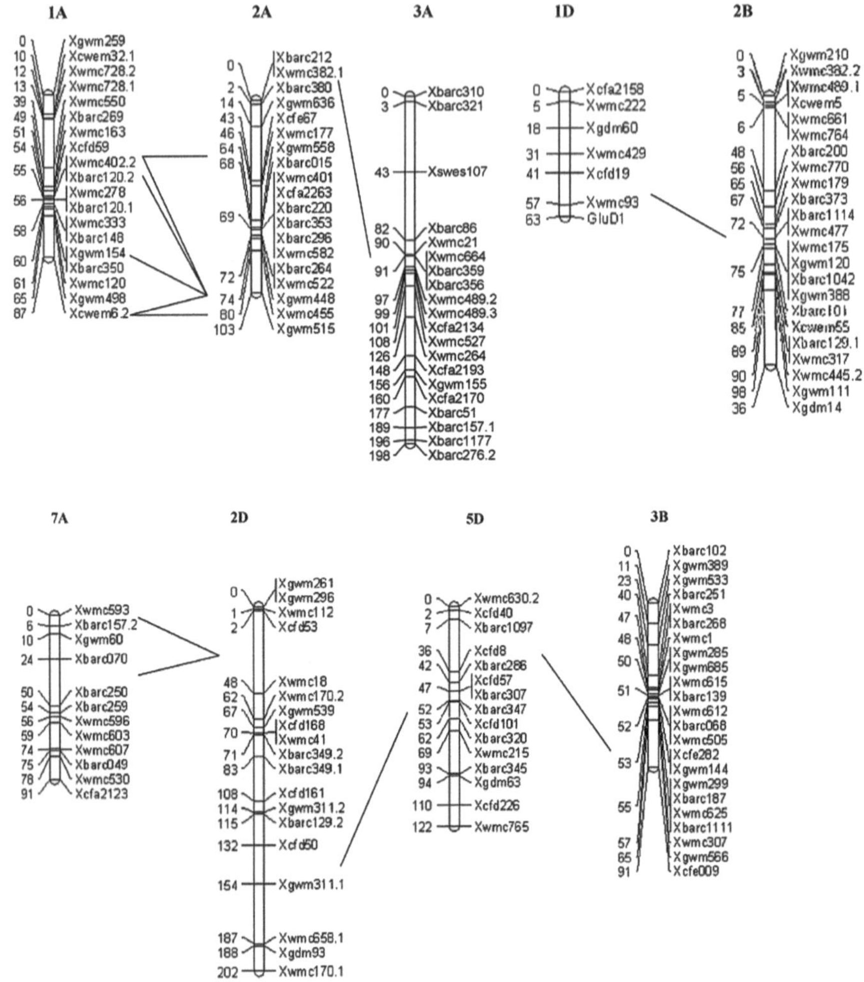

Fig. 6.13 Positions of epistatic QTL for root traits at seedling stage in the IF_2 population derived from Huapei 3 × Yumai 57

additive (AA), additive and dominance (AD), dominance and additive (AD), as well as dominance and dominance (DD) were also detected.

For RTL, one additive QTL and two pairs of epistatic QTLs were detected. The additive QTL on chromosome 2D was contributed by Yumai 57, accounting for 4.44 % of phenotypic variation, and performed overdominant effect. The two pairs of epistatic QTLs on chromosomes 2D-5D and 3B-5D explained phenotypic variation from 0.17 to 14.05 % and performed AA, DA, and DD interactions. Meanwhile, the QTL designated as *QRtl2D* performed epistatic effect.

One additive QTL and two pairs of epistatic QTLs conferring RSA were identified. The additive QTL on chromosome 2D accounted for 8.18 % of phenotypic effect and performed dominant effect. The two pairs of epistatic QTLs on chromosomes 2D-5D and 3B-5D explained 14.05 and 6.38 % of phenotypic variation, respectively. Further, the epistatic QTL of 2D-5D performed AD interaction, while the epistatic QTL of 3B-5D performed DD interaction.

One additive QTL for RAD on chromosome 2A was detected, accounting for 9.32 % of phenotypic variation, and contributed by Huapei 3. No pair of epistatic QTL was identified.

One additive QTL and five pairs of epistatic QTLs conferring RV were detected. The additive QTL on chromosome 7D, explaining 0.03 % of phenotypic variation, was contributed by Huapei 3 and showed overdominant effect. Five pairs of epistatic QTLs were all mapped on chromosomes 1A and 2A and showed different effects of AA, DA, and DD, explaining phenotypic variation from 0.04 to 5.46 %.

One additive QTL for MRL was identified on chromosome 6D, accounting for 9.98 % of phenotypic variation, and contributed by Yumai 57. And it performed partial dominant effect.

One pair of epistatic QTL for RT was mapped on chromosomes 1D-2B, and showed AD, DA, and AD effects, accounting for 0.20, 7.20, and 7.02 % of phenotypic variation, respectively. No QTL with additive effect and dominant effect was identified.

One additive QTL, one dominant QTL, and two pairs of epistatic QTLs for RDW were detected. Further, the additive QTL on chromosome 2D was contributed by Yumai 57, accounting for 3.53 % of phenotypic variation, and showed overdominant effect. The dominant QTL explained 9.81 % of phenotypic variation. The two pairs of epistatic QTLs were mapped on chromosomes 1D-2A and 2A-3A, accounting for 8.49 and 8.68 % of phenotypic variation, respectively, and performed AA interactions.

Among the seven additive QTLs for root traits detected in this study, some loci only performed additive effects, and some loci only performed dominant effects, while only a few loci showed additive and dominant effects simultaneously. Furthermore, the loci with both additive and dominant effects gave priority to dominant effects, and only one locus was detected in this study. There were differences in effect size among different loci, and their directions were not consistent. Twelve pairs of epistatic QTLs detected in this study gave priority to AA and DD effects.

6.5.2 *Research Progress of Root Traits' QTL Mapping and Comparison of the Results with Previous Studies*

6.5.2.1 Research Progress of QTL Mapping Conferring Root Traits

Growth and development of root directly affects acquiring nutrient substance, thus affecting final yield of wheat because growth of wheat was affected by environment condition such as drought, water, N, P, K, heavy metal, and salt. Therefore, in recent years, researches regarding root traits were focused on QTL mapping for related traits under abiotic stress. It was found that Wu et al. (2007), Landjeva et al. (2010), Yang (2012), Xu et al. (2012), and Ren et al. (2012) detected QTL for wheat root traits under salt stress using DH and RIL populations, and a total of 26 QTLs were identified. Most of these QTLs distributed on chromosomes 3A, 5A, 5B, and 2D, among which 16 QTLs were major QTLs with the highest R^2 of 36.06 % (Table 6.34). Ibrahim et al. (2012) detected QTL for root traits in wheat using BC2F4-6 population under drought condition, and 32 QTLs were detected. Furthermore, multiple QTLs conferring root traits were distributed on chromosomes 1D, 2A, 2D, and 7D. Liu et al. (2013) detected QTL for root traits under water stress, and 46 QTLs including 20 major QTLs were detected with the highest effect of 24.31 %. An et al. (2006) mapped five QTLs for RDW under condition with different level of N fertilizer, and four major QTLs were found with the highest effect of 19.6 %. Bail et al. (2013), Liu et al. (2011), Ren et al. (2012), Jiang (2012), and Hamada et al. (2012) identified 69 QTLs conferring root-related traits of wheat seedlings on chromosomes 2A, 2B, 5D, 2D, 3B, 4D, 3A, 6A, and 7D, using different DH population and RIL population, and 29 of 69 QTLs were major QTLs with the highest effect of 68 % (Ren et al. 2012). Sharma et al. (2011) identified 15 QTLs for root-related traits using 1RS-1BS map, and the highest effect of single QTL was 56.0 %.

6.5.2.2 Comparsion of the Results with the Previous Studies

A total of seven QTLs for root-related traits of wheat seedlings were identified in this study. Among them, three QTLs were contributed by Yumai 57, while two QTLs were contributed by Huapei 3, and *QRtl2D*, *QRv7D,* and *QRdw2D* showed overdominant effects, indicating that the parents with strong advantage should be selected to configured crosses. Among the 12 pairs of epistatic QTLs detected in this study, some QTLs interacted with other two QTLs simultaneously. For example, *QRtl5D* interacted with *QRtl2D* and *QRtl3B*, controlling RTL; while the QTL located within the interval Xgwm448–Xwmc455 interacted with both *QRdw1A* and *QRv3A*, indicating that epistasis was very important for heredity of root-related traits in wheat seedlings, but the mechanism was very complicated (Xing et al. 2002; Li et al. 2001; Mei et al. 2005), which needed further research. Most of the additive QTLs for root traits, detected in this study, were mapped on

Table 6.34 Summary of QTL of wheat root traits (PVE > 10 %)

Env.	Trait	QTL	Flanking marker	PVE (%)	Population	Reference
Salt–tolerance	RDW/SDW	*QRsrc.ipk–5D*	Xgwm1122	18.6	DILs	Landjeva et al. (2010)
			Xgwm174	18.6		
			Xgwm182	18.6		
			Xgwm3063	18.6		
			Xgdm99	17.2		
	PRL	*qTL3A*	Xwmc527–Xwmc264	17.45	DH	Yang (2012)
		qTL3D–2	Xgdm72–Xbarc1119	23.72		
		qTL7D	Xwmc630.1–Xgdm67	13.9		
	RRW	*qFRW2D–1*	Xwmc170.2–Xgwm539	36.06		
		qFRW2D–2	Xbarc349.2–Xbarc349.1	19.39		
		qFRW5B	Xgwm213–Xswes861.2	10.33		
	RDW	*qDRW5B–4*	Xgwm213–Xswes861.2	12.98		
	RL	*QRl–7B*	Xgwm297–NP43	14.75	RIL	Xu et al. (2012)
	RDW	*QRdw.sqn–3A*	Xgwm497.1–Xcfa2193	10.2	RIL	Ren et al. (2012)
	MRL	*QMrl.sqn–3A*	Xgwm497.1–Xcfa2193	25.5		
	RRDW	*QRdwr.sqn–4A*	Xbarc78–Xgwm350.1	12.2		
Normal	TRL	*5BS*	gwm213.5B4D7B–barc74.5BLT	11.44	DH	Bai et al. (2013)
	TRSA	*2D*	gwm132–wPt–9997.2DS	10.1		
		5A	gwm293b.5ASM–gwm146a.5ASM	12.89		
	RDW/SDW	*4D*	RhtMrkD1.4D–wPt–0431. 4B4D/wPt–5809.4B4D	21.08		
	NR > 30 cm	*1RS–1BS*	Sec–1	34	RL	Sharma et al. (2011)
		1RS–1BS	Xurc–1–Pm8	26		
		1RS–1BS	Pm8	57		

(continued)

Table 6.34 (continued)

Env.	Trait	QTL	Flanking marker	PVE (%)	Population	Reference
		1RS–1BS	Sr31	47		
	LRLa	*1RS–1BS*	Xurc–8–Gli–1,Glu–3	18		
	TRL	*1RS–1BS*	Xurc–4	52		
		1RS–1BS	Pm8	56		
		1RS–1BS	Xurc–2–Pm8	24		
		1RS–1BS	Sr31	40		
	SRW	*1RS–1BS*	NOR–Xurc–4	15		
		1RS–1BS	Pm8–Gli–1,Glu–3	18		
	DRW	*1RS–1BS*	Pm8–Gli–1,Glu–3	31		
		1RS–1BS	NOR–Xurc–4	11		
	TRW	*1RS–1BS*	NOR–Xurc–4	14		
		1RS–1BS	Pm8–Gli–1,Glu–3	23		
	MRL	*QMRL.cgb–4A*	Xgwm610–Xgwm397	12.37	DH	Liu et al. (2011)
	RN	*QRN.cgb–2B*	Xgwm429–Xgwm388	11.9		
	RA	*QRA.cgb–3B*	P3622–400–P2076–147	12.16		
	MRL	*qMRL–2B*	Xgwm210–Xbarc1138.2	68	RIL	Ren et al. (2012)
	PRL	*qPRL–2B1*	Xgwm210–Xbarc1138.2	59		
	LRLb	*qLRL–6A*	Xgwm570–Xgwm169.2	30.5		
	TRL	*qTRL–2B*	Xgwm210–Xbarc1138.2	20.3		
	RN	*qRN–6A*	Xgwm570–Xgwm169.2	24.5		
		qRN–6D	Xgwm55.3–Xgdm14.6	10.3		
	RL	*QRl–7A*	wpt4637–barc121	12.26	RIL	Jiang (2012)
	RRW	*QFrw–4A*	wpt671707–barc70a	14.45		
		QFrw–6A	wpt668031–wpt4229	11.69		

(continued)

Table 6.34 (continued)

Env.	Trait	QTL	Flanking marker	PVE (%)	Population	Reference
	RDW	*QDrw–4A.1*	swes147–swes624c	10.63		
		QDrw–4A.2	wpt671707–barc70a	10.05		
	RV	*QV–1A*	wpt729788–wpt667395	11.57		
		QV–3A	wpt1562–wpt2587	10.21		
	RAD	*QRdm–3B*	ubc834a–wpt5906	12.71		
	RTN	*QRn–1B*	wpt668027–swes169a	13.29		
		QRn–3A	wpt730892–barc314	11.15		
		QRn–4A	issr23b–wmc308	15.4		
	RTN	*qRN*	wmc150a	11.63	DH	Hamada et al. (2012)
	DRR	*qDRR–2*	wmc97	11.49		
	ER	*qER–1*	cfd266	12.53		
		qER–2	wmc405	14.73		
	HR	*qHR–1*	wmc278	13.44		
Water stress	MRL	*QMRL.cgb–5D*	Xgwm205.2–Xgwm68	12.2	DH	Liu et al. (2013)
		QMRL.cgb–2D	WMC453.1–WMC18	12.22		
		QMRL.cgb–5B	WMC380–Xgwm540	11.95		
	RN	*QSRN.cgb–3B*	WMC3–P6934.380	14.98		
		QSRN.cgb–5A	P2470.2–Xgwm154	19.82		
	TRL	*QTRL.cgb–1B*	P3470.2–P4133.1	11.43		
		QTRL.cgb–1B	CWM65–P8222.5	16.13		
		QTRL.cgb–3B	WMC231–Xgwm284	10.43		
		QTRL.cgb–3B	Xgwm644.2–WMC3	10.42		
		QTRL.cgb–5D	Xgwm3–Xgwm43	10.3		
		QTRL.cgb–7D	Xgwm44–Xgwm121	10.69		

(continued)

Table 6.34 (continued)

Env.	Trait	QTL	Flanking marker	PVE (%)	Population	Reference
		QTRL.cgb–3B	WMC3–P6934.380	13.81		
		QTRL.cgb–3B	P3622.4–P2076.1	14.23		
		QTRL.cgb–5B	P8143.3–P2454.1	10.9		
	PRA	*QPRA.cgb–7D*	Xgwm44–Xgwm121	11.9		
		QPRA.cgb–5B	P8143.3–P2454.1	16.24		
	RSA	*QRSA.cgb–7D*	QRSA.cgb–7D	11.93		
		QRSA.cgb–5B	QRSA.cgb–5B	16.22		
	RA	*QSRA.cgb–7D*	Xgwm44–Xgwm121	10.46		
		QSRA.cgb–2B	WMC474–Xgwm374	10.66		
		QSRA.cgb–2B	WMC179.2–P6901.2	11.16		
		QSRA.cgb–3B	WMC3–P6934.3	24.31		
Using efficiency of N	RDW		CWM70–P3474–480	19.6	DH	An et al. (2006)
			Xgwm539–P4233–175	11		
			P8422–170–CWM539.2	10.4		
			EST25–CWM88	11		

Note: RDW/SDW the ration of root dry weight to shoot dry weight; *PRL* primary root length; *RRW* root raw weight; *RDW* root dry weight; *RL* root length; *MRL* maximum root length; *RRDW* relative root dry weight; *TRL* total root length; *TRSA* total root surface area; *RN* root number; *RA* root angle; *RV* root volume; *RAD* root average diameter; NR > 30 cm, number of roots greater than 30 cm; *LRL*[a] longest root length; *SRW* shallow root weight; *DRW* deep root weight; *TRW* total root weight; *LRL*[b] lateral root length; *RTN* root tip number; *DRR* Deep root ratio; *ER* elongation rate of the primary seminal root; *HR* hydrotropic response of root; *PRA* primary root area chromosomes 2D and 7D, which were also found in the previous researches, indicating that important QTLs or genes confer root traits in wheat distributed on D genome.

6.6 QTL Mapping Conferring Leaf-Related Traits in Wheat

Leaf is the main photosynthetic organ. Among them, flag leaf of wheat, with the highest photosynthetic efficiency at late growth stage and the highest contribution to formation of grain and yield, is the main source of carbohydrates in wheat grain and can contribute to yield for one-third. At home and abroad, lots of researches related to effects of wheat flag leaf on photosynthetic efficiency and yield were conducted, but few researches focused on genetic loci conferring flag leaf. Keller et al. (1999) identified eight QTLs for flag leaf width on chromosomes 1A, 1B, 2A, 3B, 5A, 5B, and 6A, respectively, which could account for 59.5 % of phenotypic variation, using a RIL population derived from Forno/spelt Oberkulmer including 226 lines. Lohwasser et al. (2004) detected 23 QTLs for length and width of the three basal leaves by using a RIL population including 114 lines under greenhouse conditions. Identifying molecular markers closely linked with leaf morphology on the base of previous studies is very important for improving photosynthetic efficiency and yield at the molecular level.

6.6.1 QTL Mapping for Leaf Morphology of Wheat Based on a DH Population

6.6.1.1 Materials and Methods

Five plants of each line were sampled on 10 days after heading to measure the included angle between flag leaf of main stem and stem designed as flag leaf angle (FLAN). While five main stems of each line were sampled at filling stage (20 days after anthesis) to measure length and width of the upper three leaves (flag leaf, second leaf, and third leaf). And leaf area was obtained by using the formula as follows: leaf area = (leaf length × leaf width)/1.2.

6.6.1.2 QTL Mapping and Effect Analysis of Leaf-Related Traits

Leaf morphology included the traits such as FLAN, and the length, width, and area of the upper three leaves. Furthermore, 31 additive QTLs and 22 pairs of epistatic QTLs confer leaf morphology, and seven of the 31 additive QTLs involved in QTL × environment interaction (Tables 6.35 and 6.36).

Table 6.35 Estimated additive (A) and additive × environment (AE) interactions of QTL for leaf morphology

Trait	QTL	Flanking marker	Position (cM)	A	H^2 (%)	A × E1		A × E2		A × E3	
						AE1	H^2 (%)	AE2	H^2 (%)	AE3	H^2 (%)
FLAN	*qFLAn2B*	Xgwm120–Xbarc1042	75.1	−3.08	5.15						
	qFLAn2D	Xcfd53–Xwmc18	1.7	2.00	2.17						
	qFLAn4D	Xcfe254–BE293342	192.5	−5.17	14.47						
	qFLAn5Ba	Xbarc36–Xbarc140	13.0	−3.18	5.48						
FLL	*qFLLe3Aa*	Xbarc86–Xwmc21	86.5	−0.54	13.82						
	qFLLe5D	Xbarc320–Xwmc215	64.3	−0.38	6.87						
	qFLLe6D	Xcfd42–Xcfd13	35.0	0.33	5.28			0.31	4.67	–0.17	1.32
FLW	*qFLWi3A*	Xwmc264–Xcfa2193	141.9	0.11	9.19						
	qFLWi4B	Xwmc48–Xbarc1096	18.4	−0.10	7.26						
	qFLWi4D	Xbarc334–Xwmc331	2.1	0.10	8.64						
	qFLWi7D	Xgwm676–Xgwm437	123.9	0.08	5.45						
FLAR	*qFLAr2Aa*	Xbarc380–Xgwm636	1.7	−0.95	2.02						
	qFLAr3Aa	Xswes107–Xbarc86	71.1	−1.10	2.74						
	qFLAr3Ac	Xwmc264–Xcfa2193	141.9	1.82	7.47						
	qFLAr4B	Xwmc48–Xbarc1096	18.4	−1.51	5.14						
	qFLAr4D	Xwmc473–Xbarc334	0.0	1.48	4.94						
	qFLAr7D	Xgwm676–Xgwm437	122.9	0.72	1.17						
SLL	*qSLLe2A*	Xwmc401–Xcfa2263	68.9	0.52	2.66						
	qSLLe2D	Xgwm261–Xgwm296	0.0	−1.18	13.74						
	qSLLe5D	Xwmc215–Xgdm63	72.4	−1.16	13.28			–0.67	4.37	0.85	7.12
SLW	*qSLWi5D*	Xcfd101–Xbarc320	58.6	−0.06	5.29			–0.04	2.30	0.06	5.83
SLAR	*qSLAr2D*	Xcfd53–Xwmc18	1.7	8.06	2.24						
	qSLAr5D	Xbarc320–Xwmc215	66.3	32.69	12.31			–0.84	7.45	–1.98	10.33

(continued)

Table 6.35 (continued)

Trait	QTL	Flanking marker	Position (cM)	A	H^2 (%)	A × E1		A × E2		A × E3	
						AE1	H^2 (%)	AE2	H^2 (%)	AE3	H^2 (%)
TLL	*qTLLe2D*	Xcfd53–Xwmc18	13.7	−0.40	3.55					–0.52	5.91
	qTLLe4A	Xwmc718–Xwmc262	0.0	−0.48	5.12						
	qTLLe5D	Xwmc215–Xgdm63	73.4	−1.00	21.91						
TLW	*qTLWi2D*	Xwmc170.2–Xgwm539	65.5	0.04	3.46						
	qTLWi5D	Xbarc320–Xwmc215	66.3	−0.07	8.38			–0.09	15.90	0.08	13.50
TLA	*qTLAr2D*	Xwmc170.2–Xgwm539	65.5	0.62	1.94						
	qTLAr4B	Xwmc657–Xwmc48	15.7	0.27		0.80	3.33			–0.80	3.33
	qTLAr5D	Xbarc320–Xwmc215	67.3	−1.88	18.0			–1.45	10.69	1.04	5.55

E1: Suzhou, 2006; E2: Tai'an, 2006; E3: Tai'an, 2005

FLAN Flag leaf angle; *FLL* Flag leaf length; *FLW* Flag leaf width; *FLAR* Flag leaf area; *SLL* Second leaf length from the top; *SLW* Second leaf width from the top; *SLAR* Second leaf area from the top; *TLL* Third leaf length from the top; *TLW* Third leaf width from the top; *TLAR* Third leaf area from the top

Table 6.36 Estimated epistasis (AA) and epistasis × environment (AAE) interactions of QTL for leaf morphology

Trait	QTL	Flanking marker	Position (cm)	QTL	Flanking marker	Position (cm)	AA	H^2 (%)	AA × E1		AA × E2		AA × E3	
									AAE1	H^2 (%)	AAE2	H^2 (%)	AAE3	H^2 (%)
FLAN	*qFLAn1Ba*	Xbac312–Xcfe023.1	36.1	qFLAn6D	Xgwm55–Xrwm133.2	77.9	−1.31	0.92						
	qFLAn1Bb	Xgwm218–Xgwm582	43.2	qFLAn6D	Xgwm55–Xrwm133.2	77.9	−1.06	0.61						
	qFLAn2Aa	Xgwm636–Xcfe67	40.1	qFLAn5D	Xbac1097vXcfd8	14.4	−1.66	1.50						
	qFLAn2Ab	Xgwm448–Xwmc455	80.2	qFLAn3D	Xbac1119–Xcfd4	17.8	2.13	2.45						
	qFLAn4D	Xcfe254- BE293342	192.5	qFLAn7Bb	Xbarc276.1–Xwmc396	33.5	2.26	2.77						
	qFLAn5Bb	Xbarc232–Xwmc235	23.7	qFLAn7D	Xgwm428–Xcfd175	163.7	3.10	5.22						
FLL	*qFLLe2B*	Xwmc317–Xwmc445.2	89.3	qFLLe3D	Xwmc631–Xbarc071	90.1	0.34	5.42						
	qFLLe2D	Xbarc349.2–Xbarc349.1	76.0	qFLLe5Ab	Xcfe026.1–Xcwem32.2	2.0	0.40	7.43						
	qFLLe3Ab	Xbarc1177–Xbarc276.2	196.3	qFLLe4B	Xwmc125–Xwmc140	0.0	0.33	5.08						
	qFLLe5Aa	Xbarc180–Xcwem40	31.6	qFLLe5B	Xbarc36–Xbarc140	0.0	0.33	5.27						
FLW	*qFLWi1D*	Xgdm60–Xwmc429	17.5	qFLWi6D	Xcfd49–Xcfd42	7.4	−0.07	3.41						
	qFLWi2Ba	Xbarc1042–Xgwm388	75.2	qFLWi6Ba	Xcfd48–Xwmc737	51.7	−0.05	2.04						
	qFLWi2Bb	Xcwem55–Xbarc129.1	85.0	qFLWi6Bb	Xgwm58–Xwmc737	61.3	−0.04	1.21						
	qFLWi2Bb	Xcwem55–Xbarc129.1	85.0	qFLWi4B	Xwmc48–Xbarc1096	18.4	0.05	2.23						
FLA	*qFLAr1B*	Xgwm18–Xwmc57	34.5	qFLAr4A	Xbarc078–Xwmc722	41.0	0.85	1.62						
	qFLAr2Ab	Xgwm455–Xgwm515	82.7	qFLAr7B	Xwmc396–Xgwm333	40.8	1.57	5.51						
	qFLAr2Ab	Xgwm455–Xgwm515	102.7	qFLAr3Ab	Xwmc21–Xwmc664	90.3	0.87	1.71						
	qFLAr5A	Xcfe223–Xwmc273.3	103.0	qFLAr6A	Xgwm334–Xbarc023	12.5	−1.42	4.50						
	qFLAr7Aa	Xgwm60–Xbarc070	9.8	qFLAr7Ab	Xgdm67–Xwmc634	152.5	−1.89	7.98						
SLL	*qSLLe6A*	Xwmc553–Xgwm732	56.5	qSLLe6B	Xcfa2187–Xgwm219	0.0	0.83	6.82						
SLW	*qSLWi4B*	Xwmc657–Xwmc48	17.7	qSLWi7D	Xwmc42–Xswes23	211.3	0.04	3.26	0.06	6.87			−0.04	3.07
TLL	*qTLLe2A*	Xbarc264–Xwmc522	72.2	qTLLe6A	Xgwm459–Xgwm334	5.0	−0.33	2.39						

Abbreviations are the same as in Table 6.35

6.6.1.2.1 QTL Mapping and Effect Analysis of Flag Leaf Angle (FLAN)

A total of four additive QTLs for FLAN were mapped on chromosomes 2B, 2D, 4D, and 5B, accounting for 5.15, 2.17, 14.47, and 5.48 % of phenotypic variation, respectively (Table 6.35), among which qFLAn4D had the highest R^2 value, explaining 14.47 % of phenotypic variation. Furthermore, in addition to qFLAn2D, the other three additive loci were contributed by Yumai 57. And no AE was detected.

Six pairs of epistatic QTLs for FLA distributed on chromosomes 1B-6D (2 regions), 2A-5D, 2A-3D, 4D-7B, and 5B-7D, respectively, were identified, accounting for phenotypic variation from 0.61 to 5.22 % (Table 6.36). However, no interactions between epistasis and environment were detected.

6.6.1.2.2 QTL Mapping and Effect Analysis of Flag Leaf Length (FLL)

Three additive QTLs conferring FLL on chromosomes 3A, 5D, and 6D were detected, accounting for 13.82, 6.87, and 5.28 % of phenotypic variation, respectively (Table 6.35). Among them, the QTL named as *qFLLe3Aa* had the highest contribution, explaining 13.82 % of phenotypic variation. In addition to *qFLLe6D*, other two QTLs were contributed by Yumai 57. Meanwhile, *qFLLe6D* involved in AE interactions and contributed 5.99 % of phenotypic variation.

Four pairs of epistatic QTLs for FLL on chromosomes 2B-3D, 2D-5A, 3A-4B, and 5A-5B were detected, explaining 5.42, 7.43, 5.08, and 5.27 % of phenotypic variation, respectively (Table 6.36), and no AAE was detected. The total contribution of epistasis was 23.20 % of phenotypic variation.

6.6.1.2.3 QTL Mapping and Effect Analysis of Flag Leaf Width (FLW)

Four additive QTLs for FLW on chromosomes 3A, 4B, 4D, and 7D were detected, accounting for 9.19, 7.26, 8.64, and 5.45 % of phenotypic variation, respectively (Table 6.35). Among them, *qFLWi3A* had the highest contribution, explaining 9.19 % of phenotypic variation. In addition, the four loci were all contributed by Huapei 3. No AE was detected.

Four pairs of epistatic QTLs conferring FLW on chromosomes 1D-6D, 2B-6B (2 regions), and 2B-4B were detected, accounting for 3.41, 2.04, 1.21, and 2.23 % of phenotypic variation, respectively (Table 6.36). And, no AAE was detected.

6.6.1.2.4 QTL Mapping and Effect Analysis of Flag Leaf Area (FLAR)

A total of six QTLs for FLAR on chromosomes 2A, 3A (2 regions), 4B, 4D, and 7D were detected, accounting for phenotypic variation from 1.17 to 7.47 % (Table 6.35), and the QTL (*qFLAr3Ac*) had the highest contribution, accounting for

7.47 % of phenotypic variation. The three QTLs, *qFLAr3Ac*, *qFLAr4D,* and *qFLAr7D,* were contributed by Huapei 3, while other QTLs were contributed by Yumai 57. And, no AE was detected.

A total of five pairs of epistatic QTLs for FLAR on chromosomes 1B-4A, 2A-7B, 2A-3A, 5A-6A, and 7A-7A were detected, explaining 1.62, 5.51, 1.71, 4.50, and 7.98 % of phenotypic variation, respectively (Table 6.36). And, no AAE was detected.

6.6.1.2.5 QTL Mapping and Effect Analysis of Second Leaf Length (SLL)

For SLL, three additive QTLs on chromosomes 2A, 2D, and 5D were identified, explaining 2.66, 13.74, and 13.28 % of phenotypic variation, respectively (Table 6.35), and *qSLLe2D* had the highest contribution, explaining 13.74 % of phenotypic variation. In addition to *qSLLe2A*, other QTLs were contributed by Yumai 57. The QTL (*qSLLe5D*) involved in AE, explaining 11.49 % of phenotypic variation.

Only one pair of epistatic QTL for SLL on chromosomes 6A-6B was identified (Table 6.36), explaining 6.82 % of phenotypic variation, and no AAE was detected in this study.

6.6.1.2.6 QTL Mapping and Effect Analysis of Second Leaf Width (SLW)

Only one additive QTL for SLW on chromosome 5D was mapped, explaining 5.29 % of phenotypic variation (Table 6.35), whose positive alleles originated from Yumai 57. Furthermore, the QTL involved in AE, and the total contribution of additive effect was 12.42 %.

One pair of epistatic QTL for SLW was also mapped on chromosomes 4B-7D (Table 6.36), explaining 3.26 % of phenotypic variation, and showed AAE.

6.6.1.2.7 QTL Mapping and Effect Analysis of Second Leaf Area (SLAR)

For SLAR, a total of two additive QTLs on chromosomes 2D and 5D were detected (Table 6.35), accounting for 2.24 and 12.31 % of phenotypic variation. Among them *qSLAr5D* had the highest contribution, with the value of 12.31 %, and involved in AE, explaining 17.78 % of phenotypic variation. The total contribution of additive effect was 32.23 %. Furthermore, no epistatic QTL for SLAR was mapped.

6.6.1.2.8 QTL Mapping and Effect Analysis of Third Leaf Length (TLL)

A total of three additive QTLs on chromosomes 2D, 4A, and 5D were identified, accounting for 3.55, 5.12, and 21.91 % of phenotypic variation (Table 6.35), and *qTLLe5D* had the highest contribution, accounting for 21.91 % of phenotypic variation. All of the three QTLs were contributed by Yumai 57. And no AE was detected.

Only one pair of epistatic QTL for TLL was detected on chromosome 2A-6A (Table 6.36), explaining 2.39 % of phenotypic variation, and no AAE for TLL was detected.

6.6.1.2.9 QTL Mapping and Effect Analysis of Third Leaf Width (TLW)

For TLW, two additive QTLs on chromosomes 2D, 4B, and 5D were detected, accounting for 3.46 and 8.38 % of phenotypic variation, respectively (Table 6.35). Meanwhile, *qTLWi2D* was contributed by Huapei 3, while *qTLWi5D* was contributed by Yumai 57. Furthermore, *qTLWi5D* involved in AE, and the interactive effect was 29.40 %.

No AAE was detected for TLW in this study.

6.6.1.2.10 QTL Mapping and Effect Analysis of Third Leaf Area (TLAR)

A total of three additive QTLs for TLAR were identified on chromosomes 2D and 5D, respectively. And the QTL (*qTLAr5D*) had the highest contribution, explaining 18.0 % of phenotypic variation, and whose positive alleles originated from Huapei 3. Meanwhile, both *qTLAr4B* and *qTLAr5D* involved in environmental interactions, accounting for 22.92 % of phenotypic variation (Table 6.35). And no epistasis was detected for TLAR.

6.6.2 *Association Analysis for Leaf Morphology Based on a Natural Population Derived from the Founder Parent Aimengniu and Its Progenies*

6.6.2.1 Materials and Methods

6.6.2.1.1 Materials

A total of 109 wheat accessions including sister lines, parents, and derived lines of the founder parent Aimengniu. Among which, the three parents and seven sister lines were provided by Tai'an subcenter of national wheat improvement, and the others were provided by Institute of Crop Sciences, Chinese Academy of Agricultural Sciences.

6.6.2.1.2 Field Trial and Determining Phenotypic Data

Field trial was conducted continuously for four years in 2007–2010 in farm of Shandong Agricultural University (Tai'an, Shandong province). The experimental design followed a completely randomized block design with three replications in each environment. In autumn, each year, all varieties were planted in 2-m-long three-row plots (25 cm apart). Management was in accordance with local practices. At filling stage (20 days after anthesis), five flag leaves of main stems of each line were sampled to measure leaf length and width, and leaf area was calculated by the formula as follows: Leaf area = Leaf length × Leaf width ÷ 1.2. And the average value of each trait was used to analysis.

6.6.2.1.3 Analysis of DArT Marker

DNA of the 109 wheat accessions was extracted from adult plant leaves of five individuals using the cetyl trimethyl ammonium bromide (CTAB) method and then genotyped by DArT markers at the Diversity Arrays Technology Pty Limited (Canberra, Australia; http://www.triticarte.com.au). The concentration and purity of DNA were determined using 0.8 % agarose (final concentration of EB was 0. 5 μg mL^{-1}).

A total of 7000 DArT markers, exploited on wheat, were used to scan all of the DNA samples by Triticarte Pty. Ltd. (Canberra, Australia). The known map including Cranbrook × Halberd (339 DArT markers) (Akbari et al. 2006), Arina × NK93604 (189 DArT markers) (Semagn et al. 2006), Avocet × Saar (112 DArT markers) (Lillemo et al. 2008), Colosseo × Lloyd (392 DArT markers) (Mantovani et al. 2008), the map consisted of 779 DArT markers (Wenzl et al. 2004), 3B physics map (Paux et al. 2008), and the information from nine populations were integrated by Triticarte Pty. Ltd. (http://www.triticarte.com.au/). Wheat DArT marker genetic map was constructed using the software of Mapchart 2.1 (Voorrips et al. 2002).

6.6.2.1.4 Analysis of Population Structure

A total of 42 DArT markers (one marker was selected on long arm and short arm of each chromosome) were used to analyze the population structure among wheat accessions using the software Structure 2.0 (Pritchard et al. 2000) with the admixed model. Five independent runs were performed setting the number of populations (K) from 2 to 12, burn-in period 100,000, and iterations 100,000. The maximum likelihood score corresponding to the setting K as target to select appropriate K value as subgroup number was taken.

6.6.2.1.5 Analysis of Linkage Disequilibrium

LD between mapped DArT loci was calculated by the squared allele frequency correlation coefficient (r^2) implemented in TASSEL 2.0.1 (http://www.maizegenetics.net). The pairwise significance was computed by 1000 permutations after removal of loci with rare alleles ($f < 0.10$). LD was calculated separately for unlinked loci and loci on the same chromosome.

6.6.2.1.6 Association Analysis

Significant marker–trait associations were identified using a mixed linear model (MLM) in TASSEL 2.1 (http://www.maizegenetics.net/). The population structure was inferred by program Structure 2.0 and kinship matrix was calculated by software TASSEL 2.1. The significance of associations between a marker locus and a trait was indicated by the p value. And it was considered that there were associations between them, when $P \leq 0.001$.

6.6.2.2 Analysis of Marker–Trait Associations

6.6.2.2.1 Phenotypic Data

The flag leaf length (FLL), flag leaf width (FLW), and flag leaf area (FLAR) of the population under the four environments were summarized in Table 6.37. There were significant differences in flag leaf traits among different individuals, while the differences were smaller among different environments. The mean percentages of phenotypic variation explained by population structure for FLL, FLW, and FLAR were 25.43, 9.78, and 25.73 %, respectively. And broad-sense heritability for FLL, FLW, and FLAR were 76.3, 80.1, and 72.8 %, respectively.

Table 6.37 Descriptive statistics and percentage of phenotypic variation explained by population structure for flag leaf length, width, and area (FLL, FLW, FLAR)

Trait	Environment	Min	Max	Mean	SD	R^{2a} (%)	H^{2b} (%)
FLL	E1: Tai'an (2007)	14.83	35.23	21.15	4.06	30.9	76.3
	E2: Tai'an (2008)	10.55	40.55	19.31	4.60	25.1	
	E3: Tai'an (2009)	13.38	29.96	18.82	3.13	22.6	
	E4: Tai'an (2010)	11.58	29.33	18.13	2.70	23.1	
FLW	E1: Tai'an (2007)	1.33	2.40	1.72	0.21	10.6	80.1
	E2: Tai'an (2008)	1.00	2.15	1.59	0.21	9.8	
	E3: Tai'an (2009)	1.33	2.24	1.68	0.17	9.5	
	E4: Tai'an (2010)	1.10	1.97	1.52	0.17	8.9	
FLAR	E1: Tai'an (2007)	18.07	69.27	30.61	8.47	31.2	72.8
	E2: Tai'an (2008)	11.94	72.65	25.95	8.65	18.9	
	E3: Tai'an (2009)	17.04	47.44	26.41	5.86	25.6	
	E4: Tai'an (2010)	13.32	46.44	22.97	4.87	27.2	

[a]Percentage of phenotypic variation explained by population structure
[b]Broad-sense heritability; abbreviations are the same as in Table 6.35

6.6.2.2.2 Association Analysis

The associations between DArT markers and FLL, FLW, and FLAR were tested through the mixed linear model. Percentage of phenotypic variation explained by individual-associated marker and significance of association was summarized in Table 6.38. Based on the critical p value less than 0.01, we identified 61 marker-trait associations (MTAs) involving 46 DArT markers distributed on 14 chromosomes (1B, 1D, 2A, 2B, 2D, 3A, 3B, 4A, 5B, 6A, 6B, 6D, 7A, and 7B) for the three traits and the R^2 ranges from 0.1 to 16.4 % (Fig. 6.14).

A total of 13 significant MTAs for FLL were detected on chromosomes 1B, 2B, 3A, 3B, 4A, 5B, 6A, 6B, and 6D, explaining phenotypic variation from 0.1 to 14.49 %. And wPt-3109 (2B) had the highest R^2.

Thirty eight significant MTAs involving 31 markers distributed on chromosomes 1D, 2A, 2B, 2D, 3A, 4A, 5B, 6A, 6B, 7A, and 7B for FLW were identified and the R^2 ranged from 1.03 to 16.4 %. And wPt-9422 had the highest R^2 (3A, 49.3 cM). Meanwhile, several MTAs were repeatedly detected in two environments, for example, wPt-665037 (1D, 11.7 cM), wPt-664989 (1D, 12.1 cM), wPt-665204 (1D, 12.1 cM), wPt-6711 (2A), wPt-1902 (2D), wPt-3130 (6B, 39.6 cM), wPt-9990 (6B, 39.6 cM), which explained phenotypic variation 11.28, 13.21, 12.18, 10.88, 5.45, 8.77, and 7.77 %, respectively.

A total of 10 significant MTAs distributed on chromosomes 2D, 3B, 4A, 5B, 6A, and 7B for FLAR were identified, and R^2 ranged from 1.1 to 13.97 %. Meanwhile, wPt-6043 (3B) had the highest R^2.

It is worth noting that some markers associated with several traits. For example, wPt-3457 (5B, 92.3 cM) simultaneously associated with both FLL and FLW, wPt-5836 (3B, 71.6 cM) and wPt-4270 (6A) associated with both FLL and FLAR, while wPt-730744 (2D, 61.4 cM), wPt-667476 (2D, 62.3 cM), wPt-1902 (2D), wPt-5737 (5B, 69.9 cM), and wPt-5737 (7B, 56.6 cM) associated with both FLW and FLAR.

6.6.3 Research Progress of Leaf Morphology QTL Mapping and Comparison of the Results with Previous Studies

6.6.3.1 Research Progress of Leaf Morphology QTL Mapping

Few researches related to QTL for physiological characteristic of leaf morphology in wheat. Keller et al. (1999) identified eight QTLs for FLW on chromosomes 1A, 1B, 2A, 3B, 5A, 5B, and 6A, explaining up to 59.5 % of phenotypic variation, by using a RIL population consisted of 226 lines derived from Forno/spelt Oberkulmer. Lohwasser et al. (2004) detected 23 QTLs for length and width of the third basal leaf by using a RIL population including 114 lines under greenhouse conditions. Zhang et al. (2012) used a RIL population to identify QTLs for FLL,

Table 6.38 The marker loci associated with flag leaf and corresponding explained phenotypic variation

Chr.	Marker	Position	R^2 (%)		
			FLL	FLW	FLAR
1B	wPt-9605	–	7.2* (E2)		
1D	wPt-665037	11.7		9.85* (E1) 12.7* (E2)	
	wPt-664989	12.1		11.45* (E1) 14.96** (E2)	
	wPt-665204	12.1		10.79* (E1) 13.56** (E2)	
	wPt-3855	–		10.69* (E2)	
2A	wPt-669355	281.9		10.69* (E1)	
	wPt-6711	–		9.16* (E1) 12.59* (E3)	
	wPt-0568	–		9.5* (E1)	
2B	wPt-2106	22.8		13.38** (E2)	
	wPt-3109	–	14.49* (E3)		
2D	wPt-1554	7.6		10.19* (E1)	
	wPt-730744	61.4		11.43* (E1)	7.47* (E4)
	wPt-667476	62.3		9.7* (E1)	10.11* (E4)
	wPt-668120	62.3		9.9* (E1)	
	wPt-731134	62.3		9.9* (E1)	
	wPt-1902	–		9.76* (E1) 1.03* (E4)	6.58* (E4)
	wPt-3692	–		14.1* (E3)	
	wPt-6704	–		10.67* (E1)	
3A	wPt-9422	49.3		16.4* (E3)	
	wPt-0398	146.4	8.71* (E1)		
	tPt-0519	–		9.91* (E2)	
3B	wPt-5836	71.6	6.49* (E1)		11.54* (E1)
	wPt-10186	–			12.17* (E3)
	wPt-2491	–	1.27** (E1)		
	wPt-6043	–			13.97* (E3)
4A	wPt-8091	180.1	9.16* (E2)		
	wPt-2985	–			1.11* (E1)
	wPt-6900	–		14.00* (E3)	
	wPt-6757	–	8.83* (E2)		
5B	wPt-5737	69.9		10.06* (E1)	5.84* (E1)
	wPt-3457	92.3	10.66* (E3)	11.86* (E1)	
	wPt-0819	–		6.75* (E4)	
	wPt-1548	–		10.33* (E1)	

(continued)

Table 6.38 (continued)

Chr.	Marker	Position	R^2 (%)		
			FLL	FLW	FLAR
6A	wPt-666988	39.8	0.9* (E2)		
	wPt-667618	142.4		6.67* (E4)	
	wPt-4270	–	10.19* (E2)		12.63* (E2)
6B	wPt-3130	39.6		8.89* (E1) 8.64* (E2)	
	wPt-9990	39.6		8.89* (E1) 8.64* (E2)	
	wPt-0959	57.7		10.14* (E1)	
	wPt-8183	92.5		9.97* (E1)	
	wPt-2424	96.1		8.56* (E1)	
	wPt-0171	172.1	0.2** (E3)		
	wPt-3581	–	0.1** (E4)		
6D	wPt-664719	134.9	10.18* (E2)		
7A	tPt-1755	–		11.28* (E2)	
7B	wPt-5737	56.6		10.06* (E1)	5.84* (E1)

Marker position "–" indicates that this marker has no definite genetic distance. Markers with significant marker–trait associations are listed ($P < 0.001$), and the phenotypic variation explained (R^2) is marked with single asterisk (*) or double asterisks (**) denotes the p value ranging from 0.0001 to 0.0010 or smaller than 0.0001, respectively. Abbreviations are the same as in Table 6.35

FLW, FLAR, SLL, SLW, SLAR, TLL, TLW, TLAR, and 29 QTLs were detected, and none of 29 QTLs were major QTLs, with the highest effect of 13.87 %. Meanwhile, several QTLs controlling leaf morphology were found on chromosomes 4B and 4D.

6.6.3.2 Comparsion of the Results with the Previous Studies

In this study, a total of 31 additive QTLs for leaf morphology-related traits were identified, mainly distributing on chromosomes 2D, 4D, 5D, and 7D, and six of 31 QTLs were major QTLs with the highest effect of 21.91 %. Meanwhile, the QTLs for leaf morphology were mainly mapped on chromosomes 2D, 4A, 4B, 4D, 5D, and 7D based on association analysis, and within the linked marker intervals of leaf morphology, some QTLs for yield-related traits, quality traits, and important agronomic traits were also mapped. There were similar results between the partial results of this paper and those of Keller et al. (1999) and Zhang et al. (2012), indicating that important QTLs or genes controlling leaf morphology of wheat were distributed on D genome. Moreover, some associations were repeatedly detected in several environments, which could be considered to be relatively stable, and linked molecular makers with higher contribution to phenotypic variation could be used in MAS breeding programs. For example, several markers distributing on

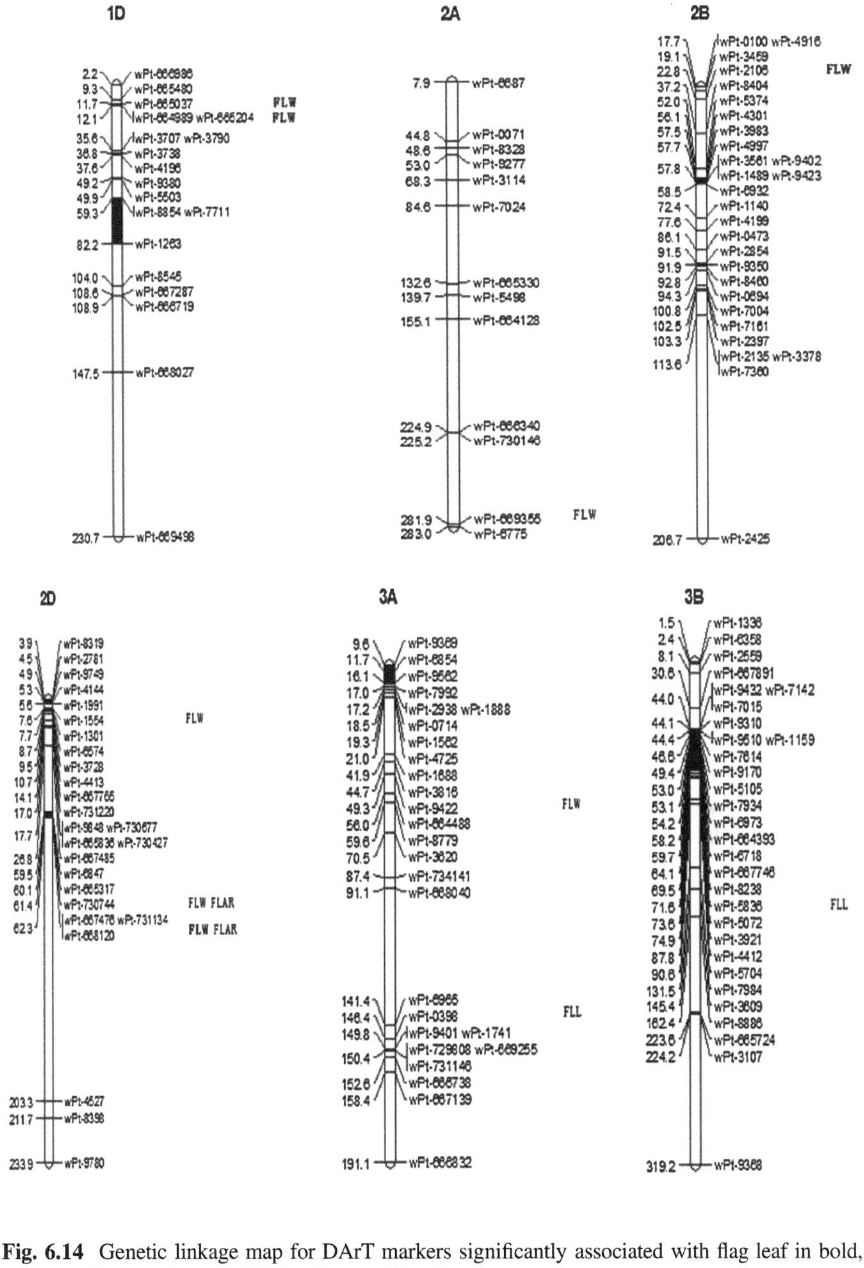

Fig. 6.14 Genetic linkage map for DArT markers significantly associated with flag leaf in bold, markers significantly associated with more than three environments

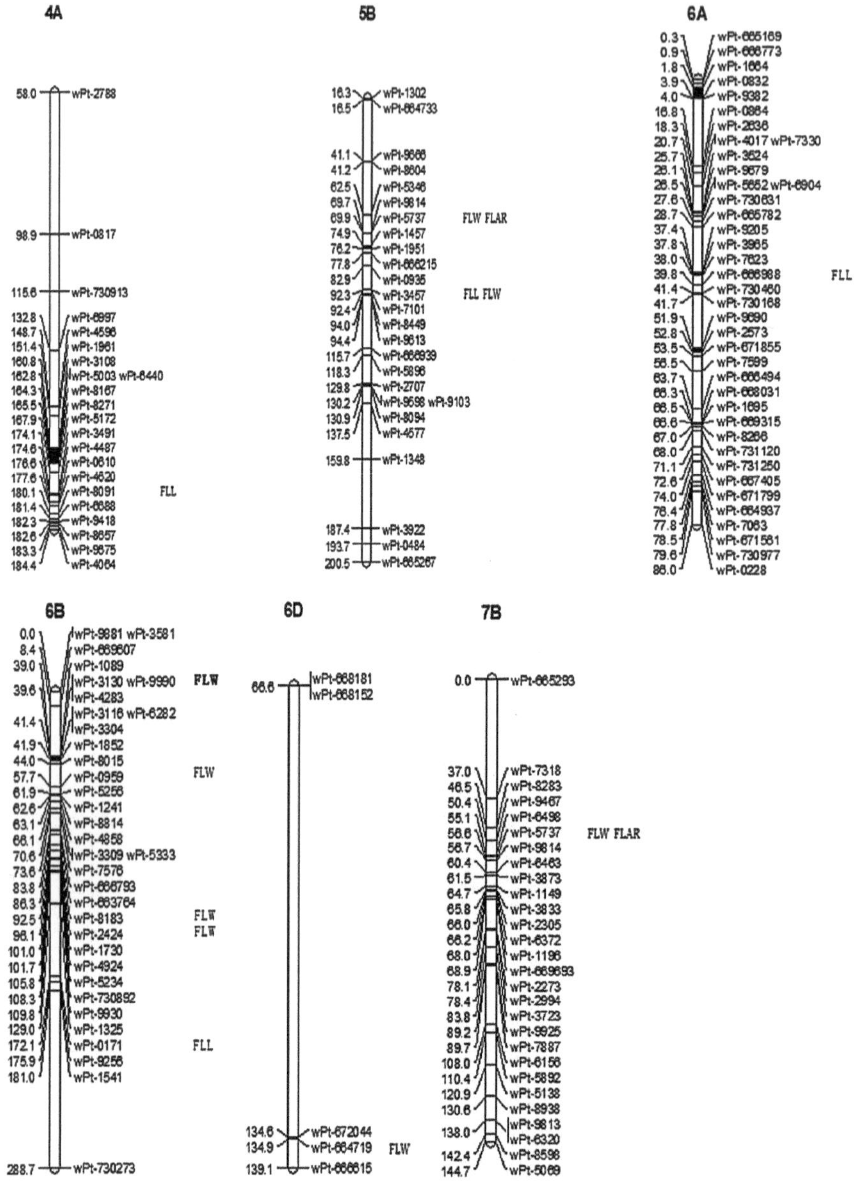

Fig. 6.14 (continued)

chromosomes 1D and 6B were associated with FLW and had the higher contributions to phenotypic variation.

The big ranges of variation of flag leaf in different varieties were in favor of mapping more key intervals linked with flag leaf traits. For example, several markers distributing on chromosomes 1D, 2D, and 6B were associated with some traits, which may be enrichment regions of genes controlling flag leaf traits, and that needs to be studied in further research.

References

Akbari M, Wenzl P, Caig V, Carling J, Xia L, Yang S, Uszynski G, Mohler V, Lehmensiek A, Kuchel H, Hoyden MJ, Howes N, Sharp P, Vaughan P, Rathmell B, Hutter E, Kilian A. Diversity arrays technology (DArT) for high-throughput profiling of the hexaploid wheat genome. Theor Appl Genet. 2006;113:1409–20.

Ali MB, Ibrahim AMH, Malla S, Rudd J, Hays D. Family-based QTL mapping of heat stress tolerance in primitive tetraploid wheat (*Triticum turgidum* L.). Euphytica. 2013;192:189–203.

An D, Su JY, Liu QY, Zhu YG, Tong YP, Li JM, Jing RL, Li B, Li ZS. Mapping QTLs for nitrogen uptake in relation to the early growth of wheat (*Triticum aestivum* L.). Plant Soil. 2006;284:73–84.

Båga M, Chodaparambil SV, Limin AE, Pecar M, Fowler DB, Chibbar RN. Identification of quantitative trait loci and associated candidate genes for low-temperature tolerance in cold-hardy winter wheat. Funct Integr Genomics. 2007;7:53–68.

Bai CH, Liang YL, Hawkesford MJ. Identification of QTLs associated with seedling root traits and their correlation with plant height in wheat. J Exp Bot. 2013;64:1745–53.

Börner A, Schumann E, Fürste A, Cöster H, Leithold B, Röder MS, Weber WE. Mapping of quantitative trait loci determining agronomic important characters in hexaploid wheat (*Triticum aestivum* L.). Theor Appl Genet. 2002;105:921–36.

Brube-Bable AL, Fowler DB. Genetic control of cold hardiness and venalization requirement in winter wheat. Crop Sci. 1988;28:879–84.

Cao WD, Jia JZ, Jin JY. Identification and interaction analysis of QTL for chlorophyll content in wheat seedlings. Plant Nutr Ferti Sci. 2004;10(5):473–8 (in Chinese with English abstract).

Caradus JR. Genetic control of phosphorus uptake and phosphorus status in plants. In: Genetic manipulation of crop plants to enhance integrated nutrient management in cropping system. Patancheru: ICRISAT Asia Centre; 1995. pp. 55–67.

Czyczylo-Mysza I, Tyrka M, Marcińska I, Skrzypek E, Karbarz M, Dziurka M, Hura T, Dziurka K, Quarrie SA. Quantitative trait loci for leaf chlorophyll fluorescence parameters, chlorophyll and carotenoid contents in relation to biomass and yield in bread wheat and their chromosome deletion bin assignments. Mol Breed. 2013;32(1):189–210.

Galiba G, Quarrie SA, Sutka J, Morgounov A, Snape JW. RFLP mapping of the vernalization (Vrn1) and frost resistance (Fr1) genes on chromosome 5A of wheat. Theor Appl Genet. 1995;90:1174–9.

Groos C, Robert N, Bervas E, et al. Genetic analysis of grain protein-content, grain yield and thousand-kernel weight in bread wheat. Theor Appl Genet. 2003;106(6):1032–40.

Guo HJ. QTL analysis for stem strength and its correlated traits in wheat. Master's thesis of Chinese Academy of Agricultural Sciences; 2002 (in Chinese with English abstract).

Guo PG, Baum M, Varshney RK, Graner A, Grando S, Ceccarelli S. QTLs for chlorophyll and chlorophyll fluorescence parameters in barley (*Hordeum vulgare* L.) under post-flowering drought. Euphytica. 2008;163:203–14.

Hai L, Guo HJ, Xiao SH, Jiang GL, Zhang XY, Yan CS, Xin ZY, Jia JZ. Quantitative trait loci (QTL) of stem strength and related traits in a doubled-haploid population of wheat (Triticum aestivum L.). Euphytica. 2005;141:1–9.

Hamada A, Nitta M, Nasuda S, Kato K, Fujita M, Matsunaka H, Okumoto Y. Novel QTLs for growth angle of seminal roots in wheat (*Triticum aestivum* L.). Plant Soil. 2012;354(1–2): 395–405.

Hanocq E, Niarquin M, Heumez E, Rousset M, Le GJ. Detection and mapping of QTL for earliness components in a bread wheat recombinant inbred lines population. Theor Appl Genet. 2004;110:106–15.

Huang XQ, Cöster H, Ganal MW, Röder MS. Advanced backcross QTL analysis for the identification of quantitative trait loci alleles from wild relatives of wheat (*Triticum aestivum* L). Theor Appl Genet. 2003;106:1379–89.

Huang XQ, Cloutier S, Lycar L, Radovanovic N, Humphreys DG, Noll JS, Somers DJ, Brown PD. Molecular detection of QTLs for agronomic and quality traits in a doubled haploid population derived from two Canadian wheats (*Triticum aestivum* L.). Theor Appl Genet. 2006;113:753–66.

Hund A, Frascaroli E, Leipner J, Jompuk C, Stamp P, Fracheboud Y. Cold tolerance of the photosynthetic apparatus: pleiotropic relationship between photosynthetic performance and specific leaf area of maize seedlings. Mol Breed. 2005;16:321–31.

Ibrahim SE, Schubert A, Pillen K, Léon J. QTL analysis of drought tolerance for seedling root morphological traits in an advanced backcross population of spring wheat. International Journal of Agri Science. 2012;2(7):619–29.

Jiang P. QTL mapping for related traits of wheat germination, Master's thesis of Shandong Agricultural University, 2012 (in Chinese with English abstract).

Jing RL, Hu RH, Zhu ZH, Chang XP. A study on heritabilities of seedling morphological traits and drought resistance in winter wheat cultivars of different genotype. Acta Bot Boreal-Occident Sin. 1997;17(2):152–7 (in Chinese with English abstract).

Ju W, Yang CF, Zhang SH, Tian JC, Hai Y, Yang XJ. Mapping QTL for cell membrane permeability of leaf treated by low temperature in Winter Wheat. Acta Agronomica Sinica. 2012;38(7):1247–52 (in Chinese).

Keller M, Karutz C, Schmid JE, Stamp P, Winzeler M, Keller B, Messmer MM. Quantitative trait loci for lodging resistance in a segregating wheat × spelt population. Theor Appl Genet. 1999;98:1171–82.

Landjeva S, Lohwasser U, Börner A. Genetic mapping within the wheat D genome reveals QTL for germination, seed vigour and longevity, and early seedling growth. Euphytica. 2010;171:129–43.

Leipner J, Jompuk C, Camp KH, Stamp P and Fracheboud Y. QTL studies reveal little relevance of chilling-related seedling traits for yield in maize (*Zea mays* L.). Theor. Appl. Genet, 2008, 116: 555–562.

Li ZK, Luo LJ, Mei HW, Shu QY, Tabien R, Zhong DB, Ying CS, Stansel JW, Khush GS, Paterson AH. Overdominance epistatic loci are the primary genetic basis of inbreeding depression and heterosis in rice: I. Biomass and grain yield. Genetics. 2001;158(4):1737–53.

Li SP, Chang XP, Wang CS, Jing RL. Mapping QTL for heat tolerance at grain filling stage in common wheat. Scientia Agricultura Sinica. 2013;46(10):2119–29 (in Chinese with English abstract).

Lillemo M, Asalf B, Singh RP, Huerta-Espino J, Chen XM, He ZH, Biornstad A. The adult plant rust resistance loci Lr34/Yr18 and Lr46/Yr29 are important determinants of partial resistance to powdery mildew in bread wheat line Saar. Theor Appl Genet. 2008;116:1155–66.

Limin AE, Danyluk J, Chauvinl LP, Fowler DB, Sarhan F. Chromosome mapping of low-temperature induced Wcs120 family genes and regulation of cold-tolerance expression in wheat. Mol Gen Genet. 1997;253:720–7.

Liu YY. Cold-resistance-index screening and QTL analysis of wheat. Master's thesis of Henan Agricultural University, 2005 (in Chinese with English abstract).

Liu TJ, Qi CH, Tang JJ. Studies on relationship between the character parameters of root and yield formation in Rice. Scientia Agricultura Sinica. 2002;35(11):1416–9 (in Chinese with English abstract).

Liu XL, Chang XP, Li RZ, Jing RL. Mapping QTLs for seminal root architecture and coleoptile length in wheat. Acta Agronomica Sinica. 2011;37(3):381–8 (in Chinese with English abstract).

Liu XL, Li RZ, Chang XP, Jing RL. Mapping QTLs for seedling root traits in a doubled haploid wheat population under different water regimes. Euphytica. 2013;189:51–66.

Lohwasser U, Röder MS, Börner A. QTL mapping of vegetative characters in wheat (*Triticum aestivum* L.). Genetic variation for plant breeding. Proceedings of the 17th EUCARPIA General Congress, Tulln, Austria, 8–11 September 2004 pp. 195–198.

Mantovani P, Maccaferri M, Sanguineti MC, Tuberosa R, Catizone I, Wenzl P, Thomson B, Carling J, Huttner E, DeAmbrogio E, Kilian A. An integrated DArT-SSR linkage map of durum wheat. Mol Breed. 2008;22:629–48.

Marza F, Bai GH, Carver BF, Zhou WC. Quantitative trait loci for yield and related traits in the wheat population Ning7840 × Clark. Theor Appl Genet. 2006;112(4):688–98.

Mei HW, Li ZK, Shu QY, Guo LB, Wang YP, Yu XQ, Ying CS, Luo LJ. Gene actions of QTL affecting several agronomic traits resolved in a recombinant inbred rice population and two backcross population. Theor Appl Genet. 2005;110:649–59.

Moudal S K, Kour K. Genetic variability and correlation coefficients of some root characteristics and yield components in bread wheat (*Triticum aestivum* L.) under rainfed condition. Environ Ecol, 2004, 22: 646–648.

Nagata K, Shimizu H, Terao T. Quantitative trait loci for nonstructural carbohydrate accumulation in leaf and culms of rice (*Oryza sativa* L.) and their effects on grain filling. Breeding Sci, 2002, 52: 275–283.

Partha DAS, Ali MN, Sarkar HK. Genetical studies on roots in bread wheat. J Interacademicia. 2004;8:166–8.

Paux E, Sourdille P, Salse J, Saintenac C, Choulet F, Leroy P, Korol A, Michalak M, Kianian S, Spielmeyer W, Lagudan E, Somers D, Kilian A, Alaux M, Vautrin S, Berges H, Eversole K, Appels R, Safar J, Simkova H, Dolezel J, Bernard M, Feuillet C. A physical map of the 1-Gigabase bread wheat chromosome 3B. Science. 2008;322:101–4.

Pelleschi S, Leonardi A, Rocher JP, Cornic G, Vienne D de, Thévenot C, Prioul JL. Analysis of the relationships between growth, photosynthesis and carbohydrate metabolism using quantitative trait loci (QTLs) in young maize (*Zea mays* L.) plants subjected to water deprivation. Mol Breeding, 2006, 17: 21–39.

Porra RJ, Thompson WA, Kriedemann PE. Determination of accurate extinction coefficients and simultaneous equations for assaying chlorophylls a and b extracted with four different solvents: verification of the concentration of chlorophyll standards by atomic absorption spectroscopy. Biochimica et Biophysica Acta (BBA)-Bioenergetics, 1989, 975 (3): 384–394.

Pritchard JK, Stephens M, Donnelly P. Inference of population structure from multilocus genotype data. Genetics. 2000;155:945–59.

Quarrie SA, Steed A, Calestani C, Semikhodskii A, Lebreton C, Chinoy C, Steele N and Pljevljakusić D. A high-density genetic map of hexaploid wheat (*Triticum aestivum* L.) from the cross Chinese Spring 9 SQ1 and its use to compare QTLs for grain yield across a range of environments. Theor Appl Genet, 2005, 110: 865–880.

Ren YZ, He X, Liu DC, Li JJ, Zhao XQ, Li B, Tong YP, Zhang AM, Li ZS. Major quantitative trait loci for seminal root morphology of wheat seedlings. Mol Breeding. 2012a;30:139–48.

Ren YZ, Xu YH, Gui XW, Wang SP, Ding JP, Zhang QC, Ma YS, Pei DL. QTLs analysis of Wheat seedling traits under salt stress. Scientia Agricultura Sinica. 2012b;45(14):2793–800 (in Chinese with English abstract).

Ritter KB, Jordan DR, Chapman SC, Godwin ID, Mace ES, Lynne MC. Identification of QTL for sugar-related traits in a sweet × grain sorghum (*Sorghum bicolor* L. Moench) recombinant inbred population. Mol Breed, 2008, 22: 367–384.

Semagn K, Bjornstad H, Skinnes AG, Marφy Y. Tarkegne, William M. Distribution of DArT, AFLP and SSR markers in a genetic linkage map of a double haploid hexaploid wheat population. Genome. 2006;49:545–55.

Sharma S, Xu SZ, Ehdaie B, Hoops A, Close TJ, Lukaszewski AJ, Waines JG. Dissection of QTL effects for root traits using a chromosome arm-specific mapping population in bread wheat. Theor Appl Genet. 2011;122(4):759–69.

Song YX, Jing RL, Huo NX, Ren ZL and Jia JZ. Detection of QTLs for heading in common wheat (*T. aestivum* L.) using different populations. Scientia Agricultura Sinica, 2006, 39(11): 2186–2193 (in Chinese with English abstract).

Song YX. QTLs analysis of heading date and other agronomic traits in wheat. Doctoral thesis of Sichuan Agricultural University, 2005 (in Chinese with English abstract).

Su JY, Tong YP, Liu QY, Li B, Jing RL, Li JY, Li ZS. Mapping quantitative trait loci for post-anthesis dry matter accumulation in wheat (*Triticum aestivum* L.). J Integr Plant Biol. 2006;48:938–44.

Sutka J. Genes for frost resistance in wheat. Euphytica. 2001;119(1–2):169–77.

Tóth B, Galiba G, Fehér E, Sutka J, Snape JW. Mapping genes affecting flowering time and frost resistance on chromosome 5B of wheat. Theor Appl Genet. 2003;107:509–14.

Vágújfalvi A, Galiba G, Cattivelli L, Dubcovsky J. The cold-regulated transcriptional activator Cbf3 is linked to the frost-tolerance locus Fr-A2 on wheat chromosome 5A. Mol Gen Genomics. 2003;269:60–7.

Vijayalakshmi K, Fritz AK, Paulsen GM, Bai GH, Pandravada S, Gill BS. Modeling and mapping QTL for senescence-related traits in winter wheat under high temperature. Mol Breeding. 2010;26(2):163–75.

Voorrips RE. MapChart: software for the graphical presentation of linkage maps and QTLs. J Hered. 2002;93:77–8.

Waldman M, Rikin A, Dovrat A, Richmond AE. Horironal regulation of morphogenesis and cold resistance. J Exp Bot. 1975;26:853–9.

Wei XY, Li SS, Jiang FS, Guo Y, Li RJ. QTL mapping for premature senescence and related physiological traits in Wheat. Acta Bot Boreal-Occident Sin. 2007;27(3):485–9 (in Chinese with English abstract).

Wenzl P, Carling J, Kudrna D, Jaccoud D, Huttner E, Kleinhofs A, Kilian A. Diversity arrays technology (DArT) for whole genome profiling of barley. Proc Natl Acad Sci USA. 2004;101:9915–20.

Wu YQ, Liu LX, Guo HJ, Zhao LZ, Zhao SR. Mapping QTL for salt tolerant traits in wheat. J Nuclear Agri Sci. 2007;21(6):545–9 (in Chinese with English abstract).

Xing YZ, Tan YF, Hua JP, Sun XL, Xu CG, Zhang QF. Characterization of the main effects, epistatic effects and their environmental interactions of QTLs on the genetic basis of yield traits in rice. Theor Appl Genet. 2002;105:248–57.

Xu SB. Construction of genetic map by SSR marker and location QTL for plant height and heading time in wheat. Master's thesis of Xinjiang Agricultural University; 2005 (in Chinese with English abstract).

Xu YF, An DG, Liu DC, Zhang AM, Xu HX, Li B. Mapping QTLs with epistatic effects and QTL×treatment interactions for salt tolerance at seedling stage of wheat. Euphytica. 2012;186:233–45.

Xue DW, Chen MC, Zhou MX, Chen S, Mao Y, Zhang GP. QTL analysis of flag leaf in barley (Hordeum vulgare L.) for morphological traits and chlorophyll content. J Zhejiang Univ Sci B. 2008;9(12):938–43.

Yan L, Helguera M, Kato K, Fukuyama S, Sherman J, Dubcovsky J. Allelic variation at the VRN-1 promoter region in polyploid wheat. Theor Appl Genet. 2004;109(8):1677–86.

Yang CF. Mapping QTL for salt tolerant traits in wheat seedling stage under different salt stress. Master's thesis of Agricultural University of Hebei; 2012 (in Chinese with English abstract).

Yang DL, Jing RL, Chang XP, Li W. Quantitative trait loci mapping for chlorophyll fluorescence and associated traits in wheat (*Triticum aestivum* L.). J Integ Plant Biol. 2007a;49:646–54.

Yang DL, Jing RL, Chang XP, Li W. Identification of quantitative trait loci and environmental interactions for accumulation and remobilization of water-soluble carbohydrates in Wheat (*Triticum aestivum* L.) Stems. Genetics. 2007b;176:571–84.

Yao Q, Zhou RH, Pan YM, Fu TH, Jia JZ. Construction of genetic linkage map and QTL analysis of agronomic important traits based on a RIL population derived from common wheat variety Yanzhan 1 and Zaosui 30. Sci Agricultura Sinica. 2010;43(20):4130–9 (in Chinese with English abstract).

Zhang KP. Construction of wheat (*Triticum aestivum* L.) molecular genetic map and QTL analysis. Doctoral thesis of Shandong Agricultural University; 2008 (in Chinese with English abstract).

Zhang Q. Genetic map construction and QTL mapping of important agronomic traits in common wheat. Master's thesis of Shandong Agricultural University; 2012 (in Chinese with English abstract).

Zhang ZB, Xu P. Reviewed on wheat genome. Hereditas. 2002;24(3):389–94 (in Chinese with English abstract).

Zhang KP, Xu XB, Tian JC. QTL mapping for grain yield and spike related traits in common wheat. Acta Agronomica Sinica. 2009;35(2):270–8 (in Chinese with English abstract).

Zhou XG, Jing RL, Hao ZF, Chang XP, Zhang ZB. Mapping QTL for seedling root traits in common wheat. Sci Agricultura Sinica. 2005;38(10):1951–7 (in Chinese with English abstract).

Chapter 7
Genetic Dissection of Stress-Tolerance Traits in Wheat

Abstract Wheat (*Triticum aestivum* L.) growth and its productivity were always affected by abiotic or biotic stress. Especially, drought, salinity, waterlogging and disease often cause severe reductions in wheat yield. Therefore, it is greatly important to discover resistant genes in wheat. In this chapter, drought resistance, heavy metals resistance (cadmium and chromium stress), pre-harvest sprouting resistance, disease resistance (adult-plant resistance to powdery mildew, fusarium head blight resistance), salt resistance and potassium resistance were genetically dissected by QTL mapping. Some major QTLs identified in this chapter could provide important genetic and molecular information for marker-assisted selection breeding in wheat.

Keywords Stress tolerance · Drought resistance · Heavy metals resistance · Pre-harvest sprouting · Powdery mildew resistance · Fusarium head blight · Salt-tolerance · Potassium-deficiency tolerance · QTL mapping

Adversity is also known as environmental stress, which has various environmental factors that exceed the scope of the normal life and not conducive to plant growth and development of the plant. The field crop will encounter a variety of environmental stress which includes not only biotic stresses, such as diseases, insects, weeds, and other adversity, but also abiotic stresses, such as water deficit, low temperature, high temperature, salinity, and environmental pollution. The plants in nature must have suitable growth characteristics and physiological mechanisms to protect themselves from or less from the from external environment. Such as, ephemeral plants in desert plants can complete their life cycle when adequate rainfall; likewise, heat-resisting plants can recognize stress signal, activate signaling pathway, regulate gene expression, and then change physiological metabolism of intact plant when they encounter high temperature. Resistance or susceptibility of plants to stress factors was affected not only by the severity of environmental stress, duration, frequency, and stress combination of exposure, but also by the species, genotype, and developmental stage (Wang 2011).

Abiotic stress causes significant crop losses. The stresses are numerous and seriously affect the normal growth of wheat. Those include drought, high salinity,

J. Tian et al., *Genetic Analyses of Wheat and Molecular Marker-Assisted Breeding, Volume 1*, DOI 10.1007/978-94-017-7390-4_7

temperature extremes, cold damage, and other adversity. Currently, with the rapid development of the molecular biology and molecular markers, quantitative trait loci (QTL) mapping under environmental stress has been rapidly developed (Mao 2010). Sutka and Snape (1989), Witkowski et al. (2008), and Poysa (1984) detected QTL for frost resistance on chromosomes 5A, 5D, and 1A; Quarrie et al. (2005) reported QTL for salt tolerance on chromosomes 5A and 5B; Quarrie et al. (1994), Hao et al. (2003), and Kirigwi et al. (2007) detected QTL for drought resisting on chromosomes 1B, 2B, 3B, 4A, 5A, 6B, 7A, and 7B. Yang et al. (2002), Francki et al. (2002), Mason et al. (2011), and Vijayalakshmi et al. (2010) detected QTL for heat resistance on chromosomes 1A, 1B, 2A, 2B, 3A, 3B, 5A, 6A, 6B, and 7A.

Compared to QTL mapping of yield and quality traits, the study about QTL mapping under environmental stress has been few reported because of its complexity of stress-tolerance conditions and the difficulty of multipoint-year phenotypic identification. At present, we only obtained some resistance QTL genes. The QTLs of stress tolerance are difficult to the genetic improvement of wheat. Therefore, analysis of genetic resistance must be a more in-depth study, in order to obtain molecules marker used to improve wheat resistance traits and to enhance the yield and quality of wheat. This chapter includes genetic analysis of drought resistance, heavy metal, sprouting resistant, disease, and salt tolerance.

7.1 QTL Mapping and Effect Analysis of Drought Resistance

Drought resistance of wheat is a quantitative trait controlled by multiple genes, and its molecular and genetic mechanism is complex. A QTL approach was applied to dissect a complex trait into different QTLs; that is to say, QTL was mapped to chromosome-specific location and the next effect of the single QTL was analyzed in a specific genetic background and environmental condition. At present, QTL mapping of drought resistance focuses on morphological indexes and physiological traits in wheat. Landjeva et al. (2008) mapped the QTLs for root length, coleoptiles length, seedling height, and root length/seedling height under drought stress simulated by PEG and control condition, respectively. Zhou et al. (2005) reported that several root traits, including root number (RN), maximum root length (MRL), root fresh weight (RFW), root dry weight (RDW), ratio of root fresh weight to shoot fresh weight (RFW/SFW), and ratio of root dry weight to shoot dry weight (RDW/SDW) per plant of hydroponic seedlings, were measured under water stress and control conditions, respectively. Zhang et al. (2000) mapped QTL for water-use efficiency on chromosomes 1A and 6D. Morgan et al. (1996) mapped the osmotic regulation gene of wheat on chromosome 7A using RFLP markers and found a marker (Xpsr119) closely linked with the gene. Furthermore, the Xpsr119 was used as genetic marker to select osmotic regulation gene in wheat breeding program. Yang et al. (2001) mapped QTL controlling proline on chromosomes 5A and 5D using the Chinese spring—Hope alien substitution lines. Lebreton et al. (1995) detected four QTLs related to stomatal conductance on chromosomes 1, 2, 7, and 10 under drought condition. Shi (2012) analyzed QTL for drought-resistant physiological character under rain-fed and irrigated conditions.

Present studies were designed to identify the QTL for coleoptiles length and main radicle length in DH group of wheat. The main aim of this study was to provide the basis for the study of drought resistance genetic mechanism and to be helpful for marker-assisted selection (MAS) in wheat drought-related traits breeding programs.

7.1.1 QTL Mapping of Drought Resistance Traits

7.1.1.1 Plant Materials

A total of 168 doubled haploid (DH) lines were developed through the cultured anthers of the F_1 hybrid derived from a cross between Huapei 3 and Yumai 57, followed by chromosome doubling. The parents differ in seedling characteristics, plant height, mature period, and several agronomic important traits.

An immortalized F_2 population derived from the DH population, which was divided into two groups, randomly matched hybridization was conducted between groups (each line was only used one time). A total of 84 cross-combinations were made after one round and 168 cross-combinations were made after two rounds, which formed a population. And each cross-combination was a F_2 strain. Meanwhile, same combinations could be made in different year, so the same population could be used in different environment and different year.

7.1.1.2 Experimental Design

High-quality seeds of the 168 IF_2 strains were soaked in 5 % H_2O_2 for 10 min, rinsed 2–3 times with tap water, and then placed in glass plates and cultured in an illuminated growth chamber at 4 °C for 2 days. Twenty healthy germinated seedlings, at the same developmental stage, from each strain were selected and planted in germination gauze disk. The germination gauze disks were soaked in distilled water (normal condition), 10, 20, and 30 % PEG-6000, respectively. Two replicates were designed for each treatment and held in a light incubator at 20 °C for 2 days, with light intensity is 2500 μmol m^{-2} s^{-1}, and photoperiodic is 12 h/12 h. When plantule stretched out to be 1.5 cm, 6–8 germinated seedlings from each treatment were selected and measured coleoptile length and radicle length, respectively.

7.1.1.3 Results and Analysis

7.1.1.3.1 Analysis of Phenotypic Data

The parent Yumai 57 showed better coleoptile length and radicle length than Huapei 3 under water stress and non-stress treatments (Table 7.1). With the increase of PEG-6000 concentration, the coleoptile length of two parents and IF_2 group increased, and then decreased, while their radicle length showed a trend of increase (Table 7.1), which indicated that water stress effects coleoptile length and radicle length. The group of this study exhibited transgressive segregations, and their absolute values of skewedness and peak were all smaller than 1.0, fitting normal distributions pattern, indicating its polygenic inheritance (Fig. 7.1). Based on these results, QTL mapping analysis of these traits can be performed.

Table 7.1 Phenotypic performance of coleoptile length and radicle length in the IF_2 population

Trait	Treatment	Parent		Immortalized F_2 population			
		Huapei 3	Yumai 57	Mean	Range	Skewness	Kurtosis
CL (cm)	Normal	2.73^{ab}	2.85^{ab}	2.46	1.65–3.74	0.70	2.07
	10 % PEG-6000	2.65^{b}	2.74^{b}	2.80	2.03–4.12	0.33	0.28
	20 % PEG-6000	2.80^{a}	3.21^{a}	2.94	1.44–3.98	−0.11	1.09
	30 % PEG-6000	2.40^{c}	2.31^{c}	2.64	1.49–3.89	0.41	0.13
RL (cm)	Normal	4.11^{c}	4.43^{c}	6.64	2.41–9.20	−0.85	0.85
	10 % PEG-6000	5.82^{b}	6.51^{b}	5.94	4.12–7.57	0.02	−0.27
	20 % PEG-6000	5.80^{b}	6.21^{b}	5.74	3.32–8.55	0.15	0.34
	30 % PEG-6000	7.52^{a}	9.52^{a}	6.03	3.21–10.70	0.17	0.58

Values followed by different letters are significantly different at 0.05 probability level. *CL* coleoptile length; *RL* radicle length

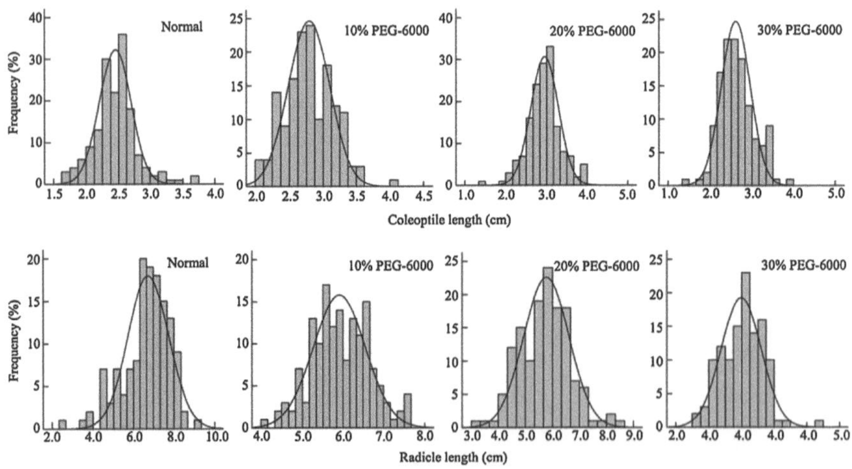

Fig. 7.1 Analysis of coleoptile length and radicle length in the IF_2 population

7.1.1.3.2 QTL Analyses

At four different drought stress conditions, a total of twenty-three QTLs (Tables 7.2, 7.3; Fig. 7.2) were detected for coleoptile length and radicle length. Among them, eleven QTLs for coleoptile length were detected and mapped on chromosomes 1A, 2A, 3D, 4B, and 6D, and single QTL accounted for 6.77–35.37 % of the phenotypic variation. Twelve QTLs for radicle length were detected and mapped on chromosomes 1A, 2D, 4B, 5A1, 6A, 6D, and 7D, accounting for 4.93–17.99 % of the phenotypic variation.

7.1.1.3.2.1 QTL for coleoptile length

Under normal condition, three additive QTLs on chromosomes 3D, 4B, and 5D were detected for coleoptile length, explaining the variances of 7.83, 35.37, and 12.42 %, respectively (Table 7.2; Fig. 7.2). One QTL *QCl4B* accounted for 35.37 % of the phenotypic variances, with the allele increasing coleoptile length QTL was inherited from the Yumai 57, showing super dominant effect. The other two QTLs were contributed by Huapei 3, showing partially dominant effect.

Under 10 % PEG-6000 condition, three additive QTLs on chromosomes 1A, 2A, and 3D were detected for coleoptile length, explaining the variances of 8.77, 9.40,

Table 7.2 Intervals, effects, and contributions of additive QTLs for coleoptile length (CL) in the IF_2 population

QTL	Flanking marker	Site (cM)	Additive[a]	Dominance[b]	H^2 (%)	Gene action[c]
Normal						
QCl3D-a	*Xcfd223–Xbarc323*	69.0	0.13	0.00	7.83	PD
QCl4B	*Xcfd39.2–Xcfd22.2*	13.0	−0.03	−0.40	35:37	OD
QCl6A	*Xwmc553–Xgwm732*	63.0	0.15	−0.12	12.42	PD
10 % PEG-6000						
QCl1A	*Xgwm259–Xcwem32.1*	2.0	−0.03	0.19	8.77	OD
QCl2A	*Xgwm448–Xwmc455*	79.0	0.04	0.21	9.40	OD
QCl3D-b	*Xwmc492–Xcfd223*	62.0	0.09	−0.15	8.02	OD
20 % PEG-6000						
QCl1A	*Xgwm259–Xcwem32.1*	0.0	0.01	0.21	6.77	OD
QCl3D-a	*Xcfd223–Xbarc323*	69.0	0.19	−0.08	11.74	PD
QCl6D	*Xswes679.1–Xcfa2129*	151.0	−0.20	0.01	10.52	PD
30 % PEG-6000						
QCl6D	*Xgwm133.2–Xswes861.1*	99.0	−0.04	−0.23	10.14	OD
QCl6D	*Xswes679.1–Xcfa2129*	139.0	0.22	−0.35	14.61	OD

[a]Alleles from Huapei 3 and Yumai 57 with positive effect are defined in positive and negative values, respectively

[b]Positive values indicate that the heterozygote has higher phenotypic values than the homozygote

[c]*PD* partial dominant (D/A < 1.00); *D* dominant (D/A = 1.00); *OD* over-dominant (D/A > 1.00)

and 8.02 %, respectively (Table 7.2; Fig. 7.2), all of them show super dominant effect. *QCl1A*, whose alleles of better coleoptile length, was originated from Yumai 57, while other two QTLs were contributed by Huapei 3.

Under 20 % PEG-6000 condition, three additive QTLs on chromosomes 1A, 3D, and 6D were detected for coleoptile length, explaining the variances of 6.77, 11.74, and 10.52 %, respectively (Table 7.2; Fig. 7.2). *QCl3D* had the largest effect and explained 11.74 % of the phenotypic variances. Positive alleles of *QCl6D* were originated from alleles of Yumai 57, which could increase the coleoptile length by 0.2 cm, showing partially dominant effect. The other two QTLs were contributed by Huapei 3.

Under 30 % PEG-6000 condition, two additive QTLs on chromosome 6D were detected for coleoptile length, explaining the variances of 10.14 and 14.61 %, respectively (Table 7.2; Fig. 7.2), showing super dominant effect. The QTL with the larger effect was contributed by Huapei 3, while the other QTL was contributed by Yumai 57.

7.1.1.3.2.2 QTLs for Radicle Length

Under normal condition, two additive QTLs on chromosomes 4B and 5A1 were detected for radicle length, explaining the variances of 7.79 and 8.88 %,

Table 7.3 Intervals, effects, and contributions of additive QTLs for radicle length (RL) in the IF_2 population

QTL	Flanking marker	Site (cM)	Additive[a]	Dominance[b]	H^2 (%)	Gene action[c]
Normal						
QRl4B	*Xwmc657–Xwmc48*	18.0	−0.04	−0.59	7.79	OD
QRl5A1	*Xgwm186–Xcfe223*	77.0	−0.42	0.22	8.88	PD
10 % PEG-6000						
QRl6A	*Xbarc1165–Xgwm82*	42.0	0.03	0.36	6.70	OD
QRl6D	*Xwmc412.1–Xcfd49*	2.0	0.03	−0.30	5.22	OD
QRl6D	*Xgwm55–Xgwm133.2*	85.0	0.07	−0.41	7.25	OD
20 % PEG-6000						
QRl1A	*Xwmc728.1–Xwmc550*	25.0	−0.23	0.34	9.69	OD
QRl16D	*Xwmc412.1–Xcfd49*	1.0	−0.21	−0.34	4.93	OD
QRl7D	*Xwmc14–Xwmc42*	199.0	−0.22	−0.64	8.68	OD
30 % PEG-6000						
QRl1A	*Xwmc550–Xbarc269*	48.0	0.66	−0.01	17.99	PD
QRl1A	*Xbarc350–Xwmc120*	60.0	−0.48	−0.09	9.47	PD
QRl2D	*Xwmc170.2–Xgwm539*	65.0	0.10	0.68	10.57	OD
QRl5A1	*Xswes45–Xbarc180*	6.0	0.01	−0.73	11.97	OD

[a]Alleles from Huapei 3 and Yumai 57 with positive effect are defined in positive and negative values, respectively

[b]Positive values indicate that the heterozygote has higher phenotypic values than the homozygote

[c]*PD* partial dominant (D/A < 1.00); *OD* over-dominant (D/A > 1.00)

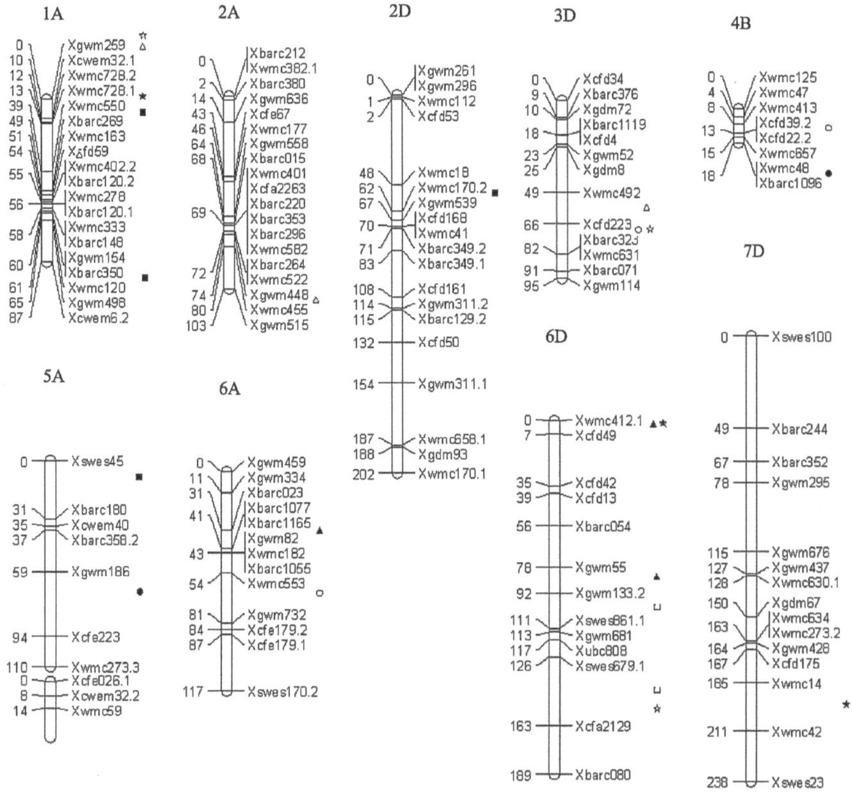

Fig. 7.2 Positions of additive QTLs associated with coleoptile length (*CL*) and radicle length (*RL*). ○, △, ☆, □: Addition QTL for coleoptile length under normal condition, 10, 20, and 30 % PEG-6000 stress. ●, ▲, ★, ■: Addition QTL for radicle length under normal condition, 10, 20, and 30 % PEG-6000 stress

respectively (Table 7.3; Fig. 7.2), showing super dominant effect and partially dominant effect, respectively. The two QTLs were originated from alleles of Yumai 57 and increased 0.04 and 0.42 cm of the radicle length, which corresponds with a larger radicle length of Yumai 57.

Under 10 % PEG-6000 conditions, three additive QTLs on chromosomes 6A and 6D were detected for radicle length and were responsible for 5.22–7.25 % of the phenotypic variation (Table 7.3; Fig. 7.2).

The QTL, *QRl6D*, mapped in marker region *Xgwm55-Xgwm133.2* had the largest effect, explaining the variance of 7.25 %. All of three QTLs show super dominant effect and were originated from alleles of Huapei 3.

Under 20 % PEG-6000 conditions, three additive QTLs on chromosomes 1A, 6D, and 7D were detected for radicle length, explaining the variances of 9.69, 4.93, and 8.68 %, respectively (Table 7.3; Fig. 7.2), all of them showing super dominant effect. All of three QTLs were originated from alleles of Yumai 57 and increased 0.21–0.23 cm of the radicle length.

Under 30 % PEG-6000 conditions, four additive QTLs on chromosomes 1A, 2D, and 5AL were detected for radicle length and were responsible for 9.47–17.99 % of the phenotypic variation (Table 7.3; Fig. 7.2). Two QTLs mapped on chromosome 1A accounted for 17.99 and 9.47 % of the phenotypic variances, respectively, showing partially dominant effect. The QTL which explained the variance of 9.47 % was originated from alleles of Yumai 57, and the other three QTLs were contributed by Huapei 3.

7.1.2 Research Progress of Drought Resistance QTL Mapping in Wheat and Comparison of the Results with the Previous Studies

7.1.2.1 Research Progress of Drought Resistance QTL Mapping

At present, the studies of QTL mapping of drought resistance in wheat mainly focused on the morphological and physiological characteristics, including roots, water-use efficiency, ABA, osmotic adjustment, photosynthetic rate, transpiration rate and stomatal conductance, and other traits.

For root drought resistance, QTL analysis of RN, MRL, shoot fresh weight, RFW, total fresh weight, shoot dry weight, RDW, and total dry weight in seedling stage were conducted under different drought stress conditions. A total of seventeen additive QTLs and fifteen pairs of QTLs with epistatic effect for root drought resistance were detected across all 21 chromosomes, except for chromosomes 5A, 4B, 2D, 6D, and 7D, with the largest effect by 37.30 % (Qi et al. 2010; Zhou et al. 2005). Hao et al. (2003) detected thirteen QTLs for bud drought resistance on chromosomes 2B, 3B, 4A, 5A, and 6B. Among them, the QTL-mapped chromosome 2B had the largest effect by 30.6 %, and the QTL for seedling drought resistance were detected on chromosomes 1A, 1B, 3A, 4A, and 7B, and among them, the QTL-mapped chromosome 7B had the largest effect by 16.1 %. About osmotic adjustment, Morgan et al. (1996) detected RFLP marker *Xpsr119* on chromosomes 7A, which were tightly linked to the osmotic adjustment genes and used in MAS breeding. In addition, there was one QTL for proline on chromosome 1A, which could increase 2.83 of proline and accounted for 10.02 % of the phenotypic variances.

7.1.2.2 Comparison of the Results with the Previous Studies

Under K-deprivation conditions, a total of eight QTLs for coleoptile length were detected in this study. The QTLs mapped to chromosomes 4B and 4D were in close range with the QTLs for plant height detected on chromosomes 4B and 4D by our group using the same population (Zhang et al. 2008; Wang et al. 2009). Gene *Rht-B1* and *Rht-D1* for both coleoptile length and plant height were detected on chromosomes 4B and 4D using different groups (Fick and Qualset 1976; Allan 1989; Richards 1992; Rebetzke et al. 1999, 2004). Rebetzke et al. (1999, 2004) detected many QTLs for coleoptile length on chromosomes 2B, 2D, 4A, 5D, and 6B, whereas QTL for coleoptile length on above chromosomes had not been detected in this study, but a new QTL for coleoptile length was found on chromosome 2A at the interval of *Xbarc380* to *Xgwm636*.

Among ten additive QTLs for radicle length, *QRl6A* mapped on chromosome 6A at the interval of *Xgwm82* to *Xwmc553* was detected under all water stress conditions, explaining the variances of 8.26 % (normal) and 9.74 % (20 % PEG-6000), respectively. Zhang and Xu (2002) detected QTL for root length on chromosome 6A at the close interval using a RIL population, which was a useful QTL for marker-assisted breeding of coleoptile length and drought resistance in wheat breeding. Under normal condition, one QTL for root length was detected on chromosome 6D, explaining 10.32 % of phenotypic variation. Li et al. (2010) also detected one QTL for root length on chromosome 6D using the same population; however, the marker intervals were different. The QTLs for root length detected in this study were different from those detected by Zhou et al. (2005) and Zhang and Xu (2002); meanwhile, the different QTLs for root length were also detected under two different treatment conditions, which may be some minor QTL or stress QTL under different conditions.

7.2 QTL Mapping and Effect Analysis of Heavy Metals Resistance

With the rapid development of the global economy, a lot of the development and utilization of mineral resources and the extensive use of various chemical products, a large number of organic and 需要在 into 前面加动词 the environment, agroecological system pollution has become an important issue as it cause serious deterioration of environmental, particularly heavy metal pollution. Heavy metals, a group of metals with density higher than 5.0 g cm^{-3}, are generic terms including more than 40 kinds of metal elements, such as cadmium (Cd), mercury (Hg), arsenic (As), lead (Pb), chromium (Cr), copper(Cu), zinc(Zn), manganese (Mn), silver (Ag), tin (Sn). In the field of environmental pollution, heavy metal general refers to some elements with strong biological toxicity such as Hg, Cd, Pb, Cr, and As, also includes other elements with mild toxicity such as Zn, Cu, Ni, Co, Sn. In soil, heavy metal pollution mainly refers that eight elements—Zn, Cu, Cr, Cd, Pb, Ni, Hg, and As—exceed their range of normal value. In our country, twenty million hm^2 cultivated land was polluted by heavy metal, accounting for about 20 % of the total arable land, which lead to heavy metal content exceeding of ten million tons of grain annually, and reduction of ten million tons of grain output (Chen et al. 1999). Heavy metal pollution is a direct or indirect product of industrialization. Compared with pesticide residues—"immediate" health killer—it has a long incubation period. Heavy metals in the soil are generally not biodegradable or chemical degradation, so they directly permeate into the food chain through seepage into groundwater or be absorbed by the plant. This can be highly dangerous to the ecological environment and people's health. Many diseases such as Alzheimer's disease, Parkinson's disease, and muscular dystrophy disorders are associated with heavy metal pollution.

There are many reports about heavy metal stress on plants growth and development—Cd stress and Cr stress have great effect on seed germination and growth, physiological metabolism of wheat. At present, QTL mapping of wheat under Cd stress and Cr stress has not been reported. The present study used a population of 168 DH derived from a cross between Huapei 3 and Yumai 57 to perform analyses for seedling and root growth of wheat under Cd stress and Cr stress, and the objective was to reveal the genetic molecular mechanism of resistance to Cd stress and Cr stress.

7.2.1 QTL Mapping for Seedling and Root Traits Under Cadmium Stress

Cadmium (Cd) stress is one of the heavy metal pollutions which is not indispensable to wheat development, but is one of the most common noxious heavy metals in wheat seedling developmental stage. Excessive accumulation in plants often results in severe signs of poisoning and even to the crops death (Patra et al. 1994). Cd is not an essential element for plant development, but is easily absorbed by plants, and can be highly toxic. Excessive accumulation of Cd in plants can result in a variety of adverse syndromes, including hindrance of plant root development, inhibition of water and mineral absorption, reduction of photosynthetic efficiency, and other physiological and metabolic disorders, which ultimately lead to severe decrease in yield and quality (Moral et al. 1994). Numerous studies have confirmed that acute cadmium stress in plants causes leaf roll and chlorosis, reduces growth both in roots and in stems in a very general way, damages the photosynthetic apparatus, and inhibits the stomatal opening (Di Toppi and Gabbrielli 1999). Since the plant roots firstly contact with cadmium in soils, its accumulation of cadmium can become dangerous to all kinds of organisms.

Previous studies on the QTL mapping in cultivated crops have been mainly focused on the rice seedling tolerance to Cd stress, thus it is not surprising that significant progresses have been made in rice, especially over the last few years. For example, Chen et al. (2010) mapped nine QTLs that are responsible for Cd tolerance in rice seedlings within 2 years and concluded that these QTLs explained 7.23–18.02 % of phenotype variance. They found two of these nine QTLs contributed to more than 15.0 %, and two other QTLs explained 10.0–15.0 %, while the remaining QTLs' contribution was less than 10 % of phenotype variance. Shen et al. (2008) used a RIL population, derived from Xieqingzao B and Zhouhui 9308, to analyze QTL for five microelements in rice grain and detect three QTLs for Cd^{2+} content in brown rice. Similarly, Xue et al. (2009) confirmed 22 QTLs conferring seedling height, root length, RDW, stem-leaf dry weight, and chlorophyll content. And under Cd-stress condition, twelve of these QTLs were mapped based on their contributions to phenotype variance, one QTL with 35.26 % (the highest) and the remaining ones ranging from 8.11 to 17.75 %. Few researches related to Cd stress in wheat were conducted. So, a DH population, derived from Huapei 3 and Yumai 57, including 168 lines was used to map QTL for some traits related to seedling and root growth in wheat. And the results can be used in molecular marker-assisted breeding for Cd-tolerant and/or resistant traits in wheat.

7.2.1.1 Materials and Methods

7.2.1.1.1 Materials

A total of 168 DH lines were developed through the cultured anthers of the F_1 hybrid derived from a cross between the high-yield, multiresistant Huapei 3 and Yumai 57, followed by chromosome doubling (Zhang et al. 2009).

7.2.1.1.2 Seedling Cultivation

High-quality seeds of the 168 DH strains, which were harvested in the same year, were soaked in 3 % H_2O_2 for 15 min, rinsed 2–3 times with tap water, and then placed in glass plates with two pieces of moisturized filter paper at 20 °C for germination. The germinated seedlings were cultivated in deionized water until the first true leaf emerged. Nine healthy seedlings, at the same developmental stage, from each strain were selected and planted in seed trays, each of which contains 128 wells with 3 cm in diameter. Seedlings were secured with sterilized sponges and cultivated in cultivation basins containing Hoagland nutrient solution with various Cd concentrations. The Hoagland nutrient solution contained the following micronutrients (in mM): $Ca(NO_3)_2 \cdot 4H_2O$ (1.0), KH_2PO_4 (2.0), $MgSO_4 \cdot 7H_2O$ (0.5), KCl (1.5), $CaCl_2$ (1.5), H_3BO_3 (1×10^{-3}), $(NH_4)_6Mo_7O_{24} \cdot 4H_2O$ (5×10^{-5}), $CuSO_4 \cdot 5H_2O$ (5×10^{-4}), $ZnSO_4 \cdot 7H_2O$ (1×10^{-3}), $MnSO_4 \cdot H_2O$ (1×10^{-3}), Fe (III) EDTA (0.1). The pH of this nutrient solution was adjusted to 6.0. The exterior of each cultivation basin was made in black in order to create a dark growth environmental condition for ideal root development. After four-week cultivation in greenhouse, these young DH plants were taken and rinsed with distilled water prior to studies on their seedling phenotypic characters.

7.2.1.1.3 Cd Treatment

The methods of Cd treatments were referred to Huang (2007), and three concentrations were used: 0, 40, and 120 mg/L. The Cd solutions were prepared using $CdCl_2 \cdot H_2O$. Each treatment was repeated three times. All experiments were conducted in the greenhouse equipped with the auto-controlled temperatures (21 °C during the day and 16 °C at night) and humidity in the State Key Laboratory of Crop Biology, Shandong Agricultural University, China.

7.2.1.1.4 Seedling Phenotype Measurements

For analyzing fresh weight of seedling stems and leaves, we used the well-cleaned and air-dried plants which were cut at the coleoptilar node of each plant, and a millesimal electronic balance for measurements. For measuring the seedling height, the tip of the longest leaf generated from the base node was measured using 30-cm ruler. For testing the stem–leaf dry weight, stems and leaves were placed in an oven at 100 °C for 15 min to remove water, followed by drying at 80 °C until weight maintained at a constant level and were finally weighed using a millesimal electronic balance.

7.2.1.1.5 Root Trait Characterization

A WinRHIZO Root Analysis System software (Regent Instruments Inc., Canada) was used to analyze the following root characters: total root length (TRL), surface area (RSA), root diameter (RD), root volume (RV), and numbers/plant. Both fresh and dry root weight analyses were conducted using a millesimal electronic balance. For dry root weight data, all roots were placed in an oven adjusted to 100 °C and dried for 15 min to remove water, followed by 80 °C until root weight unchanged.

7.2.1.1.6 Construction of Genetic Map and Data Processing

A genetic linkage map of the DH population with 368 markers was constructed by Zhang et al. at Shandong Agricultural University in 2008. The map covered a length of 3074.1 cM with an average distance of 8.35 cM between adjacent markers. Thirteen markers remained unlinked. These markers formed 24 linkage groups at LOD 4.0. The chromosomal locations and the order of the markers were in accordance with Somers et al. (2004). The recommended map distance for genome-wide QTL scanning is an interval length less than 10 cM (Doerge 2002). Thus, the map was suitable for QTL mapping.

SPSS17.0 software was used to conduct statistical analyses for the above-described seedling and root characters, while ICI mapping software (Wang 2009) based on the ICIM method (Zhang et al. 2008; Wang 2009) was used for

QTL analysis. Specifically, the mean value of three repeats of each treatment was considered as an input datum, LOD threshold was set to 2.5, and step value was set as 1 cm. QTL designations were based on (http://www.graingenes.org), e.g., "q + trait acronym + research institute acronym + chromosome No." Multiple QTLs located on the same chromosome and related to the same trait were described using an additional letter a, b... after the chromosome number.

7.2.1.2 Results and Analysis

7.2.1.2.1 Phenotype Evaluation of Various Seedling Developmental Traits of the DH Population Under Cd-Stress Conditions

To evaluate seedling phenotypes of the 168 DHs and their two parents under the exposure of Cd environments, three Cd concentrations (0, 40 and 120 mg/L) were used in this study. Data listed in Table 7.4 indicated that development of wheat seedlings treated with Cd was significantly inhibited, and such inhibitory effect was further enhanced when Cd concentration increased from 40 to 120 mg/L compared to the control (0 mg/L Cd). The two parents showed some phenotypic differences, whereas the DH strains exhibited transgressive segregations, and their absolute values of skewedness and peak were all smaller than 1.0, fitting normal distributions pattern (Table 7.4). Based on these results, QTL mapping analysis of these traits can be performed.

7.2.1.2.2 QTL Analysis

At three different Cd concentrations, 43 QTLs (Table 7.5; Fig. 7.3) related to wheat growth and development during the seedling stage were mapped and were found to be distributed on 14 chromosomes: 1A, 1B, 1D, 2A, 2B, 2D, 3A, 3B, 3D, 6A, 6B, 6D, 5D, and 7D. Each of these QTLs was accounted for 5.81–70.05 % of phenotypic variation.

7.2.1.2.2.1 QTL Analysis Based on Seedling Performance Without Cd (0 mg/L)

Two QTLs, *qSFW1A* and *qSFW1D*, were found to control the stem–leaf fresh weight and were mapped on chromosomes 1A and 1D, explaining the stem–leaf fresh weight variances of 6.21 and 6.39 %, respectively. Data showed that the two loci acted as enhancers responsible for additive effects were derived from Yumai 57, whereas QTL *qSH6Ab* for seedling height came from Huapei 3. This QTL increased seedling height by 1.02 cm, explaining 6.79 % of seedling height variance. Another QTL that was found to play a role in shoot developmental traits was *qSDW3D* which was involved in shoot dry weight. It was mapped on chromosome

Table 7.4 The effect of growth in the DH population under Cd^{2+} stresses

Treatment		Fresh weight of stem and leaf (g)	Dry weight of stem and leaf (g)	Plant height (cm)	Leaf age	Total length of root (cm)	Root area (cm^2)	Diameter of root (mm)	Root volume (cm^3)	Apical number	Root fresh weight (g)	Root dry weight (g)
0 mg/L Cd^{2+} control												
Huapei 3		0.300	0.036	23.87	3.87	232.142	22.547	0.309	0.174	622	0.087	0.0083
Yumai 57		0.219	0.034	24.23	3.87	141.112	11.894	0.269	0.117	511	0.080	0.0083
DH	Mean	0.357	0.420	24.54	3.89	247.708	21.502	0.280	0.149	919	0.108	0.110
	Range	0.193–0.604	0.210–0.620	16.13–35.40	3.00–4.37	124.457–454.606	11.849–33.624	0.220–0.350	0.063–0.231	302–1989	0.059–0.168	0.0070–0.1750
	Variation coefficient	0.232	0.019	0.159	0.074	0.261	0.214	0.089	0.208	0.334	0.231	0.018
40 mg/L Cd^{2+}												
Huapei 3		0.134	0.029	13.97	3.27	105.108	13.201	0.400	0.132	436	0.060	0.0063
Yumai 57		0.121	0.027	15.15	3.57	91.077	11.069	0.326	0.097	389	0.064	0.0060
DH	Mean	0.186	0.029	15.65	3.44	98.091	13.028	0.428	0.139	555	0.069	0.0078
	Range	0.091–0.306	0.0150–0.0420	11.93–22.55	2.83–4.00	36.726–176.418	4.516–20.104	0.337–0.534	0.044–0.217	128–1219	0.030–0.107	0.0033–0.0125
	Variation coefficient	0.231	0.172	0.134	0.084	0.234	0.215	0.082	0.230	0.341	0.232	0.256
120 mg/L Cd^{2+}												
Huapei 3		0.109	0.023	9.70	2.67	69.637	10.673	0.488	0.130	375	0.048	0.0040
Yumai 57		0.101	0.020	10.40	2.90	59.701	7.288	0.389	0.071	202	0.039	0.0023
DH	Mean	0.113	0.0242	12.41	2.89	83.844	11.656	0.447	0.131	505	0.067	0.0067
	Range	0.064–0.175	0.0120–0.0360	7.67–18.67	2.3–3.6	34.989–154.873	4.765–19.51	0.349–0.554	0.052–0.229	147–1105	0.021–0.124	0.0017–0.0113
	Variation coefficient	0.212	0.207	0.149	0.091	0.260	0.243	0.083	0.244	0.368	0.284	0.299

Table 7.5 Position, effect, and contribution of seedling and root traits in DH population

Trait	Treatment (mg/L)	QTL	Flanking marker	Site (cM)	Additive	PVE (%)
SFW	0	*qSFW1A*	*Xwmc550–Xbarc269*	77	−0.021	6.21
		qSFW1D	*Xwmc429–Xcfd19*	32	−0.0214	6.39
	40	*qSFW3B*	*Xgwm566–Xcfe009*	101	0.0164	14.94
	120	*qSFW2A*	*Xwmc455–Xgwm515*	134	0.0066	7.63
		qSFW7D	*Xwmc630.1–Xgdm67*	144	−0.0063	6.92
SL	0	*qSL6Ab*	*Xbarc1165–Xgwm82*	74	1.017	6.79
	40	*qSL1D*	*Xcfd19–Xwmc93*	53	−0.5911	7.92
		qSL2B	*Xbarc101–Xgpw3248*	119	−0.5418	6.59
	120	*qSL1A*	*Xgpw3167–Xcwem6.2*	149	−0.4724	6.44
		qSL3B	*Xgwm533–Xbarc251*	39	0.8027	17.83
		qSL6Aa	*Xgpw2265–Xgpw322*	16	0.4528	5.97
LA	40	*qLA5D*	*Xbarc1097–Xcfd8*	16	−0.1031	12.71
RTL	0	*qRTL1Ba*	*Xcfd21–Xcwem9*	37	25.9593	15.64
		qRTL1Bb	*Xgwm218–Xgwm582*	40	−33.5079	25.95
		qRTL1Bc	*Xwmc766–Xswes98*	123	−17.6198	7.33
		qRTL6D	*Xcfd13–Xbarc054*	41	15.7022	5.81
		qRTL7D	*Xgdm67–Xwmc634*	162	−20.378	9.98
RA	0	*qRSA1Ba*	*Xcfd21–Xcwem9*	37	1.8652	15.89
		qRSA1Bb	*Xgwm218–Xgwm582*	40	−2.5736	30.15
		qRSA7D	*Xgdm67–Xwmc634*	162	−1.5441	11.29
	120	*qRSA1A*	*Xgwm498–Xgpw7412*	101	−0.7234	6.57
RAD	0	*qRAD1B*	*Xgwm218–Xgwm582*	40	0.0081	10.04
		qRAD3Da	*Xgwm52–Xgdm8*	24	0.0069	7.49
		qRAD3Db	*Xwmc631–Xbarc071*	85	−0.0068	7.17
	40	*qRAD3B*	*Xbarc251–Xwmc3*	57	0.0309	70.05
	120	*qRAD6B*	*Xcfa2187–Xgwm219*	4	0.0132	11.52
RV	0	*qRV3B*	*Xgpw1148–Xgpw4075*	73	0.0084	6.59
		qRV7D	*Xgdm67–Xwmc634*	162	−0.012	15.28
	120	*qRV1A*	*Xgwm498–Xgpw7412*	101	−0.0086	7.09
RT	0	*qRT1Ba*	*Xcfd21–Xcwem9*	37	160.5598	26.45
		qRT1Bb	*Xgwm218–Xgwm582*	40	−232.1049	55.03
		qRT3A	*Xbarc310–Xbarc321*	0	−74.6941	5.86
RFW	0	*qRFW3D*	*Xcfd34–Xbarc376*	0	−0.0062	6.31
		qRFW6B	*Xgwm58–Xwmc737*	56	−0.0182	10.4
RDW	40	*qRDW3Bb*	*Xgwm566–Xcfe009*	113	0.0004	6.34
	120	*qRDW1A*	*Xcwem32.1–Xwmc728.2*	10	−0.0005	7.14
		qRDW3Ba	*Xgwm566–Xcfe009*	109	0.0007	14.69
SDW	0	*qSDW3D*	*Xgdm72–Xbarc1119*	17	−0.0024	7.58

(continued)

Table 7.5 (continued)

Trait	Treatment (mg/L)	QTL	Flanking marker	Site (cM)	Additive	PVE (%)
	40	*qSDW2D*	*Xbarc349.1–Xcfd161*	85	0.0019	10.33
		qSDW3A	*Xbarc276.2–Xgpw1108*	234	−0.0015	7.08
		qSDW3Bb	*Xbarc139–Xwmc612*	63	0.0017	7.89
	120	*qSDW3Ba*	*Xbarc268–Xwmc1*	58	0.0026	28.94
		qSDW6D	*Xswes679.1–Xcfa2129*	151	0.0016	10.57

SFW shoot fresh weight; *SL* seedling length; *LA* leaf age; *RTL* total length of root; *RA* root area; *RAD* average root diameter; *RV* root volume; *RT* root tip number; *RFW* fresh weight of root; *RDW* fresh weight of root; and *SDW* dry weight of stem. Alleles from Huapei 3 and Yumai 57 with positive effect are defined in positive and negative values, respectively

3D and explained 7.58 % of the shoot dry weight variance. Parent, Yumai 57, was the source of this additive allele.

We found that five QTLs, *qTRL1Ba*, *qTRL1Bb*, *qTRL1Bc*, *qTRL6D,* and *qTRL7D,* contributed to the TRL and were mapped to chromosomes 1B, 6D, and 7D, explaining 5.81–25.95 % of TRL variance. Of these root length QTLs, *qTRL1Bb* had the highest contribution (25.95 %), and its positive allele was derived from Yumai 57, as those of *qTRL1Bc* and *qTRL7D*. However, the positive alleles of *qTRL1Ba* and *qTRL6D* inherited from Huapei 3. As for QTLs responsible for root surface area, *qRSA1Ba*, *qRSA1Bb*, and *qRSA7D* were identified on chromosomes 1B and 7D. They contributed to root surface area variances of 15.89, 30.15, and 11.29 %, respectively. Furthermore, the positive alleles of *qRSA1Bb* and *qRSA7D* were both derived from Yumai 57, while that of *qRSA1Ba* came from Huapei 3.

Three QTLs, *qARD1B*, *qARD3 Da*, and *qARD3Db*, involving in average RD were mapped on chromosomes 1B and 3D and could explain 10.04, 7.49, and 7.17 % of average RD variances, respectively. The additive effect alleles of *qARD3Db* were derived from Yumai 57, whereas those of the other two QTLs were both from Huapei 3. With no Cd stress, two QTLs were found to be involved in controlling RV: *qRV7D* and *qRV3B* were mapped on chromosomes 3B and 7D, respectively; the former could explain high RV variance (15.28 %), and parent Yumai 57 contributed its additive allele, while parent Huapei 3 was the source of *qRV3B* additive allele. We determined several other QTLs that play important roles in root developmental traits and these include the following: (1) Three QTLs for root tip number were mapped on chromosomes 1B and 3A and explained phenotypic variances ranging from 5.86 to 55.03 %. Of these, the locus within the *Xgwm218–Xgwm582* interval of chromosome 1B had highest contribution (55.03 %). The additive allele of *qRT1Ba* was from Huapei 3, while those of *qRT3A* and *qRT1Bb* came from Yumai 57. (2) Two QTLs for RFW were determined on chromosomes 6B and 3D, respectively. Of them, *qRFW6B* contributed most to this trait, explaining the RFW variance of 10.4 %. The additive alleles of both QTLs were derived from Yumai 57.

No QTL for leaf age (LA) and RDW was detected in the absence of Cd treatment.

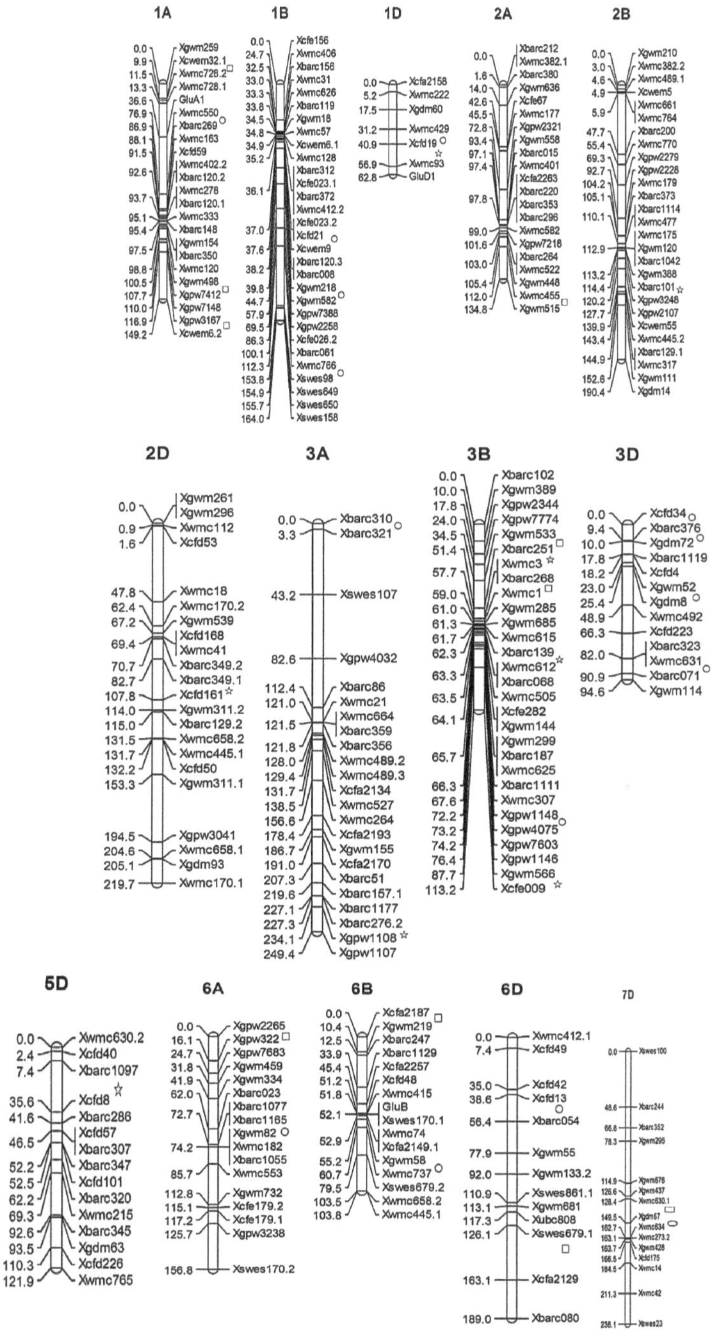

Fig. 7.3 Positions of additive QTL associated with seedling and root traits. O: Addition QTL under normal condition. ☆: Addition QTL under 40 mg/L Cd^{2+} stress. □: Addition QTL under 120 mg/L Cd^{2+} stress

7.2.1.2.2.2 QTL Analysis at 40 mg/L Cd

One QTL, *qSFW3B*, was identified as a shoot fresh weight locus on chromosome 3B at the 40 mg/L Cd level and explained 14.94 % of shoot fresh weight variance. The allele was derived from the Huapei 3 parent. The presence of 40 mg/L Cd allowed us to locate two QTLs, *qSH1D* and *qSH2B*, for shoot height, as they explained 7.92 and 6.59 % of the phenotype variances, respectively. Their additive alleles for both QTLs came from the Yumai 57. And one QTL for LA, *qLA5D*, was located on chromosome 5D as it explained 12.71 % of LA variance and its additive allele was from Yumai 57. The QTL, *qARD3B*, that showed the highest phenotype contribution (70.05 % of the variance) was determined to be associated with the average RD and was mapped on chromosome 3B, but its additive allele came from parent Huapei 3. The RDW QTL, *qRDW3Bb*, was mapped on chromosome 3B as it explained 6.34 % of the variance of this trait, and its positive allele came from Huapei 3. With phenotype variances ranging from 7.08 to 10.33 %, several important QTLs, *qSDW2D*, *qSDW3Bb* and *qSDW3A*, were found to be responsible for shoot dry weight and were mapped on chromosomes 2D, 3A, and 3B, respectively. The additive allele of *qSDW2D* (with the highest contribution of 10.33 % within these three QTLs) was derived from Yumai 57, whereas those of qSDW3Bb and qSDW3A came from Huapei 3.

No QTL associated with TRL, root surface area, RV, RFW, and root tip number was found at the 40 mg/L Cd-stress level.

7.2.1.2.2.3 QTL Analysis at 120 mg/L Cd

When the Cd concentration increased to 120 mg/L, a number of important QTLs were identified. Two QTLs, namely *qSFW2A* and *qSFW7D,* were found to be associated with shoot fresh weight and were located on chromosomes 2A and 7D, respectively. QTL of *qSFW2A* explained 7.63 % and that of *qSFW7D* explained 6.92 % of the shoot fresh weight variances. The additive allele of *qSFW2A* was derived from Huapei 3, while that of *qSFW7D* came from Yumai 57. Three QTLs for shoot height were detected on chromosomes 1A, 3B, and 6A, respectively. They explained 5.9–17.83 % of the phenotype variance. Of these, QTL located in between *Xgwm533* and *Xbarc251* loci on chromosome 3B contributed the most (17.83 %). The additive allele of *qSH1A* was derived from Yumai 57, but those of *qSH3B* and *qSH6Aa* came from Huapei 3. We have also identified two other QTLs that are associated with another shoot trait at the 120 mg/L Cd condition. QTLs for shoot dry weight, for example *qSDW3Ba* and *qSDW6D,* were mapped on chromosomes 3B and 6D, respectively. With their phenotype variances of 28.94 % (*qSDW3Ba*) and 10.57 % (*qSDW6D*), additive alleles of both QTLs were derived from Huapei 3.

Treatment with 120 mg/L Cd also resulted in the identification of a large number of QTLs that are responsible for various root developmental traits. One QTL for root surface area, namely *qRSA1A,* contributed 6.57 % phenotype variance and was located on chromosome 1A. Parent Yumai 57 was the source of its additive allele.

Table 7.5 showed that the QTL *qARD6B* for average RD contributed phenotype variance of 11.52 %, resulted in 0.013 mm increase in average RD, and was located on chromosome 6B. Parent Huapei 3 contributed to the additive allele. The RV QTL with its phenotype variance of 7.09 % was mapped on chromosome 1A in Yumai 57. Finally, QTLs for root dry weight, *qRDW1A* (explained 7.14 % phenotype variance) and *qRDW3Ba* (explained 14.69 % phenotype variance), were mapped on chromosomes 1A and 3B, respectively. It was determined that the additive allele of *qRDW1A* was derived from Yumai 57, whereas that of *qRDW3Ba* was from Huapei 3.

No QTL for LA, TRL, RFW, and root tip number was detected under the 120 mg/L Cd-stress condition.

7.2.2 QTL Mapping for Seedling and Root Traits of Wheat Under Chromium Stress

Chromium (Cr) as one of five positions for industry is a toxic teratogenic and mutagenic agent. It is Cr^{6+} that harm to the wheat development. Their presence in the root, stem, and leaf of crops—even in trace concentrations—can cause serious problems to all organisms, and heavy metal bio-accumulation in the food chain can be highly dangerous. At present, with the heavy metal cadmium (Cr) pollution increasingly serious, the yield and quality of wheat has substantial decline. Clearly, studying the QTL mapping of Cr and MAS is absolutely essential for developing wheat varieties with low or no Cr. Analysis effect of wheat development at the seedling stage under various Cr stresses can clear chromium (Cr) toxicity on wheat, chromium pollution prediction, evaluation, and prevention. These results may provide useful information for pollution-free and high-yielding production in wheat under Cr stress.

7.2.2.1 Materials and Methods

7.2.2.1.1 Materials

A total of 168 DH lines were developed through the cultured anthers of the F_1 hybrid derived from a cross between the high-yield, multiresistant Huapei 3 and Yumai 57, followed by chromosome doubling (Zhang et al. 2009).

7.2.2.1.2 Seedling Cultivation

High-quality seeds of the 168 DH strains, which were harvested in the same year, were soaked in 3 % H_2O_2 for 20 min, rinsed 2–3 times with tap water, and then placed in glass plates with two pieces of moisturized filter paper at 25 °C for germination. The germinated seedlings were cultivated in deionized water until the first true leaf emerged. Nine healthy seedlings, at the same developmental stage, from each strain were selected and planted in seed trays, each of which contains 128 wells with 3 cm in diameter. Seedlings were secured with sterilized sponges and cultivated in cultivation basins containing Hoagland nutrient solution with various Cr concentrations. The pH of this nutrient solution was adjusted to 6.0. Three treatment: (1) 0 mg/L Cr^{6+}(A); (2) 40 mg/L Cr^{6+}(B); (3) 120 mg/L Cr^{6+} (C). The exterior of each cultivation basin was made in black in order to create a dark growth environmental condition for ideal root development. After four weeks of cultivation in greenhouse, these young DH plants were taken and rinsed with distilled water prior to studies on their seedling phenotypic characters.

7.2.2.1.3 Seedling Phenotype Measurements

For analyzing fresh weight of seedling stems and leaves, we used the well-cleaned and air-dried plants which were cut at the coleoptilar node of each plant, and a millesimal electronic balance for measurements. For measuring the seedling height, the tip of the longest leaf generated from the base node was measured using a 50-cm ruler and investigation LA. For testing the stem–leaf dry weight, stems and leaves were placed in an oven at 100 °C for 15 min to remove water, followed by drying at 80 °C until weight maintained at a constant level and were finally weighed using a millesimal electronic balance. A WinRHIZO Root Analysis System software (Regent Instruments Inc., Canada) was used to analyze TRL and calculation root length inhibition rate. Root length inhibition rate/% = (contrast root length-treatment root length)/contrast root length × 100.

7.2.2.1.4 Statistical Analysis

SPSS17.0 software was used to conduct statistical analyses for the above-described seedling and root characters. QTL analyses were performed using the software of QTLNetwork version 2.0 (Yang and Zhu 2005) based on a mixed linear model (Wang et al. 1999). Composite interval analysis was undertaken using forward–backward stepwise, multiple linear regression with a probability into and out of the model of 0.05 and window size set at 10 cM. QTL was declared if the phenotype was associated with a marker locus at $P < 0.005$. And QTL designations were based on (http://www.graingenes.org), e.g., "q + trait acronym + research institute acronym + chromosome No." Multiple QTLs located on the same chromosome and related to the same trait were described using an additional letter a, b… after the chromosome number.

7.2.2.2 Results and Analysis

7.2.2.2.1 Phenotype Evaluation of Various Seedling Developmental Traits

To evaluate seedling phenotypes of the 168 DHs and their two parents under the exposure of Cr environments, three Cr concentrations (0, 40 and 120 mg/L) were used in this study (Table 7.6). The two parents showed some phenotypic differences, whereas the DH strains exhibited transgressive segregations, and their absolute values of skewedness and peak were all smaller than 1.0 except LA, fitting normal distributions pattern. Based on these results, QTL mapping analysis of these traits can be performed.

7.2.2.2.2 QTL Analysis

A total of twenty-four additive QTLs and five pairs of epistatic QTLs for wheat seedling growth and development were detected (Tables 7.7 and 7.8). Of them, seven QTLs were detected for fresh weight of stem, four QTLs for LA, six QTLs for main root length, and two QTLs for RFW. And, three pairs of QTLs with epistatic effects were detected for fresh weight of stem, one pair of epistatic QTLs for plant height, and one pair of epistatic QTLs for RFW. Furthermore, fourteen additive QTLs and two pairs of epistatic QTLs were involved in AE interactions.

7.2.2.2.2.1 QTL Analysis for Stem-Leaf Fresh Weight

Seven QTLs, *qSFW1A*, *qSFW1B*, *qSFW2D*, *qSFW5A.2*, *qSFW6A*, and *qSFW7B.2*, were found to control the stem–leaf fresh weight and mapped on chromosomes 1A, 1B, 2D, 3B, 5A.2, 6A, and 7B.2, explaining the stem–leaf fresh weight variances of 4.5, 2.0, 5.8, 11.8, 19.2, 10.1, and 145 %, respectively (Table 7.7). The QTL named as *qSFW5A.2* on chromosome 5A.2 had the highest contribution (19.2 %). We found that *qSFW1A*, *qSFW6A*, and *qSFW7B.2* mapped to chromosomes 1A, 6A, and 7B.2 were inherited from Yumai 57, while the else were inherited from Huapei 3. Among all of QTLS, *qSFW3B*, *qSFW6A* and *qSFW7B.2*, were involved in AAE interactions; at the same time, *qSFW6A* were also involved in AAE interactions with 120 mg/L Cr^{6+}, accounting for 0.55 % of the phenotypic variance.

Three pairs of epistatic QTLs for the stem–leaf fresh weight were detected in the three environments (Table 7.8). We found that *qSFW1A*/*qSFW2D*, *qSFW1*A/qSFW6A, and *qSFW4D*/*qSFW6A* were located on chromosomes 1A-2D, 1A-6A, and 4D-6A, explaining the phenotypic variance ranging from 2.0 to 5.0 %. Of them, *qSFW1A*/*qSFW6A* had a positive effect. However, the other two pairs showed negative effect on the stem–leaf fresh weight. Furthermore, both *qSFW1A*/*qSFW6A* and *qSFW4D*/*qSFW 6A* were involved in AAE_2 and AAE_3 interactions and had the contributions of 4.0, 2.0, 3.0, and 2.77 %, respectively. However, no QTL, involved in AAE1 interaction, was found.

Table 7.6 Developmental characters of parents and DH lines

Trait	Treatment	Parent		DH population			
		Huapei 3	Yumai 57	Mean ± SD	Range	Skewness	Kurtosis
SFW (g)	0 mg/L Cr^{6}	0.238	0.33	0.353 ± 0.100	0.089–0.711	0.35	0.39
	40 mg/L Cr^{6+}	0.075	0.071	0.068 ± 0.017	0.030–0.117	0.248	−0.221
	120 mg/L Cr^{6+}	0.045	0.048	0.04168 ± 0.008604	0.023–0.064	0.074	−0.224
SL (cm)	0 mg/L Cr^{6}	23.51	28.75	24.632 ± 3.362	11.10–31.63	−0.553	0.731
	40 mg/L Cr^{6+}	13.63	15.13	13.639 ± 2.16	7.70–19.67	0.058	−0.213
	120 mg/L Cr^{6+}	11	13.92	12.662 ± 1.9	8.47–18.10	0.348	−0.009
LA	0 mg/L Cr^{6}	3.34	3.95	3.741 ± 0.386	2.67–5.00	−0.409	−0.053
	40 mg/L Cr^{6+}	2.02	2.16	2.063 ± 0.146	1.67–3.00	2.38	10.79
	120 mg/L Cr^{6+}	2.04	2.13	1.982 ± 0.147	1.30–2.33	−2.396	8.878
SDW (g)	0 mg/L Cr^{6}	0.058	0.077	0.063 ± 0.013	0.015–0.104	−0.318	0.909
	40 mg/L Cr^{6+}	0.023	0.024	0.021 ± 0.004	0.010–0.032	−0.147	−0.149
	120 mg/L Cr^{6+}	0.019	0.21	0.019 ± 0.003	0.009–0.026	−0.189	−0.028
RL (cm)	0 mg/L Cr^{6}	19.95	28.27	22.093 ± 3.007	8.70–29.77	−0.712	0.787
	40 mg/L Cr^{6+}	13.67	15.5	13.707 ± 1.868	7.35–19.80	−0.197	0.888
	120 mg/L Cr^{6+}	11.77	15.53	12.886 ± 1.460	8.20–15.90	−0.38	0.308
RFW (g)	0 mg/L Cr^{6}	0.228	0.306	0.253685 ± 0.0724494	0.048–0.500	0.218	0.604
	40 mg/L Cr^{6+}	0.081	0.095	0.06997 ± 0.020595	0.01–0.144	0.238	0.831
	120 mg/L Cr^{6+}	0.052	0.086	0.05021 ± 0.013608	0.018–0.085	0.355	−0.166
RDW (g)	0 mg/L Cr^{6}	0.025	0.028	0.023 ± 0.005	0.004–0.0391	−0.095	0.941
	40 mg/L Cr^{6+}	0.011	0.011	0.009 ± 0.002	0.003–0.014	−0.131	0.307
	120 mg/L Cr^{6+}	0.008	0.012	0.008 ± 0.001	0.004–0.013	−0.066	0.249

SFW stem–leaf fresh weight; *SL* seedling length; *LA* leaf age; *SDW* dry weight of stem; *RL* root length; *RFW* root fresh weight; and *RDW* root dry weight

Table 7.7 Estimated additive (A) and additive × environment (AE) interactions of QTLs for developmental characters of wheat seedling in DH lines based on the averaged phenotypic data from three environments

Trait	QTL	Flanking marker	Site (cM)	Range	Additive		A × E_1		A × E_2		A × E_3	
					A	H^2 (%)	AE_1	*H* (%)	AE_2	H^2 (%)	AE_3	H^2 (%)
SFW	*qSFW1A*	**XGPW3167–*XCWEM6.2*	148.9	140.9–148.9	−0.006	4.5	−0.012	15.5				
	qSFW1B	**XWMC128–*XBARC312*	35.2	33.0–36.1	0.007	2.0	0.009	0.6				
	qSFW2D	**XBARC129.2–*XWMC658.2*	118.0	114.0–124.0	0.007	5.8	0.012	9.7				
	qSFW3B	**XBARC102–*XGWM389*	1.0	0.0–7.0	0.009	11.8	0.016	22.5	−0.009	7.9		
	qSFW5A.2	**XWMC59–*XGPW2109*	30.7	24.7–31.7	0.011	19.2	0.016	21.1				
	qSFW6A	**XGPW2265–*XGPW322*	0	0.0–5.0	−0.011	10.1	−0.018	15.4	0.008	2.5	0.01	5.5
	qSFW7B.2	**XGPW2224–*XGPW3256*	31.5	22.5–38.5	−0.012	14.5	−0.022	25.7	0.013	9.2		
SL	*qSL2A*	**XBARC380–*XGWM636*	1.6	0.0–5.6	−0.251	13.3						
	qSL3B	**XGWM533–*XBARC251*	48.5	40.5–56.4	0.495	19.8						
	qSL4B	**XBARC1096–*XGPW3206*	25.3	19.3–29.3	−0.485	24.5						
	qSL6A	**XGWM82–*XWMC182*	74.2	72.7–74.2	0.40	20.4						
	qSL7D	**XGDM67–*XWMC634*	156.5	151.5–162.5	−0.298	0.7						
LA	*qLA2A*	**XWMC455–*XGWM515*	112	108.4–118.0	−0.028	8.3	−0.06	22.8				
	qLA3D	**XWMC492–*XCFD223*	55.9	39.4–63.9	0.038	7.7						
	qLA5B.2	**XBARC232–*XWMC235*	33.7	31.7–33.7	0.044	15.3	0.038	6.3				
	qLA6A	**XGWM82–*XWMC182*	74.2	72.7–74.2	−0.045	17.4	−0.037	13.5				
RL	*qRL4B*	**XBARC1096–*XGPW3206*	27.3	21.3–29.3	0.304	27.7						
	qRL5A	**XBARC358.2–*XGWM186*	56.3	49.3–63.3	−0.325	5.1						
	qRL6A	**XGWM459–*XGWM334*	31.8	29.7–37.8	−0.241	27.2						
	qRL6A	**XGWM82–*XWMC182*	74.2	72.7–74.2	−0.604	63.1	−0.310	12.2				
	qRL7B.2	**XCFD22.1–*XGPW3226*	14.3	9.5–19.3	−0.214	0.7	−0.373	9.9				
	qRL7D	**XGWM676–*XGWM437*	122.9	115.9–140.4	−0.405	2.43						
RFW	*qRFW5A.2*	**XWMC59–*XGPW2109*	31.7	26.7–31.7	0.004	6.7	0.009	13.9				
	qRFW6A	**XBARC1055–*XWMC553*	74.2	72.7–79.2	−0.008	16.1	−0.01	15.2				

E_1: 0 mg/L Cr^{6+} stress; E_2: 40 mg/L Cr^{6+} stress; E_3: 120 mg/L Cr^{6+} stress; alleles from Huapei 3 and Yumai 57 with positive effect are defined in positive and negative values, respectively. *SFW* stem–leaf fresh weight; *SL* seedling length; *LA* leaf age; *RL* root length; *RFW* root fresh weight

Table 7.8 Estimated epistasis (AA) and epistasis × environment interactions of QTLs for developmental characters of wheat seedling in DH lines based on the averaged phenotypic data from three environments

Trait	QTL	Flanking marker	Site (cM)	QTL	Flanking marker	Site (cM)	AA	H^2 (%)
SFW	*qSFW1A*	**XGPW3167–*XCWEM6.2*	148.9	*qSFW2D*	**XBARC129.2–*XWMC658.2*	118	−0.008	2.2
	qSFW1A	**XGPW3167–*XCWEM6.2*	148.9	*qSFW6A*	**XGPW2265–*XGPW322*	0	0.005	5.0
	qSFW4D	**XGPW7258–*XGPW1144*	57.3	*qSFW6A*	**XGPW2265–*XGPW322*	0	−0.006	2.0
SL	*qSL2A*	**XBARC380–*XGWM636*	1.6	*qSL7D*	**XGDM67–*XWMC634*	156.5	0.389	13.0
RFW	*qRFW3A*	**XGWM155–*XCFA2170*	190.7	*qRFW6A*	**XBARC1055–*XWMC553*	74.2	0.004	10.6

Only two pairs of epistatic QTLs for SFW were involved in AAE interactions. And for *qSFW1A/qSFW6A*, the effects of AAE_2 and AAE_3 were 0.0042 and −0.0051, accounting for 0.004 and 0.002 % of phenotypic variation; for *qSFW4D/Qsfw6A*, the effects of AAE_2 and AAE_3 were 0.0058 and −0.0061, accounting for 0.003 and 0.0027 % of phenotypic variation. The QTL with epistatic effect involved in AAE1 interaction was not found. For the effect of AAE, a positive value means that the parent-type effect is greater than the recombinant-type effect, and the negative value means that the parent-type effect is less than the recombinant-type effect. Abbreviations as in Table 7.7

7.2.2.2.2.2 QTL Analysis for Plant Height

Five QTLs, *qSL2A*, *qSL3B*, *qSL4B*, *qSL6A*, and *qSL7D*, were found to control plant height and mapped on chromosomes 2A, 3B, 4B, 6A, and 7D, explaining plant height variances of 13.3, 19.8, 24.5, 20.4, and 0.7 %, respectively (Table 7.7). *qSL4B* mapped on chromosome 4B had the highest contribution (24.5 %), while *qSL7D* mapped on chromosome 7D had the smallest contribution (0.7 %). We found that *qSL3B* and *qLA6A* mapped to chromosomes 2A and 6A were inherited from Huapei 3, while the else were inherited from Yumai 57. No AE interactions were identified for plant height.

One pair of QTLs (*qSL2A/qSL7D*) with epistatic effects was detected for plant height in the three environments (Table 7.8). They explained that 13.0 % of phenotypic variance had a positive effect (0.389). No AAE interactions were identified for plant height.

7.2.2.2.2.3 QTL Analysis for Leaf Age

Four QTLs, *qLA2A*, *qLA3D*, *qLA5B.2* and *qLA6A*, were found to control LA and mapped on chromosomes 2A, 3D, 5B.2, and 6A, explaining LA variances of 8.3, 7.7, 15.3, and 17.4 %, respectively (Table 7.7). Of them, *qLA6A* locating on chromosome 6A had the highest contribution (17.4 %), while *qLA3D* mapped on chromosome 3D had the smallest contribution (7.7 %). We found that *qLA2A* and *qLA6A* mapped to chromosomes 2A and 6A were inherited from Yumai 57, while *qLA3D* and *qLA5B.2* mapped to chromosomes 3D and 5B.2 were contributed by Huapei 3. No pair of epistatic QTL for LA was detected.

7.2.2.2.2.4 QTL Analysis for Main Root Length

Six QTLs, *qRL4B*, *qRL5A*, *qRL6A* (**XGWM459-*XGWM334*), *qRL6A*(**XGWM82-*XWMC182*), *qRL7B.2,* and *qRL7D*, were found to control main root length and mapped on chromosomes 4B, 5A, 6A (**XGWM459-*XGWM334*), 6A (**XGWM82-*XWMC182*), 7B.2, and 7D, explaining main root length variances ranging from 0.70 to 63.1 % (Table 7.7). Of them, *qRL6A* (**XGWM82-*XWMC182*) mapped on chromosome 6A (**XGWM82-*XWMC 182*) had the highest contribution (63.1 %). And *qRL4B* was contributed by Huapei 3, while the other QTLs were contributed by Yumai 57. Among all of QTLs, *qRL6A* (**XGWM82-*XWMC182*) and *qRL7B.2* were involved in AE1 interactions, accounting for 1.22 and 0.99 % of the phenotypic variance, respectively. No detected QTL for root length was involved in AE_2 and AE_3. And, no pair of epistatic QTL for root length was found.

7.2.2.2.2.5 QTL Analysis for Root Fresh Weight

Two QTLs, *qRFW5A.2* and *qRFW6A*, were found to control RFW and mapped on chromosomes 5A.2 and 6A, explaining RFW variances of 6.7 and 16.1 %, respectively (Table 7.7). Of them, *qRFW6A* mapped on chromosome 6A had the highest contribution (16.1 %), and its addictive alleles were inherited from Yumai 57. Two QTLs, *qRFW5A.2* and *qRFW6A*, were involved in AE interactions, accounting for 13.9 and 15.2 % of the phenotypic variance, respectively.

One pair of epistatic QTLs for RFW was detected in the three environments (Table 7.8). Furthermore, *qRFW3A/qRFW6A*, located on chromosomes 3A-6A, explaining 10.6 % of the phenotypic variance, had a positive effect. No AAE interactions were identified for RFW.

7.2.3 Research Progress of Heavy Metals Resistance QTL Mapping and Comparison of the Results with the Previous Studies

7.2.3.1 Research Progress of Heavy Metals Resistances QTL Mapping

At present, most studies related to heavy metal focused on QTL mapping of rice under stress tolerance. Chen et al. (2010) mapped nine QTLs, accounting for 7.23–18.02 % of phenotypic variation, for Cd^{2+} stress resistance in rice seedling for two years. Among them, two QTLs contributed more than 15.0 %, and other two QTLs contributed 10.0–18.02 %. Shen et al. (2008) analyzed QTL controlling Cd, five kinds of microelement contents in rice grain, and detected three QTLs for Cd^{2+} content on chromosomes 5, 7, and 11, respectively. Xue et al. (2009) also conducted the similar research and mapped 22 QTLs for seedling height, root length, RDW, dry weight of stem, and chlorophyll. Among them, a total of 12 QTLs were detected under Cd-stress condition with the highest contribution of 35.26 %, and the others could contribute 8.11–17.75 % of phenotypic variation. However, few researches related to QTL conferring stress tolerance in wheat were conducted.

7.2.3.2 Comparison of the Results with the Previous Studies

In this study, we performed QTL analyses under three Cd^{2+} or Cr^{6+} treatments, mimicking three levels of Cd-polluted or Cr^{6+} environments. The identified QTLs can be categorized into three groups: (1) those identified under all three Cd^{2+} conditions and showed the same genetic effects, which could be considered as Cd-neutral; (2) those detected only in the absence of Cd^{2+}, but not in the presence of 40 and/or 120 mg/L Cd^{2+}, suggesting that these QTLs were Cd^{2+}-sensitive (non-tolerant); and (3) those mapped only when Cd^{2+} is present at 40 and/or 120 mg/L Cd^{2+}, indicating that these QTLs were unique loci for Cd^{2+} tolerance, and thus are valuable for wheat Cd^{2+} tolerance studies and useful for the breeding programs.

However, the QTL expressed under all three conditions has not been identified in our study; that is, the first group of QTLs does not exist. This result was in agreement with the study by Xue et al. (2009), who reported that no rice QTL expression was detected in the absence or presence of Cd. Significantly, these results strongly suggested that under Cd-stress conditions, expressions of some QTLs could be inhibited, while expressions of others could be induced at the same time. None of the 22 QTLs mapped under control condition was detected at 40 or 120 mg/L Cd. These QTLs belong to group 2 and were classified as Cd-sensitive or Cd-intolerant. However, many of these QTLs have significant effects on seedling and radicle development. Examples include the following: *qTRL1Bb* was involved in TRL, explaining 25.95 % of root length variance; *qRSA1Bb* was responsible for root surface area, explaining 30.15 % of its phenotype variance; *qRT1Ba* and *qRT1Bb* were related to the number of roots, explaining 26.45 and 55.03 % of their phenotype variances, respectively. These results suggested that these QTLs had great influence on seedling and radicle growth and development under normal condition and that could be useful in molecular marker-assisted selection. Importantly, QTLs of the third group involve in 21 QTLs which could be detected only at 40 or 120 mg/L Cd. These QTLs were considered to be Cd tolerant under Cd stress and might be very valuable for future studies. Implications of these QTLs include: seven Cd-tolerant QTLs were mapped to chromosome 3B, suggesting the possible presence of a Cd-tolerant gene on this chromosome; both *qARD3B* which is associated with the RD and *qSDW3Ba* which is involved in stem-leaf dry weight showed effect magnitudes as high as 70.05 and 28.94 %, respectively, and thus are of great value, especially the former. It is important to point out that the most obvious trait observed when seedlings under Cd stress was the change in the thickness of the radicles. Therefore, constructing the secondary population for fine mapping of this QTL should be considered as the future intensive and in-depth research. There were not the same QTL detected at 40 and 120 mg/L Cd. This suggested that Cd stress under different concentrations could produce different effects on the expression of QTLs. With the increase in concentration the expression of certain QTLs could be inhibited while the expression of other QTLs can be induced.

The present study revealed that the QTLs that we mapped were clustered on chromosomes. Several genes were located on the same chromosome, and multiple QTLs encoding different traits were found to be concentrated on a specific segment of the same chromosome. This conclusion can be evidenced by the following examples: without Cd stress, QTLs responsible for TRL, root surface area, and the RN were all detected on chromosome 1B at the interval of *Xcfd21* to *Xcwem9*; QTLs for TRL, root surface area, average RD, and the RN were also located on chromosome 1B, but between the segment of *Xgwm218* to *Xgwm582*; QTLs for TRL, root surface area, and RV were found on chromosome 7D at the interval of *Xgdm67* to *Xwmc634*; with 40 mg/L Cd stress, QTLs for shoot fresh weight and RDW were present on chromosome 3B in between *Xgwm566* and *Xcfe009*; and finally, with 120 mg/L Cd stress, QTLs for root surface area and RV were both identified on chromosome 1A at the interval of *Xgwm498* to *Xgpw7412*. Similar information from wheat has been observed by Ci (2012). These results strongly suggest that these concentrated distribution intervals with these QTLs reflected, to some extent, an important QTL inheritance relationship with respect to the Cd-tolerant traits: mutually promotive, but mutually inhibitory at the same time. Moreover, these QTLs most likely play a pleiotropy role and have a tight genetic linkage relationship. To discriminate the QTLs with promotive effects from those with inhibitory effects accurately, a near-isogenic line should be created via backcrossing so that compound QTLs can be separated from each other. This would allow us to develop lines with an individual Mendelian inheritance factor, thereby performing the fine mapping. Importantly, the use of these tightly linked QTLs in breeding programs will result in efficient molecular marker-assisted selection and maximum benefit for the genetic improvement of several linked traits simultaneously.

7.3 QTL Mapping and Effect Analysis for Wheat Preharvest Sprouting Resistance

Wheat is easily prone to preharvest sprouting (PHS) in humid environment before harvest because it has the shorter dormancy. PHS can cause a series of physiological changes inside the wheat seed, which not only affects wheat yield but also severely reduces wheat quality. In addition, even if it is not serious, it can largely reduce the seed storage life, which will cause the huge losses in agronomic production. Besides the temperature and raining amount during the wheat ripening period for harmful degree of PHS, the genotype of variety is an important factor for PHS resistance. The resistance to PHS of varieties is affected by multigenes, and their interactions seemed to be complex, which not only has the additive effect of single gene but also has the epistatic effect of different genes interactions including with environment. So, it is the effective means to select the resistant varieties for PHS using MAS. But the first thing for MAS is to find the markers closely linked with resistance to PHS, which needs to QTL mapping of PHS under natural or greenhouse environments. Therefore, a DH population derived from Huapei 3 and Yumai57 was used to study the PHS under three different environments in this study.

7.3.1 QTL Mapping for Wheat Preharvest Sprouting Resistance

7.3.1.1 Materials and Methods

7.3.1.1.1 Materials

Materials and Planting were same as one of 7.2.2.1.1 in this chapter.

7.3.1.1.2 Methods

The plant materials were grown in the experimental field in Shandong Agricultural University with random plot design in 2008. Each line of DH population and the two parents were planted with four lines, and the length of each line is 2.5 m with 0.25-m plant space. There were three treatments with different nitrogen and water irrigations under the same soil condition. E1 included four times irrigations: winter water irrigation, joint water irrigation, anthesis water irrigation, and grain filling irrigation, and nitrogen was also applied each time at jointing and grain filling stages; The irrigation condition of E2 was the same as that of E1, but no nitrogen was applied during the wheat growth period. The nitrogen application of E3 was the same as that of E1, while no water was irrigated during the wheat whole growth period. The anthesis date was investigated and then marked. The eight spikes with the length 7–8 cm peduncle were extracted at 35 days after anthesis. They were used to sprouting test. First, bind them in bundles, and then immersed in sterile water for 4–6 h to make the spikes filling water. They were inserted in a prepared foam board with some spaces between bundles and were placed in a turnover plastic box at room temperature. During 24 h, the sterile water was sprayed every 1–2 h with about three or five times a day. The number of sprouting grains and non-sprouting grains was counted after 7 days of sprouting.

7.3.1.1.3 Data Analysis

The phenotypic data were analyzed using Microsoft Excel. The QTLNetwork 2.0 software was used to analyze the QTL mapping, and QTL was named according to McIntosh et al.'s method.

7.3.1.2 Results and Analysis

7.3.1.2.1 Phenotypic Data

In three environments, the two parents had the higher germination rate (Table 7.9), and Huapei 3 showed higher than Yumai 57. In DH population, the germination rate had a wide variation ranging from 6 to 98 % (E1), 10 to 98 % (E2), and 7 to 98 % (E3), respectively.

In three environments, the trend of sprouting was similar (Figs. 7.4 and 7.5). A transgressive phenomenon was observed in the DH population. The lines with seventy percent of germination rate accounted for a large proportion, which showed be prone to the genetic effect of female parent. This result was similar to the previous researches.

Table 7.9 Analysis of preharvest sprouting in the DH population derived from Huapei 3 × Yumai 57

Environment	Parent		DH population			
	Huapei 3 (%)	Yumai 57 (%)	Mean (%)	Minimum (%)	Maximum (%)	SD
Env. 1	75	56	71	6	98	0.18
Env. 2	81	59	70	10	98	0.19
Env. 3	79	61	74	7	98	0.19

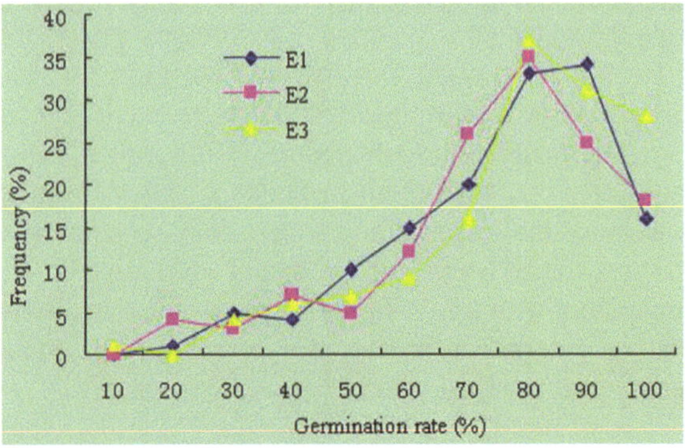

Fig. 7.4 The frequency distribution chart of preharvest sprouting derived from Huapei 3 × Yumai 57

Fig. 7.5 The performance of the parents and the part lines seven days later

7.3.1.2.2 QTL Mapping of Sprouting

7.3.1.2.2.1 Additive QTL

There were three QTLs (*qPhs2B.2*, *qPhs5D.1*, and *qPhs4A*) detected in E1 distributing on 2B, 5D, and 4A chromosomes (Table 7.10; Fig. 7.6). The additive values of three QTLs were −0.075, 0.050, and −0.043 with explaining 9.45, 7.74, and 6.09 % of phenotypic variation, respectively.

Table 7.10 Intervals, effects, and contributions of QTLs for preharvest sprouting in the DH population derived from Huapei 3 × Yumai 57

Environment	QTL	Flanking marker	Site (cM)	Additive	
				A	H^2 (%)
Env. 1	*qPhs2B.2*	*XBARC373–XBARC1114*	68.1	−0.075**	9.45
	qPhs5D.1	*XCFD40–XBARC1097*	4.4	0.050**	7.74
	qPhs4A	*XWMC-313–XWMC497*	32.5	−0.043**	6.09
Env. 2	*qPhs1B*	*XWMC766–XSWES158*	95.4	0.021**	7.62
	qPhs2B.1	*XCWEM55–XBARC129.1*	87.6	0.016**	8.88
	qPhs5D.1	*XCFD40–XBARC1097*	2.4	0.047**	5.33
Env. 3	*qPhs5D.1*	*XCFD40–XBARC1097*	2.4	0.052**	5.70
	qPhs5D.2	*XWMC215–XBARC345*	69.3	−0.039**	5.85

**Significance at level of 5 %

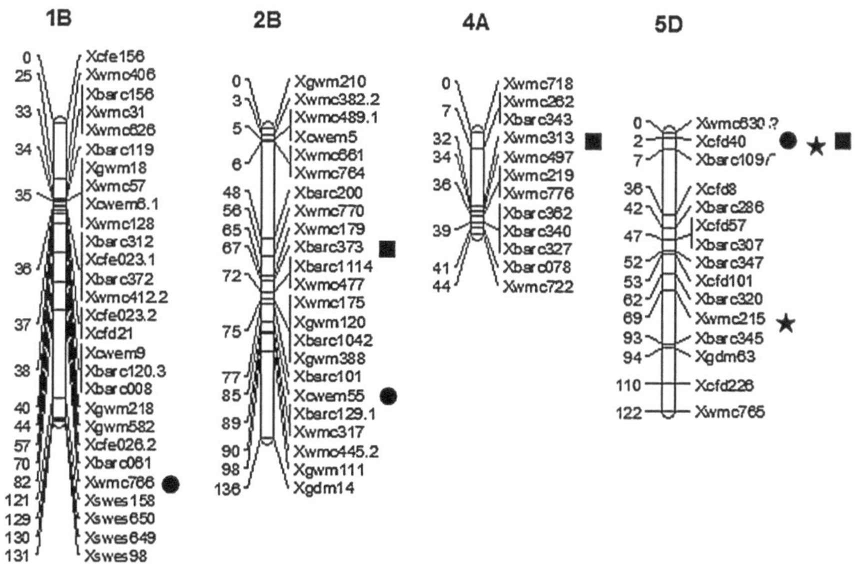

Fig. 7.6 Positions of additive QTLs associated with preharvest sprouting in the DH population derived from Huapei 3 × Yumai 57. ■: E1 QTL. ●: E2 QTL. ★ E3: QTL

In E2, three QTLs, *qPhs1B*, *qPhs2B.1* and *qPhs5D.1*, were identified on 1B, 2B, and 5D chromosomes, explaining 7.62, 8.88, and 5.33 % of phenotypic variation, respectively. And their additive values were 0.021, 0.016, and 0.047, respectively.

While in E3, two QTLs, *qPhs5D.1* and *qPhs5D.2*, were detected on chromosome 5D, explaining 5.70 and 5.85 % of phenotypic variation, respectively. And their additive values were 5.70 and 5.85 %, respectively.

In all these QTLs, only one QTL, *qPhs5D.1*, was always detected in three environments, and their additive effects showed positive; but the residual QTLs were only found in one environment with positive or negative additive value, which indicated the alleles would be from two parents and caused the transgressive separation.

7.3.1.2.2.2 Epistasis QTL

Two pairs of epistatic QTLs were detected in E1 (Table 7.11) distributing on chromosomes 2B-4A and 2B-5D, explaining 1.31 and 6.70 % of phenotypic variation, respectively. In E2, only one pair of epistatic QTL was found on chromosomes 1A-5B; while in E3, two pairs of epistatic QTLs were detected on chromosomes 1A-3D and 2A-3A, explaining 13.37 and 8.30 % of phenotypic variation, respectively. There was one major QTL. This indicated that the spike sprouting was also affected by the interaction between non-allele genes.

7.3.2 Research Progress of Preharvest Sprouting Resistance QTL Mapping and Comparison of the Results with the Previous Studies

7.3.2.1 The Research Progress of QTL Mapping of Preharvest Sprouting Resistance

Spike sprouting trait is closely related to wheat yield and quality. Many researchers had studied the QTL mapping of agronomic traits which is related to spike sprouting, such as PHS resistance, germination index, and falling number (Kulwal et al. 2004; Mohan et al. 2009; Fofana et al. 2009; Imtiaz et al. 2008; Rasul et al. 2009; Flintham et al. 2002; Sun et al. 2008). More than 90 QTLs have been identified, and of which, 28 QTLs belonged to the major QTL with more than 10 % explanation of phenotypic variation, and the maximum PVE was 52.7 % (Table 7.12). Most of them were located on 3A, 3B, 3D, and 4A chromosomes.

7.3.2.2 Comparison of the Results with One of the Previous Studies

In this study, the additive QTLs were found on 1B, 2B, 4A, and 5D chromosomes, and the epistasis QTL was also identified. The previous study also found QTL on 2B and 5D chromosomes, which indicated these two chromosomes were important for PHS; perhaps, they had some important gene or gene regions for resistance to PHS. In addition, there were some QTLs detected on 3B, 3D, 4A, 6A, and 6B chromosomes, especially the QTLs on chromosomes 4A and 3D were related to the seed germination characteristics, while that on chromosomes 4A and 6B were closely related to the falling number. These results dissected the relationship between PHS and seed germination and four falling number at molecular level.

Table 7.11 Estimated epistasis associated with preharvest sprouting in the DH population derived from Huapei 3 × Yumai57

Environment	QTL	Flanking marker	Site (cM)	QTL	Flanking marker	Site (cM)	AA	H^2 (%)
Env. 1	*qPhs2B.3*	*XBARC373–XBARC1114*	68.1	*qPh4A*	*XWMC313–XWMC497*	32.5	−0.014**	1.31
	qPhs2B.3	*XBARC373–XBARC1114*	68.1	*qPhs5D.1*	*XCFD40–XBARC1097*	2.4	0.048**	6.70
Env. 2	*qPhs1A.2*	*XBARC120.1–XWMC333*	57.2	*qPhs5B*	*XBARC232–XWMC235*	24.7	0.067**	9.50
Env. 3	*qPhs1A.1*	*XCWEM32.1–XWMC527*	9.9	*qPhs3D*	*XGDM8–XWMC492*	25.4	−0.064**	13.37
	qPhs2A	*XWMC382.1–XBARC380*	1.0	*qPhs3A*	*XCFA2134–XWMC527*	104.9	0.015**	8.30

**Significance at level of 5 %

Table 7.12 Summary of QTL results of wheat preharvest resistance (PVE > 10 %)

Trait	QTL	Flanking marker	PVE (%)	Population	Reference
Preharvest resistance	*QPhs.ccsu-3B.4*	*Xfbb283–XATPase.2*	19.84	RILs	Kulwal et al. (2004)
	QPhs.ccsu-3B.5	*Xgwm108–Xfbb378*	10.69		
	QPhs.ccsu-2A.5	*Xgwm1045–Xgdm296*	13.90	RIL	Mohan et al. 2009
	QPhs.ccsu-3A.1	*Xwmc153–Xgwm155*	15.99		
Germination index (GI)	*QGi.crc-3B*	*Xbarc77–Xwmc307*	27	DH	Fofana et al. (2009)
	QGi.crc-3D	*Xwmc552–Xwmc533*	17		
GI-7 at seven days	*QPhs.dpivic-3D.1*	*Xcfd211–XRGC*	–	BC1F7	Imtiaz et al. (2008)
	QPhs.dpivic-3D.2	*Xgdm8–Xgwm52*	–		
	QPhs.dpivic-4A.2	*Xwms937–Xwms894*	–		
GI-14 at 14 days	*QPhs.dpivic-3D.1*	*XRGC–Xwms1200*	–	BC1F7	
	QPhs.dpivic-4A.1	*Xbarc170–Xgwm269c*	–		
	QPhs.dpivic-4A.2	*Xwms937–Xwms894*	–		
GI	*QGI-4B*	*Xwmc349*	52.7	DH	Rasul et al. (2009)
Visible germinated seeds (VI)	*QPhs.dpivic-3D.1*	*XRGC–Xwms1200*	12	BC1F7	Imtiaz et al. (2008)
	QPhs.dpivic-4A.2	*Xwms937–Xwms894*	24	BC1F7	Imtiaz et al. (2008)
	QPhs.dpivic-3D.1	*Xwms1200–Xgwm3*		BC1F7	Imtiaz et al. (2008)
SI	*QSI-3A*	*Xwmc264*		DH	Rasul et al. (2009)
	QSI-3D	*Xgwm191–Xbarc71*			
	QSI-4A.2	*Xgwm494*			
	QSI-4B	*Xwmc349*			
	QSI-7D	*Xgwm482*			
	QSi.crc-3B	*Xbarc77–Xwmc307*	21	DH	Fofana et al. (2009)
	QSi.crc-3D	*Xwmc552–Xwmc533*	11		
	QSi.crc-5D	*Xgwm469–Xcfd10*	44		

(continued)

Table 7.12 (continued)

Trait	QTL	Flanking marker	PVE (%)	Population	Reference
Anti-sprouting damage	*1BS*	*Xpsp3000*		RIL	Flintham et al. (2002)
	4BL	*Xpsp3030-Xpsp3078*			
	7AS	*Xpsp3050*			
	1AS	*Xbcd1434*		RIL	Anderson et al. (1993)
	1AS	*Xcdo431*			
	3BL	*Xcdo482*		RIL	
	4AL	*Xcdo545*			
	5DL	*Xbcd450*			
	6BL	*Xbcd1426*			
	6BS	*Xwmc104*		RIL	Roy et al. (1999)
	7DL	*Xmst101*			
	1BS	*Xgwm18–Xglk483*		RIL	Zanetti et al. (2000)
	2AL	*Xplpap–Xglk684a*			
	3AS	*Xpsr304–Xpsr598*			
	3BL	*Xglk80–Xpsr1054*			
	4DL	*Xpsr1101a–Xpsr160b*			
	5AS	*Xglk163a–Xpsr945a*			
	5AL	*Xpsr1194–Xpsr918b*			
	6DL	*Xpsr167a–Xmwg684b*			
	7BL	*Xpsr350–Xpwir232b*			
	4AL	*Xcdo795–Xpsr115*		DH	Kato et al. (2001)
	4BL	*Xbcd1431*			
	4DL	*Xbcd1431*			

(continued)

Table 7.12 (continued)

Trait	QTL	Flanking marker	PVE (%)	Population	Reference
	2DL	*sub-telomeric*		DH	Mares et al. (2002)
	4AL	*Xpsr115–Ger*			
	7BL	*Xpsr547–Xpsr680b*			
	3DL	*Xwms654*		DH	Mrva and Mares (2002)
Falling number	*QFN-4A.1*	*Xwmc48*	13.7	DH	Rasul et al. (2009)
	QFN-4B	*Xwmc349*	14.9	DH	Rasul et al. (2009)
	QFn.sdau-6B	*Xgwm132b–Xwmc487*	17.9	RIL	Sun et al. (2008)
	QFn.crc-3B	*Xbarc77–Xwmc307*	33	DH	Fofana et al. (2009)

Moreover, the epistasis QTL was detected with no additive effect, which indicated these QTLs could affect the PHS by interacting with other genes.

7.4 QTL Mapping and Effect Analysis of Disease Resistance

In China, breeding for disease resistance in wheat, which focused on stripe rust, began in the early 1950s. In recent years, besides stripe rust many original secondary diseases gradually become as main diseases for limiting wheat production, such as powdery mildew, fusarium head blight (FHB), sharp eyespot, take-all disease, and so on. Physiological races of pathogenic bacteria for the most diseases in wheat were complex and changeable, which often bring about the loss of resistance in varieties. So, the task of preventing and controlling disease in wheat is very difficult. Besides existing disease-resistant genes were used to conduct conventional breeding, excavating new disease-resistant genes and multiple disease-resistant genes in the same variety was the important breeding strategy. In addition, with the wide application and development of molecular markers, molecular markers of more and more resistant genes were obtained, which provides convenience to MAS and speeding up the breeding process of polymerizing target genes.

Although some QTLs and molecular markers of main disease resistance were obtained, few were applied in actual production. Therefore, QTL mapping for adult-plant resistance (APR) to powdery mildew and FHB in wheat was conducted to find closely linked molecular markers to be used in MAS breeding for APR to powdery mildew and GDR.

7.4.1 QTL Mapping for Adult-Plant Resistance to Powdery Mildew

7.4.1.1 Materials

Materials and Planting were same as one of 7.2.2.1.1 in this chapter.

7.4.1.2 Investigation of APR to Powdery Mildew

The APR to powdery mildew was evaluated under natural infection with the native population of the pathogen. According to the visually estimated percentage covered with symptoms in the 2 upper leaves, disease severity was expressed as a disease index from 1 (high resistance) to 9 (high susceptible) scale (referring to GB/T 19557.2-2004). The plants were scored in milky ripe stage (14 days after flowering).

7.4.1.3 Results and Analysis

7.4.1.3.1 Statistical Analysis of Phenotypic Assessments

The variations of APR to powdery mildew for parents and DH population in three environments are shown in Fig. 7.7. The parent Yumai 57 showed better APR to powdery mildew with disease index of 3, whereas Huapei 3 exhibited a higher disease index of 7. APR to powdery mildew (Fig. 7.7) of the DH population segregated continuously and followed a normal distribution, indicating its polygenic inheritance and suitability of the data for QTL analysis.

7.4.1.3.2 QTL Mapping and Effect Analysis of APR to Powdery Mildew

Two additive QTLs on chromosomes 4D and 5D were detected for APR to powdery mildew. One QTL *qApr4D* accounted for 20.0 % of the phenotypic variances, which had no AE interactions. The other QTL *qApr5D* could explain 1.3 % of the phenotypic variances with AE interactions. Both alleles were originated from the resistant parent Yumai 57 to APR (Table 7.13).

Two pairs of epistatic QTL on chromosomes 1B/3A were also identified for APR to powdery mildew, which had negative effect. They explained 3.6 and 1.3 % of phenotypic variances with no AAE interactions (Table 7.14).

7.4.2 QTL Mapping for Resistance to Fusarium Head Blight

7.4.2.1 Materials

A RILF$_8$ population was used in the study, which derived from the cross between Shannong 01-35 (39-1/Hesheng 2) and Gaocheng 9411 (77546/Linzhangmai) by using a single seed descend method and consisted of 182 lines.

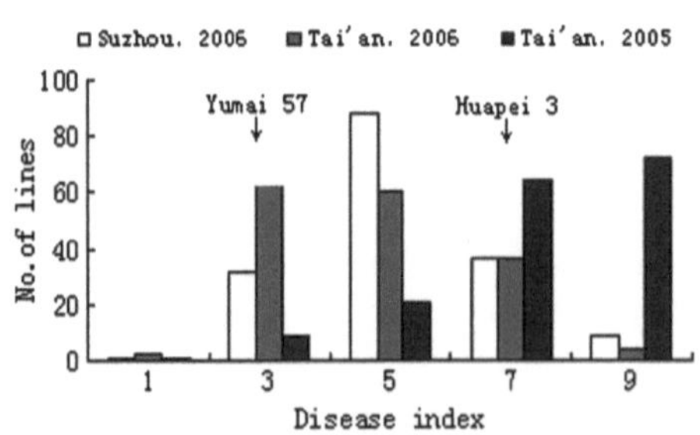

Fig. 7.7 Frequency distribution of disease index of adult plant to powdery mildew

Table 7.13 Estimated additive (A) and additive × environment (AE) interactions of QTLs for adult-plant resistance of powder mildew

Trait	QTL	Flanking marker	Site (cM)	Additive		A × E_1		A × E_2		A × E_3	
				A	H^2 (%)	AE_1	H^2 (%)	AE_2	H^2 (%)	AE_3	H^2 (%)
APR	*qApr4D*	*Xgwm194–Xcfa2173*	106.0	0.82	20.0						
	qApr5D	*Xwmc215–Xgdm63*	69.3	0.20	1.3	−0.34	3.4				

APR Adult-plant resistance; E1: Suzhou (2006); E2: Tai'an (2006); E3: Tai'an (2005)

Table 7.14 Estimated epistasis (AA) and epistasis × environment (AAE) of QTLs for adult-plant resistance of powder mildew

Trait	QTL	Flanking marker	Site (cM)	QTL	Flanking marker	Site (cM)	AA	H^2 (%)	AA × E1		AA × E2		AA × E3	
									AAE1	H^2 (%)	AAE2	H^2 (%)	AAE3	H^2 (%)
APR	*qApr1Ba*	*Xcfe023.2–Xcfd21*	37.3	*qApr3A*	*Xwmc489.2–Xwmc489.3*	59.8	−0.35	3.6						
	qApr1Bb	*Xgwm582–Xcfe026.2*	51.6	*qApr3A*	*Xwmc489.2–Xwmc489.3*	59.8	−0.21	1.3						

APR Adult-plant resistance; E1: Suzhou (2006); E2: Tai'an (2006); E3: Tai'an (2005)

7.4.2.2 Investigation on Resistance to FHB

Mixed suspension consisted of strong virulent strains F609, F301, F15, and 7136, with the spore concentration of 5000 conidia/mL, was used as the spore liquid which was supplied by Nanjing Agricultural University. Spray inoculation method was used. At flowering stage, 20 spikes of each line were selected and the spore liquid was sprayed. Three days after inoculation, spraying clear water on spikes to keep moist and ensure the spore germination. Furthermore, 21 days after inoculation, investigating the disease condition and counting pathogenetic spikelet rate (PSR). PSR = (Number of disease spikelet/Total number of spikelet) × 100 %.

7.4.2.3 Results and Analysis

7.4.2.3.1 Statistical Analysis of Phenotypic Assessments

The variation ranges of PSR of the RIL population in the three different environments were 10.15–61.3 %, 8.5–56.2 %, and 9.6–54.3 %, with the mean of 38.9, 29.6, and 27.8 %, respectively. And PSR of the population followed a normal distribution. There was a large difference between two parents. Shannong 01-35 had the better resistance to FHB, and the variation range of PSR was 17.8–38.5 % in the three environments, however, that of Gaocheng 9411 was 20.1–71.2 %. It was found that disease index of each line in the population followed a normal distribution, after analyzing the average value of three years (Fig. 7.8). And the average value of the population was 33.9 %, between Shannong 01-35 (21.6 %) and Gaocheng 9411 (62.3 %), which tended to the resistant parent.

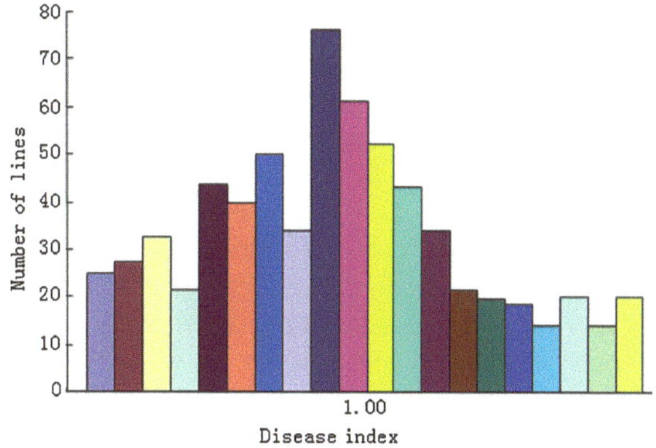

Fig. 7.8 Frequency distribution of disease index of adult plant to fusarium head blight

7.4.2.3.2 QTL Mapping and Effect Analysis of Resistance to FHB

A total of two QTLs for APR to FHB distributed on chromosomes 4B and 5A were detected (Table 7.15). Among which, *Qfhi.nau-4B* on chromosome 4B contributed 10.31 % of the total phenotypic variation, and stably expressed in each environment; however, *Qfhi.nau-5A* on chromosome 5A contributed 11.51 % of the total phenotypic variation. Meanwhile, the two loci involved in QE interaction, and both positive alleles were originated from Shannong 01-35, which was in accord with that Shannong 01-35 had the better resistance to FHB (Table 7.15).

7.4.3 Research Progress of Wheat Disease-Resistant QTL Mapping and Comparison of the Results with the Previous Studies

7.4.3.1 Research Progress of QTL Mapping Conferring Disease-Resistant

Now, more than 60 genes of APR to powdery mildew in wheat were mapped on 47 locus (*Pm1–Pm47*), distributed on each chromosome except chromosomes 3A and 4D (Table 7.16). Among them, both *Pm38* and *Pm39* were ADR genes of powdery mildew; however, other resistant genes were all major genes of physiological race of pathogenic bacteria. Because of continuous variation for physiological race, most of resistant genes have lost resistance (Lan 2010). Many researches related to APR to powdery mildew have been conducted. For example, Liu et al. (2001a, b, c) detected three QTLs conferring slow mildewing resistance in wheat, among which *QPm.vt-1B*, *QPm.vt-2A*, and *QPm. vt-2B* were mapped on chromosomes 1B, 2A, and 2B, and contributed 17, 29, and 11 % of phenotypic variation, respectively. Meanwhile, all the three loci expressed additive effect. Börner et al. (2002) detected six QTLs related to powdery mildew resistance, distributed on chromosomes 2D, 3B, 4B, 5D, 6A, and 7D, respectively, by using a RIL population derived from W7984 × Opata8 in several environments for multiyear and multisite. He et al. (2011) reviewed QTLs/genes related to APR to powdery mildew, which have been mapped in wheat, and found that 80 QTLs including 49 major QTLs were mapped on 21 chromosomes in wheat, among which the highest contribution for individual QTL was 63 % of phenotypic variation.

On the international, a total of 51 major stripe rust resistance genes, mapped on 48 loci (*Yr1–Yr48*), have been named, and, except *Yr11–Yr14*, the others were mapped on specific chromosomes (Wu 2011). Furthermore, *Yr3* locus has three multiple alleles (*Yr3a*, *Yr3b*, and *Yr3c*), *Yr4* locus has two multiple alleles (*Yr4a* and *Yr4b*), and *Yr11*, *Yr12*, *Yr13*, *Yr14*, *Yr16*, *Yr18*, *Yr29*, *Yr30*, *Yr36*, *Yr39*, *Yr46*, and *Yr48* loci were the APR genes; however, other 39 were the seedling resistance

Table 7.15 Estimated additive (A) and additive × environment (AE) interactions of QTLs for APR to FHB

Trait	QTL	Flanking marker	Site (cM)	Additive		A × E_1		A × E_2		A × E_3	
				A	H^2 (%)	AE_1	H^2 (%)	AE_2	H^2 (%)	AE_3	H^2 (%)
APR to FHB	*Qfhi.nau-4B*	*Xgwm149–Xwmc349*	1.0	0.06	10.31	0.06	16.12	0.06	16.43	0.06	17.54
	Qfhi.nau-5A	*Xwmc96–Xwmc446*	2.0	0.07	11.51	0.07	17.22	0.07	17.63	0.08	19.84

APR adult-plant resistance; *FHB* fusarium head blight; E1: Suzhou (2006); E2: Tai'an (2006); E3: Tai'an (2005)

Table 7.16 Summary of QTLs for plant resistance to disease in wheat (PVE > 10 %)

Trait	QTL	Donor	Flanking marker	PVE (%)	Reference
YR	*QYr.jirc-1B*	Fukuho-komugi	*Xwmc320*	10.4	Suenaga et al. (2003)
	QPst.jic-1B	Guardian	*Xgwm818*	18.0–22.0	Melichar et al. (2008)
	QYr.cimmyt-1BL	Pavon 76	*Xgwm259*	33.0–33.9	William et al. (2006)
	QYrtm.pau-2A	*T.monococcum*	*Xwmc407–Xwmc170*	14.0–27.0	Chhuneja et al. (2008)
	QTL 2AS	PAU14087/Kris	*Xwmc407–Xgwm071d*	27	Christiansen et al. (2006)
	QYR2	Camp Remy	*Xgwm356–Xgwm382*	10.7–15.4	Boukhatem et al. (2002)
	QYr.inra-2AL	Camp Remy	*Xgwm282a–Xgwm359*	20.0–40.0	Mallard et al. (2005)
	QTL 2AL	Deben/Kris/Soloist	*Xwmc198a–Xwmc170B*	40.0–41.0	Christiansen et al. (2006)
	QYR3	Opata 85	*Xcdo405–Xbcd152*	30.7	Boukhatem et al. (2002)
	QYr.inra-2BS	Camp Remy	*Xgpv3032–Xcfd50a*	22.0–70.0	Mallard et al. (2005)
	QYr.sgi-2B.1	Kariega	*Xgwm148–s12m60A*	17.0–46.0	Ramburan et al. (2004)
	QYrlu.cau-2BS.1	Luke or Aquileja	*Xwmc154–Xgwm148*	36.6	Guo et al. (2008)
	QYR1	Camp Remy	*Xgwm47–Xgwm501*	45.8–46.0	Boukhatem et al. (2002)
	QYr.inra-2BL	Camp Remy	*Xbarc101–Xgwm120*	42.0–61.0 c	Mallard et al. (2005)
	QTL 2BL	Deben	*Xwmc149–Xwmc317a*	21	Christiansen et al. (2006)
	QYraq.cau-2BL	Luke or Aquileja	*Xwmc175–Xwmc332*	61.5	Guo et al. (2008)
	QYr.inra-2DS	Camp Remy	*Xgwm102–Xgwm539*	24.0–69.0 d	Mallard et al. (2005)
	QYr.jirc-2DL	Fukuho-komugi	*Xgwm349*	10.1–11.4	Suenaga et al. (2003)
	QYr.cimmyt-3BS	Opata 85	*Xfba190–XksuG53*	16.0–28.0	Singh et al. (2000)
	QYrex.wgp-3BL	Express	*Xgwm299–Xwgp66*	22.1–27.4	Lin et al. (2009)
	QYr.cimmyt-3D	Opata 85	*Xfba241–Xfba91*	14	Singh et al. (2000)
	QYR6	Opata 85	*Xcdo407–XksuA6*	11.7	Boukhatem et al. (2002)
	QPst.jic-4B	Guardian	*Xwmc652–Xwmc692*	8.0–12.0	Melichar et al. (2008)
	QYr.cimmyt-4BL	Avocet S	*Xgwm495*	7.4–12.7	William et al. (2006)
	QYr.cimmyt-4DS	Synthetic	*Xbcd265–Xmwg634*	9.0–31.0	Singh et al. (2000)

(continued)

Table 7.16 (continued)

Trait	QTL	Donor	Flanking marker	PVE (%)	Reference
	QYR5	Opata 85	*Xfbb209–Xabg391*	15	Boukhatem et al. (2002)
	QYrtm.pau-5A	*T.boeoticum* PAU5088	*Xbarc151–Xcfd12*	24	Chhuneja et al. (2008)
	QYr.caas-5AL	Pingyuan 50	*Xwmc410–Xbarc261*	5.0-19.9	Lan et al. (2010)
	QYr.inra-5BL.1	Camp Remy	*Xgwm639a–Xgwm639c*	18.0–26.0	Mallard et al. (2005)
	QYr.inra-5BL.2	Camp Remy	*Xgwm234–XDuPW115a*	29.0–34.0	Mallard et al. (2005)
	QYr.jirc-5BL	Oligoculm	*Xwmc415*	2.4–16.1	Suenaga et al. (2003)
	QYrex.wgp-6AS	Express	*Xgwm344–Xwgp56*	24.5–30.9	Lin et al. (2009)
	QYrst.wgp-6BS.1	Stephens	*Xbarc101–Xbarc136*	32.0–45.0	Santra et al. (2008)
	QYrst.wgp-6BS.2	Stephens	*Xgwm132–Xgdm113*	25.0–43.0	Santra et al. (2008)
	QTL 6BL	Soloist/Kris	*Xwmc397–Xwmc105b*	25.0–29.0	Christiansen et al. (2006)
	QYr.cimmyt-6BL	Pavon76	*XPstAGGMseCGA1*	10.1–17.6	William et al. (2006)
	QYR7	W-7984	*Xbcd1510–XksuD27*	13.1	Boukhatem et al. (2002)
	QYr.nsw-7B	Tiritea	*Xgwm611*	26	Imtiaz et al. (2004)
	Yr39	Alpowa	*Xwgp36–Xwgp45*	59.1–64.2 e	Lin et al. (2007)
	YrCK	Cook	*M59/P41–215–M49/P33–280*	18.0–30.0	Navabi et al. (2005)
	QYr.jirc-7DS	Fukuho-komugi	*Xgwm295*	10.7–23.7	Suenaga et al. (2003)
	QYr.sgi-7D	Kariega	*Xgwm295–Ltn*	9.0–29.0	Ramburan et al. (2004)
	QYr.nsw-7DS	Otane	*Xgwm44*	13	Imtiaz et al. (2004)
	QYr.caas-7DS	Libellula and Strampelli	*csLV34–Xgwm295*	14.6–39.1	Lu et al. (2009)
	QYr.cimmyt-7D	Opata 85	*Xbcd1438–Xwg834*	12.0–36.0	Singh et al. (2000)
	QYR4	Opata 85	*XWg834–Xbcd1438*	13.9	Boukhatem et al. (2002)

(continued)

Table 7.16 (continued)

Trait	QTL	Donor	Flanking marker	PVE (%)	Reference
PM	*QPm.osu-1A*	2174	*Pm3a*	63.0 a	Chen et al. (2009)
	QPm.caas-1AS	Fukuho-komugi	*Xgdm33–Xpsp2999*	19.9–26.6	Liang et al. (2006)
	QPm.crag-1A	RE714	*Xcdo572–Xbad442*	39.3–43.0	Mingeot et al. (2002)
	QPm.sfr-1B	Forno	*CD9b–Xpsr593a*	11.6	Keller et al. (1999)
	QPm.vt-1BL	Massey	*Xgwm259–Xbarc80*	15.0–17.0	Tucker et al. (2007)
	QPm.vt-1B	USG3209	*WG241*	17	Liu et al. (2001a, b, c)
	Yr29/Lr46/Pm39	Saar	*Xwmc719–Xhbe248*	7.3–35.9	Lillemo et al. (2008)
	QPm.osu-1B	2174	*Xwmc134*	14	Chen et al. (2009)
	QPm.inra-1D.1	RE9001	*Xgwm106*	12.6	Bougot et al. (2006)
	QPm.crag-2A	RE714	*Pm4b-gbxG303*	22.7–33.6	Mingeot et al. (2002)
	QPm.vt-2AL	Massey	*Xgwm304–Xgwm312*	29	Liu et al. (2001a, b, c)
	QPm.vt-2A	USG3209	*Xgwm304–Xgwm294*	26.0–29.0	Tucker et al. (2007)
	QPm.crag-2B	Festin	*Xgwm148–gbxG553*	23.6–71.5	Mingeot et al. (2002)
	QPm.caas-2BS	Lumai 21	*Xbarc98-Xbarc1147*	10.6–20.6	Lan et al. (2010)
	QPm.vt-2B	Massey	*WG338–Xgwm526a*	11	Liu et al. (2001a, b, c)
	QPm.inra-2B	RE9001	*Xrtp114R–Xcfd267b*	10.3–36.3	Bougot et al. (2006)
	QPm.vt-2BL	USG3209	*Xgwm501–Xgwm191*	11.0–15.0	Tucker et al. (2007)
	QPm.caas-2BL	Lumai 21	*Xbarc1139–Xgwm47*	5.2–10.1	Lan et al. (2010)
	QPm.inra-2D-a	RE9001	*Xgwm102*	19	Bougot et al. (2006)
	QPm.inra-2D-b	RE9001	*Xcfd2e*	16.5	Bougot et al. (2006)
	QPm.sfr-2D	Oberkulmer	*Xpsr932–Xpsr331a*	10	Keller et al. (1999)
	QPm.caas-2DL	Lumai 21	*Xwmc18–Xcfd233*	5.7-11.6	Lan et al. (2010)
	QPm.sfr-3A	Forno	*Xpsr598–Xpsr570*	10.4	Keller et al. (1999)
	QPm.crag-3A	Festin	*Xpsr598–Xgwm5*	21.4–25.9	Mingeot et al. (2002)

(continued)

Table 7.16 (continued)

Trait	QTL	Donor	Flanking marker	PVE (%)	Reference
	QPm.nuls-3AS	Saar	*Xstm844tcac–Xbarc310*	8.1–20.7	Lillemo et al. (2008)
	QPm.inra-3B	Courtot	*Xgwm389*	22.7	Bougot et al. (2006)
	QPm.osu-3B	2174	*Xwmc533*	10	Chen et al. (2009)
	QPm.sfr-3D	Oberkulmer	*Xpsr1196a–Lrk10-6*	15.7	Keller et al. (1999)
	QPm.inra-3D	RE9001	*Xcfd152, Xgwm707*	9.3–15.2	Bougot et al. (2006)
	QPm.sfr-4A.1	Forno	*Xgwm111c-Xpsr934a*	14.7	Keller et al. (1999)
	QPm.sfr-4A.2	Forno	*Xmwg710b–Xglk128*	14.3	Keller et al. (1999)
	QPm.ttu-4A	*Triticum militinae*	*Xgwm232–Xgwm160*	35.0–54.0 c	Jakobson et al. (2006)
	QPm.crag-4A	RE714	*XgbxG036–XgbxG542*	22.3	Mingeot et al. (2002)
	QPm.osu-4A	2174	*Xwms160*	12	Chen et al. (2009)
	QPm.nuls-4BL	Avocet	*XwPt1505–Xgwm149*	21.0–40.2	Lillemo et al. 2008
	QPm.sfr-4D	Forno	*Xglk302b–Xpsr1101a*	14.4	Keller et al. 1999
	QPm.caas-4DL	Bainong 64	*Xbarc200–Xwmc33*	15.2–22.7	Lan et al. (2009)
	QPm.sfr-5A.1	Oberkulmer	*Xpsr644a–Xpsr945a*	22.9	Keller et al. (1999)
	QPm.sfr-5A.2	Oberkulmer	*Xpsr1194–Xpsr918b*	16.6	Keller et al. (1999)
	QPm.sfr-5A.3	Oberkulmer	*Xpsr911–Xpsr120a*	10.5	Keller et al. (1999)
	QPm.nuls-5A	Saar	*Xgwm617b–Xwmc327*	4.2–15.2	Lillemo et al. (2008)
	QPm.sfr-5B	Oberkulmer	*Xpsr580b–Xpsr143*	12.6	Keller et al. (1999)
	QPm.inra-5B.2	Courtot	*Xgwm790b*	11.1	Bougot et al. (2006)
	QPm.crag-5D.1	RE714	*Xgwm639a–Xgwm174*	30.2–38.9 b	Mingeot et al. (2002)
	QPm.crag-5D.2	RE714	*Xcfd8B9–Xcfd4A6*	24.0–37.8	Mingeot et al. (2002)
	QpmVpn.inra-5D	Courtot	*Xcfd8*	11	Bougot et al. (2006)
	QPm.inra-5D.1	RE714	*Xcfd26*	28.1–37.7	Chantret et al. (2001)
	QPm.inra-5D.2	RE714	*XgbxG083c*	37.7	Chantret et al. (2001)

(continued)

Table 7.16 (continued)

Trait	QTL	Donor	Flanking marker	PVE (%)	Reference
	QPm.inra-6A	RE714	*MIRE(Xgwm427)*	8.8–13.4	Chantret et al. (2001)
	QPm.crag-6A	RE714	*MIRE*	19.8–53.9 d	Mingeot et al. (2002)
	QPm.caas-6BS	Bainong 64	*Xbarc79–Xgwm518*	10.3–16.0	Lan et al. (2009)
	QPm.sfr-7B.1	Forno	*Xpsr593c–Xpsr129c*	11.3	Keller et al. (1999)
	QPm.sfr-7B.2	Forno	*Xglk750–Xmwg710a*	31.8	Keller et al. (1999)
	QPm.crag-7B	RE714	*XpdaC01–XgbxR035b*	22.8–33.5	Mingeot et al. (2002)
	QPm.caas-7DS	Fukuho-komugi	*Ltn-Xgwm295.1*	12	Liang et al. (2006)
	Yr18/Lr34/Pm38	Saar	*Xgwm1220–Xswm10*	19.0–56.5	Lillemo et al. (2008)
	QPm.inra-7D.1	Courtot	*Xgpw1106*	10.6	Bougot et al. (2006)

YR yellow rust; *PM* powdery mildew

genes (Wu 2011). There were 72 QTLs for APR to stripe rust mapped on each chromosome, except chromosomes 1A, 1D, 3D, and 7A (Lan 2010).

There were 68 brown leaf rust resistance genes in wheat which have been formally named, and the majority of which were major genes for specific physiological race (Liu et al. 2013). A total of 39 molecular markers linking or co-separation with brown leaf rust resistance genes have been reported (Sun et al. 2011).

In recent years, sharp eyespot in wheat was very serious in China especially in Huanghuai Winter Wheat Region, and turned into one of the main diseases of wheat in China. However, few researches related to genetic of wheat were conducted. Zhang et al. (2005) mapped QTL for sharp eyespot resistance using two RIL populations (RIL-8 and RIL-SES) and identified one SSR marker (Xgwm526) associating with the sharp eyespot resistance. And the genetic distance between the marker and sharp eyespot resistance gene *Ses1* was 27.9 cM.

Wheat FHB mainly occurs in warm, humid, and semi-humid regions. In China, FHB is very serious in Middle and Yellow Yangtze Valley Winter Wheat Region and Northeastern Spring Wheat Region. In recent years, with the global warming, improvement of irrigation, and change of farming system, there is a trend of further spread in Huanghuai Winter Wheat Region. So far, QTLs conferring FHB resistance in wheat were mapped on all chromosomes except chromosome 7D (Buerstmayr et al. 2009) (Table 7.16). Furthermore, the QTLs distributed on chromosomes 3B, 5A, and 6B were identified in Sumai 3, Wangshuibai, Wuhan, Nyubai, Frontana, CM82036 and DH181, and which were the major QTLs for FHB resistance in wheat. However, until now no wheat variety has been found to completely immune to FHB.

7.4.3.2 Comparison of the Results with One of the Previous Studies

In this study, two additive QTLs and two pairs of epistatic QTLs for APR to powdery mildew were identified, among which *qApr4D* distributed on chromosome 4D had the highest contribution (20 %) and was the major QTL. And these QTLs were mainly mapped on chromosomes 4D, 5D, 1B, and 3A. QTL for APR to powdery mildew were also detected in these chromosomes in the previous studies, and majority of these QTLs were major QTL.

A total of two additive QTLs for FHB in wheat were detected on chromosomes 4B and 5A, respectively. Among them, *Qfhi.nau-4B* with the contribution of 10.31 % was mapped on chromosome 4B, and *Qfhi.nau-5A* with the contribution of 11.51 % was mapped on chromosome 5A. Furthermore, both the two loci were involved in QE interaction. Now, QTL analysis of FHB resistance was mainly focused on its resistance to spread and initial infection, furthermore, relative QTL were mainly mapped on chromosomes 3B, 6B and 5A. Besides *Qfhi*.nau-5A (mapped on chromosome 5A), *Qfhi.nau-4B* was also mapped on chromosome 4A in this study, explaining 16.12, 16.43, and 17.54 % of phenotypic variation, respectively, which were consistent with the results of Tian et al. (2008).

7.5 QTL Mapping and Effect Analysis of Salt Resistance

Saline soils as one of the most important abiotic stresses plays an important role in the yield reduction of crop plants worldwide. With the rapid growing of population, food security has become an important issue. In addition to taking some saline soil improvement measures (such as irrigation pressure alkali), studying the plant salt resistance and developing wheat salt-tolerance varieties are the fundamental approaches to the effective use of soil salinization. Wheat salt tolerance is a quantitative trait controlled by multiple genes, and the molecular mechanisms and genetic mechanisms are more complex. Selection of salt-tolerance index is an important issue to salt identification. Previous studies are on the morphological index, physiological, and biochemical indicators of wheat salt tolerance. Over the last few years, with the rapid development of biotechnology, the genetic basis of salt tolerance in wheat, especially gene/QTL localization, has become a hot spot. Ren et al. (2012) reported that a hydroponic culture was carried out to evaluate wheat seedling traits (RDW, MRL, SDW, TDW) under control (CK) and salt stress (ST) conditions by using a recombinant inbred line (RIL) population derived from two Chinese wheat varieties, Xiaoyan 54 and Jing 411; the relative values of each trait (ratio of ST/CK) were also calculated, and a total of 25 QTLs for 4 traits were detected, which were distributed on chromosomes 1A, 2A, 2D, 3A, 4A, 4B, 5B, 5D, 6B, 7A, and 7B, and explained phenotypic variation ranging from 4.4 to 25.5 %. The gene *Knal* control of K^+/Na^+ discrimination locus was mapped on a short region in the 4DL arm with RFLP marker (Dubcovsky et al. 1996). Munnus et al. (2002) found QTLs for Na^+ absorption traits on chromosome 2A using F_2 population. The QTLs for proline accumulation were located on chromosomes 5A and 5D. Liu et al. (2001a, b, c) detected salt tolerance major gene close linkage with marker *WMS67* and *WMSZ13* using SSR marker and mapped on chromosome 5BL. A wheat-specific SSR marker *Xgwm304* was decided to link with salt-tolerant locus of wheat variety Shanrong No. 3 was located on 5AS of wheat chromosome (Shan et al. 2006).

The above-mentioned studies performed QTL analyses for different salt-tolerant traits in wheat using different wheat population. However, the simultaneous analysis on QTL of additive and epistatic effect using DH population has not been studied. In this study, we investigated seedling and root system traits in wheat with a set of DH population derived from anther culture of HP3/YM57. The purposes of this study were to detect QTLs with additive effect and epistatic effect for salt-tolerant traits and obtain closely linked molecular markers that could be used for marker-assisted selection in wheat breeding programs.

7.5.1 QTL Mapping of Salt-Resistance Traits

7.5.1.1 Materials

Materials and Planting were same as one of 7.2.2.1.1 in this chapter.

7.5.1.2 Experimental Design

7.5.1.2.1 Seedling Cultivation

A total of 100 high-quality seeds of the 168 DH strains were soaked in 3 % of H_2O_2 for 20 min, rinsed 2–3 times with tap water, and then placed in glass plates with two pieces of moisturized filter paper at 20 ± 2 °C for germination. The germinated seedlings were cultivated in deionized water until the hypocotyl extent by 1 cm. Healthy seedlings, at the same developmental stage, from each strain were selected and planted in seed trays, each of which contains 84 (5 cm × 5 cm) wells. Experiment consisted of three treatments with different NaCl concentrations in the culture solution: 0 mmol/L (normal), 50 mmol/L, and 100 mmol/L. The exterior of each cultivation basin was made in black in order to create a dark growth environmental condition for ideal root development.

7.5.1.2.2 Trait Measurements

After 25 days, six healthy seedlings, at the same developmental stage, from each strain were selected and seedling height, root length, root surface area, dry seedling weight, and dry root weight were measured, respectively. A WinRHIZO Root Analysis System software was used to analyze the following root characters: root length and root surface area.

7.5.1.2.3 Data and QTL Analysis

SPSS17.0 software was used to conduct statistical analyses for the above-described seedling and root characters, whereas ICI mapping software (Wang 2009) based on the ICIM method (Zhang et al. 2008; Wang 2009) was used for QTL analysis. Specifically, the mean value of three repeats of each treatment was considered as an input datum, LOD threshold was set to 2.5, and step value was set as 1 cm. QTL designations were based on (http://www.graingenes.org), e.g., "q + trait acronym + research institute acronym + chromosome No." Multiple QTLs located on the same chromosome and related to the same trait were described using an additional letter a, b… after the chromosome number.

7.5.1.3 Results and Analysis

7.5.1.3.1 Analysis of Phenotypic Data

The parents of the DH population, Huapei 3 and Yumai 57, were markedly different from one another with respect to most investigated traits under different salt stress conditions. The DH population showed a wide range of variation for most traits (Table 7.17; Fig. 7.9). The difference between normal condition and 50 mmol/L condition for seedling height and dry seedling weight did not reach the significance level, whereas the difference with 100 mmol/L condition reached significance level and the difference of three different conditions for root length, root surface, and dry root weight all reached the significance level. Five traits of the DH population segregated continuously and followed a normal distribution, and both absolute values of skewness and kurtosis were less than 1.0, indicating its polygenic inheritance and suitability of the data for QTL analysis.

7.5.1.3.2 QTL Analyses

At three different conditions, a total of thirty-eight QTLs and ten pairs of QTLs with epistatic effect were detected for seedling height, root length, root surface area, dry seedling weight, and dry root weight and mapped to chromosomes 1A, 2A, 2B, 2D, 4B, 4D, 5D, and 6D (Tables 7.18, 7.19; Fig. 7.10). These QTLs accounted for 3.38–22.96 % of the phenotypic variation.

7.5.1.3.2.1 QTLs for Seedling Height

At three different conditions, a total of eight QTLs were detected for seedling height (Table 7.18; Fig. 7.10). Under normal condition, three QTLs were found and mapped to chromosomes 4D, 6B, and 7B, explaining the variances of 17.9, 5.22, and 5.43 %, respectively. Under 50 mmol/L NaCl conditions, two QTLs were found and mapped to chromosomes 4D and 7B, explaining the variances of 3.38 and 22.96 %, respectively. Under 100 mmol/L NaCl conditions, three QTLs were found and mapped to chromosomes 2A, 2D, and 3B, explaining the variances of 6.39, 5.37, and 4.49 %, respectively. *QSH4D* and *QSH7B* were detected under normal and 50 mmol/L NaCl conditions and originated from alleles of Yumai 57 increase 0.58–1.47 cm of seedling height, which explained 40.86 and 8.81 % of the phenotypic variances, respectively. The above two QTLs were detected simultaneously at two conditions indicating that the additive effect plays an important role in the genetic height wheat.

Ten pairs of epistatic QTL on chromosomes 6A, 7D, 1D, and 2D were also identified for seedling height (Table 7.19). They explained 7.52–16.26 % of phenotypic variances. Three pairs of epistatic QTL were identified under normal

Table 7.17 Phenotypic performance of seedling height, root length, root surface area, dry seedling weight, and dry root weight in the DH population

Trait	Treatment	Parent		DH population			
		Huapei 3	Yumai 57	Mean	Range	Skewness	Kurtosis
SH (cm)	0 mmol/L NaCl	19.76	17.36	18.07A	11.30–26.87	0.162	−0.303
	50 mmol/L NaCl	20.63	19.80	18.18A	11.47–28.20	0.154	−0.129
	100 mmol/L NaCl	7.30	14.4	13.45B	7.30–19.37	−0.569	0.755
RL (cm)	0 mmol/L NaCl	251.48	326.16	275.86A	101.03–561.47	0.257	0.524
	50 mmol/L NaCl	224.14	139.91	146.30B	46.02–323.82	0.653	0.582
	100 mmol/L NaCl	16.88	61.75	64.34C	16.88–139.20	0.439	0.192
RSA (cm^2)	0 mmol/L NaCl	26.58	26.87	25.77A	9.56-45.18	-0.038	0.077
	50 mmol/L NaCl	22.83	14.44	15.70B	6.19–38.03	0.012	0.408
	100 mmol/L NaCl	2.43	8.32	8.36C	2.43-17.14	0.454	0.408
DSW (mg)	0 mmol/L NaCl	0.0230	0.0200	0.0237A	0.0133–0.0300	−0.081	−0.237
	50 mmol/L NaCl	0.0300	0.0267	0.0235A	0.0133–0.0333	−0.093	−0.308
	100 mmol/L NaCl	0.0040	0.0167	0.0212B	0.0040–0.0300	−0.422	0.027
DRW (mg)	0 mmol/L NaCl	0.0077	0.0087	0.0091A	0.0034–0.0151	0.144	0.078
	50 mmol/L NaCl	0.0103	0.0082	0.0072B	0.0037–0.0104	−0.103	−0.439
	100 mmol/L NaCl	0.0027	0.0040	0.0048C	0.0020–0.0098	0.579	0.333

SH seedling height; *RL* root length; *RSA* root surface area; *DSW* dry seedling weight; and *DRW* dry root weight. Capital letters mean significant at 1 % level

condition, which had positive effect. They explained totally 37.21 % of phenotypic variances. Three pairs of epistatic QTL were identified under 50 mmol/L NaCl condition. Among them, two pairs of epistatic QTLs had positive effect and one pair of epistatic QTL had negative effect, and they explained jointly 25.68 % of phenotypic variances. Four pairs of epistatic QTLs were identified under 100 mmol/L NaCl condition. Among them, three pairs of epistatic QTLs had positive effect and one pair of epistatic QTL had negative effect, and they explained totally 42.85 % of phenotypic variances. Results suggested that both additive effects and epistatic effects controlled wheat seedling height in different concentrations of salt stress.

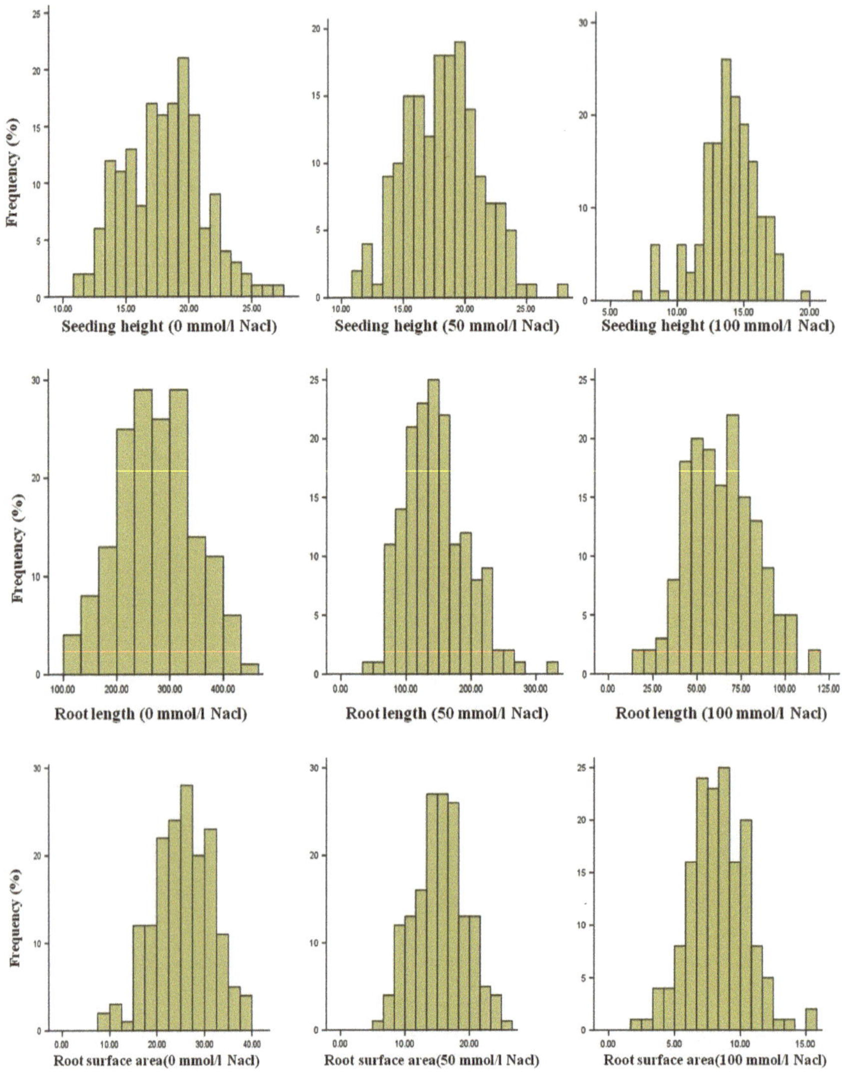

Fig. 7.9 Distribution of seedling height, root length, root surface area, dry seedling weight, and dry root weight in the DH Population

7.5.1.3.2.2 QTLs for Root Length

At three different conditions, a total of eight QTLs were detected for root length (Table 7.18; Fig. 7.10). Under normal condition, three QTLs were found and mapped to chromosomes 2D, 5B, and 5B2, explaining the variances of 5.66, 10.36, and 5.22 %, respectively. Under 50 mmol/L NaCl conditions, three QTLs were found and mapped to chromosomes 2A and 2B, explaining the variances of 5.40,

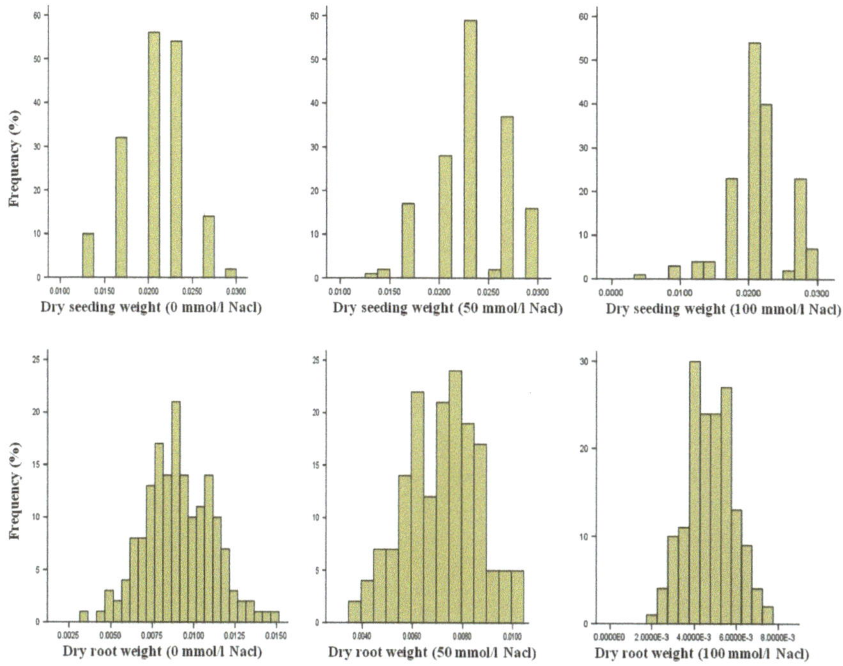

Fig. 7.9 (continued)

6.13, and 8.49 %, respectively. Under 100 mmol/L NaCl conditions, two QTLs were found and mapped to chromosomes 4A and 5B2, explaining the variances of 9.37 and 4.03 %, respectively.

Nine pairs of epistatic QTLs on chromosomes 6A, 4A, and 3D were also identified for seedling height (Table 7.19). They explained 12.29–20.24 % of phenotypic variances. Three pairs of epistatic QTLs were identified under normal condition. Among them, two pairs of epistatic QTLs had positive effect and one pair of epistatic QTL had negative effect, and they explained totally 47.64 % of phenotypic variances. Three pairs of epistatic QTL were identified under 50 mmol/L condition. Among them, two pairs of epistatic QTL had positive effect and one pair of epistatic QTL had negative effect, and they explained totally 49.65 % of phenotypic variances. Three pairs of epistatic QTLs were identified under 100 mmol/L NaCl condition. Among them, two pairs of epistatic QTLs had positive effect and one pair of epistatic QTL had negative effect, and they explained totally 48.87 % of phenotypic variances. Results suggested that both additive effects and epistatic effects controlled wheat root length in different concentrations of salt stress.

Table 7.18 Positions, effects, and contribution rates of additive QTLs for seedling height (SH), root length (RL), root surface area (RSA), dry seedling weight (DSW), and dry root weight (DRW)

Trait	QTL	Site (cM)	Flanking marker	A	LOD	H^2 (%)
SH[a]	*QSH4D*	0	*Xbarc334–Xwmc331*	−1.32	8.04	17.9
	QSH6B	0	*Xcfa2187–Xgwm219*	0.74	2.52	5.22
	QSH7B	3	*Xswes625–Xbarc72*	−0.74	2.28	5.43
SH[b]	*QSH4D*	0	*Xbarc334–Xwmc331*	−1.47	9.51	22.96
	QSH7B	0	*Xswes625–Xbarc72*	−0.58	1.64	3.38
SH[c]	*QSH2A*	43	*Xcfe67–Xwmc177*	0.54	2.26	6.39
	QSH2D	102	*Xbarc349.1–Xcfd161*	−0.51	1.51	5.37
	QSH3B	47	*Xbarc268–Xwmc1*	−0.48	1.57	4.49
RL[a]	*QRL2D*	54	*Xwmc18–Xwmc170.2*	−17.94	1.94	5.66
	QRL5B	58	*Xgwm213–Xswes861.2*	25.31	3.94	10.36
	QRL5B2	25	*Xbarc232–Xwmc235*	17.23	1.98	5.22
RL[b]	*QRL2A*	86	*Xwmc455–Xgwm515*	11.72	1.57	6.13
	QRL2B	5	*Xcwem5–Xwmc661*	−13.80	3.07	8.49
	QRL2B	68	*Xbarc373–Xbarc1114*	11.01	2.13	5.40
RL[c]	*QRL4A*	2	*Xwmc718–Xwmc262*	−6.38	2.95	9.37
	QRL5B2	22	*Xgdm116–Xbarc232*	−4.14	1.55	4.03
RSA[a]	*QRSA2D*	69	*Xgwm539–Xcfd168*	−1.41	1.87	4.88
	QRSA5B	58	*Xgwm213–Xswes861.2*	2.08	3.78	9.97
RSA[b]	*QRSA1A*	12	*Xwmc728.2–Xwmc728.1*	1.19	2.25	5.99
	QRSA2B	5	*Xcwem5–Xwmc661*	−1.25	2.72	7.48
RSA[c]	*QRSA2A*	17	*Xgwm636–Xcfe67*	0.53	1.62	5.21
	QRSA4A	2	*Xwmc718–Xwmc262*	−0.56	1.77	5.92
	QRSA6B	1	*Xcfa2187–Xgwm219*	−0.55	1.70	5.22
DSW[a]	*QDSW1A*	71	*Xgwm498–Xcwem6.2*	0.01	2.21	7.78
	QDSW2D	95	*Xbarc349.1–Xcfd161*	−0.01	1.89	8.54
	QDSW5B	58	*Xgwm213–Xswes861.2*	0.01	1.98	5.34
DSW[b]	*QDSW2B*	17	*Xwmc764–Xbarc200*	−0.01	1.93	7.37
	QDSW2D	110	*Xcfd161–Xgwm311.2*	−0.01	2.73	8.03
	QDSW4D	3	*Xbarc334–Xwmc331*	−0.01	2.75	9.10
DSW[c]	*QDSW1A*	61	*Xbarc350–Xwmc120*	0.01	2.84	7.87
	QDSW1B	130	*Xswes649–Xswes98*	−0.01	1.77	4.53
DRW[a]	*QDRW2D*	67	*Xwmc170.2–Xgwm539*	−0.10	1.89	5.16
	QDRW5D	58	*Xgwm213–Xswes861.2*	0.10	2.34	6.29
	QDRW6D	0	*Xwmc412.1–Xcfd49*	−0.10	2.24	5.95
DRW[b]	*QDRW2A*	52	*Xwmc177–Xgwm558*	0.01	2.40	8.35
	QDRW4B	18	*Xwmc657–Xwmc48*	−0.01	4.01	10.94
DRW[c]	*QDRW2A*	29	*Xgwm636–Xcfe67*	0.01	1.78	7.66
	QDRW6B	49	*Xcfa2257–Xcfd48*	0.02	2.15	12.07

SH seedling height; *RL* root length; *RSA* root surface area; *DSW* dry seedling weight; *DRW* dry root weight; *A* additive effect, the positive value indicates that the corresponding QTL is from Huapei 3, and the negative value indicates that the corresponding QTL is from Yumai 57

[a]0 mmol/L NaCl treatment

[b]50 mmol/L NaCl treatment

[c]100 mmol/L NaCl treatment

Table 7.19 Epistatic effects of QTLs for seedling height (SH), root length (RL), root surface area (RSA), dry stem weight (DSW), and dry root weight (DRW)

Trait	QTL	Flanking marker	Site (cM)	QTL	Flanking marker	Site (cM)	AA	LOD	H^2 (%)
SH[a]	*QSH1A*	*Xcwem32.1–Xwmc728.2*	10	*QSH6A*	*Xbarc023–Xbarc1077*	40	0.88	3.51	7.52
	QSH1B	*Xwmc766–Xswes158*	85	*QSH7D*	*Xwmc14–Xwmc42*	190	1.29	4.52	13.43
SH[b]	*QSH1B*	*Xbarc061–Xwmc766*	80	*QSH7D*	*Xwmc14–Xwmc42*	190	1.42	5.13	16.26
	QSH1D	*Xwmc429–Xcfd19*	40	*QSH6A*	*Xwmc553–Xgwm732*	65	1.01	3.28	9.87
	QSH5D	*Xbarc1097–Xcfd8*	10	*QSH7D*	*Xwmc14–Xwmc42*	185	0.95	3.73	7.62
	QSH6A	*Xbarc023–Xbarc1077*	40	*QSH7A*	*Xgwm60–Xbarc070*	15	−0.87	3.45	8.19
SH[c]	*QSH1D*	*Xcfa2158–Xwmc222*	5	*QSH2D*	*Xcfd161–Xgwm311.2*	110	−0.69	3.91	10.55
	QSH2D	*Xbarc349.2–Xbarc349.1*	80	*QSH3B*	*Xgwm285–Xgwm685*	50	−0.85	3.54	12.20
	QSH3D	*Xgdm72–Xbarc1119*	10	*QSH7D*	*Xgwm295–Xgwm676*	80	0.67	3.33	8.85
	QSH5D	*Xbarc1097–Xcfd8*	35	*QSH6A*	*Xcfe179.1–Xswes170.2*	90	−0.73	3.52	11.25
RL[a]	*QCL5D*	*Xcfd53–Xwmc18*	25	*QCL5D*	*Xwmc331–Xgwm194*	30	−31.76	3.03	18.56
	QCL5D	*Xwmc527–Xwmc264*	120	*QCL5D*	*Xswes100–Xbarc244*	15	33.31	3.42	16.59
	QCL5D	*Xbarc1125–Xgwm213*	5	*QCL5D*	*Xwmc553–Xgwm732*	70	−26.45	3.85	12.49
RL[b]	*QCL2B*	*Xbarc1042–Xgwm388*	75	*QCL4A*	*Xbarc343–Xwmc313*	15	−17.17	4.04	14.34
	QCL2B	*Xbarc101–Xcwem55*	80	*QCL4D*	*Xgwm194–Xcfa2173*	75	21.37	3.31	20.07
	QCL5D	*Xwmc215–Xbarc345*	75	*QCL6D*	*Xubc808–Xswes679.1*	125	−26.18	3.96	15.24
RL[c]	*QCL1B*	*Xswes649–Xswes98*	130	*QCL7D*	*Xgwm295–Xgwm676*	95	8.24	3.42	14.65
	QCL3A	*Xcfa2170–Xbarc51*	170	*QCL3D*	*Xcfd223–Xbarc323*	70	−7.15	3.20	13.98
	QCL4A	*Xbarc343–Xwmc313*	10	*QCL5D*	*Xbarc1097–Xcfd8*	30	−1.33	5.16	20.24
RSA[a]	*QRS1A*	*Xgwm259–Xcwem32.1*	0	*QRS7D*	*Xwmc14–Xwmc42*	205	−2.67	4.57	11.51
	QRSA3A	*Xwmc527–Xwmc264*	125	*QRSA7D*	*Xswes100–Xbarc244*	35	2.98	6.93	19.30
	QRSA3B	*Xgwm566–Xcfe009*	70	*QRSA5B2*	*Xbarc36–Xbarc140*	0	1.79	3.73	7.84

(continued)

Table 7.19 (continued)

Trait	QTL	Flanking marker	Site (cM)	QTL	Flanking marker	Site (cM)	AA	LOD	H^2 (%)
RSA[b]	*QRSA5A*	*Xswes45–Xbarc180*	15	*QRSA6D*	*Xubc808–Xswes679.1*	120	5.03	3.73	20.71
	QRSA5B	*Xgwm213–Xswes861.2*	30	*QRSA6D*	*Xubc808–Xswes679.1*	120	5.46	4.81	21.83
	QRSA6D	*Xubc808–Xswes679.1*	120	*QRSA7D*	*Xswes100–Xbarc244*	25	−5.08	3.77	21.27
RSA[c]	*QRSA1B*	*Xgwm218–Xgwm582*	40	*QRSA5A*	*Xbarc358.2–Xgwm186*	50	0.93	3.80	15.46
	QRSA4A	*Xbarc343–Xwmc313*	10	*QRSA5D*	*Xbarc1097–Xcfd8*	30	−1.05	5.07	20.19
	QRSA5D	*Xwmc630.2–Xcfd40*	0	*QRSA7D*	*Xswes100–Xbarc244*	10	−1.03	3.86	19.13
DSW[a]	*QDSW1B*	*Xbarc061–Xwmc766*	75	*QDSW7B*	*Xswes625–Xbarc72*	5	−0.01	3.70	13.90
	QDSW3A	*Xwmc527–Xwmc264*	125	*QDSW7D*	*Xbarc244–Xbarc352*	50	0.01	4.13	12.01
DSW[b]	*QDSW1A*	*Xgwm259–Xcwem32.1*	5	*QDSW5A*	*Xcfe223–Xwmc273.3*	95	0.01	3.49	10.54
	QDSW2B	*Xgwm111–Xgdm14-6D*	100	*QDSW6D*	*Xswes679.1–Xcfa2129*	130	−0.01	3.22	11.59
DSW[c]	*QDSW1A*	*Xwmc728.1–Xwmc550*	25	*QDSW3B*	*Xgwm285–Xgwm685*	50	0.01	3.04	13.12
	QDSW2D	*Xgwm311.1–Xwmc658.1*	160	*QDSW5D*	*Xbarc320–Xwmc215*	65	0.01	3.54	12.51
	QDSW5D	*Xbarc1097–Xcfd8*	30	*QDSW6A*	*Xcfe179.1–Xswes170.2*	90	−0.01	3.84	14.10
DRW[a]	*QDRW2A*	*Xwmc177–Xgwm558*	55	*QDRW5A*	*Xgwm186–Xcfe223*	80	0.01	3.34	19.42
	QDRW3A	*Xbarc157.1–Xbarc1177*	195	*QDRW5A*	*Xswes45–Xbarc180*	20	0.01	5.30	19.83
	QDRW3B	*Xgwm566–Xcfe009*	90	*QDRW6D*	*Xcfa2129–Xbarc080*	175	−0.01	3.69	13.95
DRW[b]	*QDR2A*	*Xwmc455–Xgwm515*	85	*QDRW5D*	*Xcfd40–Xbarc1097*	5	−0.01	3.84	12.27
	QDRW5A	*Xgwm186–Xcfe223*	85	*QDRW6D*	*Xwmc412.1–Xcfd49*	0	0.01	4.44	13.51
DRW[c]	*QDRW4A*	*Xbarc343–Xwmc313*	10	*QDRW5D*	*Xbarc109—Xcfd8*	25	−0.01	3.25	13.31

SH seedling height; *RL* root length; *RSA* root surface area; *DSW* dry seedling weight; *DRW* dry root weight; *AA* epistatic effect, the positive AA value indicates that effect of the parent-type is greater than that of the recombinant-type, and the negative AA value indicates that effect of the parent-type is less than that of the recombinant-type

[a]0 mmol/L NaCl treatment

[b]50 mmol/L NaCl treatment

[c]100 mmol/L NaCl treatment

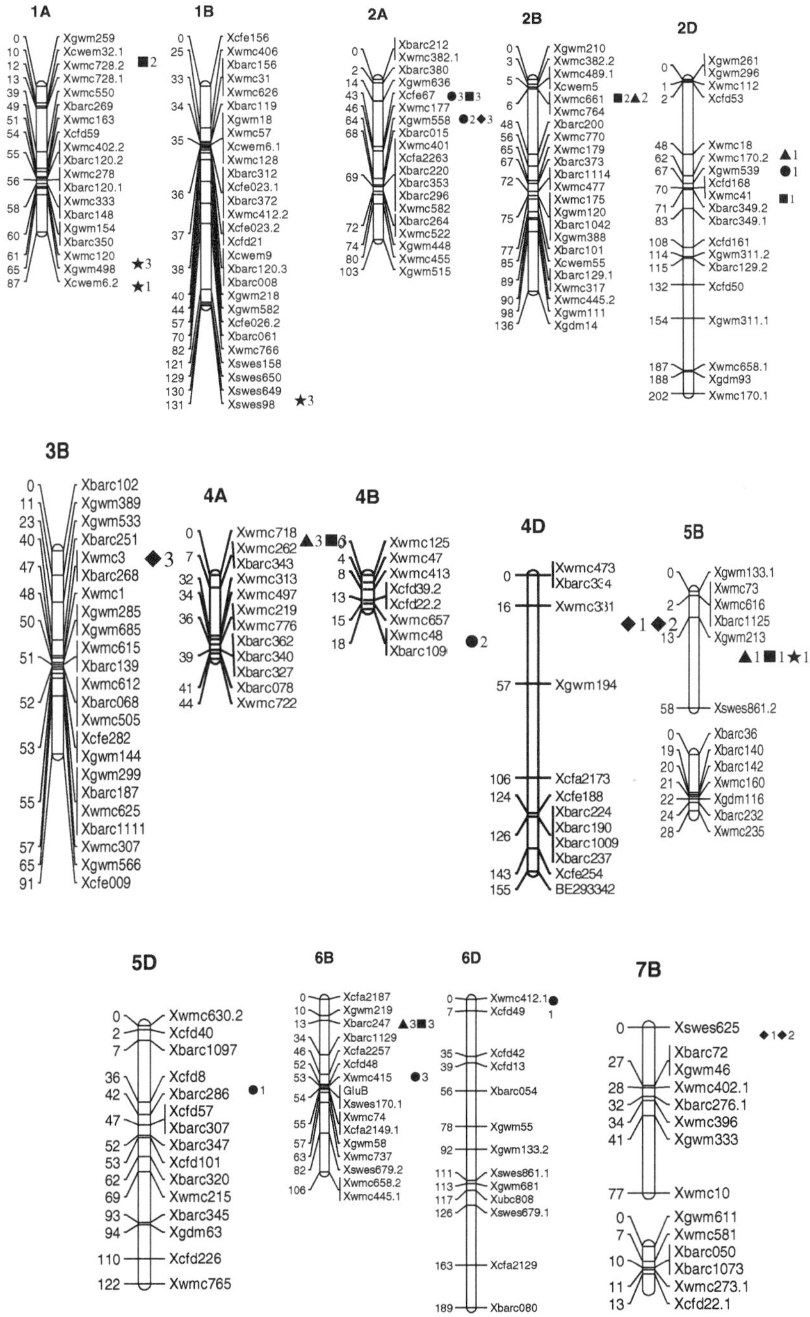
1A
Xgwm259
Xcwem32.1
Xwmc728.2
Xwmc728.1
Xwmc550
Xbarc269
Xwmc163
Xcfd59
Xwmc402.2
Xbarc120.2
Xwmc278
Xbarc120.1
Xwmc333
Xbarc148
Xgwm154
Xbarc350
Xwmc120
Xgwm498
Xcwem6.2
1B
Xcfe156
Xwmc406
Xbarc156
Xwmc31
Xwmc626
Xbarc119
Xgwm18
Xwmc57
Xcwem6.1
Xwmc128
Xbarc312
Xcfe023.1
Xbarc372
Xwmc412.2
Xcfe023.2
Xcfd21
Xcwem9
Xbarc120.3
Xbarc008
Xgwm218
Xgwm582
Xcfe026.2
Xbarc061
Xwmc766
Xswes158
Xswes650
Xswes649
Xswes98
2A
Xbarc212
Xwmc382.1
Xbarc380
Xgwm636
Xcfe67
Xwmc177
Xgwm558
Xbarc015
Xwmc401
Xcfa2263
Xbarc220
Xbarc353
Xbarc296
Xwmc582
Xbarc264
Xwmc522
Xgwm448
Xwmc455
Xgwm515
2B
Xgwm210
Xwmc382.2
Xwmc489.1
Xcwem5
Xwmc661
Xwmc764
Xbarc200
Xwmc770
Xwmc179
Xbarc373
Xbarc1114
Xwmc477
Xwmc175
Xgwm120
Xbarc1042
Xgwm388
Xbarc101
Xcwem55
Xbarc129.1
Xwmc317
Xwmc445.2
Xgwm111
Xgdm14
2D
Xgwm261
Xgwm296
Xwmc112
Xcfd53
Xwmc18
Xwmc170.2
Xgwm539
Xcfd168
Xwmc41
Xbarc349.2
Xbarc349.1
Xcfd161
Xgwm311.2
Xbarc129.2
Xcfd50
Xgwm311.1
Xwmc658.1
Xgdm93
Xwmc170.1
3B
Xbarc102
Xgwm389
Xgwm533
Xbarc251
Xwmc3
Xbarc268
Xwmc1
Xgwm285
Xgwm685
Xwmc615
Xbarc139
Xwmc612
Xbarc068
Xwmc505
Xcfe282
Xgwm144
Xgwm299
Xbarc187
Xwmc625
Xbarc1111
Xwmc307
Xgwm566
Xcfe009
4A
Xwmc718
Xwmc262
Xbarc343
Xwmc313
Xwmc497
Xwmc219
Xwmc776
Xbarc362
Xbarc340
Xbarc327
Xbarc078
Xwmc722
4B
Xwmc125
Xwmc47
Xwmc413
Xcfd39.2
Xcfd22.2
Xwmc657
Xwmc48
Xbarc109
4D
Xwmc473
Xbarc334
Xwmc331
Xgwm194
Xcfa2173
Xcfe188
Xbarc224
Xbarc190
Xbarc1009
Xbarc237
Xcfe254
BE293342
5B
Xgwm133.1
Xwmc73
Xwmc616
Xbarc1125
Xgwm213
Xswes861.2
Xbarc36
Xbarc140
Xbarc142
Xwmc160
Xgdm116
Xbarc232
Xwmc235
5D
Xwmc630.2
Xcfd40
Xbarc1097
Xcfd8
Xbarc286
Xcfd57
Xbarc307
Xbarc347
Xcfd101
Xbarc320
Xwmc215
Xbarc345
Xgdm63
Xcfd226
Xwmc765
6B
Xcfa2187
Xgwm219
Xbarc247
Xbarc1129
Xcfa2257
Xcfd48
Xwmc415
GluB
Xswes170.1
Xwmc74
Xcfa2149.1
Xgwm58
Xwmc737
Xswes679.2
Xwmc658.2
Xwmc445.1
6D
Xwmc412.1
Xcfd49
Xcfd42
Xcfd13
Xbarc054
Xgwm55
Xgwm133.2
Xswes861.1
Xgwm681
Xubc808
Xswes679.1
Xcfa2129
Xbarc080
7B
Xswes625
Xbarc72
Xgwm46
Xwmc402.1
Xbarc276.1
Xwmc396
Xgwm333
Xwmc10
Xgwm611
Xwmc581
Xbarc050
Xbarc1073
Xwmc273.1
Xcfd22.1

◀ **Fig. 7.10** Positions of QTLs associated with seedling height, root length, root surface area, fresh root weight, dry stem weight, and dry root weight. ◆1, ◆2, and ◆3 indicate the QTL for seedling height under 0, 50, and 100 mmol L^{-1} NaCl stress, respectively; ▲1, ▲2, and ▲3 indicate the QTL for root length under 0, 50, and 100 mmol L^{-1} NaCl stress, respectively; ■1, ■2, and ■3 indicate the QTL for root surface area under 0, 50, and 100 mmol L^{-1} NaCl stress, respectively; ★1, ★2, and ★3 indicate the QTL for dry seedling weight under 0, 50, and 100 mmol L^{-1} NaCl stress, respectively; ●1, ●2, and ●3 indicate the QTL for dry root weight under 0, 50, and 100 mmol L^{-1} NaCl stress, respectively

7.5.1.3.2.3 QTLs for Root Surface Area

At three different conditions, a total of seven QTLs were detected for root surface area (Table 7.18; Fig. 7.10). Under normal condition, two QTLs were found and mapped to chromosomes 2D and 5B, explaining the variances of 4.88 and 9.97 %, respectively. Under 50 mmol/L NaCl conditions, two QTLs were found and mapped to chromosomes 1A and 2B, explaining the variances of 5.99 and 7.48 %, respectively. Under 100 mmol/L NaCl conditions, three QTLs were found and mapped to chromosomes 2A, 4A, and 6B, explaining the variances of 5.21, 5.92, and 5.22 %, respectively.

Nine pairs of epistatic QTLs were also identified for root surface (Table 7.19). They explained 7.84–21.83 % of phenotypic variances. Three pairs of epistatic QTLs were identified under normal condition. Among them, two pairs of epistatic QTLs had positive effect and one pair of epistatic QTL had negative effect, and they explained totally 38.65 % of phenotypic variances. Three pairs of epistatic QTLs were identified under 50 mmol/L condition. Among them, two pairs of epistatic QTLs had positive effect and one pair of epistatic QTL had negative effect, and they explained totally 63.81 % of phenotypic variances. Three pairs of epistatic QTLs were identified under 100 mmol/L NaCl condition. Among them, two pairs of epistatic QTL had positive effect and one pair of epistatic QTL had negative effect, and they explained totally 54.78 % of phenotypic variances. Results suggested that both additive effects and epistatic effects controlled root surface area in different concentrations of salt stress.

7.5.1.3.2.4 QTLs for Dry Seedling Weight

At three different conditions, a total of eight additive QTLs were detected for dry seedling weight (Table 7.18; Fig. 7.10). Under normal condition, three QTLs were found and mapped to chromosomes 1A, 2D, and 5B, explaining the variances of 7.78, 8.54, and 5.43 %, respectively. Under 50 mmol/L NaCl conditions, three QTLs were found and mapped to chromosomes 2B, 2D, and 4D, explaining the variances of 7.37, 8.03, and 9.10 %, respectively. Under 100 mmol/L NaCl conditions, two QTLs were found and mapped to chromosomes 1A and 1B, explaining the variances of 7.78 and 4.53 %, respectively.

Seven pairs of epistatic QTLs were also identified for dry seedling weight (Table 7.19). They explained 10.54–14.10 % of phenotypic variances. Two pairs of epistatic QTL were identified under normal condition. Among them, one pair of epistatic QTL had positive effect and one pair of epistatic QTL had negative effect, and they explained totally 25.91 % of phenotypic variances. Two pairs of epistatic QTLs were identified under 50 mmol/L condition. Among them, two pairs of epistatic QTLs had positive effect and one pair of epistatic QTL had negative effect, and they explained totally 22.13 % of phenotypic variances. Three pairs of epistatic QTLs were identified under 100 mmol/L NaCl condition. Among them, two pairs of epistatic QTL had positive effect and one pair of epistatic QTL had negative effect, and they explained totally 39.73 % of phenotypic variances. Results suggested that both additive effects and epistatic effects controlled wheat dry seedling weight in different concentrations of salt stress.

7.5.1.3.2.5 QTLs for Dry Root Weight

At three different conditions, a total of seven QTLs were detected for dry root weight (Table 7.18; Fig. 7.10). Under normal condition, three additive QTLs were found and mapped to chromosomes 2D, 5D, and 6D, explaining the variances of 5.16, 6.29, and 5.95 %, respectively. Under 50 mmol/L NaCl conditions, two QTLs were found and mapped to chromosomes 2A and 4B, explaining the variances of 8.35 and 10.94 %, respectively. Under 100 mmol/L NaCl conditions, two QTLs were found and mapped to chromosomes 2A and 6B, explaining the variances of 7.65 and 12.07 %, respectively.

Six pairs of epistatic QTLs were also identified for dry root weight (Table 7.19). They explained 12.27–19.83 % of phenotypic variances. Three pairs of epistatic QTL were identified under normal condition. Among them, two pairs of epistatic QTL had positive effect and one pair of epistatic QTL had negative effect, and they explained totally 53.20 % of phenotypic variances. Two pairs of epistatic QTL were identified under 50 mmol/L condition. Among them, one pair of epistatic QTL had positive effect and one pair of epistatic QTL had negative effect, and they explained totally 25.78 % of phenotypic variances. One pair of epistatic QTL was identified under 100 mmol/L NaCl condition, which had negative effect, and they explained 13.31 % of phenotypic variances. Results suggested that both additive effects and epistatic effects controlled wheat dry root weight in different concentrations of salt stress.

7.5.2 Research Progress of Salt-Tolerance Traits QTL Mapping and the Comparative Analysis with this Study

7.5.2.1 Research Progress of Salt-Tolerance Traits QTL Mapping

The difference between normal condition and 50 mmol/L condition for seedling height and dry seedling weight did not reach the significance level, whereas the difference with 100 mmol/L condition reached significance level, and the difference of three different conditions for root length, root surface, and dry root weight all reached the significance level, which indicated that 100 mmol/L NaCl treatment reached salt stress for seedling height and dry seedling weight and two conditions (50 mmol/L NaCl treatment and 100 mmol/L NaCl treatment) reach salt stress to root length, root surface, and dry root weight.

At three different conditions, eight QTLs were detected for seedling height. Among them, five QTLs were detected on chromosomes 4D, 6B, and 7B under non-salt stress condition (normal and 50 mmol/L NaCl treatment); *QSH4D* and *QSH7B* were identified under non-salt stress condition (normal and 50 mmol/L NaCl treatment). *QSH4D* was a major QTL, explaining the variances of 5.43 and 3.38 %, respectively. *QSH7B* explained 5.43 and 3.38 % of its phenotype variance under two conditions, respectively. Three new QTLs were detected on chromosomes 2A, 2D, and 3B, respectively, under salt stress (100 mmol/L NaCl), which may be related to salt-tolerance adaptation.

At three different conditions, eight QTLs were detected for root length. Among them, three QTLs were detected on chromosomes 2D, 5B, and 3B under non-salt stress condition (normal treatment); *QRL5B* was a major QTL, explaining the variance of 10.36 %. Five new QTLs were detected on chromosomes 2A, 2B, 4A, and 5B2 under salt stress (50 and 100 mmol/L NaCl). *QRL5B2* was detected under non-salt stress condition and salt stress condition, explaining the variances of 5.22 and 4.03 %. The other four QTLs were new QTLs for root length under salt stress conditions.

At three different conditions, seven QTLs were detected for root surface. Among them, two QTLs were detected on chromosomes 2D and 5B under non-salt stress condition (normal treatment). Five new QTLs were detected on chromosomes 1A, 2A, 2B, 4A, and 6B under salt stress (50 and 100 mmol/L NaCl).

At three different conditions, eight QTLs were detected for dry seedling weight. Among them, six QTLs were detected on chromosomes 1A, 2B, 2D, 4D, and 7B under non-salt stress condition (normal and 50 mmol/L NaCl treatment); *QDSW2D* was detected under non-salt stress condition and salt stress condition, explaining the variances of 8.54 and 8.03 %. Two QTLs were detected on chromosomes 1A and 1B under salt stress (100 mmol/L NaCl). *QDSW1A* was detected under non-salt stress condition and salt stress condition, explaining the variances of 7.78 and 7.87 %. *QDSW1B* was a new QTLs for dry seedling weight under salt stress conditions.

At three different conditions, seven QTLs were detected for dry root weight. Among them, three QTLs were detected on chromosomes 2D, 5D, and 6D under non-salt stress condition (normal treatment). Five new QTLs were detected on chromosomes 2A, 4A, and 6B under salt stress (50 and 100 mmol/L NaCl). *QDRW2A* was detected under non-salt stress condition and salt stress condition, explaining the variances of 8.35 and 7.66 %.

It was thought that the QTLs detected both in non-salt stress condition and in salt stress condition, such as *QRL5B2* and *QDSW1A*, which were contributing to salt tolerance, whereas the QTLs detected only in salt stress condition may be related to salt adaptive response.

7.5.2.2 Comparison of the Results with One of the Previous Studies

The identified QTLs of salt-tolerance traits in this study agreed with the previous findings. *QRL5B* detected on chromosome 5B was a major QTL (explaining 10.36 % of the phenotypic variance) conferring root length. At the same time, QTLs for root surface and dry root weight were detected in the same position. This finding agrees with the finding of Shi et al. (1999), Liu et al. (2001a, b, c), and Wu et al. (2007). In this study, two QTLs mapped on chromosomes 4D and 7B for seedling height were found under two conditions. Among them, the QTL mapped on chromosome 4D accounted for 17.9 and 22.96 % of the phenotypic variances, which was a major QTL. This may be the first report of QTLs for seedling height in this genetic location.

This study reveals that the QTLs that we mapped are clustered on chromosomes. The QTLs were detected for seedling height, root surface, and dry root weight on chromosome 2A at the interval of *Xgwm636* to *Xgwm558*, and the QTLs were detected for seedling height, root length, root surface, and dry seedling weight on chromosome 2D at the interval of Xwmc18 to Xgwm311.2.

7.6 QTLs Mapping of Potassium-Deficiency Tolerance at the Seedling Stage in Wheat

Potassium (K) is an important mineral nutrient required for plant growth and development. Numerous studies have confirmed that K deficiency has a strong influence on plant development and crop yield (Rengel and Damon 2008; Fageria et al. 2011). In many agricultural soils, K supply is insufficient to sustain rapid crop growth. In both intensive and extensive agricultural systems, K fertilizer application is thus an effective technique for improving crop production (Fageria et al. 2011). On the other hand, the high input and low use efficiency of K fertilizers increase financial costs and cause environmental problems (Lægreid et al. 1999). To reduce the use of K fertilizers, crop cultivars with enhanced K acquisition or utilization efficiency are consequently being developed as a more sustainable solution (Baligar et al. 2001; Pettigrew 2008).

In this study, we used a doubled haploid (DH) wheat population to investigate morphological traits and physiological parameters in roots and shoots subjected to two K supply treatments (normal K and K deficiency) at the seedling stage. The goals of this study are (1) determination of the relationship between investigated morphological traits and physiological parameters under K-deficiency conditions; (2) identification of QTLs with major additive effects controlling plant tolerance to K deprivation; and (3) elucidation of genetic mechanisms associated with cooperative uptake of K, Na, Ca, and Mg in wheat. Our results should be helpful for MAS in wheat K-deficiency tolerance breeding programs.

7.6.1 QTL Mapping of Potassium-Deficiency Tolerance

7.6.1.1 Materials and Methods

7.6.1.1.1 Materials

Materials and Planting were same as one of 7.2.2.1.1 in this chapter.

7.6.1.1.2 Experimental Design

The 168 DH lines and their parents were hydroponically grown in a greenhouse at Hebei Agricultural University, Baoding, China. Two separate experiments were set up in 2011: one determining root activity, in November, and the other determining the remaining traits, in December. The nutrient solution used for culture of DH lines and parents was a modified Hoagland's solution (Hoagland and Arnon 1950). Each experiment consisted of two treatments with different K concentrations in the culture solution: a normal K treatment [Control group (N), 2 mmol L^{-1} K] and a K-deprivation treatment [Low K group (L), 0.01 mmol L^{-1} K]. K_2SO_4 was used as the K source in the treatments. A complete randomized design block was employed, with three replicates per treatment.

Seeds of each DH line and parent (1500 grains each) were sterilized in 10 % of H_2O_2 for 5 min, rinsed three times with distilled water, and immersed in distilled water for 24 h. The uniformly germinated seeds were then transferred onto filter paper in Petri dishes containing distilled water, and cultured in an illuminated growth chamber at 20 °C for 5 days. The 180 uniform and vigorous seedlings [30 (plants) × 2 (treatments) × 3 (replications)] of each genotype were selected and their endosperms were removed. For each replication, 30 plants of each genotype were then transferred into opaque plastic containers whose covers were held in place by rubber bands. The distances between different lines were 3 × 3 cm. Each container contained 25 L of nutrient solution, which was continuously aerated by a mini-pump and maintained at a temperature of 15–30 °C. The solutions were replaced every 5 days, at which time the pH was readjusted to 6.0.

7.6.1.1.3 Trait Measurements

After 4 weeks of growth, seedlings of DH lines and parents were harvested. Roots of nine plants (three replicates) were rinsed with distilled water and divided into primary roots and lateral roots. To measure root parameters such as length (TRL), surface area (TRS), volume (TRV), and diameter (RAD), these roots were scanned to obtain 800-dpi-resolution images, which were further analyzed using WinRHIZO software (Regent Instruments, Canada). After root image acquisition, root and shoot samples were oven-dried at 105 °C for 30 min, and then at 80 °C until completely dry (approximately 48 h). RDW and shoot dry weight (SDW) were measured on an electronic balance. The ratio of root dry weight to shoot dry weight (RSDW) was calculated by dividing RDW by SDW. To measure levels of nutrients K^+, Na^+, Ca^{2+}, and Mg^{2+} in each DH line and parent under the two treatment conditions, roots, and shoots were digested with H_2SO_4 and H_2O_2. Concentrations of these inorganic ions were determined using an Analyst 800 atomic absorption spectrophotometer (PerkinElmer, USA).

For measurement of physiological and biochemical parameters, leaves from each DH line and parent (about 0.5 g) were crushed into fine powder in a mortar with 5.0 mL ice-cold 0.05 mol L^{-1} phosphate buffer (pH 7.8). After centrifugation at 10,000 × *g* for 20 min at 4 °C, the supernatant was used for enzyme [superoxide dismutase (SOD) and peroxidase (POD)] and malondialdehyde (MDA) analysis. SOD activity was assayed according to the method described by Li and Mei (1989) using a UV-240 UV/V spectrophotometer (Shimadzu, Tokyo, Japan). One unit of SOD activity was defined as the amount of enzyme causing 50 % inhibition of NBT reduction at 560 nm. POD activity was analyzed according to the method of Zhang (1992). Each sample was assayed in 2.91-mL phosphate buffer (pH 7.0) containing 0.05 mL of 20 mmol L^{-1} guaiacol, 0.02 mL of 40 mmol L^{-1} H_2O_2, and 0.02 mL supernatant. The increase of absorbance at 470 nm was recorded after addition of 0.02 mL 20 % chloroacetic acid. MDA content was assayed following.

7.6.1.1.4 Data and QTL Analysis

QTLs with additive effects in the DH population were mapped under a mixed linear model using IciMapping v2.2 (Wang et al. 2009).

7.6.1.2 Results and Analysis

7.6.1.2.1 Seedling Growth-Related Phenotypes

Seedling growth-related phenotypes of DH and parental plants subjected to control (N) and K-deprivation (L) treatments are listed in Table 7.20. The parents of the DH population, Huapei 3 and Yumai 57, were markedly different from one another with respect to most investigated traits under either treatment condition. The DH population showed a wide range of variation for most traits, with the coefficient of variation (CV) higher than 15 % for all 20 traits. DH population mean values for all traits were higher under N than under L conditions; in addition, significant differences were observed in RSDW, POD, SOD, MDA, shoot Na^+ concentration (NCS), root Na^+ concentration (NCR), shoot Ca^{2+} concentration (CCS), root Ca^{2+} concentration (CCR), shoot Mg^{2+} concentration (MCS), and root Mg^{2+} concentration (MCR) between treatments (Table 7.20). Frequency distributions of all traits showed continuous variation and significant transgressive segregation in both

Table 7.20 Phenotypic performance for traits related to seedling growth in a doubled haploid (DH) population and its parents under N and L treatments

Trait	Treatment	Parent		DH population				
		Huapei 3	Yumai 57	Mean	Minimum	Maximum	SD	CV (%)
Morphological traits								
RAD (mm)	*N*	0.28 ± 0.02	0.22 ± 0.01*	0.34	0.23	0.55	0.07	21.38
	L	0.25 ± 0.01	0.22 ± 0.02*	0.26	0.17	0.41	0.05	21.26
TRL (cm)	*N*	577.24 ± 13.22	504.44 ± 13.33*	477.60	160.83	1152.72	202.16	42.33
	L	546.56 ± 16.44	304.94 ± 6.61*	351.14	110.25	727.03	154.67	44.05
TRS (cm)	*N*	42.30 ± 4.02	41.19 ± 2.04	42.42	18.13	70.07	11.45	26.99
	L	35.07 ± 0.61	31.68 ± 0.52*	31.79	11.19	56.73	10.04	31.59
TRV (cm^2)	*N*	0.22 ± 0.007	0.13 ± 0.007*	0.21	0.09	0.44	0.07	35.02
	L	0.20 ± 0.009	0.13 ± 0.010*	0.17	0.04	0.34	0.07	39.60
SDW (mg)	*N*	76.83 ± 4.92	75.43 ± 3.16*	80.35	50.90	125.37	16.38	20.38
	L	56.57 ± 0.50	45.40 ± 1.25*	57.69	30.57	87.73	12.87	22.31
RDW (mg)	*N*	27.98 ± 1.44	27.84 ± 3.91*	24.31	10.20	44.88	7.21	29.64
	L	18.34 ± 0.34	16.90 ± 0.47	20.42	10.03	35.87	5.81	28.45
RDW/SDW	*N*	0.31 ± 0.044	0.24 ± 0.022*	0.31	0.12	0.51	0.08	24.97
	L	0.36 ± 0.009	0.40 ± 0.010*	0.35	0.22	0.56	0.06	17.92
Physiological traits								
RA (ug TTF/g FW/h)	*N*	217.73 ± 10.16	238.88 ± 3.99	260.75	103.09	357.67	61.51	23.59
	L	93.29 ± 1.02	80.62 ± 0.44*	188.14	38.02	311.99	69.81	37.11
POD OD470/(g min) FW	*N*	95.88 ± 1.44	105.98 ± 1.99*	99.19	42.41	164.31	30.82	31.07
	L	142.40 ± 6.98	119.74 ± 10.78*	109.29	48.01	204.50	34.72	31.77
SOD (u mol/g FW)	*N*	27.87 ± 3.90	35.14 ± 3.16*	31.85	7.77	48.83	9.03	28.34
	L	53.83 ± 1.66	45.36 ± 3.72*	39.13	12.40	60.00	8.90	22.75
MDA (u mol/g FW)	*N*	1.56 ± 0. 39	1.50 ± 0.32*	1.52	0.44	3.20	0.65	42.34
	L	2.05 ± 0.24	1.66 ± 0. 25*	2.29	0.65	5.15	0.94	41.19

(continued)

Table 7.20 (continued)

Trait	Treatment	Parent		DH population				
		Huapei 3	Yumai 57	Mean	Minimum	Maximum	SD	CV (%)
CHL (mg/g)	*N*	2.23 ± 0.002	1.97 ± 0.007*	1.78	1.00	2.58	0.32	18.26
	L	1.75 ± 0.026	1.47 ± 0.032*	1.51	0.83	2.20	0.30	19.59
K^+ Na^{2+} Ca^{2+} Mg^{2+} concentration								
KCS (mg/g)	*N*	34.27 ± 0.08	29.19 ± 0.08*	40.22	10.35	70.56	16.47	40.95
	L	11.45 ± 0.18	9.75 ± 0.17*	13.43	3.47	24.21	5.50	40.92
KCR (mg/g)	*N*	38.19 ± 0.61	33.98 ± 0.62*	32.99	10.42	65.93	14.30	43.33
	L	25.36 ± 0.61	15.38 ± 0.43*	17.71	5.07	40.42	9.35	52.76
NCS (mg/g)	*N*	2.98 ± 0.11	5.52 ± 0.16*	4.09	1.17	8.28	0.88	16.89
	L	4.47 ± 0.18	5.47 ± 0.10*	5.19	2.19	8.55	1.43	35.02
NCR (mg/g)	*N*	1.78 ± 0.07	1.24 ± 0.09*	1.33	0.61	2.30	0.28	20.70
	L	1.88 ± 0.08	2.13 ± 0.10*	1.81	1.20	2.92	0.33	18.49
CCS (mg/g)	*N*	4.36 ± 0.91	2.58 ± 0.32	4.85	1.44	10.48	2.12	43.76
	L	5.32 ± 0.18	3.11 ± 0.11*	5.60	1.62	10.85	2.61	46.51
CCR (mg/g)	*N*	3.47 ± 0.05	2.06 ± 0.10*	2.90	0.24	6.17	1.33	46.05
	L	3.97 ± 0.06	2.80 ± 0.10*	3.27	0.71	6.99	1.55	47.36
MCS (mg/g)	*N*	1.35 ± 0.04	2.49 ± 0.13	3.53	1.51	7.86	1.42	40.33
	L	2.04 ± 0.08	4.18 ± 0.07*	3.92	1.57	8.96	1.78	45.52
MCR (mg/g)	*N*	3.18 ± 0.14	2.28 ± 0.13*	2.57	0.60	5.29	0.88	34.23
	L	3.48 ± 0.06	5.08 ± 0.05*	4.15	1.16	7.76	1.54	37.15

The parents were significantly different at $P \leq 0.05$. *N* indicates normal K^+ concentration (2 mmol L^{-1}); *L* indicates low K^+ concentration (0.01 mmol L^{-1}); *RAD* root average diam; *TRL* total root length; *TRS* total root surface area; *TRV* total root volume; *SDW* shoot dry weight; *RDW* root dry weight; *RA* root active; *POD* peroxidase; *SOD* superoxide dismutase; *MDA* malondialdehyde; *CHL* chlorophyll concentration; *KCS* shoot K^+ concentration; *KCR* root K^+ concentration; *NCS* shoot Na^+ concentration; *NCR* root Na^+ concentration; *CCS* shoot Ca^{2+} concentration; *CCR* root Ca^{2+} concentration; *MCS* shoot Mg^{2+} concentration; *MCR* root Mg^{2+} concentration

*Significant at 0.05 level

directions, indicating that these traits were under polygenic control and suitable for QTL analysis.

Correlation analyses were performed among all traits under different K supply treatments. Most of the 20 traits were found to be positively significantly correlated with one another at the $P \leq 0.05$ level. Significant negative correlations were detected between Na^+ concentration and the other traits under L conditions. In addition, 84 correlations were not significant at the $P \leq 0.05$ level.

7.6.1.2.2 QTL Analyses

A total of 65 QTLs for 20 traits were detected across 18 chromosomes, except for chromosomes 2B, 5A, and 7B (Table 7.21; Fig. 7.11). Individual QTLs in the two K supply treatments explained between 5.35 % (NCR) and 39.64 % (RAD) of the observed phenotypic variations. The highest LOD value for a single QTL was 13.76 (SDW) identified under the N treatment.

7.6.1.2.2.1 Morphological Traits

A total of 29 QTLs for morphological traits were detected (Table 7.21; Fig. 7.11). Four QTLs for RAD were detected and mapped to chromosomes 3B, 5B, and 7B. These QTLs accounted for 7.36–39.64 % of the phenotypic variation. The allele increasing RAD at the major QTL *QRAD-7B.1* was inherited from the parental line Huapei 3. Five QTLs for TRL were detected and mapped to chromosomes 1A, 1D, 5B, and 6B and accounted for 5.94–11.63 % of the phenotypic variation. Higher TRL at major QTLs *QTRL-3B*, *QTRL-6B,* and *QTRL-5B.2* was conferred by Huapei 3 alleles. Three QTLs for TRS, responsible for 6.72–7.68 % of the phenotypic variation, were detected and mapped to chromosomes 1D, 5B, and 6D. *QTRS-5B* was detected under both N and L treatments, with the allele increasing TRS QTL was inherited from the parental line Huapei 3. Five QTLs for TRV mapped to chromosomes 1A, 1B, 2A, 4B, and 5B and accounted for 9.47–24.61 % of the phenotypic variation.

The allele increasing RAD at the major QTL *QTRV-1A* was inherited from the parental line Huapei 3. For SDW, four QTLs were detected and mapped to chromosomes 1B, 3B, 3D, and 6A. These QTLs accounted for 8.14–31.55 % of the phenotypic variation. Higher SDW at major *QSDW-1B* and *QSDW-6A* was conferred by Huapei 3 alleles, one of which, *QSDW-6A*, was detected under both N and L treatments. Four QTLs for RDW were mapped to chromosomes 1A, 3A, and 5B and explained 6.48–20.61 % of the phenotypic variation. Higher RDW at major QTLs *QRDW-1A.1*, *QRDW-1A.2*, and *QRDW-3A* was conferred by Huapei 3 alleles, with QTL *QRDW-3A* detected under both treatments. For RSDW, four QTLs were detected and mapped to chromosomes 1A, 1B, 1D, and 6A. These QTLs accounted for 6.40–10.86 % of the phenotypic variation. The allele increasing RSDW at the major QTL *QRSDW-1B* was inherited from the parental line Huapei 3.

Table 7.21 Detected QTLs with additive effects at the seedling stage under N and L treatments

Trait	QTL	Flanking marker	LOD		Additive effect		PVE (%)	
			N	L	N	L	N	L
Morphological traits								
RAD	*QRAD-3B*	*Xgpw7774–Xgwm533*	3.15		−0.023		7.36	
	QRAD-5B	*Xbarc232–Xwmc235*		3.18		0.022		7.52
	QRAD-7B.1	*Xwmc273.1–Xcfd22.1*		13.53		0.082		39.64
	QRAD-7B.2	*Xgwm611–Xwmc581*	4.62		−0.071		9.58	
TRL	*QTRL-1D*	*Xcfa2158–Xwmc222*		3.04		−14.65		5.94
	QTRL-3B	*Xgwm566–Xcfe009*		4.01		10.96		10.77
	QTRL-5B.1	*Xbarc1125–Xgwm213*		3.01		6.25		6.85
	QTRL-5B.2	*Xbarc232–Xwmc235*	5.24		22.11		10.66	
	QTRL-6B	*Xcfa2149.1–Xgwm58*	4.21		27.19		11.63	
TRS	*QTRS-1D*	*Xgdm60–Xwmc429*		3.02		3.18		6.78
	QTRS-5B	*Xbarc1125–Xgwm213*	3.32	3.12	3.11	3.47	7.38	7.68
	QTRS-6D	*Xswes679.1–Xcfa2129*	3.09		−2.04		6.72	
TRV	*QTRV-1A*	*GluA1–Xwmc550*	8.11		0.048		24.61	
	QTRV-1B	*Xgwm582–Xgpw7388*	4.56		0.027		9.86	
	QTRV-2A	*Xcfa2263–Xbarc264*	4.54		0.026		9.65	
	QTRV-4B	*Xcfd22.2–Xwmc657*		3.54		−0.22		7.93
	QTRV-5B	*Xbarc232–Xwmc235*		4.33		0.25		9.47
SDW	*QSDW-1B*	*Xwmc406-Xbarc156*	13.76		9.12		31.55	
	QSDW-3B	*Xgpw7774–Xgwm533*	4.6		−8.34		9.68	
	QSDW-3D	*Xcfd34–Xbarc376*		3.37		4.23		8.14
	QSDW-6A	*Xgwm334–Xbarc023*	3.63	3.83	5.22	4.11	10.85	10.29

(continued)

Table 7.21 (continued)

Trait	QTL	Flanking marker	LOD		Additive effect		PVE (%)	
			N	L	N	L	N	L
RDW	*QRDW-1A.1*	*Xbarc120.2–Xwmc278*	4.3		−1.01		10.1	
	QRDW-1A.2	*Xbarc148–Xgwm154*		8.4		3.11		20.61
	QRDW-3A	*Xswes107–Xgpw4032*	3.13	4.32	2.02	2.17	8.21	10.01
	QRDW-5B	*Xbarc1125–Xgwm213*		3.68		1.88		6.48
RSDW	*QRSDW-1A*	*Xgpw3167–Xcwem6.2*	3.51		−0.023		8.06	
	QRSDW-1B	*Xwmc406–Xbarc156*		4.15		0.031		10.86
	QRSDW-1D	*Xcfa2158–Xwmc222*		3.19		−0.015		6.40
	QRSDW-6A	*Xgpw2265–Xgpw322*		3.15		−0.018		7.47
Physiological traits								
RA	*QRA-1B*	*Xwmc406–Xbarc156*	3.8	3.83	20.08	20.14	10.09	10.21
	QRA-3A	*Xbarc157.1–Xbarc1177*	3.14		−3.26		5.95	
	QRA-4B	*Xcfd22.2–Xwmc657*		3.77		−19.17		7.85
POD	*QPOD-1B*	*Xwmc406–Xbarc156*	9.06		16.29		21.10	
	QPOD-4D	*BE293342–Xgpw7258*	4.92	3.51	−11.29	−7.74	10.55	5.54
SOD	*QSOD-3A*	*Xgpw1108–Xgpw1107*	3.01		2.55		6.72	
	QSOD-7B	*Xcfd22.1–Xgpw3226*	3.28		3.01		8.82	
	QSOD-7D	*Xgwm676–Xgwm437*		3.68		2.33		5.17
MDA	*QMDA-1D*	*Xcfa2158–Xwmc222*	4.64	3.94	−0.34	−0.31	11.27	10.25
	QMDA-3A	*Xwmc264–Xcfa2193*	3.51		−0.23		8.38	
	QMDA-4D	*Xcfa2173–Xcfe188*		3.56		−0.14		6.48
	QMDA-6D	*Xwmc412.1–Xcfd49*		3.99		0.21		7.74
CHL	*QCHL-6D*	*Xwmc412.1–Xcfd49*	3.66		0.09		6.93	

(continued)

Table 7.21 (continued)

Trait	QTL	Flanking marker	LOD		Additive effect		PVE (%)	
			N	L	N	L	N	L
K^+ Na^{2+} Ca^{2+} Mg^{2+} concentration								
KCS	*QKCS-4A*	*Xwmc219–Xwmc776*		3.37		3.24		8.56
	QKCS-5B	*Xgdm116–Xbarc232*	4.66		2.11		9.75	
	QKCS-7B	*Xgwm611–Xwmc581*		3.79		−3.21		6.91
KCR	*QKCR-4A.1*	*Xwmc718–Xwmc262*	3.69	3.24	5.39	1.81	10.51	8.51
	QKCR-4A.2	*Xbarc327–Xbarc078*	3.68		0.24		6.35	
	QKCR-7B	*Xgwm611–Xwmc581*	3.43		−2.43		7.94	
NCS	*QNCS-3B*	*Xgpw7774–Xgwm533*	7.68		−0.74		20.36	
	QNCS-5D	*Xcfd226–Xwmc765*		3.89		0.25		9.88
	QNCS-6A	*Xbarc023–Xbarc1077*	3.22		−0.12		7.21	
NCR	*QNCR-1A*	*GluA1–Xwmc550*	3.93		−0.43		8.88	
	QNCR-1D	*Xcfa2158–Xwmc222*		4.38		−0.23		12.08
	QNCR-6A	*Xgwm334–Xbarc023*	3.18		0.32		5.35	
CCS	*QCCS-1A*	*GluA1–Xwmc550*		6.32		−0.76		8.51
	QCCS-1D	*Xwmc429–Xcfd19*		4.83		−0.68		7.93
	QCCS-6D	*Xswes861.1Xgwm681*		8.94		0.41		23.15
CCR	*QCCR-1A*	*Xbarc350–Xwmc120*		5.21		−0.37		10.23
	QCCR-4A	*Xbarc327–Xbarc078*		3.08		-0.23		7.06
	QCCR-5D	*Xcfd226–Xwmc765*		6.04		0.33		14.65
MCS	*QMCS-2D*	*Xgwm539–Xcfd168*		7.58		0.27		7.05
	QMCS-4A	*Xbarc327–Xbarc078*	3.98		−0.22		6.76	
	QMCS-6B	*Xcfa2257–Xcfd48*		3.36		0.58		16.48
MCR	*QMCR-6D*	*Xgwm133.2–Xswes861.1*	3.68		0.44		11.83	
	QMCR-7B	*Xgwm611–Xwmc581*	3.71	3.79	−0.39	−0.40	6.68	6.86

N indicates normal K^+ concentration; *L* indicates low K^+ concentration. Abbreviations as in Table 7.20

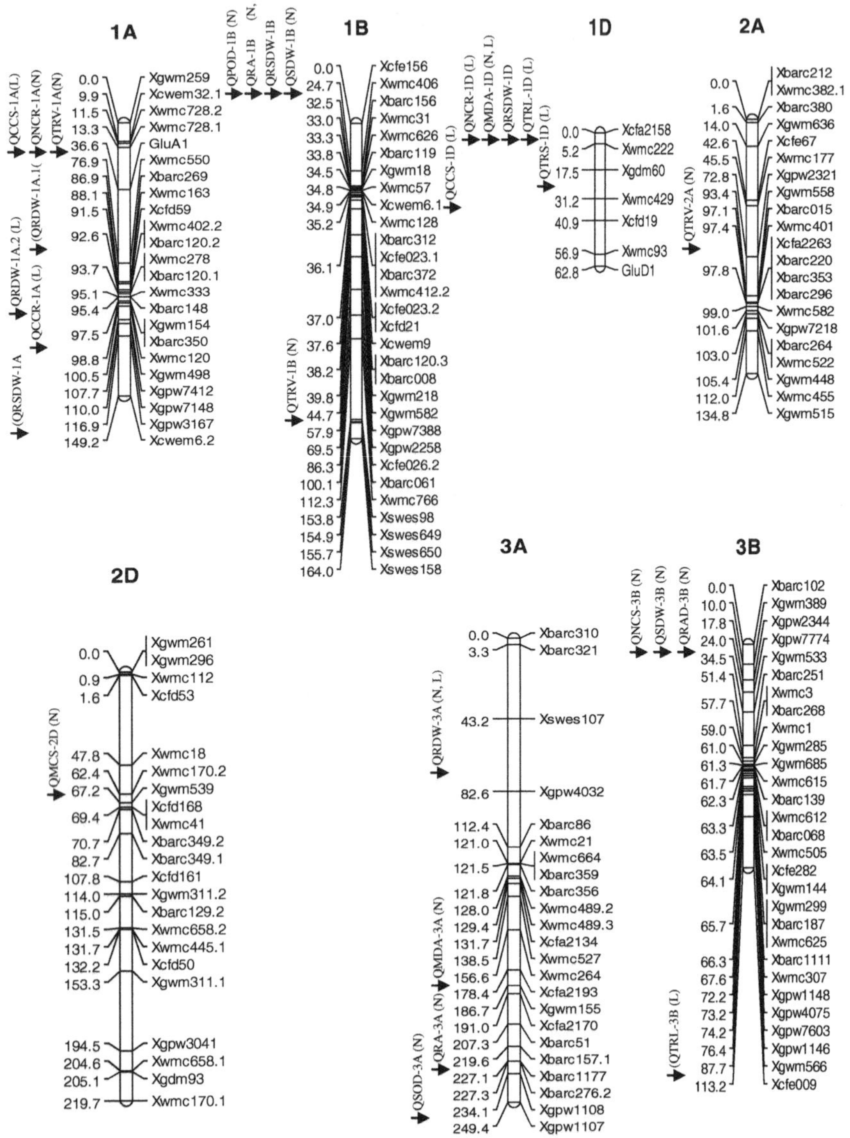

Fig. 7.11 Location of QTLs for traits related to seedling growth under different K^+ concentration treatments

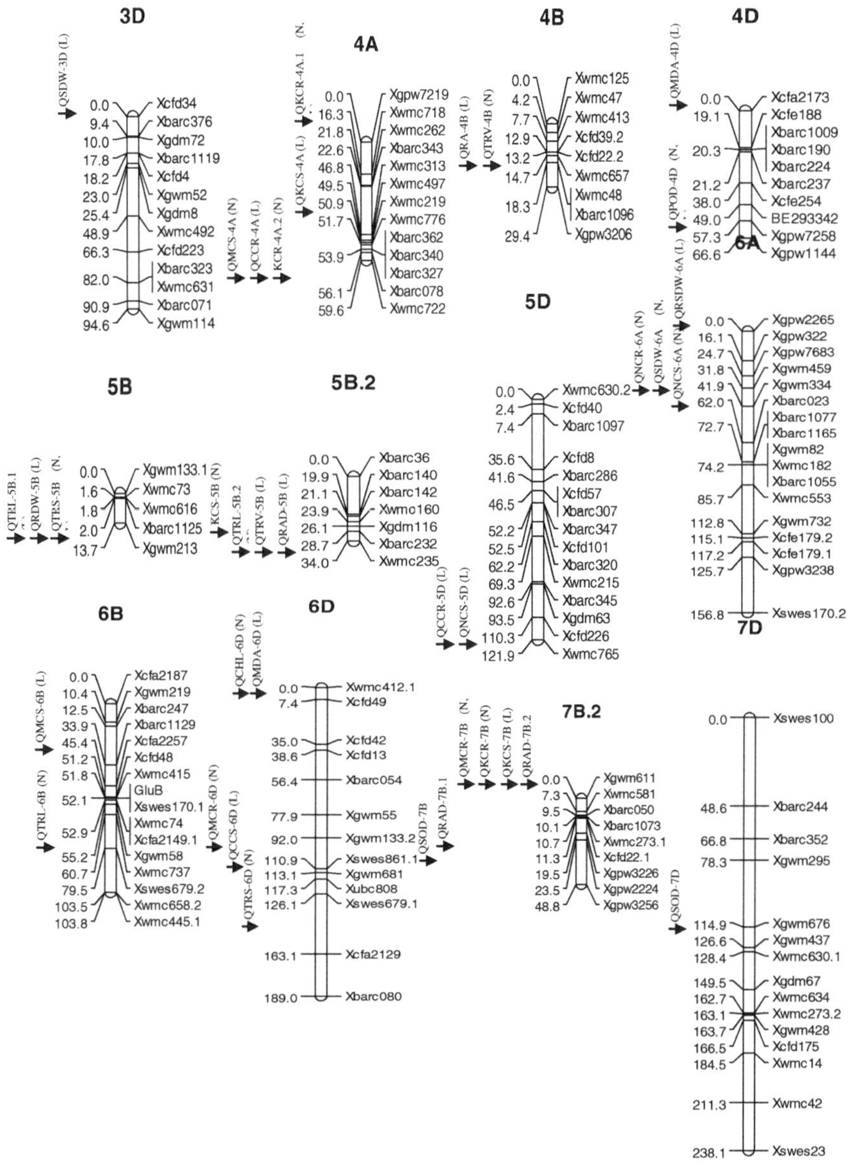

Fig. 7.11 (continued)

7.6.1.2.2.2 Physiological Traits

Thirteen QTLs for physiological traits were detected (Table 7.21; Fig. 7.11). Three QTLs for RA were detected and mapped to chromosomes 1B, 3A, and 4B. These QTLs accounted for 5.92–10.21 % of the phenotypic variation. Two major QTLs (*QRA-1B*) detected under both treatments were conferred by Huapei 3 alleles. Two QTLs for POD, mapped to chromosomes 1B and 4D, were responsible for 5.54–21.10 % of the phenotypic variation. Two major QTLs (*QPOD-1B* and *QPOD-4D*) were contributed by Huapei 3 or Yumai 57 alleles; *QPOD-4D* was detected under both treatments. For SOD, three QTLs were mapped to chromosomes 3A, 7B, and 7D and were responsible for 6.72–8.82 % of the phenotypic variation. Four QTLs for MDA were identified on chromosomes 1D, 3A, 4D, and 6D and accounted for 6.48–11.27 % of the phenotypic variation. Two of these (*QMDA-1D*), contributed by Huapei 3 alleles, were detected under both treatments. One QTL for CHL was detected on chromosome 6D and was contributed by Huapei 3 alleles.

7.6.1.2.2.3 K^+, Na^+, Ca^{2+}, and Mg^{2+} Concentration

We uncovered 23 QTLs for K^+, Na^+, Ca^{2+}, and Mg^{2+} concentrations (Table 7.21; Fig. 7.11). For KCS, three QTLs were mapped to chromosomes 4B, 7B, and 5B. These QTLs accounted for 6.91–9.75 % of the phenotypic variation. Three QTLs for KCS, mapped to chromosomes 4A and 7B, were associated with 6.35–10.51 % of the phenotypic variation; the allele increasing KCS at the major QTL *QKCR-4A.1* was detected under both treatments and inherited from the parental line Huapei 3. Three QTLs for NCS were detected and mapped to chromosomes 3B, 5D, and 6A. These QTLs accounted for 7.21–20.36 % of the phenotypic variation. The allele increasing NCS at the major QTL *QNCS-3B* was inherited from the parental line Huapei 3. For NCR, three QTLs were detected and mapped to chromosomes 1A, 1D, and 6A. These QTLs accounted for 5.35–12.08 % of the phenotypic variation. The allele increasing NCR at the major QTL *QNCR-1D* was inherited from the parental line Huapei 3. Three QTLs for CCS were mapped to chromosomes 1A, 1D, and 6D and explained 7.93–23.15 % of the phenotypic variation. The allele increasing CCS at the major QTL *QCCS-6D* was inherited from the parental line Huapei 3.

For CCR, three QTLs were detected and mapped to chromosomes 1A, 4A, and 5D. These QTLs accounted for 7.06–14.65 % of the phenotypic variation. Two major QTLs, *QCCR-1A* and *QCCR-5D,* were contributed by Yumai 57 or Huapei 3 alleles, respectively.

For MCS, three QTLs were detected and mapped to chromosomes 2D, 4A, and 6B. These QTLs accounted for 6.76–16.48 % of the phenotypic variation. The allele increasing MCS at the major QTL, *QMCS-6B,* was inherited from the parental line Huapei 3.

Two QTLs for MCR were mapped to chromosomes 6D and 7B and explained 6.68–11.83 % of the phenotypic variation. The allele increasing MCR at the major QTL, *QMCR-6D,* was inherited from the parental line Huapei 3.

7.6.2 Research Progress of Potassium-Deficiency Tolerance QTL Mapping and the Comparative Analysis with this Study

7.6.2.1 Research Progress of Potassium-Deficiency Tolerance QTL Mapping

Numerous studies have revealed that K-deficiency tolerance in plants is a complicated quantitative trait, with strong interactions occurring between the genotypes and the environment (Wu et al. 1998; Atienza et al. 2003; Harada and Leigh 2006; Kanter et al. 2010; Prinzenberg et al. 2010). QTL analysis has thus proved to be an effective approach for uncovering genetic loci that control plant tolerance to K deprivation.

Over the past several years, QTL analysis has been successfully used to dissect quantitative traits into single components and to investigate the related effects of a specific trait (Doerge 2002). A number of QTLs have been determined to be associated with K deficiency in various plant species such as rice (Wu et al. 1998), *Arabidopsis thaliana* (Ghandilyan et al. 2009), *Brassica oleracea*, and *Miscanthus sinensis* (Atienza et al. 2003). To the best of our knowledge, however, only one study focusing on QTLs associated with K-deficiency tolerance has been carried out in wheat (*Triticum aestivum* L.). In that study, 380 QTLs for morphological traits, nutrient (N, P, K) contents, and utilization efficiencies of the aforementioned nutrients were identified using a set of RILs derived from a cross between Chuan 35050 and Shannong 483 (Guo et al. 2012). K-deficiency tolerance in wheat was associated with growth traits, such as root architecture, physiological and biochemical parameters, such as cellular protective enzyme activities and ion concentrations, but requires further characterization.

7.6.2.2 Comparison of the Results with One of the Previous Studies

A previous QTL analysis for K tolerance in hydroponically cultured seedlings focused on K content and K utilization efficiency (Guo et al. 2012). No reports have yet appeared of QTL analyses for K tolerance in relation to root architecture, protective enzyme, and ion (Na^+, Ca^{2+}, and Mg^{2+}) concentration traits. In this study, 65 QTLs for 20 such traits were detected on all 21 wheat chromosomes, except for chromosomes 2B, 5A, and 7B (Fig. 7.11). Most of the uncovered QTLs mapped to new marker regions except two loci were located in adjacent marker regions: a QTL for MDA, TRL, RSDW, and NCR at *Xcfa2158-Xwmc222* on 1D, and a QTL for KCS, KCR, MCR, MCR, and RAD at *Xgwm611-Xwmc581* on 7B. QTLs for K content and K utilization efficiency traits have also been reported in comparable regions of these two loci (Guo et al. 2012). These results indicated that these two loci are possibly associated with other stress-tolerance loci. The *GluA1-Xwmc550* region has been found to be linked with salt tolerance in wheat seedlings

(Xu et al. 2012). Using our DH population, we also found that *Xwmc406-Xbarc156* is associated with cold resistance in wheat (unpublished data). These results suggest that QTLs controlling different aspects of stress tolerance may be combined through MAS in wheat breeding programs.

Under K-deprivation conditions, other cations can replace K to exercise its role (Rengel and Damon 2008; White and Karley 2010). Potassium–cation interaction mechanisms are thus very important for normal wheat growth and worthy of study, especially with respect to cooperative uptake relationships. In our investigation, correlation coefficients among K, Mg, Ca, and Na ion content traits were found to be significantly positive under the two K supply conditions. This result indicated the existence of a cooperative uptake relationship among K, Mg, Ca, and Na. In addition, we also found similar evidence for an ion cooperative uptake relationship at the QTL level. We considered a cooperative uptake locus to be present when QTLs were detected for at least two K, Mg, Ca, or Na concentration traits in roots or shoots. In this study, clusters C1, C5, and C9 were associated with cooperative uptake. This result may provide further evidence that similar genetic mechanisms are involved in cooperative uptake among K, Mg, Ca, and Na.

References

Allan RE. Agronomic comparisons between *Rht1* and *Rht2* semidwarf genes in winter wheat. Crop Sci. 1989;29(5):1103–8.

Anderson JA, Sorrells ME, Tanksley SD. RFLP analysis of genomic regions associated with resistance to pre-harvest sprouting in wheat. Crop Sci. 1993;33:453–9.

Atienza SG, Satovic Z, Petersen KK, Dolstra O, Martín A. Identification of QTLs influencing combustion quality in Miscanthus sinensis Anderss. II. Chlorine and potassium content. Theor Appl Genet. 2003;107(5):857–63.

Baligar VC, Fageria NK, He ZL. Nutrient use efficiency in plants. Commun Soil Sci Plant. 2001;32:921–50.

Börner A, Schumann E, Furste A, Coster H, Leithold B, Röder MS, Weber WE. Mapping of quantitative trait loci determining agronomic important characters in hexaploid wheat (*Triticum aestivum* L.). Theor Appl Genet. 2002;105:921–36.

Bougot Y, Lemoine J, Pavoine MT, Guyomar'ch H, Gautier V, Muranty H, Barloy D. A major QTL effect controlling resistance to powdery mildew in winter wheat at the adult plant stage. Plant Breeding. 2006;125(6):550–6.

Boukhatem N, Baret PV, Mingeot D, Jacquemin JM. Quantitative trait loci for resistance against Yellow rust in two wheat-derived recombinant inbred line populations. Theor Appl Genet. 2002;104(1):111–8.

Buerstmayr H, Ban T, Anderson J. QTL mapping and marker-assisted selection for fusarium head blight resistance in wheat: a review. Plant Breed. 2009;128:1–26.

Chantret N, Mingeot D, Sourdille P, Bernard M, Jacquemin JM, Doussinault G. A major QTL for powdery mildew resistance is stable over time and at two development stages in winter wheat. Theor Appl Genet. 2001;103(6–7):962–71.

Chen TB. Contamination of heavy metals on soil. Metal Word. 1999;3:10–11 (in Chinese with English abstract).

Chen YH, Hunger RM, Carver B, Zhang HL, Yan LL. Genetic characterization of powdery mildew resistance in U.S. hard winter wheat. Mol Breeding. 2009;24:141–52.

Chen ZD, Zhou WG, Wang J, Wang ZF, Zhang HS. Mapping QTL of tolerance to Cd^{2+} stress at seedling stage in rice (*Oryza sativa* L.). J Nanjing Agric Univ. 2010;33(3):1–5 (in Chinese with English abstract).

Chhuneja P, Kaur S, Garg T, Ghai M, Kaur S, Prashar M, Bains NS, Goel RK, Keller B, Dhaliwal HS, Singh K. Mapping of adult plant stripe rust resistance genes in diploid A genome wheat species and their transfer to bread wheat. Theor Appl Genet. 2008;116(3):313–24.

Christiansen MJ, Feenstra B, Skovgaard IM, Andersen SB. Genetic analysis of resistance to yellow rust in hexaploid wheat using a mixture model for multiple crosses. Theor Appl Genet. 2006;112(4):581–91.

Ci DW, Jiang D, Li SS, Wollenweber B, Dai TB, Cao WX. Identification of quantitative trait loci for cadmium tolerance and accumulation in wheat. Acta Physiologiae Plantarum. 2012;34 (1):191–202.

Di Toppi LS, Gabbrielli R. Response to Cadmium in higher plants. Environ EXP Bot. 1999;41:105–30.

Doerge RW. Mapping and analysis of quantitative trait loci in experimental populations. Nat Genet. 2002;3:43–52.

Dubcovsky J, Maria GS, Epstein E, Luo MC, Dvořák J. Mapping of the K^+/Na^+ discrimination locus Knal in wheat. Theor Appl Genet. 1996;92:448–54.

Fageria NK, Baligar VC, Jones CA. Growth and mineral nutrition of field crops. Boca Raton: CRC Press; 2011.

Fick GN, Qualset CO. Seedling emergence, coleoptile length, and plant height relationships in crosses of dwarf and standard-height wheats. Euphytica. 1976;25(1):679–84.

Flintham J, Adlam R, Bassoi M, Holdsworth M, Gale M. Mapping genes for resistance to sprouting damage in wheat. Euphytica. 2002;126:39–45.

Fofana B, Humphreys DG, Rasul G, Cloutier S, Brû lé-Babel A, Woods S, Lukow OM, Somers DJ. Mapping quantitative trait loci controlling pre-harvest sprouting resistance in a red × white seeded spring wheat cross. Euphytica. 2009;165:509–21.

Francki MG, Berzonsky WA, Ohm HW, Anderson JM. Physical location of a HSP70 gene homologue on the centromere of chromosome 1B of wheat (*Triticum aestivum* L.). Theor Appl Genet. 2002;104:184–91.

Ghandilyan A, Ilk N, Hanhart C, Mbengue M, Barboza L, Schat H, Koornneef M, El-Lithy M, Vreugdenhil D, Reymond M, Aarts MG. A strong effect of growth medium and organ type on the identification of QTLs for phytate and mineral concentrations in three Arabidopsis thaliana RIL populations. J Exp Bot. 2009;60:1409–25.

Guo Y, Kong FM, Xu YF, Zhao Y, Liang X, Wang YY, An DG, Li SS. QTL mapping for seedling traits in wheat grown under varying concentrations of N, P and K nutrients. Theor Appl Genet. 2012;124:851–65.

Guo Q, Zhang ZJ, Xu YB, Li GH, Feng J, Zhou Y. Quantitative trait loci for high-temperature adult-plant and slow-rusting resistance to puccinia stritformis f. sp. tritici in wheat cultivars. Phytopathology. 2008;98(7):803–9.

Hao ZF, Jing RL, Jia JZ, Jing RL, Li YZ, Jia JZ. QTL mapping for drought tolerance at stages of germination and seedling in wheat (*Triticum aestivum* L.) using a DH population. Agr Sci China. 2003;2(9):943–9.

Harada H, Leigh RA. Genetic mapping of natural variation in potassium concentrations in shoots of *Arabidopsis thaliana*. J Exp Bot. 2006;57:953–60.

He ZH, Lan CX, Chen XM, Zou YC, Zhuang QS, Xia XC. Progress and perspective in research of adult-plant resistance to stripe rust and powdery mildew in wheat. Sci Agric Sin. 2011;44 (11):2193–215 (in Chinese with English abstract).

Hoagland DR, Arnon DI. The water culture method for growing plants without soil. California Agricultural Experiment Station, Circular. 1950;347:4–32.

Huang PX. Response and mechanism on different wheat varieties to lead and cadmium. Master Dissertation of Graduate School of Yangzhou University. 2007 (in Chinese with English abstract).

Imtiaz M, Ogbonnaya FC, Oman J, Ginkel M. Characterization of QTL controlling genetic variation for pre-harvest sprouting in synthetic backcross derived wheat lines. Genetics. 2008;178:1725–36.

Imtiaz M, Ahmad M, Cromey MG, Griffin WB, Hampto JG. Detection of molecular markers linked to the durable adult plant stripe rust resistance geneYr18in bread wheat (*Triticum aestivum* L.). Plant Breeding. 2004;123(5):401–4.

Jakobson I, Peusha H, Timofejeva L, Järve K. Adult plant and seedling resistance to powdery mildew in a *Triticum aestivum* × *Triticum militinae* hybrid line. Theor Appl Genet. 2006;112 (4):760–9.

Kanter U, Hauser A, Michalke B, Draexl S, Schaeffner AR. Caesium and strontium accumulation in shoots of *Arabidopsis thaliana*: genetic and physiological aspects. J Exp Bot. 2010;61:3995–4009.

Kato H, Taketa S, Ban T, Iriki N, Murai K. The influence of a spring habit gene, Vrn-D1, on heading time in wheat. Plant Breeding. 2001;120(2):115–20.

Keller M, Keller B, Schachermayr G, Winzeler M, Schmid JE, Stamp P, Messmer MM. Quantitative trait loci for resistance against powdery mildew in a segregating wheat × spelt population. Theor Appl Genet. 1999;98(6–7):903–12.

Kirigwi FM, Van Ginkel M, Brown-guedira G, Gill BS, Paulsen GM, Fritz AK. Markers associated with a QTL for grain yield in wheat under drought. Mol Breed. 2007;20:401–13.

Kulwal PL, Singh R, Balyan HS, Gupta PK. Genetic basis of pre-harvest sprouting tolerance using single-locus and two-locus QTL analyses in bread wheat. Func Integr genomic. 2004;4(2):94–101.

Lan CX. QTL Mapping for Adult-plant Resistance to Stripe Rust and Powdery Mildew in Common Wheat. PHD dissertation of Graduate School of Chinese Academy of Agricultural Sciences, 2010 (in Chinese with English abstract).

Lan CX, Ni XW, Yan J, Zhang Y, Xia XC, Chen XM, He ZH. Quantitative trait loci mapping of adult-plant resistance to powdery mildew in Chinese wheat cultivar Lumai 21. Mol Breeding. 2010;25(4):615–22.

Landjeva S, Neumann K, Lohwasser U, Börner A. Molecular mapping of genomic regions associated with wheat seedling growth under osmotic stress. Biol Plant. 2008;52(2):259–66.

Lægreid M, Bøckman OC, Kaarstad O. Agriculture, fertilizers, and the environment, CABI Publishing in association with Norsk Hydro ASA, New York, 1999.

Lebreton C, Lazić-Jančić V, Steed A, Pekić S, Quarrie SA. Identification of QTL for drought responses in maize and their use in testing causal relationships between traits. J Exp Bot. 1995;46(7):853–65.

Li BL, Mei HS. Relationship between oat leaf senescence and activated oxygen metabolism. Acta Phytophysiol Sin. 1989;15:6–12 (in Chinese with English abstract).

Li ZK, Peng T, Zhang WD, Xie QG, Tian JC. Analysis of QTLs for root traits at seedling stage using an "Immortalized F_2" population of wheat. Acta Agron Sin. 2010;36(3):442–8 (in Chinese with English abstract).

Liang SS, Suenaga K, He ZH, Wang ZL, Liu HY, Wang DS, Singh RP, Sourdille P, Xia XC. Quantitative trait loci mapping for adult-plant resistance to powdery mildew in bread wheat. Phytopathology. 2006;96(7):784–89.

Lillemo M, Asalf B, Singh RP, Huerta-Espino J, Chen XM, He ZH, Bjørnstad Å. The adult plant rust resistance loci *Lr34/Yr18* and *Lr46/Yr29* are important determinants of partial resistance to powdery mildew in bread wheat line Saar. Theor Appl Genet. 2008;116(8):1155–66.

Lin F, Chen XM. Quantitative trait loci for non-race-specific, high-temperature adult-plant resistance to stripe rust in wheat cultivar Express. Theor Appl Genet. 2009;118(4):631–42.

Lin ZS, Cui ZF, Zeng XY, Ma YZ, Zhang ZY, Nakamura T, Ishikawa G, Nakamura K, Yoshida H, Xin ZY. Analysis of wheat-Thinopyrum intermedium derivatives with BYDV-resistance. Euphytica. 2007;158(1–2):109–18.

Liu J, Liu D, Tao W, Li W, Wang S, Chen P, Cheng S, Gao D. Molecular marker-facilitated pyramiding of different genes for powdery mildew resistance in wheat. Plant Breed. 2001a;119:21–4.

Liu SX, Griffey CA, Saghai Maroof MA. Identification of molecular markers associated with adult plant resistance to powdery mildew in common wheat cultivar Massey. Crop Sci. 2001b;41:1268–75.

Liu X, Shi J, Zhang XY, Ma YS, Jia JZ. Screening salt tolerance germplasms and tagging the tolerance gene(s) using microsatellite (SSR) markers in wheat. Acta Bot Sin. 2001c;43(9):948–54 (in Chinese with English abstract).

Liu ZJ, Liang Y, Zhu J, Li GQ, Yang ZM, Sun QX, Liu ZY. Genetic improvement and marker assisted selection of wheat with leaf rust resistance gene *Lr9*. J Triticeae Crop. 2013;33:236–42 (in Chinese with English abstract).

Mallard S, Gaudet D, Aldeia A, Abelard C, Besnard AL, Sourdille P, Dedryver F. Genetic analysis of durable resistance to yellow rust in bread wheat. Theor Appl Genet. 2005;110(8):1401–9.

Mao SL. Meta-analysis for some important traits and QTL analysis for photosynthesis and waterlogging tolerance in wheat. PhD dissertation of Graduate School of Sichuan Agricultural University, 2010 (in Chinese with English abstract).

Mares D, Mrva K, Tan MK, Sharp P. Dormancy in white-grained wheat: Progress towards identification of genes and molecular markers. Euphytica. 2002;126(1):47–53.

Mason Esten R, Mondal S, Beecher FW, Hays DB. Genetic loci linking improved heat tolerance in wheat (*Triticum aestivum* L.) to lower leaf and spike temperatures under controlled conditions. Euphytica. 2011;180:181–94.

Melichar JPE, Berry S, Newell C, MacCormack R, Boyd LA. QTL identification and microphenotype characterisation of the developmentally regulated yellow rust resistance in the UK wheat cultivar Guardian. Theor Appl Genet. 2008;117(3):391–99.

Mingeot, Chantret N, Baret PV, Dekeyser A, Boukhatem N, Sourdille P, Doussinault G, Jacquemin JM. Mapping QTL involved in adult plant resistance to powdery mildew in the winter wheat line RE714 in two susceptible genetic backgrounds. Plant Breeding. 2002;121 (2):133–40.

Mohan A, Kulwal P, Singh R, Kumar V, RoufMir R, Kumar J, Prasad M, Balyan HS, Gupta PK. Genome-wide QTL analysis for pre-harvest sprouting tolerance in bread wheat. Euphytica. 2009;168:319–29.

Moral R, Gomez I. Navarro Pedreno J, Mataix J. Effect of cadmium on nutrient distribution, yield and growth of tomato grown in soilless culture. J Plant Nutr. 1994;17(6):953–62.

Morgan JM, Tan MK. Chromosomal location of a wheat osmotic regulation gene using RFLP. Aust J Plant Physiol. 1996;23:803–6.

Mrva K, Mares D. Screening methods and identification of QTLs associated with late maturity alpha-amylase in wheat. Euphytica. 2002;126(1):55–9.

Munnus R, Husain S, Rivelli RA. Avenues for increasing salt tolerance of crops and the role of physiologically based selection traits. Plant Soil. 2002;247:93–105.

Navabi A, Tewari JP, Singh RP, McCallum B, Laroche A, Briggs KG. Inheritance and QTL analysis of durable resistance to stripe and leaf rusts in an Australian cultivar, *Triticum aestivum* "Cook". Genome. 2005;48:97–107.

Patra J, Lenka M, Panda BB. Tolerance and co-tolerance of the grass *Chloris barlata* Sw. to tercury, cadmium and zinc. New Phytol. 1994;128:165–71.

Pettigrew WT. Potassium influences on yield and quality production for maize, wheat, soybean and cotton. Plant Physiol. 2008;133(4):670–81.

Poysa VW. The genetic control of low temperature, ice encasement and flooding tolerance by chromosome 5A, 5B and 5D in wheat. Cereal Res Commun. 1984;12:135–41.

Prinzenberg AE, Barbier H, Salt DE, Stich B, Reymond M. Relationships between growth, growth response to nutrient supply, and ion content using a recombinant inbred line population in Arabidopsis. Plant Physiol. 2010;154(3):1361–71.

Qi ZL, Zhu H, Li SS. Comprehensive evaluation of drought resistance of wheat RIL population at seedling stage. Shandong Agric Sci. 2010;9:5–9 (in Chinese with English abstract).

Quarrie SA, Gulli M, Calestani C, Steed A, Marmiroli N. Location of a gene regulation drough-induced abscisic acid production on the long arm of chromosome 5A of wheat. Theor Appl Genet. 1994;89:794–800.

Quarrie SA, Steed A, Calestani C, Semikhodskii A, Lebreton C, Chinoy C, Steele N, Pljevljakusić D. A high-density genetic map of hexaploid wheat (*Triticum aestivum* L.) from the cross Chinese Spring XSQ1 and its use to compare QTLs for grain yield across a range of environments. Theor Appl Genet. 2005;110:965–90.

Ramburan VP, Pretorius ZA, Louw JH, Boyd LA, Smith PH, Boshoff WHP, Prins R. A genetic analysis of adult plant resistance to stripe rust in the wheat cultivar Kariega. Theor Appl Genet. 2004;108(7):1426–33.

Rasul G, Humphreys DG, Brû lé-Babel A, McCartney CA, DePauw RM, Knox RE, Somers DJ. Mapping QTLs for pre-harvest sprouting traits in the spring wheat cross 'RL4452/AC Domain'. Euphytica. 2009;168:363–78.

Rebetzke GJ, Botwright TL, Moore CS, Richardsa RA, Condona AG. Genotypic variation in specific leaf area for genetic improvement of early vigour in wheat. Field Crop Res. 2004;88 (2):179–89.

Rebetzke GJ, Richards RA, Fischer VM, Mickelson BJ. Breeding long coleoptile, reduced height wheats. Euphytica. 1999;106(2):159–68.

Ren YZ, Xu YH, Gui XW, Wang SP, Ding JP, Zhang QC, Ma YS, Pei DL. QTLs analysis of wheat seedling traits under salt stress. Sci Agric Sin. 2012;45(14):2793–800 (in Chinese with English abstract).

Rengel Z, Damon PM. Crops and genotypes differ in efficiency of potassium uptake and use. Physiol Plant. 2008;133:624–36.

Richards RA. The effect of dwarfing genes in spring wheat in dry environments. II. Growth, water use and water-use efficiency. Crop Pasture Sci. 1992;43(3):529–39.

Roy JK, Prasad M, Varshney RK, Balyan HS, Blake TK, Dhaliwal HS, Singh H, Edwards KJ, Gupta PK. Identification of a microsatellite on chromosomes 6B and a STS on 7D of bread wheat showing an association with preharvest sprouting tolerance. Theor Appl Genet. 1999;99 (1–2):336–40.

Santra DK, Chen XM, Santra M, Campbell KG, Kidwell KK. Identification and mapping QTL for high-temperature adult-plant resistance to stripe rust in winter wheat (*Triticum aestivum* L.) cultivar Stephens. Theor Appl Genet. 2008;117(5):793–802.

Shan L, Zhao SY, Chen F, Xia GM. Screening and localization of SSR markers related to salt tolerance of somatic hybrid wheat Shanrong No. 3. Sci Agric Sin. 2006;9(2):225–30 (in Chinese with English abstract).

Shen XH, Cao LY, Shao GS, Zhan XD, Chen SG, Wu WM, Cheng SH. QTL mapping for the content of five trace elements in brown rice. Mol Plant Breed. 2008;6(6):1061–7 (in Chinese with English abstract).

Shi J. Screening salt tolerance germplasm and tagging salt tolerance gene(s) using microsallite (SSR) markers in wheat. Master Dissertation of Graduate School of Chinese Academy of Agricultural Sciences. 1999 (in Chinese with English abstract).

Shi W. Genetic dissection of quantitative loci for traits associated with drought tolerance in Wheat (*Triticum Aesitvum* L.) Backcrossing introgression lines. Master dissertation of Graduate School of Chinese Academy of Agricultural Sciences, 2012 (in Chinese with English abstract).

Singh S, Singh M. Genotypic basis of response to saliniity stress in some crosses of spring wheat *Triticum aestivum* L. Euphytica. 2000;115(3):209–14.

Somers DJ, Isaac P, Edwards K. A high-density microsatellite consensus map for bread wheat (*Triticum aestivum* L.). Theor Appl Genet. 2004;109(6):1105–14.

Suenaga K, Singh RP, Huerta-Espino J, William HM. Microsatellite markers for genes *Lr34/Yr18* and other quantitative trait loci for leaf rust and stripe rust resistance in bread wheat. Phytopathology. 2003;93(7):881–90.

Sun HY, Lu JH, Fan YD, Zhao Y, Kong F, Li RJ, Wang HG, Li SS. Quantitative trait loci (QTLs) for quality traits related to protein and starch in wheat. Prog Nat Sci. 2008;18:825–31.

Sun Y, Hu YY, Yang WX, Liu DQ. Evaluation of the resistance to leaf rust of 6 wheat lines. J Triticeae Crop. 2011;31:762–8 (in Chinese with English abstract).

Sutka J, Sanpe JW. Location of a gene for frostreistance on chromosome 5A of wheat. Euphytica. 1989;42:41–4.

Tian DG, Lin F, Zhang CQ, Zhang ZZ, Xue SL, Cao Y, Li CJ, Ma ZQ. Mapping QTLs associated with resistance to fusarium head blight using an "immortalized F_2" population. Acta Agron Sin. 2008;34(4):539–44 (in Chinese with English abstract).

Tucker DM, Griffey CA, Liu S, Brown-Guedira G, Marshall DS, Saghai Maroof MA. Confirmation of three quantitative trait loci conferring adult plant resistance to powdery mildew in two winter wheat populations. Euphytica. 2007;155:1–13.

Vijayalakshmi K, Fritz AK, Paulsen GM, Bai GH, Pandravada S, Gill BS. Modeling and mapping QTL for senescence-related traits in winter wheat under high temperature. Mol Breeding. 2010;26:163–75.

Wang JK. Inclusive composite interval mapping of quantitative trait genes. Acta Agron Sin. 2009;35:239–45 (in Chinese with English abstract).

Wang CX. Isolation and functional analysis of stress-response genes TaABC1 and TaSAP1/2 from Wheat (*Triticum Aestivum* L.). PHD dissertation of Graduate School of Chinese Academy of Agricultural Sciences, 2011 (in Chinese with English abstract).

Wang Y, Fu XM, Gao GJ, He YQ. Improving the grain quality of Minghui63, a restorer line of rice with good quality through marker-assisted selection. Mol Plant Breed. 2009;7(4):661–5 (in Chinese with English abstract).

Wang DL, Zhu J, Li ZK, Paterson AH. Mapping QTLs with epistatic effects and QTL × environment interactions by mixed linear model approaches. Theor Appl Genet. 1999;99:1255–64.

White PJ, Karley AJ. In: Hell R, Mendel R-R (eds) Plant cell monographs 17, cell biology of metals and nutrients. Berlin: Springer; 2010.

William HM, Singh RP, Huerta-Espino J, Palacios G, Suenaga K. Characterization of genetic loci conferring adult plant resistance to leaf rust and stripe rust in spring wheat. Genome. 2006;49:977–90.

Witkowski E, Waga J, Witkowska K, Rapacz M, Gut M, Bielawska A. Association between frost tolerance and the alleles of high molecular weight glutenin subunits present in polish winter wheat. Euphytica. 2008;159:377–84.

Wu L. QTL mapping and molecular characterization of adult-plant resistance to stripe rust in common Wheat. PHD dissertation of Graduate School of Sichuan Agricultural University, 2011 (in Chinese with English abstract).

Wu P, Ni JJ, Luo AC. QTLs underlying rice tolerance to low-potassium stress in rice seedlings. Crop Sci. 1998;38:1458–62.

Wu YQ, Liu LX, Guo HJ, Zhao LS, Zhao SR. Mapping QTL for salt tolerant traits in wheat. Acta Agric Nucl Sin. 2007;21(6):545–9 (in Chinese with English abstract).

Xu YF, An DG, Liu DC, Zhang AM, Xu HX, Li B. Mapping QTLs with epistatic effects and QTL × treatment interactions for salt tolerance at seedling stage of wheat. Euphytica. 2012;186 (1):233–45.

Xue DW, Chen MC, Zhang GP. Mapping of QTLs associated with cadmium tolerance and accumulation during seedling stage in rice. Euphytica. 2009;165:587–96.

Yang K, Chang XP, Hu RH, Jia JZ. Chromosomal localization of genes associated with proline accumulation under drought stress in wheat (*Triticum aestivum* L.). Acta Agron Sin. 2001;27 (3):363–66 (in Chinese with English abstract).

Yang J, Sears RG, Gill BS, Paulsen GM. Quantitative and molecular characterization of heat tolerance in hexaploid wheat. Euphytica. 2002;126:275–82.

Yang J, Zhu J. Predicting superior genotypes in multiple environments based on QTL effects. Theor Appl Genet. 2005;110:1268–74.

Zanetti S, Winzeler M. Genetic Analysis of pre-harvest sprouting resistance in a wheat × Spelt Cross. Crop Sci. 2000;40(5):1406–17.

Zhang XZ. Crop physiology research method. Beijing: China Agricultural Press; 1992.

Zhang ZB, Xu P. Reviewed on study of wheat QTLs. World Sci-tech R&D. 2002;24(1):52–8 (in Chinese with English abstract).

Zhang KP, Fang ZJ, Liang Y, Tian JC. Genetic dissection of chlorophyll content at different growth stages in common wheat. J Genet. 2009;88(2):183–9.

Zhang ZB, Shan L, Xu Q. Background analysis of chromosome controling genetic of water use efficiency of Triticum. J Genet Genomics. 2000;27(3):240–6 (in Chinese with English abstract).

Zhang XC, Li SS, Zhao XH, Fan YD, Li RJ. QTL and molecular markers for resistance gene of wheat sharp eyespot. 2005;6:276–9 (in Chinese with English abstract).

Zhang K, Tian J, Zhao L, Wang SS. Mapping QTLs with epistatic effects and QTL× environment interactions for plant height using a doubled haploid population in cultivated wheat. J Genet Genomics. 2008;35(2):119–27.

Zhao XH. QTL mapping for physiological traits related to drought-resistance in wheat. Master Dissertation of Graduate School of Shandong Agricultural University. 2004 (in Chinese with English abstract).

Zhou XG, Jing RL, Hao ZF, Chang XP, Zhang ZB. Mapping QTL for seedling root traits in common wheat. Sci Agric Sin. 2005;38(10):1951–7 (in Chinese with English abstract).

Afterword

After more than one year of work, this book is finally in its stage of completion and can be submitted for publication. As a scientist who has experienced several decades of social changes, I am very grateful to the state science policy. Without the reform and opening-up policy, it would be impossible for scientists to focus and advance their research. Similarly, without all the financial supports we received from the State "973" Project, the Natural Science Foundation of China and the Science and Technology Department of Shangdon Province over the years, it would be inconceivable for my research team to be able to complete systematic research in wheat genetic map construction, QTL analysis, and development of a number of new wheat varieties. Next, I would like to thank the strong support and assistance that we have received from our counterparts at home and abroad, including Dr. Walker (Kansas State University, USA) who came to my laboratory three times and gave lectures, Dr. Mingcheng Luo (University of California at Davis, USA) who assisted in the SNP whole genome scanning, Dr. Ji-Ping Zhao (senior scientist at Ball Horticultural Company, USA) who kindly assistance in language improvement of several chapters of this book, and appreciate his helpful advices on some contents presented in this book. Wang, Daowen and Zhang, Aimin (chief experts of the "973" Project, Institute of Genetics and Developmental Biology, Chinese Academy of Sciences); Jia, Jizeng, He, Zhonghu, Wang, Jiankang and Xia, Xianchung (renowned wheat experts from Chinese Academy of Agricultural Sciences); professors Sun, Qixin (China Agricultural University); Zheng, Youliang (Sichuan Agricultural University); Yan, Yueming (Beijing Normal University); and Yang, Xueju (Hebei Agricultural University). Very special thanks go to Li Zhengsheng (academician and receiver of the National Supreme Scientific and Technological Award) who inscribed this book and Liu Xu (academician and renowned expert in wheat genetic breeding and germplasm innovation) who wrote the preface for this book.

The main contents of this book consist of my research team's comprehensive summaries. Despite of our great effort in writing this book, there will be, inevitably, insufficient data in some areas due to the rapid development of the molecular biology. It is my intention to summarize our current research data so that the readers will be able to utilize and share the research information. Most importantly, we

J. Tian et al., *Genetic Analyses of Wheat and Molecular Marker-Assisted Breeding, Volume 1*, DOI 10.1007/978-94-017-7390-4

welcome feedback and comments from our counterparts at home and abroad in order to further improve upon our work. I am confident that the wheat genetic breeding and the development of new varieties in China will undoubtedly achieve many new breakthroughs under the guidance and support of such notable academicians such as Li Zhensheng and Yu Songlie.

August 1, 2015

Jichun Tian
In Taian, Shangdong Province
People's Republic of China

FSC
www.fsc.org
MIX
Papier | Fördert
gute Waldnutzung
FSC® C083411

Kolibri 360 GmbH
Ferdinand-Jühlke-Straße 7
99095 Erfurt, Deutschland
produktsicherheit@kolibri360.de